D0164463

LEADERSHIP: Personal Development and Career Success

Second Edition

Join us on the web at
www.Agriscience.Delmar.com

Delmar is proud
to support FFA activities

LEADERSHIP: Personal Development and Career Success

Second Edition

Cliff Ricketts

DELMAR
THOMSON LEARNING™

Australia Canada Mexico Singapore Spain United Kingdom United States

DELMAR
THOMSON LEARNING™

Leadership: Personal Development and Career Success, Second Edition
Cliff Ricketts

Business Unit Director:
Susan L. Simpfenderfer

Executive Editor:
Marlene McHugh Pratt

Acquisitions Editor:
Zina M. Lawrence

Developmental Editor:
Andrea Edwards Myers

Editorial Assistant:
Elizabeth Gallagher

Executive Production Manager:
Wendy A. Troeger

Production Manager:
Carolyn Miller

Production Editor:
Kathryn B. Kucharek

Executive Marketing Manager:
Donna J. Lewis

Channel Manager:
Nigar Hale

Cover Images:
Left: National FFA Organization
Right: PhotoDisc

Cover Design:
Dutton & Sherman Design

Printed in the United States of America
6 7 8 9 XXX 08 07 06 05 04

For more information contact Delmar,
3 Columbia Circle, PO Box 15015, Albany, NY 12212-5015.

Or find us on the World Wide Web at www.thomsonlearning.com, www.delmar.com, or www.Agriscience.delmar.com

Library of Congress Cataloging-in-Publication Data
ISBN: 0-7668-2536-1

Ricketts, Cliff
Leadership : personal development and career success / Cliff Ricketts. — 2nd ed.
p. cm.
Includes bibliographical references and index.
ISBN 0-7668-2536-1
HD57.7.R523 2001
658.4'092—dc21
2001042225

NOTICE TO THE READER

Publisher does not warrant or guarantee any of the products described herein or perform any independent analysis in connection with any of the product information contained herein. Publisher does not assume, and expressly disclaims, any obligation to obtain and include information other than that provided to it by the manufacturer.

The reader is expressly warned to consider and adopt all safety precautions that might be indicated by the activities herein and to avoid all potential hazards. By following the instructions contained herein, the reader willingly assumes all risks in connection with such instructions.

The Publisher makes no representation or warranties of any kind, including but not limited to, the warranties of fitness for particular purpose or merchantability, nor are any such representations implied with respect to the material set forth herein, and the publisher takes no responsibility with respect to such material. The publisher shall not be liable for any special, consequential, or exemplary damages resulting, in whole or part, from the readers' use of, or reliance upon, this material.

This book is dedicated to my family: my parents, Hall and Louise Ricketts, who have been tremendous leaders and role models in personal development; my wife, Nancy, also a leader, supporter, and role model; and my three children, John, Mitzi, and Paul, who are products of the objectives of this book, which started at a very early age.

Contents

Section 2 Communication and Speaking to Groups

Section 3 Leading Individuals and Groups

Preface

The mission statement for agricultural education and, specifically, the National FFA Organization includes the enhancement of leadership, personal growth, and career success of its students and members. Educating the total person is one of the objectives of this book. Besides the cognitive (academic) and technical (psychomotor) skill development secondary schools are presently teaching students, employers want employees with leadership and human relations skills. Schools, as a rule, are not teaching affective skills except through youth organizations. One of the purposes of this book, *Leadership: Personal Development and Career Success, Second Edition* is to provide a formal text to help teach affective skills for a stand-alone course or a supplementary text to existing agricultural education courses. As a result, employees will be better prepared for the workplace. Why? Employers tell us most people lose their jobs not because of cognitive and technical skills, but because of affective (leadership and human relations) skills.

The intent of this book is to serve as an instrument to help students become more successful in life and the workplace. *Leadership: Personal Development and Career Success, Second Edition* achieves this by teaching students to learn and enhance personal development and communication skills; by helping students select a job, get a job, and attain career success; and by helping students attain any desired leadership positions both in their careers and community. In short, the intent of this book is to help the student attain professional and personal success.

Leadership: Personal Development and Career Success, Second Edition is divided into six sections. Section 1 covers understanding leadership. Leadership categories and styles, personality types and their relationship to leadership and personal development, learning styles, and developing leaders are covered in Chapters 1 through 5.

Section 2 includes communicating to individuals and groups. Personal communication, overcoming stage fright by saying the FFA Creed, extemporaneous speaking, and public speaking are covered in Chapters 6 through 9. The FFA CDE rules are included in each chapter.

Section 3 is about leading individuals and groups. Topics include parliamentary procedure, and conducting successful meetings. These topics are included in Chapters 10 through 12. The FFA parliamentary procedure CDE rules are included.

Section 4 is about managerial leadership skills. The topics in this section include problem solving and decision making, goal setting, time management, and positive reinforcement and motivation. These topics are covered in Chapters 13 through 16.

Section 5 is about personal development. The topics in this area include: self-concept, attitude, and ethics in the workplace. These topics are covered in Chapters 17 through 19.

Section 6 includes transition to work skills. Topics in this section include selecting a career, procedures for applying and interviewing for a job, and employability skills necessary to prepare the student for transition to the work world.

PEDAGOGICAL FEATURES

Each chapter begins with a list of performance objectives, which include the main topics for each chapter. The objectives are followed by terms to know to enable the students to better understand the chapter content. The definitions of all the terms are found in the glossary and in some cases within the unit.

Each chapter ends with a summary. At the end of each chapter are review questions, which encourage the student to read the whole chapter. Fill-in-the-blank and matching exercises are designed to reinforce the student's understanding of important ideas. Activities enable students to apply newly learned concepts and practices, as well as draw on their own experiences and share those with others. Critical thinking is also a component of the activities section.

About the Author

Dr. Cliff Ricketts came through the ranks of agricultural education and the FFA and is a product of the program. He has achieved every degree offered by the FFA, including the Greenhand, Chapter, State, and American FFA (Farmer) degrees, Honorary Chapter, Honorary State, and Honorary American FFA, which was received the same year by former President George Bush and Lee Iacocca.

As a high school agricultural education instructor, his students won the state FFA Parliamentary Procedure Contest and state FFA Creed Speaking (four times); one student was runner-up in Public Speaking and Extemporaneous Speaking. Ricketts was also a recipient of the NVATA's National Outstanding Teacher Award while a high school teacher.

Ricketts received his B.S. and M.S. degrees from the University of Tennessee and his Ph.D. from Ohio State University. While teaching at Middle Tennessee State University, Ricketts received both the MTSU Foundation's Outstanding Teacher Award and Outstanding Public Service Award. The National Association of Agricultural Educators (NAAE) selected the agricultural education program at Middle Tennessee State University as the most outstanding in the southeastern United States at the beginning of the new millenium.

Dr. Ricketts coordinated the Agriscience for science credit with the Agricultural Education and Science Teachers for the Tennessee Department of Education. This was the first such effort in the country and the integration of academic and vocational education is now part of federal legislation. Ricketts was the editor of over 35 curricular guides in Agricultural Education for the Tennessee Department of Education.

Dr. Ricketts is a leader in alternative fuel research. He has worked with alternative fuels for over 20 years. Ricketts and his students have built engines to run on ethanol from corn, methane from cow manure, soybean oil, and hydrogen from water. Ricketts and his students have held the world land speed record for a hydrogen fuel vehicle since 1991 at Bonneville Salt Flats in Wendover, Utah. They also hold the land speed record for a vehicle that runs on soybean oil.

Acknowledgments

The author is appreciative of several people who had input into this book either as a resource for course content, contributing author, consultant, or reviewer:

Jay Stetzel, Education Consultant, Tennessee Department of Children's Services
Dr. Joe Townsend, Texas A & M University
William H. Coley, Retired Tennessee Agricultural Education Supervisor
Taft Davis, Fairview High School
Jill Martz, University of Tennessee Extension Service
Jerry Spivey, Tri-County Vocational School
Dawn Fox, Tennessee Tech University
Courtney Nichols, Richland High School
Sheri Smith and Robert Best, Middle Tennessee State University
Dr. Joyce Harrison, Department of Human Sciences, Middle Tennessee State University
Barry Baker, East Robertson High School
Kay Webb, Humphrey County Vocational School
Eric Swafford, Tennessee Farmer's Cooperative
Kim Elmore, Middle Tennessee State University
Jeremy Childers, Middle Tennessee State University
Benny McDonald, Tennessee Farmer's Cooperative
Steve Lomax, Middle Tennessee State University
Byron Myerhoff, Middle Tennessee State University
Harlan Dabney, Middle Tennessee State University
Dana McKay, Middle Tennessee State University
John Ricketts, University of Florida
Ricky Mabry, Hampshire High School
Jennifer Young, Smyrna Middle School
Debbi Weston, Ely & Walker
Randy Kersey, Mt. Pleasant High School
Harold Collins, Page High School
Regina Beaty, Lincoln County High School
Brad Mosley, White House High School
Leslie Dyer, Tennessee Department of Education
Dale R. Carpenter, Monroe Area High School
David N. Coile, Lake Gibson Senior High School
Ronnie A. Thomas, Park View Senior High School
Michael Barry, University of Tennessee Extension Service
Susan Pile, Graduate Assistant, Middle Tennessee State University
Glenn Ross, McEwen High School
Sharon Skiles, Buchholz High School, Gainesville, Florida
Robert W. Clark, Dauphin County Technical School, Harrisburg, Pennsylvania
Carl Vivaldi, Bucks County Technical High School, Yardley, Pennsylvania
Mike Davis, Chappell High School, Chappell, Nebraska

SECTION 1

Understanding Leadership

CHAPTER 1

Leadership Categories and Styles

Objectives

After completing this chapter, the student should be able to:

- Analyze various definitions of leadership
- Discuss the contributions of agricultural education and the FFA to leadership development
- Evaluate myths about leaders and leadership
- Describe the various leadership categories
- Explain "the big picture" relationship between leadership categories, behavior, and employment
- Describe democratic, authoritarian, and laissez-faire behavioral leadership styles
- Describe the situational (contingency) leadership behavioral style

Terms to Know

leadership
democratic
gleaned
trait leadership
skills
power leadership
influence leadership
behavioral leadership
authoritarian
situational leadership
traditional leadership
popularity (perceived) leadership
combination leadership
structural frame
human resource frame
political frame
laissez-faire leadership
continuum

Leadership is the ability to move or influence others toward achieving individual or group goals. Leadership is not a mystical trait that one individual has and another does not. It is learned behavior that anyone can improve by study and application. You can be a leader if you have the determination to develop the abilities that make a leader.

Leaders are needed in organizations, communities, states, and nations. Leaders serve. This is the key to developing your own leadership abilities. Benefits to yourself are incidental when assisting others.

Persons trained to lead have the characteristics most young people want. Leaders have respect, poise, confidence, and the ability to think and take on responsibility. They are mature because development as a leader helps develop maturity. Learning leadership helps you prepare to take your place in society as a useful, productive, well-adjusted citizen (Figure 1–1).

Every member of a group participates in leadership when he or she contributes an idea. Leadership passes from person to person as each contributes to achievement of group goals. A democratic group reaches its height when leadership is spread throughout the membership. There is no limit to the number of leaders within a group. In fact, the more leaders the better because the very act of leadership develops the initiative, creativity, and responsibility that the group needs from each of the members.

A true leader helps the group achieve what it believes is important. Leaders' thoughts and feelings are consistent with their speech and actions. They have ideas and can make suggestions, modifications, or expansions. They can receive and implement others' ideas, too. Sometimes people confuse leadership with being bossy, but there is a difference.

Figure 1–1 Community groups are excellent places to exhibit leadership skills. This student is giving a speech before she competes at the state FFA convention.

The boss assigns tasks—
the leader sets the pace.

The boss says, "I"—
the leader says, "We."

The boss says, "Go"—
the leader says, "Let's go."

The boss drives people—
the leader guides them.

The boss depends on authority—
the leader on good will.

The boss creates fear—
the leader develops confidence.[1]

DEFINITIONS OF LEADERSHIP

Researchers have given us over 350 definitions of leadership within the last seventy-five years. Even though leadership is hard to define and explain, people know it when they see it. For example, leadership researchers Warren Bennis and Burt Nanus stated, "Leadership is like the Abominable Snowman, whose footprints are everywhere but who is nowhere to be seen."[2] Leadership is like love. We cannot define it or teach it, yet we can always identify it. We know when it is present and when we need it but cannot ensure its expression or continuation. Warren Bennis also said, "To an extent, leadership is like beauty: it's hard to define, but you know it when you see it."

The term *leadership* means different things to different people. People usually define leadership according to their individual perspective or the context that has meaning to them. Warren Bennis used the assumption that leaders are people who are able to express themselves fully. They know who they are, what their strengths and weaknesses are, and how to fully display their strengths and compensate for their weaknesses. They also know what they want, why they want it, and how to communicate what they want to others to gain their cooperation and support. Also, leaders know how to achieve their goals.[3]

AGRICULTURAL EDUCATION, FFA, AND LEADERSHIP

It is within this context that agricultural education and the FFA enter the leadership picture. Bennis used the words *communication*, *cooperation*, and *support* to describe those who have become leaders. The aims and purposes of the FFA are the development of leadership, citizenship, and cooperation. Agricultural education and the FFA can expose students to a variety of experiences similar to those they may encounter in their chosen occupations.

Many FFA members define leadership as public speaking and parliamentary procedure. Although leadership is more than these skills, they are a first step toward leading groups and influencing others. We do many other things in agricultural education and the FFA that contribute to leadership development, such as communication; leading group discussions; setting goals, including ways and means; conducting meetings; solving problems and making decisions; developing social skills; and planning careers.

Importance of Youth Organizations

The National Association of Agricultural Educators (NAAE) stated that to ensure quality, vocational programs must be responsive to the needs of the individual for job-entry skills and the compatible skills of communication, citizenship, leadership, decision making, positive attitude toward learning, and personal occupational responsibility. It has been found that students who were more active or participated more frequently in vocational educational student organizations were rated higher by the students' instructors at the time of graduation and were rated higher by parents and employers six months after graduation, on leadership, citizenship, character, willingness to accept responsibility, confidence in self and work, and cooperative spirit and effort.[4]

Leadership and Personal Development Abilities

Agricultural education and the FFA can provide students with opportunities to relate positively to others in a variety of situations, participate in activities that develop leadership, and participate in activities that increase occupational competency. Youth organizations, especially the FFA, can be one of the most effective ways of maximizing learning opportunities.

A well-balanced program of FFA activities supplementing formal instruction in agricultural education can lead to a desirable educational program for students. Agricultural education and the FFA can expose students to a variety of experiences similar to those they may encounter in agribusiness and agriscience occupations.

The FFA can encourage leadership and personal development through developing and exercising leadership responsibilities. In the process of exercising initiative, accepting responsibility, and evaluating progress, members can develop leadership and personal development skills.[5] You are probably wondering how. Ten areas of the FFA that contribute to leadership and personal development follow.

Parliamentary Procedure The effective use of parliamentary procedure is an ability needed by members and leaders who want to take part in and guide group meetings. The rules of parliamentary procedure make meetings more democratic. Learning parliamentary procedure has practical applications. Practice in parliamentary procedure helps one learn to stay on the subject, respect the rights of others, and move to a decision. Parliamentary procedure also aids in developing communication skills and public speaking ability. While learning parliamentary procedure, you learn to be a **democratic** leader.

Leading Individuals and Groups The FFA organization provides many opportunities for members to practice and improve their skills in democratic leadership. One function of a leader in American society is to help groups work in a democratic way to ensure that individuals participate in making decisions. Participation and learning-by-doing are the keys to FFA leadership programs. Certainly, FFA members are not born leaders. It takes study, practice, mistakes, and successes to develop chapter, state, and national officers. The FFA gives its members an opportunity to become leaders at an early age by providing a test ground where they can practice and become experienced leaders.

Conducting Meetings Conducting and participating in meetings offers many opportunities for developing leadership skills. These skills will be valuable when members graduate and participate in professional, business, and civic organizations. Members can practice public speaking, communication with groups, planning for effective meetings, developing meeting agenda, establishing rapport to obtain group cooperation, following democratic procedures, supporting the majority decision of the group, and respecting the opinions of others.

Duties of Officers and Selection of Officers All members of an organization are responsible for electing capable officers. Among other things, FFA members learn what is expected from the officers, what the necessary qualifications are, and how to properly nominate and elect people to office. By learning the duties of officers and the proper ways to select them, FFA members can become better contributors to clubs and civic and private organizations, as well as advisors to youth clubs and activities (Figure 1–2).

Managing Financial Resources The knowledge and abilities developed in FFA financing will help members perform their responsibilities as members of other groups and organizations. Earning

Figure 1–2 Being a chapter officer and speaking in front of groups may lessen a student's fear of public speaking.

money and managing its distribution can serve as a meaningful educational experience. This is particularly true for members who keep official record books as a result of their supervised agricultural experience programs. Well-planned and executed fund-raising activities give members a chance to use their membership and leadership abilities and learn how an organization manages money (Figure 1–3).

Developing Good Work Habits Supervised agricultural experience programs help members become more responsible people by developing the good work habits necessary for advancement in the job market. Supervised agricultural experience in agricultural education is the foundation for advancement in the FFA and for help in choosing the FFA member's agricultural career. Members can explore and practice agricultural skills that prepare them for the future. Some of these skills are responsibility, personal integrity, initiative, willingness to take supervision, a positive attitude toward work, dependability, and taking pride in one's work.

Figure 1–3 Being involved in agricultural education and the FFA can include keeping records for the supervised agricultural experience program. This will help in managing financial resources in the future.

"Learning to do" and "doing to learn" (from the FFA motto) applies to supervised agricultural experience programs that can develop good work habits. The FFA degree program is based on the member's supervised agricultural experience programs and leadership development. These degrees of membership are contingent on definite accomplishment in connection with the agricultural education program of the school. Specific levels of achievement with respect to agricultural work experience cannot be attained unless good work habits are developed.

Individual Adjustment Used in this context, *individual adjustment* refers to personal development abilities that are a part of the maturation process facilitated through FFA contests and other activities. These personal development skills include, but are not limited to, following established procedures and regulations, developing competitive goals, developing the ability to perform under pressure, and learning to accept success and failure. FFA contests provide practical experience for what is learned in agricultural education courses, as well as opportunities and challenges for personal growth and leadership development. All FFA contests are to be a natural outgrowth of the instructional program in agricultural education. Other measurable skills include accepting success and failure, developing appropriate public relations skills, managing use of time, and demonstrating poise (Figure 1–4).

Participating in a Committee No matter what a student's plans may be, he or she needs practice in committee work. Learning to work with others in accomplishing tasks is an important part of an individual's education. Committee participation provides a means of getting every member involved in some area of interest for maximum chapter effectiveness and member growth. Participating in committees and groups requires cooperation. Committee activities should be designed to provide the best possible experience for members, whether they plan to become production agriculturists, enter nonfarm agricultural business,

Figure 1–4 As these students compete in the Soils Judging Contest, they learn to follow established rules and procedures, accept success and failure, and develop many other skills. Classroom instruction prepares students for theory and after school work based learning opportunities as important parts of Supervised Agricultural Experience Programs.

enter industry, or go into advanced training in a vocational-technical institution or university program. Skills developed include identifying different kinds of committees, identifying purposes of committees, selecting members for committees, being an effective committee chairperson, presenting a committee report, promoting committee member participation, and delegating responsibility.

Developing Social Skills Members must become confident, responsible people before they can become confident, responsible leaders. Social as well as recreational activities are necessary in the development of well-integrated and healthy people. Although it is not particularly the responsibility of the FFA to teach social skill development, the FFA can supplement and perhaps make more meaningful some of the learning that has been **gleaned** elsewhere and assist members in preparation for adult life. There is social value in almost every FFA activity that provides members with opportunities to practice real-life situations for personal growth. This includes properly presenting an award, selecting a meal from a menu and properly ordering, dressing appropriately for the occasion, meeting and greeting people, and making proper introductions (Figure 1–5).

Developing Citizenship Skills Community service activities help members learn to develop and practice good citizenship and leadership abilities needed now and in the future. The focus is on making the community a better place to live and work. Personal development can be analyzed in terms of desirable personal characteristics, accepted social behavior, and good citizenship.

Citizenship development helps members become informed of civic responsibilities, such as voting, paying taxes, and abiding by the laws of society. FFA activities strongly emphasize learning

Figure 1–5 By attending the FFA awards banquet, students can learn the importance of public relations, dressing appropriately for the occasion, and proper eating etiquette.

and personal development for responsible citizens. The (Building Our American Communities (BOAC) award program is an example of how the FFA emphasizes developing citizenship skills. The BOAC awards program is designed to develop interested, experienced, and knowledgeable community leaders and citizens. The BOAC program seeks to develop a sense of pride in the community and the initiative to make it a better place to work and live.

Measurable skills are the ability to plan and establish community service projects, develop a respect for national symbols and customs, develop an understanding that "if we belong, we pay dues," respect the rights and views of others, and cooperate with others in group activities.

Does Agricultural Education and the FFA Make a Difference?

Agricultural education and FFA members and students who had never taken any agricultural education classes were evaluated to measure leadership and personal development abilities. The evaluation was conducted by the Ohio State and Middle Tennessee State Universities.[6] The agricultural education students and FFA members scored significantly higher (57 percent) than nonagricultural education students (41 percent). The evaluation also revealed that the more active students were in FFA activities, the higher the level of leadership and personal development abilities.

MYTHS ABOUT LEADERS AND LEADERSHIP

There are several myths or misunderstandings regarding leadership and leaders.

Myth 1 Little is known about leadership.

Myth 2 All leaders are born with unique qualities.

Myth 3 Leaders make all the group's decisions.

Myth 4 All leaders are popular, charismatic individuals.

Myth 5 To lead requires election or appointment.

Myth 6 Leadership is a rare skill.

Myth 7 Leaders control, direct, and produce.

Myth 8 Leadership exists only at "the top" (a high position in an organization).

Information on leaders and leadership is readily available. Business, industry, educators, volunteer groups, and others use leadership information in improving their organizations. Today, leadership authorities reject theories that propose that leaders are born with special leadership qualities. Everyone can learn and develop leadership skills. Recent studies of leadership have demonstrated the need to involve members in the decision-making process. New techniques help leaders determine how to involve others in making decisions. Leadership is not a place but rather a process; therefore, those at the top are not the only leaders in a group. When members participate in group decisions, they are leading.[7]

LEADERSHIP CATEGORIES

Major categories of leadership are (a) trait, (b) power and influence, (c) behavioral, (d) situational, (e) traditional, (f) popularity, and (g) combination. These categories help explain why there are so many different definitions of leadership.[8] Different cultures, groups, organizations, and nations have different perceptions of leadership. Just as people have different views of a religion or political party, they have different views of leadership. After all these categories are discussed, a schematic illustrates how various views of leadership fit into "the big picture" or the broad range of leadership categories.

Trait Leadership

One of the earliest approaches for studying leadership was the trait approach. Traits are distinctive qualities. The **trait leadership** approach assumes that some people are "natural" leaders who possess certain traits that others don't. Trait theories search for the exact ingredients leaders exhibit in an attempt to prove that people are born leaders. An individual's physical attributes, personality, social background, abilities, and skills are among the traits examined.[9]

Traits Scholars have researched the personalities of leaders, focusing on traits such as age, weight, height, appearance, physique, energy, health, speech fluency, intelligence, scholarship, knowledge, insight, judgment and decision, originality, adaptability, introversion, extroversion, dominance, initiative, persistence, ambition, responsibility, integrity, and conviction. Although certain traits increase the likelihood that a leader will be effective, they do not guarantee effectiveness. The relative importance of different traits depends on the nature of the leadership situation.[10]

Skills Remember, leaders are not born. Therefore, traits alone do not determine leadership. **Skills**, which are learned, are different from traits. Skills are practiced abilities (Figure 1–6). Therefore, intelligence, social skills, speaking ability, group dynamics, and organizational ability are skills most frequently linked to leadership ability. Other skills or traits are discussed in detail in Chapter 2.

Figure 1–6 Dr. Fred Shively demonstrates his skill in public speaking. Verbal and public speaking ability have a high correlation with career advancement. (Courtesy of Anderson University, a four-year liberal arts university located in Anderson, IN.)

Power and Influence

The power and influence theories explain leadership effectiveness in terms of the amount and type of power a leader possesses and how he or she exercises power.

Power Leadership Power is important not only for influencing subordinates but also for influencing peers, supervisors, and people outside the organization. Many times, a person's power depends to a considerable extent on how the person is perceived by others.

There are many sources and types of power.

Formal authority. Power stemming from formal authority is sometimes called legitimate power. Authority is based on perceptions about the goals, obligations, and responsibilities associated with particular positions in an organization.[11]

Control over resources and rewards. Potential influence based on control over rewards is sometimes called reward or position power. This control stems in part from formal authority. The higher a person's position in the organization, the more control over scarce resources the person is likely to have. An example of reward power is influence over wages and promotions.

Control over punishment. This form of power is sometimes called coercive power. It is the capacity to prevent someone from obtaining deserved rewards.

Control over information. This control involves access to vital information and the distribution of information to others.[12]

Ecological control. This form of power is sometimes called situational engineering. It is control over the physical environment, technology, and any organization of the workplace. Manipulation of these physical and social conditions allows one to influence indirectly the situations of others.

Expertise. This form of power is sometimes called expert power. Experts solve problems and perform important tasks. Expertise is a source of power for a person only if others are dependent on the person for the advice or assistance they need.

Friendship and loyalty. Sometimes called referent power, this source of power is the desire of others to please a person toward whom they feel strong affection. Success in developing and maintaining referent power depends on interpersonal skills such as charm, tact, diplomacy, empathy, and humor. Referent power relates to informal power structures. People with this sort of power are not in elected or appointed positions, but due to their earned respect (reference), they can exert great power over those who are appointed or elected. Referent power is similar to influence.

Charisma. Charismatic leaders have insight into the needs, hopes, and values of followers. They are able to create a vision that motivates commitment to the leader's policies and strategies.[13] It is not clear whether these leaders have a variation of referent power or a distinct form of power. Attributes include qualities such as personal magnetism; a dramatic, persuasive manner of speaking; strong enthusiasm; and strong convictions. These qualities make the charismatic leader appear somewhat mysterious, a person who can be trusted to lead followers to victory or success.

Political power. This is a pervasive process within groups that involves efforts by members of a group to ensure power or protect existing power sources. This involves remaining in control of your group, whether it is a political party or an organization. People do many things to solidify their power base (Figure 1–7).

Influence The exercise of influence is different from the exercise of raw power. When you exercise raw power, you force a group to submit, perhaps against its will. When you influence others, you show them why an idea, a decision, or a means of achieving a goal is superior in such a way that they follow your lead of their own free will. Members will continue to be influenced as long as they are convinced that what they agreed to is right or is in their best interest as individuals or as a group.

The essence of leadership is influence over others; however, influence between a leader and followers flows in two directions: leaders influence followers, but followers also have influence over leaders. For example, in an FFA chapter, the effectiveness of the president depends on his or her influence over the advisor and fellow officers (peers). Yet, the effectiveness of the president depends on the influence of the members.

Leadership means exerting influence. **Influence leadership** is the ability to bring about changes in the attitudes and actions of others. Influence can be indirect (unconscious) or direct (purposeful). We often influence others without being aware of it. If you have a new hairstyle, wear a new suit, or purchase a flashy new car, you may influence someone who sees you to copy your hairstyle or buy a similar suit or car. In a group, a leader can influence members indirectly by serving as a role model.

Behavioral Leadership

Behavioral leadership assumes that there are distinctive styles that effective leaders use consistently. The basic two styles are **authoritarian** (theory X) and democratic (theory Y). All other behavioral styles revolve around the authoritarian and democratic styles of leadership. Due to the

Figure 1–7 Former Tennessee Governor Ned Ray McWherter, a former FFA member, participates in a group groundbreaking ceremony. Besides being a noble gesture, this type of activity helps protect his power base.

importance of behavioral styles of leadership, these are discussed later in this chapter.

Situational Leadership

Some people refer to **situational leadership** as contingency leadership. The term *contingency* used for this leadership theory comes from the fact that the emergence or effectiveness of any style is contingent on the situation in which the leader is operating. Similarly, one's assumptions, behavior, and attitudes toward people are difficult to change. Leaders can change their styles to meet the needs of a particular group. For example, you may have democratic assumptions, but if those you are leading are a low-motivated group that needs close supervision, you can use an authoritarian supervisory style.

As a leader, you must be aware of your attitudes, but you must use the appropriate leadership style for the situation. The leader who acquires a large number of leadership skills can use different abilities to meet the needs of different situations. Situational leadership is discussed in greater detail later in this chapter.

Traditional Leadership

Some people refer to **traditional leadership** as cultural and symbolic.[14] One way to sum up this leadership style is "this is the way we do things here." We do many things out of tradition; it may not be the only way, but it is the group, organization, or company way. Another way to describe this style is "playing the game." If one wants to become a leader in some groups, certain games have to be played. For example, although fighting ability is not a criterion for leading an organization, unless the organization is a gang, the game of fighting most often be played by a leader who hopes to be effective in some places. Understanding organizational culture is vital to leadership effectiveness. In some cultures, the leader may be the most intelligent, the best speaker, the hardest worker, or the best dresser. Some examples follow to illustrate traditional, cultural, and symbolic leadership.

Public Officials Public officials are victims of cultural and symbolic leadership. For example, if the mayor of a city did not attend ribbon-cutting ceremonies, give keys to the city, or speak often to clubs, convention groups, and fund-raisers, she might not be perceived as a leader even though she has been an expert manager of finances and promoter of city growth. Certain things are done because of tradition.

FFA Tradition and Culture The FFA is rich with traditional symbols and organizational culture. Many FFA advisors and members believe leadership consists of being in official dress, knowing the opening and closing ceremonies, reciting the FFA Creed, and participating in contests. It is true that some of these components do foster leadership development; much of what we do in the FFA is cultural and symbolic. FFA members who appear at official contests, especially speaking contests, without official dress are seen as not participating fully and not giving the organization the proper respect.

Does this mean that people cannot become leaders if they do not wear the official FFA dress? No, but they may not be perceived as leaders due to our FFA culture and symbols. Remember, perception is reality; if we are not perceived as leaders by our group, we will not become leaders. This is not intended to be negative. The explanation is simply to illustrate the cultural and symbolic nature of the FFA.

FFA Symbols There are many symbols and cultures within the FFA. The FFA emblem, banner, and paraphernalia are symbolic (Figure 1–8). These things can instill pride, motivation, and goal setting. If cultural symbols can enhance these characteristics, we can again thank the FFA. What about the name FFA? It doesn't stand for anything since dropping Future Farmers of America as its name. However, due to organizational culture and symbolic tradition, we cannot change FFA to another name. Should we? It is for you to decide. You are a part of the culture and symbols.

Corporate Symbols Coca-Cola® destroyed a national as well as a corporate symbol when they introduced a new Coke® formula and design several

Figure 1–8 How many FFA symbols can you find in this picture?

years ago. The power of the symbol was demonstrated by the public's refusal to change. The symbol seemed to be owned by the public in addition to the Coca-Cola® corporate shareholders.[15]

American Flag Another example is our stars and stripes, the American flag. No matter what may be perceived to be wrong with our country, when we raise this symbol and sing the national anthem, people stand with allegiance. During the song some choke back tears and feel chills of emotion, but all seem to unite.

Educational Symbols Our high school and universities exhibit symbols that can elicit strong feelings of loyalty. During graduation ceremonies, retirement banquets, inaugurations, homecomings, and athletic events, for example, symbols are everywhere. Graduation caps and gowns, crowns, and mums at homecomings are just a few of the many examples.

Motivational Symbols A person in a newspaper office invented a "taking off" symbol to catch up with the competition. Posters of space ships were used to create the image, and the president of the newspaper installed and used an airplane seat belt in his office chair. The symbol and the behaviors it inspired encouraged a more cohesive group at the newspaper office. Symbols prove a powerful tool to inspire and motivate organizational members. The FFA symbol has a similar motivational effect on many present and past members.

Popularity (Perceived) Leadership

Some people refer to **popularity (perceived) leadership** as cognitive leadership. The popularity leadership category relates to symbolic leadership in that this theory is based on perceived leadership. In other words, leadership is bestowed on those who really don't deserve it. The decision as to who should lead within an organization may be irrational.[16]

Falsely Positive Self-Concept The popularity or perceived leadership category contends that leader effectiveness is bestowed by others. This relates to follower perceptions and not necessarily to the leader's ability. Followers often acknowledge leadership where none exists. False praise of the leader reinforces a false self-concept within the leader. Self-concept is great, but a falsely positive self-concept can destroy a person's image. A person could develop a falsely positive self-concept by being elected to an office because he is good-looking but is not actually qualified to hold the office.

FFA Members FFA members should be aware of the popularity or perceived leadership category. For example, just because a boy is captain of the football team does not mean he will be a good FFA president. However, if football is "king" at a particular school, the football captain might be perceived as a good leader, and students may come to believe that good leaders are FFA presidents who are football captains. A tradition can begin based on false assumptions. Similarly, a preacher who delivers great sermons and is charismatic does not automatically qualify as a church leader or administrator except by tradition.

Organizations Organizations find ways to legitimize the choice of leaders. Organizational members might determine leaders based on salary or experience. In this leadership example, experience, salary, and tenure do not determine leadership.

Combination Leadership

The **combination leadership** category unites the previous leadership models and styles. The combination category shall be viewed through four frames, or images, of leadership: authoritarian, democratic, political, and tradition (Figure 1–9).[17]

The dynamics of leadership include all four frames or categories; problems and solutions are

Combination Leadership Category	
Authoritarian Frame Authority Goals Task oriented Rules Roles Formal relationship Focus on implementation Policy and organization charts	**Democratic Frame** Share decision making Meeting needs Provide support Feeling oriented Motivation Believe in people Visible and accessible procedure
Political (Power and Influence) Frame Power Conflict Coalitions Backroom politics Bargaining Negotiation Compromise Cooperate with rivals	**Tradition (Symbolic) Frame** "This is the way we do it here" Culture (the best fighter is the leader) Ceremonies (opening ceremony) Rituals (ribbon-cutting ceremonies) Symbols (FFA symbols) American flag

Figure 1–9 Leadership is a combination of various categories and styles. (Adapted from work by Tom Burks, Director of the Leadership Institute at Middle Tennessee State University, and from *Reframing Organizations*, by Lee Bolman and Terrence Deal.)

not restricted to any one. This leads to some assumptions about leadership effectiveness. The leader who can make positive use of the frames or categories will succeed where others may fail. Leaders who do not respond to the appropriate frame or category can jeopardize their leadership ability. Problems can arise and make solutions seem impossible.

Authoritarian or Structural Frame The authoritarian or **structural frame** relates to relationships and formal roles in the organization. Organizational charts, policies, procedures, authority, and responsibility guide the leader's decisions and behavior. The emphasis is on goals, roles, and formal relationships. Leaders

- develop a strategy
- focus on implementation
- continually experiment, evaluate, and adapt
- do their homework[18]

How does the FFA compare to the authoritarian structural frame? Structure does not imply a rigid hierarchy without flexibility. The FFA is a structured organization, yet it is flexible within boundaries, as members demand. It is structured from the local to the state and national levels, but the organization is flexible since it uses democratic principles.

Democratic or Human Resource Frame The democratic or **human resource frame** relates to the needs of members without the strong emphasis on production and policy found in the structural frame. This frame or leadership style is viewed as follows.

- Democratic leaders believe in people and communicate that belief.
- Democratic leaders are visible and accessible.
- Effective democratic leaders delegate. They increase participation, provide support, share information, and spread decision making among as many members in the organization as possible.[19]

Fitting the organization to the people and meeting the needs of followers becomes the key to effectiveness for the democratic frame. You may recognize this as the democratic style, which states that organizations should design conditions that allow people to accomplish their own goals along with organizational objectives.

How does the FFA adhere to the democratic human resource frame? Programs, projects, goals, objectives, and tasks are important, but not at the expense of people. Zig Ziglar has stated, "People don't care how much we really know until they know how much we really care." The FFA is a democratic human resource (people) organization.

Political Frame The **political frame**, not to be confused with political leaders in a democracy, focuses on the struggle for scarce resources in an organization. Power and influence are a part of it. Going is so tough that it's "every person for him- or herself." Special interest groups band and disband as the need arises. Negotiation, conflict, and compromise are part of everyday life; the ability to persuade is essential. Whether we like it or not, due to scarce resources, leadership is sometimes political. When this occurs, the organization demands political leaders who

- Clarify what they want and what they can get
- Assess the distribution of power and interests
- Cooperate with rivals
- Persuade first, negotiate second, and use coercion only if necessary

However, a politically effective leader often persuades others to follow his self-interest. Thus, self-interest rather than organizational interests can greatly affect the organization. Also, back-room political deals are made that serve the leader's own interest, resulting in bargaining with interest groups, which can impose goals that prove to be constraints on the organization. Therefore, due to the leader's bargaining, organizational goals can become the special interest group's goals, thus degrading the organization. For organizations to diminish political influence, they must have shared vision and organizational values.

How can the FFA or you as part of another organization overrule a person who has too much political influence? To have shared vision and organizational values, the FFA can fully utilize its program of activities, which is prepared by the executive committee and voted on by the members. There are times when leaders have to do what is necessary to get the job done. However, if autocratic rule becomes more than situational, the members of the group can ensure that democracy is the form of leadership to which they will adhere.

Tradition Frame Refer to the previously discussed tradition leadership category for further explanation. This frame emphasizes, among other things, shared vision and values. Organizational ceremonies, traditions, and symbols are part of this frame. Traditional and symbolic leaders interpret experience, use symbols to capture attention, and discover and communicate a vision.

Figure 1–10 summarizes the categories of leadership.

Categories of Leadership (Traditional Understanding and Agricultural/FFA Relationship)

Leadership Categories	Traditional Understanding	Agricultural Education/ FFA Relationship
A. Trait	Leaders are not born, they are made. Some people assume certain traits make you a leader. Leadership skills are learned through FFA activities.	Public speaking Parliamentary procedure Leading group discussions Conducting meetings
B. Power and Influence	Forms of power: Formal or legitimate Coercive Referent Influence Position Expert Charisma	FFA officers Chairperson, Program of Activities Committee Influence through contest and awards recognition
C. Behavior	Authoritarian (theory X) Democratic (theory Y) Laissez-faire	Democratic leadership learned through parliamentary procedure and conduct of meetings.
D. Situational	Using the appropriate leadership style for the situation.	Parliamentary procedure: Suspend the rules in certain situations; rules have to be suspended to get immediate action.
E. Tradition	"This is the way things are done here."	Official FFA dress, FFA paraphernalia, opening and closing ceremonies, FFA banner, FFA awards banquet, participating in contests, FFA emblem
F. Popularity	Leadership is bestowed on those who don't deserve it.	Spur-of-the-moment FFA officer elections without applications and interviews. Result? Most popular student elected regardless of leadership ability.
G. Combination —Structural —Human Resource —Political —Symbolic	Unites or combines the previous leadership categories. No leadership category stands alone.	Structural: FFA constitution and bylaws Human Resource: Democracy through parliamentary procedure and officer elections Political: Voting Symbolic: Refer to E above

Figure 1–10 Leadership can be divided into seven categories. The traditional understanding of each category is provided, along with its relationship to agricultural education and the FFA. (Adapted from work by Tom Burks, Director of the Leadership Institute at Middle Tennessee State University.)

The Big Picture

Figure 1–11 provides an overview of the leadership categories and their connections with each other. All these areas are discussed in this book. "The big picture" illustrates the connection, relationship, and importance of leadership and personal development to employment and leadership in your home and community. It also suggests why leadership has so many different definitions and why it can become complicated. Refer to Figure 1–11 as you read this book; it may help you fit the pieces into the leadership puzzle.

Democratic Leadership Is in the Center of the Big Picture

There are many leadership styles, but the basic style is democratic, which is closely connected to situational leadership. Although democratic leadership is the preferred style for most, there are situations when authoritarian leadership must be used.

Leaders Are Not Born, They Are Made

People are born with certain traits, such as height, color, innate intelligence, and personality type. Our traits have to be channeled toward goals. Most of the skills we possess are taught either informally through our environment or formally through our schools.

Personality Types and Leadership Styles

Basically, sanguine (influencing) personality types tend to be democratic leaders, while choleric (dominating) personality types tend to be authoritarian leaders. The other two personality types tend to be situational leaders. However, through leadership training, any personality type can be taught to use the appropriate leadership style for the situation. Personality types are discussed in Chapter 2.

Leadership Skills Are Learned in Agricultural Education and the FFA

Public speaking, parliamentary procedure, leading a group, conducting a meeting, and other skills learned in agricultural education classes and the FFA promote the growth of leadership and personal development skills. Parliamentary procedure is a key element. When parliamentary procedure is mastered, democratic leadership is mastered. All views are presented, everyone has an equal chance for input, but the wishes of the group take precedence over the desires of any one individual.

Employers Want People with Good Leadership and Personal Development Skills

Employers want people who can think, solve problems, make decisions, and be persistent and dedicated. They want people who can communicate and manage their time wisely. Employers want people who are honest, hardworking, and cooperative; who have positive attitudes and values, as well as good social skills. They also want people with a positive self-concept. If you don't like yourself, you will not like other people. You may find yourself putting others down with the misconception that it will lift you up.

Communication Style Relates to Personality Type and Leadership Style

Briefly, aggressive communicators tend to be authoritarian leaders with a choleric personality. Passive communicators tend to be laissez-faire leaders with a phlegmatic personality. Assertive communicators tend to be democratic leaders with a sanguine personality. (See Chapter 5 for discussion of these personality types.)

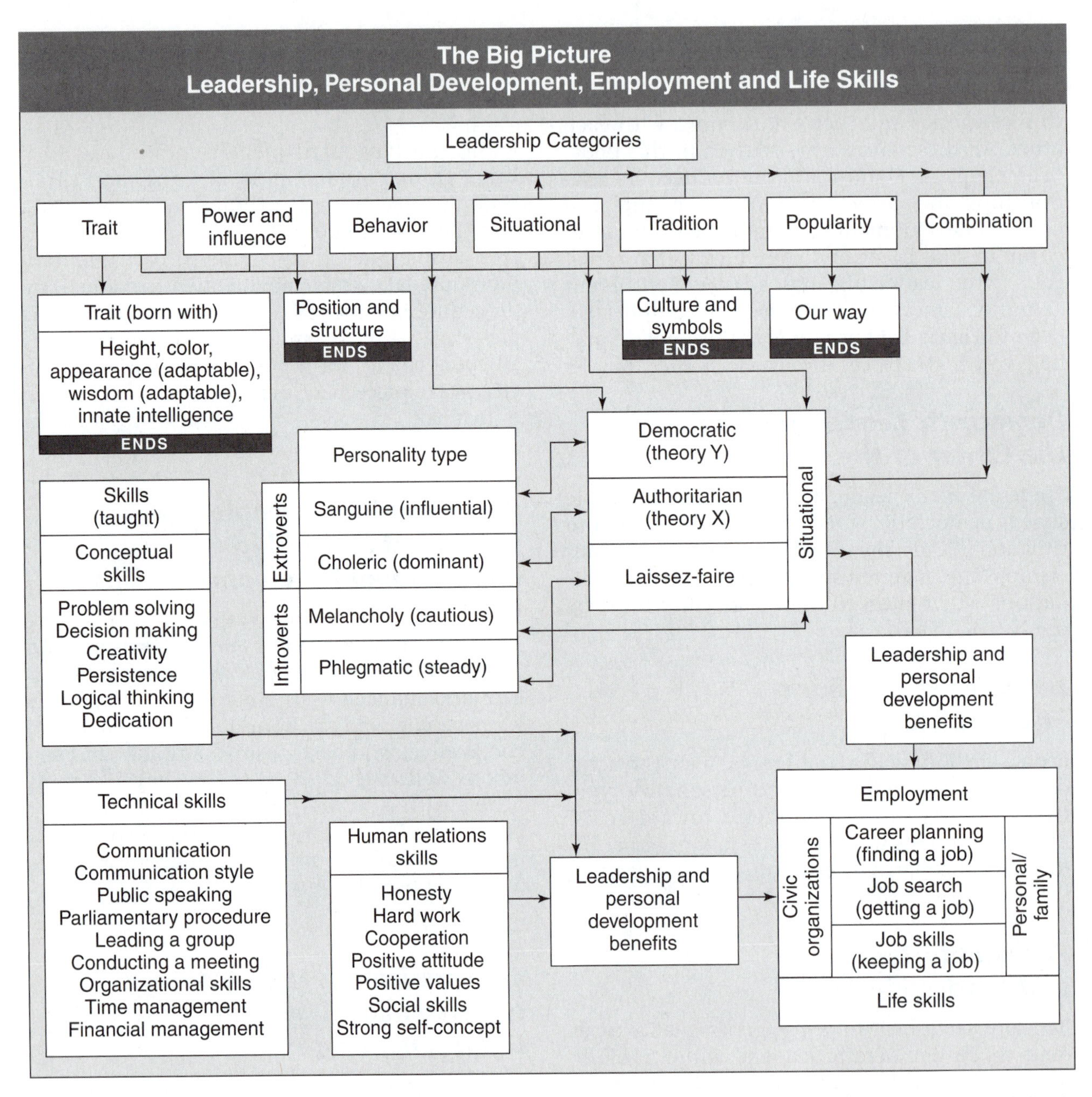

Figure 1–11 Defining leadership is not easy. Some leadership categories are related to personality type, personal development, employment, and life in general; other leadership categories are not related to anything.

Power, Tradition, and Popularity Leadership Styles Stand Alone

Power and influence, tradition, and popularity leadership styles basically stand alone, without direct connections to other leadership categories: in some cases leadership is attained by knowing somebody, tradition, or "playing the game." Obtaining the position of leader is not necessarily connected to skills and abilities formally developed for democratic leadership. There are always informal connections between categories; for example, human relations skills can be used to get appointed to a leadership position for which one does not qualify.

Leadership and Personal Development Skills Help the Total Person

Besides leading to employment, leadership and personal development skills are important for family and personal relationships. Leadership and personal development skills also help you become a better member of your community and civic organizations.

DEMOCRATIC, AUTHORITARIAN, AND LAISSEZ-FAIRE BEHAVIORAL LEADERSHIP STYLES

Democratic and authoritarian leaders are within the behavioral leadership category mentioned earlier. Most leadership styles, categories, and types center around whether a person has an authoritarian (theory X) or democratic (theory Y) philosophy. The late Douglas McGregor, while a professor at MIT, published a book in 1960 called *The Human Side of Enterprise*. In this book, McGregor set forth what were to become his famous theory X and theory Y. Although some people have generalized these ideas inappropriately, a brief review at this point will be helpful in achieving a better understanding of leaders.

Theory X (Authoritarian)

Each theory, as McGregor defined it, is a set of assumptions made about people in general—a way some of us look at humanity. Theory X is traditional and quite familiar. Here are the main characteristics of a theory X person:

- The average human being has an inherent dislike of work and avoids it if possible.
- Because of this human characteristic of dislike for work, most people must be coerced, controlled, directed, or threatened with punishment to get them to work toward the achievement of organizational objectives.
- The average human being prefers to be directed, wishes to avoid responsibility, has relatively little ambition, and wants security above all.

Those who hold such views expect their coworkers to be lazy, to require close control, and to "goof off" at every opportunity. The gentleman in Figure 1–12 appears by his facial expression to be a theory X individual.

Theory Y (Democratic)

Theory Y reflects a totally different set of values and expectations of people. Here is McGregor's theory Y description.

- The expenditure of physical and mental effort in work is as natural as in play or rest.
- External control and threat of punishment are not the only means for bringing about effort toward organizational goals.
- The average human being learns, under proper conditions, not merely to accept but to seek responsibility.
- Under the conditions of modern industrial life, the intellectual potential of the average human being are only partially used.

Figure 1–12 What characteristics of this gentleman would lead one to believe he has a theory X philosophy? (Courtesy of USDA #K4830-2.)

Figure 1–13 What characteristics of these two people would lead one to believe they have a theory Y philosophy? (Courtesy of USDA #K-4563-14.)

Theory Y states that people work because it is natural for them and, under the proper conditions, they want to achieve the goals of the group. It is generally true that the theory Y approach offers a leader more options in working with people. By their nonverbal expressions, the two workers in Figure 1–13 appear to be theory Y individuals.

Theory X and theory Y may seem somewhat simplistic. However, they are important in that they dramatize how we may feel about co-workers. If you view your co-workers as lazy people who must be coerced and controlled so that you can perform your leadership functions properly, this attitude will affect your behavior toward them.

You may unconsciously be causing mistrust, suspicion, rebellion, and many other forms of nonproductive behaviors. On the other hand, acceptance, confidence, and trust of co-workers shows that you expect them to want to do a good job. It adds to their self-esteem, self-respect, and motivation to do a good job, thus contributing to group effectiveness. Refer to Figure 1–14 for other assumptions of theory X and theory Y leaders.

Authoritarian Leaders

The authoritarian leader makes the decisions and closely supervises or instructs people. This can be related to theory X assumptions. Authoritarian leaders use power and position to force their decisions on their co-workers. Co-workers are expected simply to do as they are told. They are not asked for any suggestions, and they could even be reprimanded for making suggestions.

Saddam Hussein of Iraq is an authoritarian leader. Hitler was an authoritarian leader. Joseph Stalin ruled the Soviet Union in an authoritarian manner for 30 years, using force, power, and fear to get his way. However, there are times when even democratic leaders have to exert authoritarian leadership, as in time of war. The military has to use an authoritarian style of leadership even though we are a democracy. It wouldn't be

Motivating Subordinates (Two Sets of Assumptions about People)	
Traditional X (Authoritarian)	**Progressive Y (Democratic)**
People are naturally lazy; they prefer to do nothing.	People are naturally active; they set goals and enjoy striving.
People work mostly for money and status rewards.	People seek many satisfactions in work: pride in achievement, enjoyment of process, sense of contribution, pleasure in association, and the stimulation of new challenges.
The main force keeping people productive in their work is fear of being demoted or fired.	The main force keeping people productive is the desire to achieve their personal and social goals.
People remain children; they are naturally dependent on leaders.	People normally mature beyond childhood; they aspire to independence, self-fulfillment, and responsibility.
People expect and depend on direction from above; they do not want to think for themselves.	People close to the situation see and feel what is needed and are capable of self-direction.
People need to be told, shown, and trained in proper methods of work.	People who understand and care about what they are doing can devise and improve their own methods of doing work.
People need supervisors who will watch them closely enough to be able to praise good work and reprimand errors.	People need a sense that they are respected and are mature enough to do good work without constant supervision.
People have little concern beyond their immediate, material interests.	People seek to give meaning to their lives by identifying with nations, communities, churches, unions, companies, or causes.
People need specific instruction on what to do and how to do it; larger policy issues are not any of their business.	People need ever-increasing understanding; they need to grasp the meaning of the activities in which they are engaged; they have cognitive hunger as extensive as the universe.
People appreciate being treated with courtesy.	People crave genuine respect from their peers.
People are naturally compartmentalized; work demands are entirely different from leisure activities.	People are naturally integrated; when work and play are too sharply separated, both deteriorate.
People naturally resist change; they prefer to stay in the old ruts. *(cont'd)*	People naturally tire of monotonous routine and enjoy new experiences; everyone is creative to some degree. *(cont'd)*

Figure 1–14 Traditional X represents the old beliefs of viewing and leading others. It is the authoritarian way. Potential Y represents the new way of leading, the democratic way. Some situations, however, may require the traditional X perspective. (Adapted from *Horace Small Manufacturing Training Manual*, Nashville, TN: Horace Small Mfg. Co., 1992.)

Motivating Subordinates *(Continued)*	
Traditional X (Authoritarian)	**Progressive Y (Democratic)**
Jobs are primary and must be done; people are selected, trained, and fitted to predefined jobs.	People are primary and seek self-realization; jobs must be designed, modified, and fitted to people.
People are formed by heredity, childhood, and youth; as adults, they remain static; old dogs don't learn new tricks.	People constantly grow; it is never too late to learn; they enjoy learning and increasing their understanding and capability.
People need to be "inspired" (pep talk), pushed, or driven.	People need to be encouraged and assisted.

Figure 1–14 *(concluded)*

appropriate to have a vote from the soldiers on when to charge or retreat in the midst of a battle.

Democratic Leadership

Except when the situation demands other means, most of us prefer theory Y or democratic leadership. American democracy is founded on the concept of shared leadership and wide participation by many members of a group in making and carrying out group decisions. This concept is essential to the successful operation of the FFA or any other club, group, or organization.

Democratic leadership is based on the participation of members of a group in making decisions. Each person is recognized as important by the leader, whose job is to help the group get things done. The democratic leader also protects the rights of the minority while the wishes of the majority are being developed and pursued. If there are procedures to be followed, it is the democratic leader's duty to see they are followed.

Most Americans accept certain beliefs about leadership, just as they do about other areas in life. The democratic style of leadership is a part of our heritage and has been important to the success of American democracy. Most of us believe democratic leadership is superior to autocratic leadership because we think that each individual has something to contribute to the group or the nation. We believe the collective judgment of a group is better and wiser than the opinion of any one individual.

Consider the following in comparing authoritarian and democratic leadership.

> ***Khrushchev, Hitler, and Napoleon used leadership to attain power over people and to further their personal interests, whereas Kennedy, Churchill, and Lincoln used their leadership abilities to help people. The authoritarian leader talks about "my program" and says, "I have decided." The democratic leader discusses "our program" and says, "We have decided."***[20]

Sociologists who study how groups function report that participation in making group decisions increases the morale of individuals in the group. Group members are more likely to accept and carry out decisions they help make than those that are imposed on the group.

Some leaders act autocratically in one situation and democratically in another. It cannot be said that democratic leaders do not use authority. To accept leadership means to accept the responsibility to use authority for the best interest of the group, but the use of authority is tempered by democratic beliefs.

Group Involvement Since democratic leaders cannot simply order that things be done, they must involve group members. They are responsible for directing the activities of the organization. An effective democratic leader works with the group to accomplish its objectives.

Democratic leaders must share leadership with others. They need to encourage the growth of informal leadership within their groups. In helping this process along, the democratic leader ensures that all opinions are heard, all sides of an issue are discussed, and all members are encouraged to express their ideas. If the leader fulfills these functions, the action taken by the group will reflect the will of the majority.

Characteristics A democratic leader encourages the following:

- Participation by group members
- Group cooperation
- The full expression of opinions
- The use of accurate and adequate information
- Careful thinking and evaluation
- The expression of all relevant suggestions
- The expression of differences of opinion
- A fair hearing for all sides of an issue
- Proper behavior by members
- Progress by the group
- Direction and purpose
- Faith in democracy
- Willingness to assume responsibility
- Open-mindedness
- Service above self
- Respect and confidence of fellow members
- Belief in the authority of the group[21]

Understanding authoritarian and democratic philosophies is the key to understanding leadership. Several sophisticated models of leadership exist, but they all center around authoritarian and democratic attitudes. In Chapter 2, you will see how these philosophies relate to personality types.

Laissez-Faire Leadership Style

A leader who uses the **laissez-faire leadership** style believes the group can make its own decisions without the leader or, at least, with very little input. Although it may appear that these leaders are avoiding responsibility, they recognize that members are often in a better position to make a decision than the leader.

Elementary teachers, for example, use an authoritarian style of leadership. High school teachers tend to use a situational style of leadership. College professors teaching undergraduate students can take a more democratic approach, and graduate students can be led using the laissez-faire leadership style because of their intelligence, experience, and competence.

It is appropriate to use the laissez-faire leadership style only when the group's level of maturity and intelligence is close to that of the leader. Whatever the situation, the leader makes the final decision about which leadership style to use.

SITUATIONAL (CONTINGENCY) LEADERSHIP

A common myth related to leadership style is that one style is always superior to the others. Although the preferred style for most is democratic, all styles may be appropriate, depending on the situation in the group. Leaders need to answer three basic questions before selecting a leadership style.

- Is there an obvious solution to the task?
- Is it important for the leader to make the decision?
- Does the decision have to be made immediately?

If the leader answers yes to any of these questions, the leader should make the decision. Some input from members might be helpful, but the leader should be directive. If the answer is no to these questions, the leader assesses the situation and selects a style.

How Leaders Assess a Situation

Leaders can assess the situation by determining the members' ability and willingness to complete the activity. If it is low, the leader uses a telling style; if it is moderate to low, the leader uses a selling style; if it is moderate to high, the leader uses the participating style; and if it is high, the leader uses a delegating style.[22] Refer to Figure 1–15 for a visual explanation.

Leaders need to evaluate the individuals involved for each activity or task. An individual might be able and willing to perform one task but unable to perform another task. In this situation, the leader selects two different leadership styles to use with the same individual. Therefore, when a leader's leadership style is used appropriately with the corresponding level of follower readiness (willing and able), it is the optimum use of situational leadership. There are four steps in assessing a situation.

Step 1

Determine your preferred leadership style.
There are several evaluations to help you determine which leadership style you prefer. Check the Internet sites listed at the end of the chapter. The style can then be applied as follows.

Telling—Provides specific instructions and supervises performance. Synonyms are guiding, directing, and establishing. This style tends to be authoritarian.

Selling—Explains decisions and provides opportunities for clarification. Synonyms are explaining, clarifying, and persuading. This style tends to be democratic.

Participating—Shares ideas and facilitates the making of decisions. Other descriptors are encouraging, collaborating, and committing. This style tends to be democratic.

Delegating—Turns over responsibilities for decision and implementation. Other descriptors are observing, monitoring, and fulfilling. This style tends to be laissez-faire.[23]

Step 2

Once you have determined your preferred leadership style, determine when to use each style according to the task.
As mentioned earlier, there is no "best" supervisory style for all situations. Instead, the effective

How Leaders Assess a Situation		
Assessment of Follower	**Which Leadership Style Do I Use?**	**Who Makes the Decision?**
Not able and not willing	Telling	Leader (authoritarian)
Willing, not able	Selling	Leader; consults group (democratic)
Able, not willing	Participating	Jointly—leader and group (democratic)
Able and willing	Delegating	Group (laissez-faire)

Figure 1–15 For the most effective leadership, each situation should be assessed. The leader assesses the followers, decides which leadership style to use, and decides whether it should be a leader or group decision. (Adapted from *Personal Skill Development*, College Station, TX: Texas A & M University Instructional Materials Service, 1989.)

leader adapts his or her style to meet the capabilities of the individual or group.

Step 3

Determine the capabilities of your co-workers. Co-worker capability may be measured on indicators of willingness and ability, which you, as a leader, must determine. You select the capability level that best describes a co-worker's ability and motivation for the specific task.

There are several ways to determine the co-worker's willingness and ability. Leaders who know their co-workers are more able to assess their willingness and ability accurately.[24]

Indicators of Willingness—The individual

- Has an interest in the activity
- Volunteers for the activity
- Discusses the activity with others
- Displays a positive attitude toward the group
- Follows through with commitments

Indicators of Ability—The individual

- Has experience in the activity
- Has skills in related activities
- Is intelligent and can think through problems
- Can find and use resources effectively
- Is self-directed

Step 4

Use the appropriate leadership style. Once again, the "correct" leadership style depends on the situation and the individual; the situation is a function of co-worker capability. The better you are at matching your leadership style to co-workers' capabilities, the greater your chances of being a successful leader.

Continuum of Leadership Behavior

Another approach to describing leadership styles is to use a **continuum** rather than placing specific titles or names on styles. On the continuum, styles vary from leader-centered decisions to group-centered decisions.

All styles on the continuum are appropriate and effective, depending on the situation. Leader-centered or autocratic does not mean the leader is a dictator. Similarly, group-centered or laissez-faire does not mean the leader does not care and is avoiding responsibility by leaving the decision to the group.[25]

Conclusion

We can change or alter our leadership style. People are born with natural tendencies, but through education we learn what works best in particular situations. Just as we can alter our eating habits to our advantage, we can alter our leadership style to match the situation even though it is not our natural tendency. Great leaders care more than others think is wise, risk more than others think is safe, dream more than others think is practical, and expect more than others think is possible.

Summary

Leadership is the ability to move or influence others toward achieving individual or group goals. Leaders are needed in the workplace, organizations, communities, states, and nations. There is no limit to the number of leaders within a group. A true leader helps the group achieve what it believes is important.

Leadership is many things to many people, however, leaders know who they are, what their strengths and weaknesses are, and how to fully display their strengths and compensate for their weaknesses. They also know what they want, why they want it, and how to communicate what they want to others to gain their cooperation and support. Leaders know how to achieve their goals.

We do many things in agricultural education and the FFA that contribute to leadership and personal development. These include leading group discussions, setting goals, developing social skills, following parliamentary procedure, and managing financial resources.

Leaders are not born, they are made. There are seven leadership categories: trait, power and influence, behavior, situational, traditional, popularity, and combination.

Of the many leadership styles, categories, and types, most center around whether a person has an authoritarian (theory X) or democratic (theory Y) philosophy. Authoritarian leaders believe people have to be pushed, whereas democratic leaders believe people can be pulled or led.

Except when the situation demands other means, most of us prefer democratic leadership. American democracy was founded on the concepts of shared leadership and wide participation by many members of a group in making and carrying out group decisions. These concepts are essential to the successful operation of the FFA or any other club, group, or organization.

END-OF-CHAPTER EXERCISES

Review Questions

1. Define the Terms to Know.
2. What is the difference between a leader and a boss?
3. Do students really learn leadership and personal development skills by taking agricultural education courses and being active FFA members? Support your answer.
4. List eight myths about leaders and leadership.
5. List the seven categories of leadership.
6. Distinguish between traits and skills.
7. Distinguish between power and influence.
8. Name three FFA symbols that are part of the traditional organizational culture.
9. Give examples of three other symbols in our society (corporate, public, or educational) that either directly or indirectly signify leadership.
10. Why should your FFA chapter be aware of the popularity leadership category?
11. List 15 assumptions of theory X leaders and 15 assumptions of theory Y leaders.
12. Compare the democratic and authoritarian leadership styles.
13. List four steps leaders use in assessing a situation so they can use the appropriate leadership style.
14. What are the four leadership style possibilities to be selected in step 1 of question 13?
15. List five indicators of co-worker willingness and five indicators of co-worker ability.
16. Briefly explain the continuum of leadership behavior.

Fill-in-the-Blank

1. Power stemming from formal authority is sometimes called ________________ power.
2. Potential influence based on control over rewards is sometimes called ________________ or power.
3. Control over punishment is sometimes called ________________ power.

4. Control over the physical environment, technology, and organization of the work within a group is called __________________ control or __________________ engineering.
5. __________________ power is the type of power a person has who is the only one with the ability to solve a special problem or perform important tasks.
6. Friendship and loyalty are sometimes called __________________ power.
7. __________________ power structures involve people who are not elected or appointed to positions, but due to their earned respect or reference, exercise power over appointed or elected leaders.
8. __________________ power is a pervasive process that involves efforts by members of a group to ensure their power and protect existing power sources.
9. The combination leadership category includes four frames or images of leadership. They are __________________, __________________, __________________, and __________________.
10. A leader who uses the __________________ leadership style believes the group can make its own decisions with little or no leader input.

Matching

_____ 1. Tall and distinguished looking.
_____ 2. Comes in many forms, such as legitimate, reward, conceived, expert, referent.
_____ 3. Opening ceremony, FFA official dress.
_____ 4. Leadership is bestowed on those who may not deserve it.
_____ 5. Use whatever or as many resources as needed to lead.
_____ 6. Authoritarian or democratic leaders.
_____ 7. Different leadership style for different occasions.

A. cultural and symbolic
B. behavior
C. power and influence
D. trait
E. popularity
F. structural
G. combination

Activities

1. After reading the different definitions of leadership, write a definition that has the most meaning to you.
2. List 10 areas of the FFA that contribute to leadership and personal development. Select an activity for each area that you could perform or experience while in agricultural education and the FFA to enhance your leadership and personal development skills.
3. Write a one-page essay on the importance of learning and following parliamentary procedure in developing democratic leaders.
4. Study "the big picture" (Figure 1–11) of leadership, personal development, employment, and life skills. Write a one-page essay explaining at least five connections, observations, or relationships between leadership, personal development, employment, and life skills.
5. Describe and defend an example of when a leader would need to be each of the following: authoritarian, democratic, laissez-faire, and situational.
6. Prepare a one-page case study or scenario describing a leader who uses theory Y the democratic style of leadership.

7. Prepare a one-page case study or scenario describing a leader who uses theory X or the autocratic style of leadership.
8. Prepare a one-page case study or scenario describing a leader who uses the laissez-faire style of leadership.
9. Prepare a one-page case study or scenario describing a leader who uses the situational (contingency) style of leadership.
10. Select one of the following leadership categories and write a one-page case study or scenario describing a person you know who fits the category. Don't mention the person's name.
    - Trait leadership
    - Power or influence
    - Traditional leadership
    - Popularity leadership
11. **Take it to the Net.**

    Listed below are some Web sites that contain articles on leadership. Browse the Web sites and the articles. Try to find an article related to leadership styles. Print the article and write a summary of it. If you are having problems with the Web sites or if you cannot find an article on leadership styles, try some of the search terms listed below in the search field.

    Web sites

    <http://www.bpubs.com/Management_Science/Leadership/>

    <http://www.mapnp.org/library/ldrship/ldrship.htm>

    Search terms

    leadership styles

    leadership styles articles

    leadership

Notes

1. *Leadership of Youth Groups* (Kansas City, Missouri: Farmland Industries, [no date available]), p. 12.
2. W. G. Bennis and B. Nanus, *Leaders: Strategies for Teachers* (New York: Harper & Row, 1997).
3. W. G. Bennis, *On Becoming a Leader* (Menlo Park, CA: Addison-Wesley, 1995).
4. L. R. Rathburn, *The Relationship Between Participation in Vocational Student Organizations and Student Success* (Columbus, OH: Unpublished Dissertation, Ohio State University, 1974).
5. Samuel C. Ricketts, *Leadership and Personal Development Abilities Possessed by High School Seniors Who Are FFA Members in Superior FFA Chapters, Non-Superior Chapters, and by Seniors Who Were Never Enrolled in Vocational Agriculture* (Columbus, OH: Dissertation, Ohio State University, 1982). Note: Can be found in ERIC.
6. Ibid.
7. *Personal Skill Development in Agriculture* (College Station, TX: Instructional Materials Service, 1989), 87370A, p. 1.
8. These seven categories were adapted from work by Dr. Tom Burks, Director of the Leadership Institute at Middle Tennessee State University.
9. Tom Burks, *Leadership in Higher Education* (Nashville, TN: Unpublished Dissertation, Vanderbilt University, 1992).
10. E. Bensiman, "The Meaning of Good Presidential Leadership: A Frame Analysis," *The Review of Higher Education*, December 1989, pp. 107–123.

11. J. French and B. H. Raven, "The Bases of Social Power," in *Studies of Social Power*, ed. D. Cartwright (Ann Arbor, MI: Institute of Social Research, 1959).
12. A. M. Pettigrew, *The Politics of Organizational Decision Making* (London: Tavistock, 1973).
13. Gary A. Yukl, *Leadership in Organizations*, Third Edition (Englewood Cliffs, NJ: Prentice Hall, 1997).
14. T. Deal and A. Kennedy, *Corporate Cultures* (New York: Addison-Wesley, 1982).
15. Burks, *Leadership in Higher Education.*
16. Ibid.
17. The terms used by Burks were *structural, human resource, political,* and *symbolic.*
18. Burks, *Leadership in Higher Education.*
19. Ibid.
20. R. E. Bender, R. E. Taylor, C. K. Hansen, and L. H. Newcomb, *The FFA and You* (Danville, IL: Interstate Printers and Publishers, 1979), p. 137.
21. Ibid., pp. 139–141.
22. P. Hersey and K. Blanchard, *Management of Organizational Behavior: Utilizing Human Resources*, 5th edition (Englewood Cliffs, NJ: Prentice-Hall, 1996).
23. Ibid.
24. *Personal Skill Development in Agriculture*, 8737-B, p. 4.
25. Robert N. Lussier, *Human Relations in Organizations: Application and Skill Building* (Columbus, OH: McGraw-Hill Higher Edu., 1998).

CHAPTER

2

Personality Types and Their Relationship to Leadership and Human Behavior

Objectives

After completing this chapter, the student should be able to:

- Describe the four personality types
- Explain the importance of learning personality types
- Compare personality types with extroverts and introverts
- Show the relationship between personality type and leadership style
- Describe the communication styles of the four personality types
- Identify ways to adapt your communication style to that of others
- Explain the communication continuum
- Show the relationship between personality types and group decision making, learning styles, and career selection

Terms to Know

sanguine
choleric
melancholy
phlegmatic
impetuousness
perpetuating
conceptualize
self-deprecating
subtlety
authenticity
vivacity
self-esteem
socializers
directors
thinkers
relaters
reprimanding

Have you ever wondered why you loved agriculture while your friend hated it, felt inspired by one particular teacher while your classmates thought she was boring? Have you ever been uncomfortable at a party where others seem to be having a wonderful time? Do you ever wonder why some people miss deadlines and never seem to have things organized? All these are connected to our personality type and our preferred way of doing and viewing things.

Try this experiment. Hold out your arms as wide apart as you can. Bring them together as though you were clapping, and clasp your hands together. Look at which thumb is on top. Is it your right or left thumb? Now do the same thing, but this time change thumbs so that the other one is on top. It usually feels awkward, even uncomfortable. Do it again with your favorite thumb on top. If you clasped your hands together thousands of times you would probably place your favorite thumb on top every time. This is what we call a preferred way of acting.

Besides personality type, we also have personal preferences in leadership style, communication style, learning style, and careers. When you are in a classroom environment that matches your learning style, everything feels right. The teacher is stimulating, the material exciting, the work enjoyable. If the environment does not match your preferred learning or communication style, you feel out of place, uncomfortable, and unable to do your best.

There are no right or wrong, good or bad styles, just preferences. Understanding learning styles and personality preferences has helped people succeed in class and the workplace. It provides an important dimension of self-discovery and personal growth throughout life.

The leader who recognizes his or her own personality type and other leadership characteristics as well as the personality type, leadership style, communication style, and learning style of his co-workers can channel each person's strengths into productive areas while minimizing or eliminating their weaknesses. In this chapter, we explore each personality type and communication style and discuss how each type correlates to leadership styles, learning styles, and career selection.

Personality Types

People behave differently: some people are usually self-motivated, whereas others can't be motivated. Some people are outgoing; others are not. Some people take pride in their dress and grooming, and others do not. We tend to use trait adjectives such as warm, aggressive, and easygoing to describe people's behavior. Personality consists of a relatively stable set of behavioral traits or characteristics that aids in explaining and predicting individual behavior. Our personalities are the product of both genetics and environmental factors. Knowing someone's personality traits can help us understand and predict that person's behavior in a given situation. Our personality affects our human relations.

Can we predict our or someone else's leadership style based on personality type? I believe we can. If we know ourselves and the personalities of others, we can determine how a leader will react to a given situation.

Origin of the Identification of Four Personality Types

Hippocrates, the Greek physician and philosopher, believed there were four types of temperament. He mistakenly believed these four temperaments were the result of body liquids that predominate in each individual. He identified the liquids as follows.

- **Sanguine** (blood)—lively temperaments
- **Choleric** (yellow bile)—active temperaments
- **Melancholy** (black bile)—dark temperaments
- **Phlegmatic** (phlegm)—slow temperaments

Obviously, the idea that temperament is determined by body liquid has been discounted, but the four classifications of temperament are still widely used. Although modern psychology has presented many new suggestions for classifications of temperaments or personality, few have found more acceptance than those of Hippocrates.[1] Other names for each of the four basic personalities are shown in Figure 2–1.

Personal Profiles Personality profiles measure a person's personality. After completing a survey and analyzing the results, you can better understand yourself and others and how they will react in certain situations. All of us have developed behavioral patterns or distinct ways of thinking, feeling, and acting. Our behavioral patterns tend to remain stable because they reflect our individual identities.

Understand Personality Type The demands of everyday living often require different responses that evolve into a behavior style or personality type. You should heighten understanding of your personality type and identify situations most conducive to your success. At the same time, you should learn about the differences of others and the situations they require for maximum productivity and harmony in the workplace and community. Research evidence supports the conclusion that the most effective people are those who know themselves, know the demands of the situation, and adapt strategies to meet those needs. Therefore, a personality profile enables you to

- Identify your behavior style or personality type
- Create the situation most conducive to success
- Increase your appreciation of different personality types
- Identify and minimize potential conflicts with others[2]

The FFA's Made for Excellence program uses a "Team Discovery Profile" from the Carlson Learning Company in Minneapolis, Minnesota. Personality Profiles can be secured from them or from one of the Web sites in Activity 9 at the end of this chapter.

Ratio and Blend of Personality Types

The four temperaments or personality types are basic temperaments. No person has a single temperament type; we all display a mixture of temperaments, although usually one predominates. There are varying degrees of temperament: some may be 60 percent sanguine and 40 percent choleric. Some are a mixture of more than two. Many people have a mixture of all four personality types, such as 30 percent sanguine, 40 percent choleric, 20 percent melancholy, and 10 percent phlegmatic. In reality, determining ratios and blends is not important. The important thing is determining your basic personality type as well as the basic personality type of others. If we know our personality as well as those we lead, we can become better leaders. We can lead and direct according to the situation. Refer to Figure 2–2.

Hippocrates	Myers Briggs Type Inventory	Carlson Learning Company	True Colors	Author
Sanguine	Intuitive feelers	Influencing	Orange	Charismatic/influencer
Choleric	Intuitive thinkers	Dominant	Gold	My way
Melancholy	Sensation thinkers	Cautious	Green	Methodical
Phlegmatic	Sensation feelers	Steady	Blue	Laid-back/carefree

Figure 2–1 Although most people agree that there are four basic personality types, different people have different names for them. Five ways of characterizing each personality type are presented here.

General Characteristics of the Sanguine Personality Type

Characteristics Sanguines need the freedom of immediate action. A zest for life and a desire to test the limits characterize these people. They take pride in being highly skilled in a variety of fields. They are master negotiators. They like adventure. A hands-on approach to problem solving and a direct line of reasoning creates the excitement and immediate results sanguines admire.

Sanguines value freedom and excitement. They think that being skillful is more important than structure, logic, and feelings. They like being spontaneous and want to enjoy what they are doing. Planning things sometimes takes the fun out of it. Sanguines like games and competition. They also like to learn things that they can put to use immediately.

Strengths Sanguines have a warm, lively, and "enjoying" temperament; they are extroverted. They are receptive by nature, and external impressions easily influence them. Often they rely on feelings rather than rational thinking. They tend to be cheerful people. Sanguines often become salespeople, teachers, politicians, actors, and public speakers.

Sanguine Students Sanguines perform well in competition, especially when there is a lot of action. They love games and hands-on activities. Since sanguines love fun and excitement, they have difficulty with routines or structured presentations. Sanquines get a kick out of putting what they learn to immediate use. Sanguine students like classes that have contests, changes of pace, and variety. They learn by doing, like immediate results, and like tools; they are impulsive, physical, and competitive.

Sanguines as Friends Planning ahead bores sanguines because they never know what they want to do until the moment arrives. They like to excite their dates with new and different things, places to go, and romantic moments.

Sanguines with Family Sanguines need plenty of space and freedom. They want everyone to have fun. It is hard for them to follow rules and feel they should all just enjoy one another.

Sources of Personal Success for Sanguine Personalities

- The impulse to really live
- Testing the limits
- The need for variation
- Excitement, light-heartedness
- Charged adventure
- Being a natural entertainer
- Spontaneous relationships
- Taking off for "somewhere else"
- Being able to act in a crisis
- A love of tools
- Charm, wit, and fun
- Taking defeats only temporarily
- Considering waiting an "emotional death"[3]

Sanguine Uniqueness

Dream of:	being free, spontaneity, **impetuousness**
Value:	skills, grace, finesse, charisma
Regard:	opportunities, options, competition
Dislike:	rigidness, authority, forcefulness
Express:	optimism, impatience, eagerness, confidence
Foster:	recreation, fun, enjoyment
Respect:	skill, artistic expression
Promote:	stimulation, risk[4]

Weaknesses When studied carefully, the boundless activity of the sanguine temperament proves to be little more than restless energy. Such people are often impractical and disorganized. Their emotional nature can get them instantly excited. Often, they do not make good students because of this restless energy. Their lifelong pattern of restless activity usually proves

Strengths of the Four Personality Types				
	Sanguine	**Choleric**	**Melancholy**	**Phlegmatic**
Emotions	Appealing personality Talkative, storyteller Life of the party Good sense of humor Memory for color Physically holds on to listener Emotional and demonstrative Enthusiastic and expressive Cheerful and bubbling over Curious Good on stage Wide-eyed and innocent Lives in the present Changeable disposition Sincere at heart Always a child	Born leader Dynamic and active Compulsive need for change Must correct wrongs Strong-willed and decisive Unemotional Not easily discouraged Independent and self-sufficient Exudes confidence Can run anything	Deep and thoughtful Analytical Serious and purposeful Genius prone Talented and creative Artistic or musical Philosophical and poetic Appreciative of beauty Sensitive to others Self-sacrificing Conscientious Idealistic	Low-key personality Easygoing and relaxed Calm, cool, and collected Patient, well-balanced Consistent life Quiet but witty Sympathetic and kind Keeps emotions hidden Happily reconciled to life All-purpose person
Work	Volunteers for jobs Thinks up new activities Looks great on the surface Creative and colorful Has energy and enthusiasm Starts in a flashy way Inspires others to join Charms others to work	Goal-oriented Sees the whole picture Organizes well Seeks practical solutions Moves quickly to action Delegates work Insists on production Makes the goal Stimulates activity Thrives on opposition	Schedule-oriented Perfectionist, high standards Detail-conscious Persistent and thorough Orderly and organized Neat and tidy Economical Sees the problems Finds creative solutions Needs to finish what he/she starts Likes charts, graphs, figures, lists	Competent and steady Peaceful and agreeable Has administrative ability Mediates problems Avoids conflicts Good under pressure Finds the easy way
Friends	Makes friends easily Loves people Thrives on compliments Seems exciting Envied by others Doesn't hold grudges Apologizes quickly Prevents dull moments Likes spontaneous activities	Has little need for friends Will work for group activity Will lead and organize Is usually right Excels in emergencies	Makes friends cautiously Content to stay in background Avoids attracting attention Faithful and devoted Will listen to complaints Can solve others' problems Deep concern for other people Moved to tears with compassion Seeks ideal mate	Easy to get along with Pleasant and enjoyable Inoffensive Good listener Dry sense of humor Enjoys watching people Has many friends Has compassion and concern

Figure 2–2 Each personality type has strengths and weaknesses. Remember that no person is all one personality type but a combination. (Adapted from *Spirit-Controlled Temperament*, by Tim LaHaye. Wheaton, IL: Tyndale House Publishers, 1966.)

Weaknesses of the Four Personality Types (continued)

	Sanguine	Choleric	Melancholy	Phlegmatic
Emotions	Compulsive talker Exaggerates and elaborates Dwells on trivia Can't remember names Scares others off Too happy for some Has restless energy Egotistical Blusters and complains Naive, gets taken in Has loud voice and laugh Controlled by circumstances Gets angry easily Seems phony to some Never grows up	Bossy Impatient Quick-tempered Can't relax Too impetuous Enjoys controversy and arguments Won't give up when losing Comes on too strong Inflexible Is not complimentary Dislikes tears and emotions Is unsympathetic	Remembers the negatives Moody and depressed Enjoys being hurt Has false humility Off in another world Low self-image Has selective hearing Self-centered Too introspective Guilt feelings Persecution complex Tends to hypochondria	Unenthusiastic Fearful and worried Indecisive Avoids responsibility Quiet will of iron Selfish Too shy and reticent Too compromising Self-righteous
Work	Would rather talk Forgets obligations Doesn't follow through Confidence fades fast Undisciplined Priorities out of order Decides by feelings Easily distracted Wastes time talking	Little tolerance for mistakes Doesn't analyze details Bored by trivia May make rash decisions May be rude or tactless Manipulates people Demanding of others End justifies the means Work may become his/her god Demands loyalty in the ranks	Not people-oriented Depressed over imperfections Chooses difficult work Hesitant to start projects Spends too much time planning Prefers analysis to work Self-deprecating Hard to please Standards often too high Deep need for approval	Not goal-oriented Lacks self-motivation Hard to get moving Resents being pushed Lazy and careless Discourages others Would rather watch
Friends	Hates to be alone Needs to be center stage Wants to be popular Looks for credit Dominates conversations Interrupts and doesn't listen Answers for others Fickle and forgetful Makes excuses Repeats stories	Tends to use people Dominates others Decides for others Knows everything Can do everything better Is too independent Possessive of friends and mate Can't apologize May be right, but unpopular	Lives through others Insecure socially Withdrawn and remote Critical of others Holds back affection Dislikes those in opposition Suspicious of people Antagonistic and vengeful Unforgiving Full of contradictions Skeptical of compliments	Dampens enthusiasm Stays uninvolved Is not exciting Indifferent to plans Judges others Sarcastic and teasing Resists change

Figure 2–2 (*concluded*)

unproductive in the long run. They seldom live up to their potential. Frequently, their lives are spent running from one tangent to another, and unless disciplined, they are not lastingly productive.[5]

General Characteristics of the Choleric Personality Type

Characteristics Cholerics value order and cherish the traditions of home and family. They provide and support the structure of our society. Steadfastness and loyalty are their trademarks. Generous and parental by nature, cholerics show they care by making sure everyone does the right thing. It never occurs to them to disregard responsibility of any kind.

Being responsible and following the rules is more important than excitement and feelings. They like family life and saving money, and they plan to really make something of themselves. They enjoy belonging to groups and want to help make them run smoothly. They enjoy learning about things that are useful to them.

Strengths The choleric has a bossy, quick, active, practical, and strong-willed temperament. These people are often self-sufficient and very independent. They tend to be decisive and opinionated and find it easy to make decisions for themselves as well as for others.

Choleric Students As students, cholerics do their best when course content is structured and clearly defined. Abstract ideas and concepts should not be introduced until the foundations of a subject are plainly presented. Cholerics always want to know when they are on the right track; rules and direction are a great help to them. These people believe students should share in the responsibilities and duties of the classroom. Choleric students prefer useful subjects, thrive on routine and orderliness, are punctual and dependable, think problems through before making a decision, respect rules, have a strong sense of right and wrong, and respect the school rules.

Cholerics as Friends Cholerics prefer people who are careful with their money and who make plans ahead of time. They like their dates to be loyal, dependable, and on time. Cholerics are serious about love and show it in many practical ways.

Cholerics with Family Stability and security are important to this type, and they enjoy traditions and frequent celebrations. They like to spend holidays with family members, and they plan such gatherings for months.

Sources of Personal Success for Choleric Personalities

- Generosity
- The work ethic
- Parental nature
- Ceremony
- Sense of history
- Dignity, culture
- **Perpetuating** heritage
- Steadfastness
- Value of order
- Predictability
- Home and family
- Establishing and organizing institutions[6]

Choleric Uniqueness

Dream of:	assets, wealth, influence, status, security
Value:	dependability, accountability, responsibility
Regard:	service, dedication
Dislike:	disobedience, nonconformity, insubordination
Express:	concern, stability, purpose
Foster:	institutions, traditions
Respect:	loyalty, obligation
Promote:	groups, ties, bonds, associations, organizations[7]

Weaknesses The admirable strengths of the choleric carry with them some serious weaknesses. The most prominent are their hard, angry, impetuous, self-sufficient traits. Their bossy, authoritarian ways are not appreciated by most coworkers. They do not give up when they are losing. They cannot admit to mistakes and cholerics cannot apologize. A very unfortunate trait is that when friends cross them, they become bitter and end the relationship, even when the choleric is wrong.

General Characteristics of the Melancholy Personality Type

Characteristics Those with melancholy personalities feel best about themselves when they are solving problems and their ideas are recognized, especially when they feel ingenious. The melancholy personality seeks to express him- or herself through the ability to be an expert in everything. A melancholy's idea of a great day is to use knowledge to create solutions, because these people are complex individualists of great analytical ability. Although melancholies do not express their emotions openly, they experience deep feelings.

Melancholy personalities value knowledge—knowledge is their strength. Discovering solutions and using their brains are more important than are feelings, rules, and nonstop excitement. They like to know how and why things work in a certain way. Melancholies prefer to work on their own and also need room to think.

Strengths The melancholy is often referred to as the dark temperament. A stereotype of this personality would be a researcher, scientist, lab technician, or mathematician. They are analytical, self-sacrificing, gifted perfectionists with very sensitive emotions. No one gets more enjoyment from the fine arts than the melancholy.

Melancholy Students Melancholies perform best when exposed to the driving force or overall theory behind a subject. They prefer to work independently. New ideas and concepts arouse their curiosity, and they enjoy interpreting ideas before adding them to their bank of knowledge. Melancholies are gratified by probing abstract concepts. Melancholies respond positively to the recognition and appreciation of their competence in a subject. Students with a melancholy personality are logical, theoretical, and curious. Melancholy students **conceptualize**, are driven to understand, learn best independently, and need to be immediately challenged.

Melancholies as Friends These personalities may seem cool and without emotion. They are uneasy about being controlled by their emotions and do not want relationships to be complex. Once they have expressed their feelings, talking about their emotions causes doubt.

Melancholies with Family In the family, the melancholy personality is often seen as a loner because this type likes a lot of private time to think. Sometimes they find family activities boring and have difficulty following family rules that don't make sense to them.

Sources of Personal Success for Melancholy Personalities

- Developing models
- Abstract thinking
- Analytical processes
- Exploring ideas
- Variety of interests
- Striving for competency
- Admiring intelligence
- Storing wisdom and knowledge
- Being a perfectionist
- Abhorring redundancy
- Using precise language
- Handling complexity[8]

Melancholy Uniqueness

Dream of:	truth, perfection, accuracy
Value:	answers, resolutions, intelligence, explanations
Regard:	efficiency, increased output, reduced waste
Dislike:	injustice, unfairness
Express:	coolness, calm, reserve
Foster:	inventions, technology
Respect:	knowledge, capability
Promote:	effectiveness, competence, know-how[9]

Weaknesses Because of the perfectionist and analytical traits, the melancholy is prone to be pessimistic. No one can be more critical than the melancholy. They are more self-centered than any of the other personality types. The melancholy personality type is also more **self-deprecating** than any of the others. They are inclined to strict self-examination, which paralyzes their will and energy.[10]

General Characteristics of the Phlegmatic Personality Type

Characteristics The phlegmatic type seeks to express his or her inner self with authenticity and honesty above all else. Sensitive to **subtlety**, phlegmatics create roles in life's drama with a special flair. They enjoy close relationships with those they love, and there is a strong spirituality in their nature. Making a difference in the world comes easily to phlegmatics because they cultivate the potential in themselves and others.

Phlegmatics value people. Being liked and having everyone around them getting along is more important than are facts, rules, adventure, or logic. They are sensitive to others and become uncomfortable when there is conflict or competition. They like socializing and working with people.

Strengths Phlegmatic personality types get their name from what Hippocrates thought was the body fluid that produced a calm, cool, slow, easygoing, and well-balanced temperament. Life for them is a happy, unexcited, pleasant experience in which they avoid as much involvement as possible. They are extreme introverts and passive.

Phlegmatic Students These students feel best in an open, interactive atmosphere. They like to feel that their teacher really cares about them and gives the classroom a personal touch. Phlegmatics appreciate a lot of genuine human feedback. They thrive in a "humanistic," people-oriented environment. Phlegmatics turn off when conflicts arise, and they flourish in an atmosphere of cooperation. It is important that their teachers value them as people and respect their feelings. Phlegmatics are verbal and social, work best in a group setting, are sensitive to rejection and conflicts with teachers, and need to feel valued and reassured.

Phlegmatics as Friends Phlegmatic types look for perfect love. They are very romantic and enjoy touching, holding hands, love poems and notes, flowers, and quiet talks.

Phlegmatics with Family In family life, phlegmatic people like to be happy and loving. They are very sensitive to rejection from family and to family conflicts. They really like to be well thought of and need frequent reassurance. Phlegmatics love intimate talks and warm feelings.

Sources of Personal Success for Phlegmatic Personalities

- **Authenticity** as a standard
- Seeking reality
- Devotion to relationships
- Cultivating potential in others
- Assuming creative roles in life's drama
- Writing and speaking with poetic flair
- Self-searching
- Having a life of significance
- Sensitivity to subtlety
- Spirituality

- Making a difference in the world
- Seeking harmony[11]

Phlegmatic Uniqueness

Dream of:	love, affection, authenticity
Value:	compassion, sympathy, rapport
Regard:	meaning, significance, identity
Dislike:	hypocrisy, deception, insincerity
Express:	**vivacity**, enthusiasm, inspiration
Foster:	potential growth in people, harmony
Respect:	nurturing, empathy, shared feelings
Promote:	growth and development of others[12]

Weaknesses The outstanding weakness of phlegmatics is that they tend to be slow and lazy. They often appear to be dragging their feet because they resent having been stimulated to action against their will, so they go along as slowly as they can.[13]

Importance of Learning Personality Types

Understanding the four personality types will help you be more accepting and tolerant of others. Things that people do will then amuse you rather than hurt you, offend you, or make you angry, once you realize that they are victims of their personality type. As we mature, we learn not to take things personally. Once we learn why we do things the way we do, we can take steps to curb our undesirable behaviors.

Even though our temperaments or attitudes cannot be eliminated, we learn that it is not good to exhibit certain behaviors in particular situations. For example, a person might have an autocratic (choleric) attitude but see that it is more appropriate to follow the democratic process when conducting meetings and in many other situations. A president of a major university once had this dilemma. Being very educated, he knew that the acceptable way to run a university was in a democratic style. The faculty and students appreciated him because they had a voice. However, one-on-one with midlevel administrators, his authoritarian attitude surfaced; he became hostile when challenged. Nevertheless, since he knew his undesirable personality behaviors and generally was able to curb them, he was a somewhat successful administrator.

Although we are born with certain temperaments, we can be taught how to become leaders by adapting our personality types to varying situations.

Learning Personality Types in the Workplace

Personality tests exist in greater and greater numbers. Companies have become aware that mere warm bodies filling positions is not enough, rather, they must hire the right person, and employees must perform in order to maintain productivity and the competitive edge. Thus, tests are being used increasingly for hiring staff and for building teams among current staff.[14]

Many companies find that applying tests promotes mutual understanding.[15] When they are well used, tests make staff aware of each others' styles so they can work better together. They also improve communication and productivity. Indeed, these tests must be doing something right. In 1986, 1.5 million people in the United States took the Myers-Briggs Type Indicator (MBTI). Of these test administrations, 40 percent were business related. Corporations such as Apple, AT&T, Citicorp, Exxon, GE, Honeywell, and 3M have all used personality profile tests for management development.

Results of Using Personality Tests in the Workplace

It is often argued that people of different personality types have difficulty working together. Through personality testing and the sharing of results, people become aware of their own and others' personality traits. As a consequence, their communication and productivity usually improve. Knight-Ridder's *Charlotte Observer* used the MBTI as a basis for team building in a fractious newsroom with positive results.[16] As Rome says, "Comfort comes from being with people similar to us; growth comes from the differences of an alternative approach. Different styles can be uncomfortable, but often are essential to good management."[17]

Transamerica Corporation offers a one-week course on MBTI at considerable cost, but it has found the investment invaluable. "We've used the theory to help us change our corporate culture; it has turned out to be one of the most meaningful things we've done," one company executive said.[18] The company has used MBTI to avoid placing people in positions of potential conflict with each other. Personality tests make "people aware of which types they and their co-workers are, thus, communication improves and with it productivity."[19] Group feedback is vital and managers gain a composite of a team or work group to help in their decision making.[20]

Infusion of the Personality Types

All four of the basic personality types are needed to give variety and purposefulness in this world. No personality type can be said to be better than another. Each one contains strengths and advantages, yet each one has weaknesses and disadvantages.

Consider the following sequence of positive events involving the four personality types: The hard-driving choleric produces the inventions of the genius-prone melancholy, which are sold by the personable sanguine and enjoyed by the easygoing phlegmatic. On the other hand, consider the shortcomings of the four personality types in their relationships to other people: The sanguine type enjoys people and then forgets them. The melancholic is annoyed with people but lets them go their own crooked ways. The choleric makes use of people for his or her own benefit, then ignores them. The phlegmatic regards people with indifference.

COMPARISON OF PERSONALITY TYPE WITH EXTROVERTS AND INTROVERTS

Generally, personality types are divided into two categories: extroverts and introverts. From the previous discussion, it can be seen that the sanguine and choleric personality types tend to be extroverted, and the melancholy and phlegmatic personality types tend to be introverted. There are gray areas, however. For example, some people in comfortable or familiar surroundings may exhibit an extroverted personality, but in a less comfortable and familiar environment, that same person may act introverted.

Activities through the FFA, such as leadership camps, Made for Excellence programs, the Washington Conference, and classes or lessons on leadership and personal development, can help students develop the skills to become more outgoing (extroverted) when their natural tendencies are reserved (introverted). Of course, extroverts have to learn that there are times when it is better to be introverted: part of wisdom is to speak when you need to speak and remain silent when you should remain silent.

Although we all have inadequacies, we still control our destiny. Our attitude controls our altitude. How far we go in life or what we accomplish depends on our attitude, along with how hard we are willing to work to overcome any natural weaknesses.

RELATIONSHIP BETWEEN PERSONALITY TYPE AND LEADERSHIP STYLE

Some personality types correlate with either a democratic or an authoritarian leadership style. For example, sanguine personalities have a more democratic attitude, whereas choleric personalities have a more authoritarian attitude. Melancholy personality types tend to be situational leaders, and phlegmatic types tend to be laissez-faire leaders.

Can we change? Absolutely. Thus arises the confusion about whether leaders are born or not. People are born with natural tendencies, but through education we learn what works best in particular situations. Even though this is a democratic nation, there are times when authoritarian decisions must be made. For example, once it is democratically decided that an interstate highway is needed, authoritarian actions are needed to obtain property from land owners. Remember, no one personality type is best. All types are needed for a complete, well-rounded society, group, or team. Figure 2–3 provides a summary of the four types, along with affirmation statements for each to build **self-esteem**.

Changing or altering personality types can be compared to eating and weight control. Some people, due to metabolism, can eat as much of whatever they want without gaining weight. Others must constantly watch what they eat or they will gain weight. Similarly, we can alter our personality types to match the best leadership style even though it is not our natural tendency.

Communication Styles of the Four Personality Types

We have discussed personality types and their relationship to leadership styles. Remember that sanguine personality types tend to be democratic leaders, and choleric personality types tend to be autocratic leaders. To take this one step further, leadership style and personality type are closely related to communication style. We can become better communicators if we adapt our communication style to the leadership and personality type of the person with whom we are speaking.

Tony Alessandra and Phil Hunsaker in their book, *Communicating at Work*, have identified four styles of communicators: **socializers**, **directors**, **thinkers**, and **relaters**. Each type of communicator has particular personality characteristics that affect communication with other people. The reason for examining communication styles is, of course, to improve communication. This may be done by changing your own communication style so you can relate to any of the other communication styles; or it may mean adapting your style to mesh with the style of one person with whom you are communicating.

In this section, I describe each communication style so you may determine your own category. Then I suggest how to make your style more adaptable to the other three styles as a whole. Finally, I offer ways to adapt yourself to each of the four styles individually.

Socializers

Socializers are relationship-oriented people who appear to need the approval of those around them. They move, act, and speak quickly and avoid details when possible. They are also risk takers who want excitement and change. They enjoy the spotlight and have good persuasive skills (Figure 2–4). Socializers may be public relations specialists, talk show hosts, trial attorneys, social directors on cruise ships, or hotel personnel.[21] You may recognize this as the sanguine personality.

If you are a socializer, you can adapt your own style by reducing your need for approval from other people or groups and concentrating on developing more directive skills and actions, such as self-assertion, conflict resolution, and negotiation.[22]

If you find you must deal with a person who is a socializer, you may adapt to that person by

- Supporting his or her opinions, ideas, and dreams
- Allowing the discussion to flow and occasionally go off the topic

Summary of Four Personality Types (Strategies for Strengthening Your Self-Esteem)

Personality Type	Strengths	Weaknesses (Limitations)	Affirmation Statements*
Sanguine (influencing)	Enthusiastic Good communicator Optimistic Involved Spontaneous Persuasive People person Imaginative	Excitable, emotional Talks too much Unrealistic Disorganized Impulsive, undisciplined Manipulative Goes along with peers Daydreamer	I am enthusiastic about life. I am good at expressing my thoughts, opinions, and ideas. I am a positive person. I am eager to participate in everything that is going on. I join in on things quickly. I have a unique ability to motivate people. I like people and want them to like me. I have a lot of creative ideas.
Choleric (dominance)	Goal-oriented Confident Gets results Competitive Decisive, determined Courageous Direct, straightforward Responds quickly	Impatient Self-reliant Never slows down Attacks first Stubborn Reckless Blunt, harsh Lacks empathy	I am able to set my mind on something and then go after it. I am capable of handling many things on my own. I can accomplish a lot when I make up my mind. I am a winner. I know what I want and go after it. I am courageous. I honestly express how I feel about things. I am quick to respond to a situation and seek a solution.
Melancholy (conscientiousness)	Analytical Cautious, intense Conscientious Sensitive, intuitive Strives for excellence Does things correctly High personal standards Curious, questioning	Critical Unsociable Worries too much Easily hurt Perfectionist Fears criticism of work Judgmental Nosy	I carefully think things through. I focus a lot of energy on getting things right. I am a hard worker. I always do my best. I am attentive to what others say and feel. I am inspired by excellence. I take my time to do things correctly and without mistakes. I am a person with high standards. I like to understand all I can about what I am planning to do.
Phlegmatic (steadiness)	Steadfast Stable Systematic Easygoing Agreeable Good listener Soft-hearted Reliable	Resists change Boring Slower paced Lacks initiative, Indecisive Too accommodating Uncommunicative Easily manipulated Overly dependent	I like things to stay the same. I stick with things that I know work well. I take my time to do something well. I am relaxed and pleasant to be around. I make it a point to get along with everyone. I am a good listener. I am a compassionate person. I follow through on things I begin.

* An affirmation statement is a positive declaration about who we are and what we can become.

Figure 2–3 No person is a single temperament type, but most people reveal a pattern of behavior indicating that they lean toward one. If you know your own personality as well as that of your co-workers, you can become a better leader. (Adapted and used with permission of Carlson Learning Company.)

Figure 2–4 Former FFA member, comedian, cowboy poet, and weekend team roper Baxter Black delivers a speech. Mr. Black has a socializer communication style. (Courtesy of *FFA New Horizons Magazine*.)

- Being entertaining and fast moving
- Being interested in him or her
- Avoiding conflict and arguments
- Agreeing and making notes of the specifics of any agreement
- Complimenting his or her appearance, creative ideas, persuasiveness, and charisma
- Allowing him or her to "get things off his or her chest."[23]

Increasing the Self-Esteem of Socializers (Sanguines) There are several ways to increase the esteem of socializers by praising and rewarding them in ways most meaningful to them:

- Making immediate and timely favorable responses
- Focusing on their behavior and performance more than the finished product
- Appreciating their cleverness and spontaneity
- Commenting on skills they demonstrate
- Noting the quickness of their actions
- Recognizing the impact of their performance
- Giving a variety of tangible rewards[24]

Directors

Directors tend to be task-oriented. They move, act, and speak quickly. They want to be in charge and seek results through others. They make decisions quickly and demonstrate good administrative skills. Directors may be hard-driving newspaper reporters, stockbrokers, independent consultants, corporate CEOs, or drill sergeants.[25] You may recognize this as the choleric or dominance personality. Would you agree that the gentleman in the lab coat in Figure 2–5 could very well have a director communication style?

Figure 2–5 What nonverbal characteristics could lead someone to believe that the man on the left has a director communication style? (Courtesy of USDA.)

If you have a director style, you may wish to adapt to the other styles by lowering your emphasis on control of other people and conditions and by focusing on supportive skills and actions, such as listening, questioning, and positive reinforcement.

If you must deal with a director style, you may adapt to that person by

- Supporting his or her goals and objectives
- Talking about the desired results
- Keeping your communications businesslike
- Recognizing his or her ideas rather than recognizing him or her
- Being precise, efficient, and well organized
- Providing him or her with clearly described options with supporting analysis
- Arguing on facts, not feelings, when disagreements occur[26]

Increasing the Self-Esteem of Directors There are several ways to increase the esteem of directors by praising and rewarding them in ways most meaningful to them:

- Being honest and sincere
- Specifically mentioning what they have actually accomplished
- Noting accuracy, efficiency, and thoroughness in their performance
- Acknowledging their sense of responsibility
- Noting how the task they completed affects the well-being of others
- Recognizing how their effort makes a significant contribution[27]

Thinkers

Thinkers, like directors, tend to be task-oriented, but they move, act, and speak slowly. They enjoy solitary, intellectual work and are greatly concerned with accuracy. They are cautious decision makers and demonstrate good problem-solving skills. Thinkers may enter professions such as accounting, engineering, computer programming, the hard sciences (chemistry, physics, math), systems analysis, and architecture. You may recognize this as the melancholy or cautious personality type. The researcher shown in Figure 2–6 in all probability has a thinker communication style.

If you have a thinker style, try decreasing your need for unnecessary perfectionism and the tendency to focus on weakness in favor of developing supportive skills and actions, such as listening, positive reinforcement of others, and involvement with others of complementary strengths.

You can adapt to a person with a thinker style by

- Being thorough and well prepared
- Supporting his or her organized, thoughtful approach
- Complimenting his or her efficiency, thought processes, and organization
- Demonstrating through actions rather than words
- Being systematic, exact, organized, and prepared

Figure 2–6 What nonverbal characteristics could lead someone to believe that this man has a thinker communication style? (Courtesy of USDA.)

- Describing a process in detail and explaining how it will produce results
- Asking questions and letting him or her show you how much he or she knows
- Answering questions and providing details and analysis
- Listing advantages and disadvantages of any plan
- Providing solid, tangible, factual evidence[28]

Increasing the Self-Esteem of Thinkers There are several ways to increase the esteem of thinkers by praising and rewarding them in ways most meaningful to them:

- Acknowledging their ideas and competent performance with sincerity and appreciation
- Stating your recognition of the value and usefulness of their work
- Recognizing their specific knowledge and skills, especially in dealing with abstractions
- Noting their creativity and ingenuity
- Only giving deserved positive feedback
- Acknowledging their intellectual capabilities to analyze and give precise explanations
- Acknowledging their ability to complete tasks independently
- Providing more opportunities to exhibit competence as the best reward for a job well done
- Avoiding fake hoopla[29]

Relaters

Relaters are relationship-oriented. They move, act, and speak slowly. They avoid risk and seek tranquility (calmness) and peace. They enjoy teamwork and show good counseling skills. Relaters may enter the helping professions, such as counseling, teaching, social work, the ministry, psychology, nursing, and human resource development. You may recognize this personality type as being the phlegmatic or steady personality type. The person in Figure 2–7 appears to have a relater communication style.

Figure 2–7 What nonverbal characteristics could lead someone to believe that this man has a relater communication style? (Courtesy of USDA.)

If you have a relater style, you may adapt to the other styles by decreasing your resistance to trying or seeking out new opportunities and by trying to develop directive skills and actions, such as negotiation and divergent thinking.

In communicating with a relater, the following is helpful:

- Be warm and sincere.
- Support his or her feelings by showing personal interest.
- Assume he or she will take everything personally.
- Allow him or her time to develop trust in you.
- Move along in an informal, slow manner.
- Listen actively.

- Discuss personal feelings in the event of a disagreement.
- Discuss and support the relationship.
- Compliment his or her teamwork, relationships with others, and ability to "get along."[30]

Increasing the Self-Esteem of Relaters There are several ways to increase the esteem of relaters by praising and rewarding them in ways most meaningful to them:

- Frequently telling them how good you feel about their achievements and contributions
- Frequently acknowledging their unique personal characteristics
- Demonstrating that you care how they feel
- Clarifying their importance
- Responding to their honesty and sincerity in like manner
- Openly reflecting their participation in successful group sessions[31]

When you learn the four communication styles and how to adapt to them, you are learning to improve your communication skills. You are not imitating another style but simply relating to the other person in the style that person prefers. By doing this, you are better able to convey your messages, ideas, and beliefs while at the same time reducing the possibility of being misunderstood or experiencing conflict. Refer to Appendixes B, C, and D for further insight into the four personality and communication styles.

The Communication Continuum

Your behavior often communicates many things to those around you. Think of passive, assertive, and aggressive as points on a continuum; try to spend less time on the passive and aggressive styles (Figure 2–8). Aggressive behavior can show disrespect for others. Passive behavior can show a holding back of true feelings and ideas. Assertive behavior is honestly expressing your feelings and thoughts without threatening others or experiencing anxieties. Assertive behavior is desirable. Nonverbal assertiveness includes having a confident body position, positive eye contact, and using comfortable hand gestures.[32,33]

An example of aggressive behavior showing disrespect for others is glaring at someone or using uncomplimentary hand gestures. Passive behaviors include laughing at a question that seems to pry too deeply into your personal beliefs or feelings.

Passive

When a leader moves to the passive end of the continuum, the characteristics and consequences shown in Figure 2–9 occur. The goals of a passive leader are to be liked, to be nice and friendly, and to appease others. Permeating these goals is the desire to avoid conflict at all costs. They want to be liked rather than admired. The basic message of passive leaders is this: "What I think doesn't matter, what I feel is unimportant. I don't respect

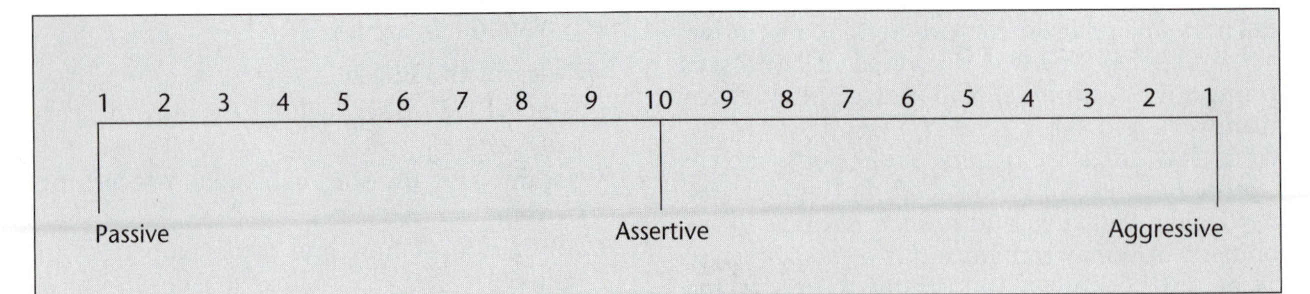

Figure 2–8 In most situations, leaders should be assertive communicators. Passive and aggressive styles are typically undesirable.

Characteristics	Consequences
Indecisive	Things don't get done: frustration
Not clear about expectations	Lack of vision
Unable to delegate	Things don't get done
Won't take a stand	Lack of leadership
Never gives correction	Don't know what to change

Figure 2–9 Passive communicators do not take charge of the moment, which results in things not getting done. Passive communicators exhibit little self-esteem, deny or do not express feelings, keep things to themselves, and allow others to make choices for them.

Payoffs	Costs
Avoid confrontation	Fear
Occasionally is liked	Doesn't meet needs of people
Few open conflicts	Perceived as weak

Figure 2–10 Because passive communicators do not take charge or deal with situations directly, they do not meet people's needs and are perceived as weak.

myself and I don't expect you to either." People accept our own subconscious evaluation of ourselves.

The thought patterns of passive leaders include these:

- I'm not okay, but I'm not sure sure about you.
- Everyone has rights but me.
- I can survive if everyone likes me and approves of what I do, say, and feel.
- Nice people don't disagree.
- Don't make waves, don't rock the boat; people won't like you.
- Peace at any price.
- What I think is unimportant.
- I won't offer my opinion. People might laugh at me.
- It's not my place to speak up.

The consequences of being a passive leader are listed in Figure 2–10.

Aggressive

When leaders move to the aggressive end of the continuum, the characteristics and consequences listed in Figure 2–11 occur.

The goals of aggressive leaders are to win, dominate, intimidate, overpower, and get what they want when they want it. They often obtain their goals by belittling, **reprimanding**, and humiliating others. The basic message of aggressive leaders is this: "You will never have to wonder what I think because I am going to tell you! You will never have to wonder what I feel because I am going to tell you! I guarantee that you are going to do what I want you to do, even if I have to use fear and intimidation to get you to do it! You are even more stupid than I thought if you disagree!"

The thought patterns of aggressive leaders include these:

- I'm okay. You're not!
- I have rights. You don't.
- People should do what I want without questioning me!
- Personnel doesn't send me good people any more!

Characteristics	Consequences
My way—PERIOD!	Blunts creativity
Humiliates others	Avoidance, sabotage
Thinks people need constant surveillance	People perceive it as nontrust
Emphasizes what people do that's wrong	Frustration, rejection

Figure 2–11 Aggressive communicators like to show authority. They go overboard when they communicate, wanting things their way. Aggressive communicators also diminish self-esteem in others, express their own negative feelings, say what they think regardless of who it hurts, like to be in control, and make choices for others. They also reach goals at the expense of others.

- If more people were like me, we wouldn't have the problems we have.
- I am never wrong!
- My feelings are more important than yours!
- I don't need to listen to them. They have nothing to offer me.

The consequences for being an aggressive leader are listed in Figure 2–12.

Situations in which aggressiveness might be your most effective response follow.

- When there is confusion
- When it is time to move . . . and no one is moving
- When counterproductive ethics are involved
- To get attention when you need to emphasize the importance of a matter
- To clear the air when tension has built up

Payoffs	Costs
Win a lot in the short run	Fear
Get what they want	Distance
Protect self and space	Anger
Dominate	Lack of teamwork

Figure 2–12 Because aggressive communicators are authoritarians, they tend to get what they want. The result, however, is fear, anger, and lack of teamwork from others.

Assertiveness

When leaders communicate assertively, the characteristics and consequences in Figure 2–13 occur.

The goals of assertive leaders are to get the work done at a high level of excellence while enhancing the growth and development of those doing the work, to communicate in a style that is accurate and respectful of the dignity of all persons involved, and to encourage those they work with to do the same.

The basic message of assertive leaders is this: "You will never have to wonder what I think because I will tell you. You will never have to wonder what I feel because I will share that with you. I guarantee you that I have no interest in being critical of you for what you think or feel. Indeed, I invite to share these things with me. We are here to get the job done and to contribute to a positive work environment."

Thought patterns of assertive leaders include these:

- I'm okay, and you're okay.
- I have rights, and so do others.
- It is alright to learn from mistakes.

Characteristics	Consequences
Competent	Trusted
Communicates goals	Knows what is expected
Listens	Cooperation
Doesn't put down	Good feelings about relationship
Helps others feel good about themselves	Teamwork—working together rather than merely alongside one another
Ability to praise and correct	You know where you stand

Figure 2–13 Leaders should strive to be assertive communicators, a style that is accurate and respectful of others' dignity. Assertive communicators have high self-esteem, which contributes to others' esteem for them; they make their own choices; and they accept responsibility for their decisions.

- I am a valuable and worthy person, and so are you.
- I have choices in nearly all situations, and I am responsible for the consequences of my choices.
- I trust you, and you can trust me.
- I am not a helpless victim.
- I will not allow others to decide for me how I will behave.
- Conflicts provide opportunities to grow and are not something to be avoided.
- I want to find a way we can all win.

The consequences for being an assertive manager are listed in Figure 2–14. Situations in which assertiveness might not be your most effective response are when it is more appropriate to be passive or aggressive

Payoffs	Costs
Little fear	More conflict due to freedom of expression
Live your own life	Less fear of ridicule and rejection
Improved confidence	Some may be threatened or intimidated by your confidence
You do what needs to be done	Unpopular with some

Figure 2–14 Because assertive communicators take charge by doing what needs to be done and are free and open, they permit those around them to be creative. Creative people are expressive and have less fear of ridicule and rejection. Remember, assertive people cannot please everyone.

Relationship Between Personality Type, Group Decision Making, Learning Style, and Career Selection

Once your personality type is identified, you can determine many other things about yourself. We have discussed how leadership styles relate to personality type and how communication style and preference relate to personality type. In other chapters I discuss how group decision making, learning styles, and the careers we select relate to personality type.

Group Decision Making

Group leaders, whether they are conscious or unconscious of it, have leadership styles they use in directing the group. How leaders involve the group in the problem-solving and decision-making processes reveals their leadership style. In Chapter 13, I show the connection between leadership styles and group problem solving and decision making. Figure 2–15 shows that autocratic decision makers tend to be choleric, consultative decision makers tend to be sanguine, participating decision makers tend to be melancholy, and laissez-faire decision makers tend to be phlegmatic.

Learning Styles

In the next chapter, people are placed in one of four groups. Your learning style correlates to your personality type. You learn why we behave and react to

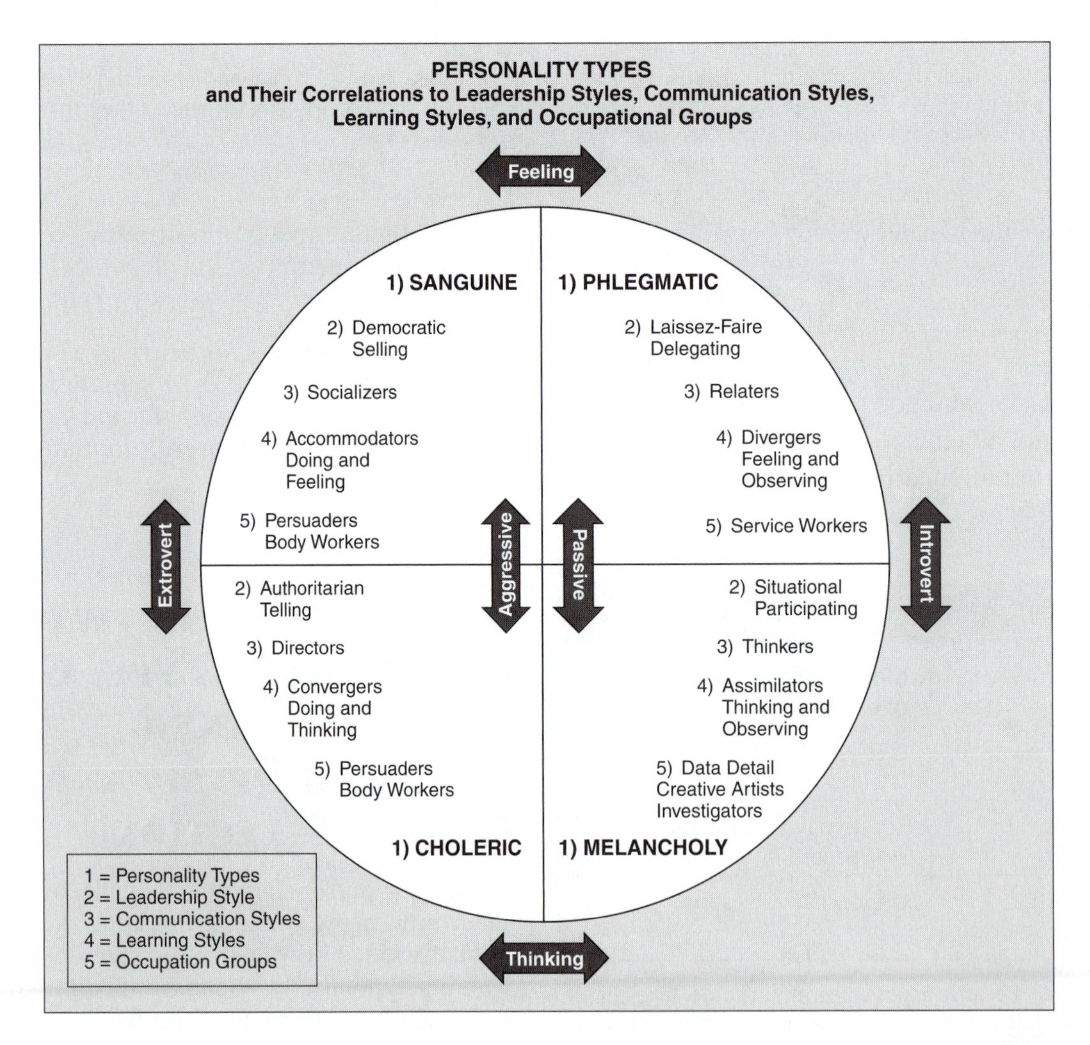

Figure 2–15 Personality types and their correlations with leadership styles, communication styles, learning styles, and occupational groups.

the things we do. By referring to Figure 2–15, we can see the following correlations: those with accommodator learning styles are sanguine, those with the converger learning style are choleric, those with an assimilator learning style are melancholy, and those with a diverger learning style are phlegmatic. Knowing co-workers' learning styles and placing them in a compatible situation should lead to higher productivity for the company.

Career Selection

In Chapter 20, I show the correlation between occupational groups and personality type. In Figure 2–15, you can see that body workers tend to be sanguine and choleric, data detailers tend to be melancholy, persuaders tend to be sanguine and choleric, service workers tend to be phlegmatic, and artists and investigators tend to be melancholy.

Your career choice should correlate with your personality type and learning style. Otherwise, you may choose the wrong occupation and wonder why you are not happy with your job.

Conclusion

Which came first, the chicken or the egg? Did personality come from learning style or did learning style come from personality? Whatever the situation, there are close correlations between personality type, communication style, learning style, leadership type, group decision making, and even the careers we choose. Once we learn these correlations, we can communicate better, enhance our personal development, and become better leaders.

Summary

There is a relationship between personality type and leadership style. The four personality types are sanguine (influencing), choleric (dominant), melancholy (cautious), and phlegmatic (steady). Sanguine personalities have a more democratic attitude, whereas choleric personalities have a more authoritarian attitude. Melancholy personality types tend to be more situational leaders, and phlegmatics tend to be laissez-faire. Understanding the personality types will help you be more accepting and tolerant of others, as well as understand leadership styles.

Leadership style can be changed. People are born with natural tendencies, but through education we learn what works best in particular situations. Just as we can alter our eating habits to our advantage, we can alter our personality types to match the appropriate leadership style, even though it is not our natural tendency.

Understanding the four personality types will help you be more accepting and tolerant of others. Things that people do may amuse you rather than hurt you, offend you, or make you angry, once you realize that they are victims of their personality type. This knowledge is also being used more and more in hiring decisions and in building teams.

Generally, people are considered either extroverted or introverted. Sanguine and choleric personality types are extroverted; melancholy and phlegmatic types are introverted.

Leadership style and personality type are closely related to communication style. The four styles of communicators—socializers, directors, thinkers, and relaters—have personal characteristics and work habits that affect their communication with others. Learning how to adapt to each style is essential to being an effective communicator. Your behavior often communicates many things to those around you. Think of passive, assertive, and aggressive as points on a continuum. Try to spend less time on the passive and aggressive styles and more on the assertive style.

Just as personality relates to leadership and communication styles, there are correlations with psychological and personal development. Learning styles, group decision making, and career choice also relates to personality type.

END-OF-CHAPTER EXERCISES

Review Questions

1. Define the Terms to Know.
2. List five personal success traits of each of the four personality types.
3. Give three key characteristics each of sanguine, choleric, melancholy, and phlegmatic students.
4. Why is it important to learn personality types?
5. Which personality types are extroverts?
6. Which personality types are introverts?
7. Match the four personality types with the four leadership styles.
8. Briefly discuss the four communication styles.
9. Match the four communication styles with the four personality styles.
10. Briefly describe the three communication behaviors on the communication continuum.
11. Which of the three communication behaviors is the most desirable? Explain.
12. What are three other things (areas of life) that correlate to our personality and communication styles?

Fill-in-the-Blank

1. __________________ have a warm, lively, and "enjoying" temperament.
2. __________________ have a bossy, quick, active, practical, and strong-willed temperament.
3. __________________ are analytical, self-sacrificing, gifted perfectionists.
4. Life for __________________ is a happy, unexcited, pleasant experience in which they avoid as much involvement as possible.
5. __________________ and __________________ personality types tend to be extroverts.
6. __________________ and __________________ personality types tend to be introverts.
7. __________________ personalities have a more democratic attitude.
8. __________________ personalities have a more authoritarian attitude.
9. __________________ personalities tend to be laissez-faire leaders.
10. __________________ behavior along the communication continuum consists of honestly expressing your feelings and thoughts without threatening others or experiencing anxieties.

Matching

_____ 1. Communication style that enjoys being in charge.
_____ 2. Communication style that favors solitary intellectual work.
_____ 3. Communication style that tends to enter the helping professions.
_____ 4. Communication style that describes risk-takers who enjoy the spotlight.
_____ 5. Avoid conflict at all costs.
_____ 6. Goals are to win, dominate, intimidate, overpower, and get what they want when they want it.

A. relaters
B. aggressive
C. directors
D. passive
E. assertive
F. thinkers
G. socializers

_____ 7. Goals are to get the work done at a high level of excellence while enhancing the growth and development of those doing the work.

_____ 8. I'm okay, and you're okay.

_____ 9. I'm okay, you're not.

_____ 10. I'm not okay, but I'm not sure about you.

Activities

1. Determine your primary personality type. You can do this by one of three methods.
   a. Complete a survey, supplied by your teacher, from Personality Plus, Florence Littauer, or Fleming H. Revell Publishing or a similar one found on the Internet.
   b. Complete a survey, supplied by your teacher, from Team, Discovery Profile™, or Carlson Learning Company, Minneapolis, MN or a similar one found on the Internet.
   c. Study the four personality types summarized in Figure 2–3. Evaluate your own personality type and/or blend as well as each member of your class. Compare your self-evaluation with the evaluation made by others.
2. Give each personality type a fictitious name that correlates to the name and personality. Write a one page case study or scenario describing each.
3. Select four people (friends, acquaintances, celebrities, public figures, or historical) to match with each personality type: Write a brief description of each person describing why you believe they match the personality type you selected.
4. Study the four personality types described in Figure 2–3. Identify your areas of strength and weakness. Write five strengths that you can build on and five weaknesses of which you need to be aware.
5. Explain how knowing the strengths and weaknesses of each personality type will help you become a better leader.
6. On a piece of paper, make four columns. Label one column "socializer," one "director," one "thinker," and one "relator." Under each column, list the characteristics of each from the text. Try to identify which style of communication you feel you practice most. Compare your results with a classmate. See if they agree or disagree with you.
7. Study Figure 2–8. Describe where you believe you are on the continuum. Where would you like to be? Name the five biggest disadvantages that you will have to overcome to be on the most appropriate assertive communication level.
8. Draw a chart, picture, or matrix showing the relationship between personality types, extroverts and introverts, leadership styles, communication styles, learning styles, and career (occupational) groups.
9. **Take it to the Net.**

   Explore personality styles on the Internet. The Web sites listed below contain personality tests. Browse as many of the sites as you want and take as many of the tests as you want. Print the results of two tests. Write whether or not you agree with the results and explain why. If you are having problems with the Web sites, try using some of the search terms provided below.

   Web sites

   <http://users.rcn.com/zang.interport/personality.html>

   <http://www.queendom.com/typea.html>

<http://www.geocities.com/Area/1303/personality.html>
<http://www.9types.com/newtest>
<http://www.advisorteam.com/user/kts.asp>

Search Terms

personality types
personality test
personality test online
free personality test

Notes

1. T. LaHaye, *Spirit Controlled Temperament* (Wheaton, IL: Tyndale House Publishers, 1973), p. 10.
2. *Personality Profile* (Minneapolis, MN: Carlson Learning Company, 1992).
3. D. Lowry, *True Colors: Keys to Successful Teaching Profile Booklet* (Laguna Beach, CA: Communication Companies International, 1988).
4. Ibid.
5. LaHaye, *Spirit Controlled Temperament.*
6. Lowry, *True Colors: Keys to Successful Teaching Profile Booklet.*
7. Ibid.
8. Ibid.
9. Ibid.
10. LaHaye, *Spirit Controlled Temperament.*
11. Lowry, *True Colors: Keys to Successful Teaching Profile Booklet.*
12. Ibid.
13. LaHaye, *Spirit Controlled Temperament.*
14. K. Evans and R. Brown, "Reducing Recruitment Risk Through Pre-employment Testing," *Personnel* 65(9):55–64 (1988).
15. P. Thome, "Psychological Testing in Manager Assessment," *International Management* 44:50 (July/August, 1989).
16. T. Moore and W. Woods, "Personality Tests Are Back," *Fortune* 115(7):74–82 (1987).
17. L. Rome, "Myers-Briggs: A Tool for Building Effective Work Teams," *Wilson Library Bulletin* 42–47 (May, 1990).
18. Moore and Woods, *Fortune*, p. 75.
19. Ibid., p. 74.
20. J. Pepper, "What Kind of Manager Are You? Your Computer Knows," *Working Woman* 15(S):45–52 (1990).
21. T. Alessandra and P. Hunsaker, *Communicating at Work* (New York: Fireside, 1993), p. 44.
22. Ibid., p. 46.
23. Ibid., p. 47.
24. Lowry, *True Colors: Keys to Successful Teaching Profile Booklet.*
25. Alessandra and Hunsaker, *Communicating at Work*, p. 38.
26. Ibid., p. 47.
27. Lowry, *True Colors: Keys to Successful Teaching Profile Booklet.*
28. Alessandra and Hunsaker, *Communicating at Work*, p. 48.
29. Lowry, *True Colors: Keys to Successful Teaching Profile Booklet.*
30. Alessandra and Hunsaker, *Communicating at Work*, p. 48.
31. Lowry, *True Colors: Keys to Successful Teaching Profile Booklet.*
32. Notes from Dr. Joe Townsend's leadership class at Texas A&M University, College Station.
33. Appreciation is extended to my friend and associate John Grogan of John M. Grogan Company, 49 Chandler Road, Mt. Juliet, TN, 37122, for use of information on passive, assertive, and aggressive communication.

Learning Styles and Leadership

CHAPTER

Objectives

After completing this chapter, the student should be able to:

- Explain the misconceptions of intelligence
- Discuss the various factors of the cognitive learning domain
- Discuss various factors of the affective learning domain
- Discuss various factors of the psychomotor learning domain
- Explain how we learn
- Describe characteristics of the four learning styles

Terms to Know

intelligence
cognitive learning
linguistic intelligence
logical-mathematical intelligence
spatial intelligence
affective learning
interpersonal intelligence
intrapersonal intelligence
psychomotor learning
bodily kinesthetics
musical intelligence
diverger
assimilator
converger
accommodator

As leaders, we need to know how to get the best out of the people we lead. Some people can learn and do some things easily while others of equal **intelligence** have difficulty with the same task. Why is this? We will attempt to answer this question so that you can assign tasks appropriately to people when you lead.

There are some people, for example, who, because of their personality and learning style are naturally good at starting conversations and making people feel at ease. Others can interpret that behavior as aggressive, however, which can lead to miscommunication, bruised feelings, and mistrust. Similarly, individuals who have a natural passive behavior can be interpreted as disinterested or discriminatory. By understanding personality, behavior, and learning styles, we can often avoid misreading people by recognizing their natural way of doing things.[1] By understanding learning styles, you can harness the rich differences between people once you are in a leadership position. Here are a few things that you will be better able to do in a leadership role as you understand the various learning styles:

- Match individual potential with job requirements by understanding individual strengths and weaknesses.
- Resolve conflicts, realizing that at the root is a learning style problem rather than a personal one.
- Make work flow more smoothly by allowing each person to work according to his or her own style.
- Reduce stress levels by understanding that what can excite and energize one person can stress and drain another.
- Meet deadlines better by realizing that different types deal with time in different ways.[2]

As a future leader, you should understand your own learning style. This allows you to identify your personal preferences and how you are similar to and different from those in your group. Although we may think we prefer differences, in reality, few of us really make allowances for them. Once you understand learning styles, you will tend to appreciate the advantages of the differences and allow them to exist for the benefit of individuals as well as for the entire group.[3]

Knowing how people learn and their psychological strengths is key to unlocking the great potential in every individual. Unless leaders know the different areas of learning, they will confuse styles of learning with intelligence. I discuss three major areas of learning: cognitive, affective, and psychomotor. Cognitive is typically the area of learning recognized in schools. First, let us take a deeper look at intelligence.

MISCONCEPTIONS OF INTELLIGENCE

Intelligence Test

Psychologist Robert Sternberg of Yale University said that if we look at the people who make the greatest difference to our society, they are often not the people with the highest IQs.[4] IQ stands for intelligence quotient. Test criteria tend to mislead teachers, said Sternberg. Just as tests are not good at measuring workers in the work force, they also are not good at measuring practical intelligence. The intelligence being measured could very well be perception and cultural conditioning. This is not to say that it is not good to score high on tests and be a good student as measured by test scores. However, if a person does not score high on tests, it does not mean he or she is not intelligent. People are not all of one style of intelligence, and leaders need to foster all aspects of intelligence.[5]

Intelligent People Find a Way

Smart people find a way to succeed. We need to recognize people who are smart (intelligent) and people who figure out what they are good at, what they are not good at, and make the most of their strengths while compensating for their weaknesses. Leaders need to realize that the most practically intelligent people are not necessarily the ones who make high grades on tests but rather, those who figure out what they are good at and then capitalize on it. Being smart in the real world means making the most of what you have, not conforming to a stereotypical pattern of what others may consider smart.[6]

Intelligence Defined

We have grown accustomed in the twentieth century to associating high intelligence with studious or academic behavior. However, by definition, intelligence is the ability to respond successfully to new situations and the capacity to learn from one's past experiences. Intelligence depends on the context, the tasks, and the demands that life presents to us, not only on an IQ score.[7]

Intelligence Tests and Real-World Success

Although intelligence tests consistently predict school success, they fail to indicate how students will do after they get out into the real world. One study of highly successful professional people indicated that a third of them had low IQ scores.[8] IQ tests have been measuring something that might be more properly called "schoolhouse giftedness," whereas real intelligence takes in a much broader range of skills: cognitive, affective, and psychomotor.

Intelligence and Self-Concept

Why do we need to know this? Why is this important? The answer lies in the perception of intelligence and self-concept. Many people have been told they are dumb when, in reality, they are very smart. They may be weak in the cognitive areas that schools use to measure intelligence but brilliant in the affective and psychomotor areas.

Self-concepts are damaged due to lack of understanding of intelligence. Many people are, in fact, brilliant. However, they have been either directly or indirectly told they were not very smart. As a result, they don't achieve their full potential due to a damaged self-concept. For example, a former graduate student of mine had to have a special letter written to get into graduate school because his admissions scores (from a cognitive test) were not high enough. Yet this individual, with his brilliance in other areas, helped create knowledge and brought more positive attention to the university than did many of the professors.

People Possess Many Areas of Intelligence

Howard Gardner, a Harvard professor, released compelling evidence that each human possesses many intelligences. Each of those intelligences appear to be housed in a different part of the brain. Although other intelligence areas may exist, Gardner and co-author Hatch have identified seven.[9] Each of these is discussed within the cognitive, affective, and psychomotor domains of learning.

COGNITIVE LEARNING

The **cognitive learning** domain is usually thought of as book smart or test smart and is where most people make their judgment as to whether a person is intelligent (Figure 3–1). Students who are cognitive learners are analytical. They are considered bright, but they are not always good at coming up with their own ideas. This type of learning includes symbols, such as words, numbers, and graphic symbols, and sensory stimuli. Cognitive learning also includes the thought patterns and reasoning processes an individual uses most in reaching conclusions and solving problems.[10] Cognitive learning includes three of the intelligences mentioned earlier: linguistic, logical-mathematic, and spatial.

Figure 3–1 Cognitive learners are "book smart" and are good at taking tests. This student is absorbed in her studies, which should lead to good test scores.

Linguistic

Linguistic intelligence has to do with verbal abilities and sensitivity to language, meanings, and the relations among words. Those who possess great amounts of this kind of intelligence tend to be very good at writing, reading, speaking, and debating. Because conventional IQ tests place a great deal of value on linguistic abilities, these people usually are considered very smart.[11]

Linguistic job skills include talking, informing, giving instructions, writing, verbalizing, speaking a foreign language, interpreting, translating, teaching, lecturing, discussing, debating, researching, listening to words, copying, proofreading, editing, word processing, filing, and reporting.

Vocations that use linguistic intelligence include librarian, archivist, curator, editor, translator, speech pathologist, writer, radio/TV announcer, journalist, legal assistant, lawyer, secretary, typist, proofreader, and English teacher.

Logical-Mathematical

Logical-mathematical intelligence has to do with an individual's abilities in numbers, patterns, abstract thought, precision, counting, organization, and logical structure. Those who are naturally gifted in this area may be scientists, mathematicians, and philosophers.[12] For most, it means successfully balancing a checkbook or grasping the significance of basic economics. Logical-mathematical job skills include financing, budgeting, doing economic research, estimating, accounting, calculating, using statistics, reasoning, analyzing, classifying, and sequencing. Examples of vocations include auditor, accountant, purchasing agent, underwriter, mathematician, scientist, statistician, computer analyst, economist, and bookkeeper.

Spatial

Spatial intelligence gives you the ability to think in vivid mental pictures, keen observation, and recreating or restructuring a given image or situation. People with good spatial intelligence can look at something and instantly pinpoint areas that could change or alter its appearance. Highly spatial professions include architecture, drafting, and mechanical drafting.[13] Job skills include drawing, painting, visualizing, creating visual presentations, designing, inventing, imagining, illustrating, coloring, drafting, mapping, photography, decorating, and filming. Examples of vocations include engineer, surveyor, architect, urban planner, graphic artist, interior decorator, photographer, art teacher, inventor, pilot, and sculptor.

Most of the standardized tests used in schools measure only these three areas of cognitive learning. Very seldom is learning style analyzed further. Jean Thorpe in her *Learning Style Inventory Manual*, uses 26 additional facets in determining learning style. Most intelligence tests (cognitive tests) use only four of the 30 intelligence areas identified by Thorpe to measure intelligence. Admittedly, these tests do measure something, but do these scores transfer to success (intelligence) in life and the workplace?

AFFECTIVE LEARNING

Affective learning has to do with intelligence in the personality or human relations arena. People who are affective learners have characteristics such as persistence, curiosity, risk-taking, and social motivation.[14] Basically, they have learned to be "people persons." These affective learners would probably score high on a personality test, but they may not score particularly high on an IQ test. Affective learning includes communicating, body language, appreciating beauty, persuasiveness, knowing one's own strengths and weaknesses, commitment, and individuality (Figure 3–2). Two other types of affective intelligence are interpersonal and intrapersonal intelligence.

Figure 3–2 Communicating is an affective skill. This student is practicing her affective skill of public speaking.

Interpersonal

Interpersonal intelligence is the ability to understand, appreciate, and get along with other people. Interpersonal intelligence also includes sensitivity to others, ability to read the intentions and desires of others, and being considerate. This intelligence is not usually measured in the traditional academic setting. People with highly developed interpersonal intelligence have a sixth sense when it comes to reading another person. They can tell when something is wrong even if no words have been spoken.[15] Workplace skills include hosting, communicating, trading, tutoring, coaching, counseling, assessing others, motivating, selling, recruiting, publicizing, encouraging, supervising, coordinating, delegating, negotiating, and interviewing. Appropriate vocations include administrator, manager, school principal, personnel worker, arbitrator, social worker, counselor, psychologist, nurse, public relations person, salesperson, travel agent, and social director.

Intrapersonal

Intrapersonal intelligence involves understanding ourselves, being sensitive to our own values, and knowing who we are and how we fit into the world. People with strong intrapersonal intelligence enjoy reflection, meditation, and time alone. They seem to possess more positive self-concepts than most, and they don't rely on others' opinions to determine their life goals and aspirations.[16] Job skills include carrying out decisions, setting goals, attaining objectives, initiating, evaluating, appraising, planning, organizing, and discerning opportunities. Vocations include psychologist, clergy, psychology teacher, therapist, counselor, theologian, program planner, and entrepreneur.

As you can see, many of the job skills in the affective learning area are not measured in a traditional school setting. However, these skills lead to successful lives.

PSYCHOMOTOR LEARNING

Psychomotor learning refers to intelligence in the area of manual dexterity or using our hands in performing a mechanical skill. It includes a wide variety of talents or abilities, such as overhauling an engine, performing various carpentry, plumbing, electrical, and other skills usually associated with a trade. Playing a piano, typing, and using a computer are also in this category.

Leaders need to be aware of the people who are talented in this area. Whereas the cognitive learner may research technical aspects of a project, and the affective learners may sell or promote the product, the psychomotor learner is the one who actually constructs the product (Figure 3–3).

Gardner identified two types of intelligence in the psychomotor area: bodily kinesthetics and musical intelligence.

Figure 3–3 Potting plants is a necessary psychomotor skill for work in the floriculture industry. This is just one of the many floriculture skills this person can perform.

Bodily Kinesthetics

Bodily kinesthetics is the control of one's body and of objects, timing, and trained responses that function like reflexes. Although schools are highly enthusiastic about physical education and sports activities, the bodily kinesthetic intelligence is not often valued as a way of being smart. We should recognize and value kinesthetic intelligence, as well as academic (cognitive) intelligence, as an indicator of accomplishment.[17] The students in Figure 3–4 are learning by bodily kinesthetics as they do mechanical work. Workplace skills include crafting, restoring, shipping, delivering, manufacturing, repairing, assembling, installing, operating, performing, signing, dramatizing, and modeling (clothes). Some vocations that use these abilities are dancer, athlete, actor, surgeon, physical therapist, recreational worker, model, farmer, mechanic, carpenter, craftsperson, physical education teacher, factory worker, choreographer, forest ranger, and jeweler.

Musical

Musical intelligence does not clearly fit into one of the three learning domains. However, if you play a musical instrument, it would fit here in the psychomotor domain. If you appreciate music, that could be an affective skill. Writing music could fit in the cognitive domain.

Musical intelligence is the sensitivity to pitch, rhythm, emotional power, and complex organization of music. Musical intelligence also resides in the mind of any individual who has a good ear for music, can sing in tune, keep time to music, and listen to different musical selections with some degree of discernment.[18] Job skills include singing, playing an instrument, recording, conducting, composing, transcribing, arranging,

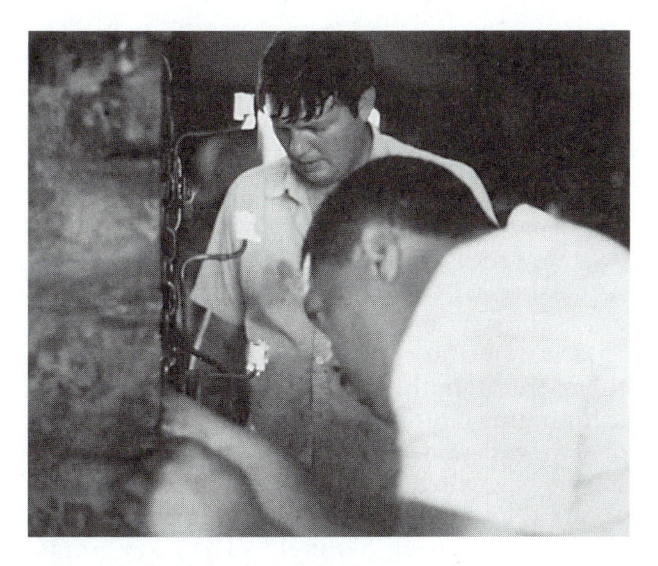

Figure 3–4 Learning by doing has always been a philosophy in agricultural education. These two students are learning how to overhaul a tractor by doing it.

listening, tuning, and orchestrating. Examples of vocations are performer, composer, conductor, recording engineer, maker of musical instruments, disc jockey, musician, piano tuner, music therapist, instrument salesperson, songwriter, and music teacher.

The Whole Person

The seven areas of intelligence must be considered together. Only by looking at all areas can a learning style be formulated or judgment be made on intelligence. A single characteristic should never be used to make assumptions about the intelligence of a person, or more specifically, no one should be labeled unintelligent because she is weak in some area. When you broaden your learning style to include cognitive, affective, and psychomotor learning, you become a totally intelligent person. This is referred to as "educating the whole person." Once people have been made aware of all three major areas of intelligence and their subcategories, they can identify and tap into their own natural resources.

Building a Team

As a leader builds a team, that person must be aware of the three major learning domains and the different ways of being smart. As a leader, study your group and unleash its full potential. Use people where they perform best and build the best team possible.

Synergy is the power of a group of people working together, in which their total output as a team is greater than their output would be if they worked individually. Part of the reason for this is that when individual abilities or learning areas (cognitive, affective, and psychomotor) are used, each individual is able to concentrate on the area in which he performs best. A person working alone must do many tasks that are not in his ability or learning area, so his productivity is not as great. The key is teamwork. You, as a leader, must assemble that team.

Genetic and Environmental Factors' Effect on Learning Style

Over the years some psychologists have argued that intelligence is genetically determined and exists at birth. Research has shown that intelligence develops through a series of stages. Other psychologists believe that intelligence is determined by the environment in which an individual is raised. People brought up in a rich environment—a home where learning is encouraged and readily available, good schools, and a highly differentiated social and physical setting—have an ideal setting for the development of intelligence.[19] It is generally agreed that intelligence is a consequence of both genetics and the environment. Both genetic and environmental factors also account for multiple intelligences. Leaders should consider this information as they work with their groups.

How We Learn

Our capacity to learn new things is important. However, we do not all learn things in exactly the same way. There are two major differences in how we learn: how we perceive and how we process.

Perceiving

The way you perceive reality, whether you favor sensing and feeling (concrete) or thinking things through (abstract), is one of the two major determinants of your learning style (Figure 3–5).

Those who perceive in a *sensing* or *feeling* way connect the experience and the information to meaning. They learn through empathy.[20] People who sense and feel tend more to the actual experience itself and immerse themselves in the concrete reality of things. They are not looking for hidden meanings or trying to make relationships between ideas and concepts. People like this place value on relating to people, being involved in real situations, and taking an open-minded approach to life.

Those who *think* through the experience tend more to the abstract dimension of reality. They analyze what is happening, and their intellect makes the first appraisal. They evaluate their experience and perceive with a logical approach.[21] Thinkers look beyond what they can't actually see. The key phrase for the abstract is "It's not always what it seems." People with this orientation value precision, the rigor and discipline of analyzing ideas, and the quality of a neat and organized system.

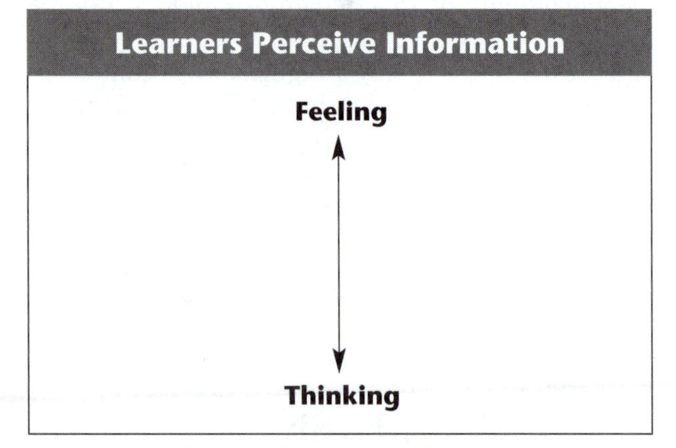

Figure 3–5 All people hover somewhere on a continuum from feeling to thinking. Both feelers and thinkers have their own strengths and weaknesses.

Process Experience and Information

The second major difference in how we learn is how we process experience and information.[22] In processing experience and information, some of us are *observers* and others are *doers* (Figure 3–6).

Observers reflect on new things. They filter them through their own experience to create meaningful connections. People reflect on and observe experiences from many perspectives. Observers have a sequential method of ordering that allows their minds to organize information in a linear, step-by-step manner.[23] Their key phrase is "follow the steps." People with this orientation value patience, impartiality, and thoughtful judgment.

Doers act out new information immediately and reflect only after they have tried it out. To make it theirs, they need to do it.[24] Doers work with unorganized information in no particular sequence, sometimes even skipping steps in a procedure, and still produce the desired result. Doers may seem impulsive or spontaneous. They are concerned with what works as opposed to what is absolute truth. Their key phrase is "Just get it done."[25]

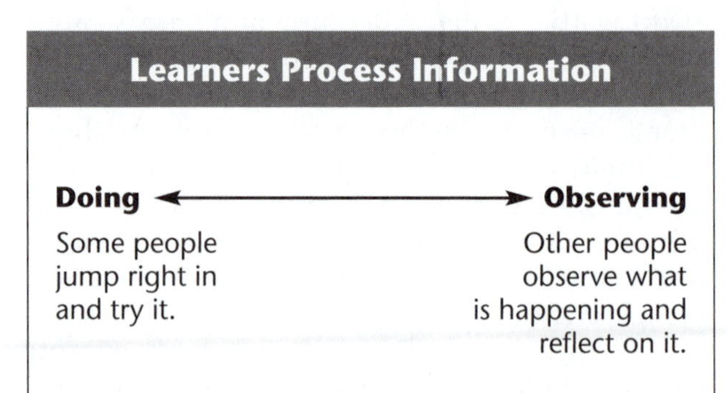

Figure 3–6 In processing information, some are observers, and some are doers. Both ways of processing information and experience are equally valuable.

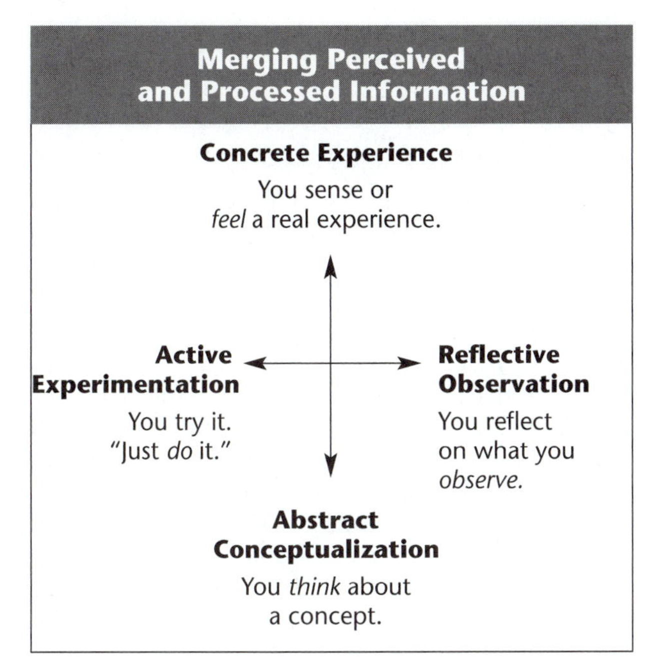

Figure 3–7 Once feeling, thinking, observing, and doing are merged, the basis for the four learning styles is established.

Merging Perceiving and Processing

Both kinds of perceiving (feeling and thinking) are equally valuable, as are both kinds of processing (observing and doing). We must learn to feel and to think, appreciating both modes of perceiving. We also must learn to observe and to do. Once you merge perceiving and processing, the way you learn begins to take shape as shown in Figure 3–7.

LEARNING STYLES

When we combine perceiving and processing, we get four distinct learning styles. Remember, no individual learns using only one style. Each of us has a dominant style or styles that provide a unique blend of strengths and abilities.[26] The four learning styles that emerge from these patterns are these:

- **Diverger** (combines observing and feeling)
- **Assimilator** (combines observing and thinking)
- **Converger** (combines doing and thinking)
- **Accommodator** (combines doing and feeling)

By learning some of the common characteristics of each of the four learning styles, we can recognize and value what we like to do best and what comes naturally to us. We can also identify our weaknesses. As leaders, we must first recognize our own natural learning style(s). As we begin to understand how to perceive and process information, we can better understand what comes naturally to us and to members of our group. We can then identify the differences between leaders and group members that cause frustration and misunderstanding (Figure 3–8).

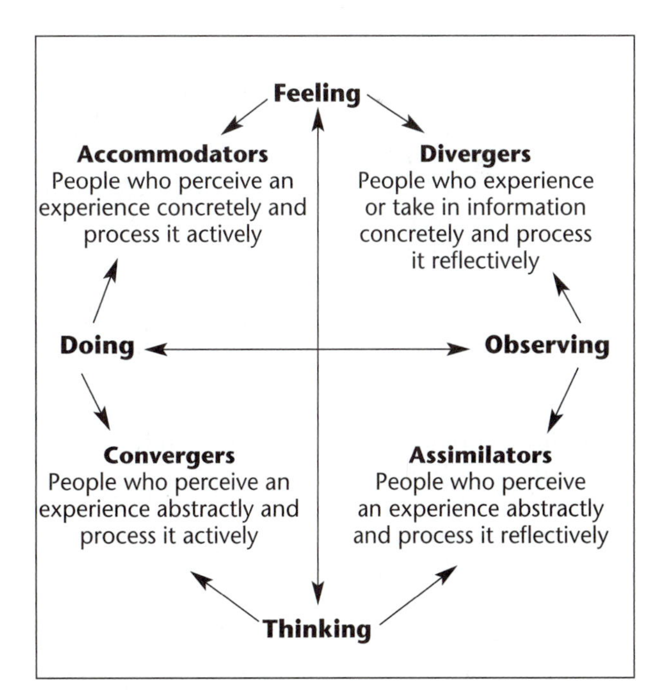

Figure 3–8 The combination of how we perceive (feeling and thinking) and how we process (observing and doing) forms the uniqueness of the four learning styles: divergers, assimilators, convergers, and accommodators.

The following discussion of these four learning styles characterizes each as learners and leaders and highlights the advantages and disadvantages of each style.

Diverger

As learners, divergers prefer learning by observing and feeling. This style is called *diverging* because people of this type perform better in situations that call for generating alternative ideas and implications, such as brainstorming sessions. They take their time and analyze many alternatives. They tend to have broad cultural interests and like to gather information. Divergers are imaginative and sensitive to the needs of other people.[27] People displaying a strong aptitude for diverging are typically interested in people, tend to be creative and feeling oriented, and tend to specialize in the liberal arts and humanities. They often seek careers in counseling, teaching, the service sector, design, social work, nursing, consulting, and personnel management.

As leaders they thrive on developing good ideas. They handle problems by first reflecting alone and then seeking the advice of others. They exercise authority with trust and participation. They work to keep the organization unified and need staff who support their goals. They lead best with frequent, honest praise; reassurance of worth, opportunities to work together; opportunities to use creativity and imagination; and acceptance of personal feelings and emotions.

Divergers tend to be imaginative and able to recognize problems. They listen sincerely to others, understand feelings and emotions, focus on themes and ideas, bring harmony to group situations, have good rapport with almost anybody, and recognize the emotional needs of others.

They tend to overanalyze problems, however, and are slow to act. Divergers often miss opportunities. It is hard for them to explain or justify feelings, be competitive, work with unfriendly people, give exact details, accept even positive criticism, and focus on more than one thing at a time.[28] They become stressed when they do not feel liked or appreciated.

Assimilator

As learners, assimilators prefer learning by observing and thinking. The learning style is less focused on people and more concerned with ideas and abstract concepts. Assimilators are effective at consolidating information and putting it into concise, logical form. They seek continuity and need to know what the experts think. They need details, are thorough, and seek intellectual competence and personal effectiveness. It is more important to them that an idea or theory have logical soundness than practical value.[29] This learning style is more characteristic of individuals in the basic sciences and mathematics rather than the applied sciences. In organizations, assimilators generally are found in the research and planning departments. They often seek careers as mathematicians, scientists, researchers, and planners.

As leaders they thrive on accumulating facts into theories. They attack problems with rationality and logic. Assimilators lead by principles and procedures.[30] They exercise authority with assertive persuasion and by knowing the facts. They need staff who are well organized, write things down diligently, and follow through on agreed decisions.[31] Assimilators lead best with organization, logical outcomes, plenty of time to work, credible sources of information, opportunities for analysis, and appreciation for their input.

They are skilled at creating models and developing theories and plans. Assimilators work through an issue thoroughly, use facts to prove or disprove theories, and analyze the means to achieve goals.

They tend to be too idealistic, however, and not practical enough. It is hard for them to rush a decision, deal with rules and regulations, express emotions, be diplomatic when convincing someone else of their point of view, and not monopolize a conversation about a subject that interests them.[32] They become stressed by being rushed through anything, not having hard questions answered, abiding by sentimental decisions, or being asked to express emotions or feelings.

Converger

Convergers learn by doing and thinking. They seek practical uses for information and integrate theory and practice; they learn by testing theories and applying common sense. The greatest strength of this learning style lies in problem solving, decision making, and the practical application of ideas: if it works, use it. They do not stand on ceremony but get right to the point. They prefer doing things the same way over and over. This learning style is called converging because a person with this style seems to do best in such situations as conventional intelligence tests because there is a single correct answer or solution to a question or problem. Convergent persons are controlled in their expression of emotion. Sometimes they seem bossy and impersonal. They prefer dealing with technical tasks and problems rather than with social and interpersonal issues. Convergers often specialize in the physical sciences. This learning style is characteristic of many engineers, technical specialists, surgeons, production supervisors, computer workers, and managers.

As leaders, they thrive on plans and timelines and are production oriented. They lead by personal forcefulness, inspiring quality, and tend to be authoritarian. Convergers work hard to make the organization productive and stable. They need employees or group members who are task oriented and won't hesitate to take immediate action.[33] They lead best with organization, routines, predictability, tangible rewards, and schedules.

Convergers are usually very good at deductive reasoning, solving problems, and making decisions. They fine-tune ideas to make them more efficient and economical. They produce concrete products from abstract ideas and work well within time limits.

On the other hand, convergers tend to make hasty decisions without reviewing all the possible alternatives. They often implement ideas without testing them first. It is hard for convergers to work in groups, discuss things with no specific point, work in a disorganized environment, work with abstract ideas, and attempt to answer questions with no right or wrong answer. They become stressed by not knowing expectations, by vague or general directions, or by not seeing an example.[34]

Accommodator

Accommodators prefer learning by doing and feeling. They tend to learn primarily from "hands-on" experience or act on "gut" feelings rather than logical analysis. They also learn by trial and error. They are believers in self-discovery and are enthusiastic about new things. They prefer to learn only what is necessary to get the job done. Accommodators believe that the strength of their competence lies in getting things done, carrying out plans and tasks, and seeking new experiences. Opportunity-seeking, risk-taking, and action characterize this competency. This style is called *accommodation* because it is best suited for situations in which one must adapt oneself to quickly changing circumstances.

In situations in which theory or plans do not fit the facts, those with an accommodative style will most likely discard the plan or theory. Accommodators often reach accurate conclusions in the absence of logical justification. They act on the spur of the moment. When making decisions they tend to rely more heavily on people for information than on their own technical analysis. They enrich reality by taking what is and adding something of themselves to it. Accommodators are generally at ease with people and may be viewed as impatient and aggressive. In organizations, people with this learning style are found in action-oriented jobs, such as marketing, sales, public relations, management, politics, social professions, and entertainment. It is not unusual for accommodators to have several careers in a lifetime or sometimes even two careers at once.

Accommodators tend to be democratic leaders. They are not deterred by the word *impossible* if they have determined the goal is a worthy one. On the other hand, they may not attempt even the most accessible goal if they have decided it is just not worth the trouble. As leaders, accommodators thrive on crisis and challenge; they lead by energizing and exercise leadership by holding up a vision of what might be. They work hard to establish their organizations as pacesetters, but they need staff who can follow up and implement details.

Accommodators are usually good leaders. They are willing to take necessary risks, and they get things done. They inspire others to action, contribute unusual and creative ideas, visualize the future, accept many types of people, and think fast on their feet.

Accommodators do not always set clear goals and develop practical plans. They often waste time on unimportant activities. It is hard for accommodators to have restrictions and limitations, complete formal reports, develop routines, redo anything once it's done, keep detailed records, show how they got an answer, and have no options. They become stressed when they have excessive restrictions and limitations, are not appreciated as unique, and are not given credit for knowing the right thing to do.[35]

The ability to learn is the most important skill we can acquire.[36] When faced with new experiences or learning situations we may need to change our learning style from getting involved (feelings), to listening (observing), to creating an idea (thinking), or to making a decision (doing). However, since we have preferred learning styles, we tend to be stronger and weaker in some of these areas.[37] Figure 3–9 provides a summary of important characteristics of each learning style.

Learning Styles and Personality Types

In Chapter 2, we compared personality types and leadership styles. We can also find some similarities between learning styles and personality types. Accommodators, being people-oriented risk-takers and operating more on feelings than theory, tend to be sanguine personality types. Like choleric personality types, convergers seek a single correct answer or solution to a question or problem. Assimilators are similar to melancholy personality types by creating models, analyzing, and developing plans. Divergers are similar to the phlegmatic personality since both are often slow to act and therefore miss opportunities.

As with personality types, you should realize that there is no best learning style; each has its own pros and cons. You probably realize that you have one preferred learning style, but you also have characteristics in other styles as well. Figure 3–9 compares learning style with personality type.[38]

Leadership and Learning Styles

Leadership styles relate to learning styles. Accommodators tend to have a democratic style of leadership. Democratic leaders are people oriented, as are accommodators; generally, the interest of the group is more important than the task. Convergers are task oriented and tend to be authoritarian; the task is much more important than people's feelings. Assimilators and divergers tend to be situational leaders with no overwhelming democratic or autocratic tendencies. However, due to where they are placed on the perception and processing learning chart (Figure 3–9), divergers slightly favor democratic leadership and assimilators slightly favor autocratic leadership.

Leadership, Learning Styles, and Decision Making

People with similar learning styles tend to behave in the same way, and people with different learning styles tend to act differently. Group productivity and harmony depend on how people are grouped as they work together. When working with a person who has a compatible learning style, you tend to get along well. However, when

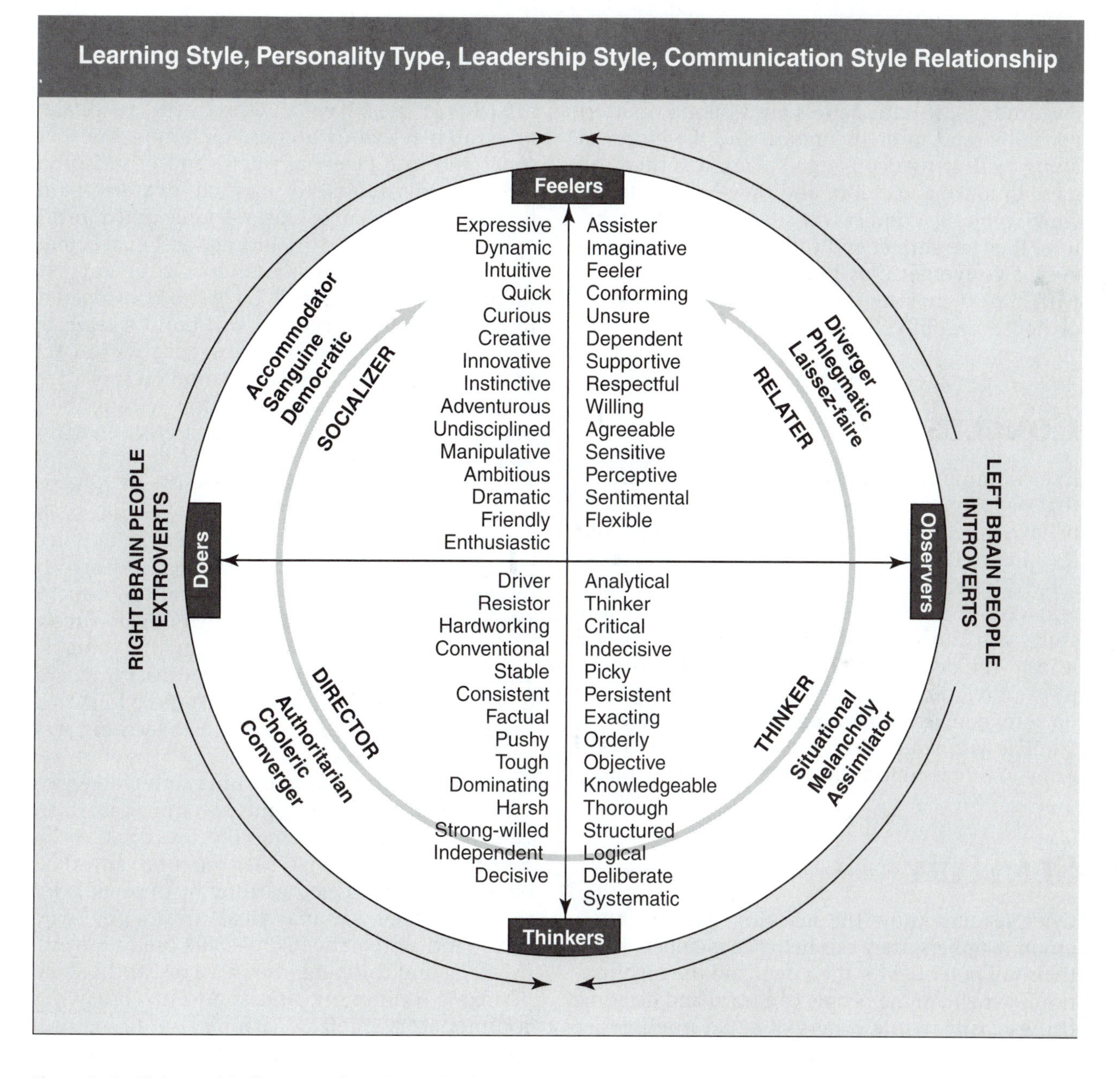

Figure 3–9 This graphic illustrates the relationship between learning style, personality type, leadership style, and communication style. Most people tend to be predominantly one or two of the styles or types.

working with a person with an incompatible learning style you may have a conflict. Conflicts can be prevented if people are grouped correctly. For example, if both parties are assimilators they generally tend to be theoretical and therefore get along well. If two convergers work together and have to make a decision, and they have different views, typically conflict will arise because both believe they are correct and will not negotiate. However, a converger (bossy) and a diverger (happy with others making decisions) should work together successfully.

CONCLUSION

In examining the four learning styles one can see that people have characteristics that lead them to behave differently. An understanding of the four learning styles helps one to realize why people behave differently. If a decision needs to be made and you know the person's preferred learning style, you can anticipate the behavior that will lead to the decision.[39] Remember, no person displays only one learning style, but most people do have tendencies toward one.[40] Once you become familiar with your learning style and those of others, you can assume a successful leadership role.

SUMMARY

Once leaders know the learning styles of their group members, they can help the members reach their full potential for their own and the organization's benefit. Some people can learn and do some things easily, while others of equal intelligence have difficulty with the same task.

Understanding learning styles and psychological strengths are critical to unlocking the great potential in every individual. Unless leaders recognize different styles of learning, they will confuse learning styles with intelligence. The three major areas of learning are cognitive, affective, and psychomotor. Cognitive learning is usually thought of as "book smart" and includes linguistic, logical-mathematical, and spatial intelligence. Affective learners have abilities in the personality or human relations arenas, including interpersonal and intrapersonal intelligence. Psychomotor learning refers to manual dexterity and performing mechanical tasks. People gifted in this area have kinesthetic intelligence and may be naturally adept at playing a musical instrument. The totally educated person is one who is competent in all three areas. A leader can build a team by using all three areas of intelligence within the group. Genetic and environmental factors affect learning style. Since we cannot do anything about our genetics, we must concentrate on environmental learning factors.

There are two major differences in how we learn: how we perceive and how we process. In certain situations, some people sense and feel their way while others think things through. In processing experience and information, some of us observe while others act. Both kinds of perceiving (feeling and thinking) are equally valuable, and both kinds of processing (observing and doing) are equally valuable. Once these four areas are merged, our intelligence area or learning style begins to take shape.

There are four major learning styles: divergers, assimilators, convergers, and accommodators. Divergers learn by observing and feeling; they are sensitive, compassionate, sentimental, and flexible. Assimilators prefer learning by observing and thinking; they are analytical, thorough, structured, and systematic. Convergers prefer learning by doing and thinking; they are hardworking, dependable, dominant, and somewhat bossy. Accommodators prefer learning by doing and feeling; they are creative, innovative, instinctive, and adventurous.

There are similarities between learning styles and personality types. Leadership styles and learning styles are also similar. By knowing all group members' learning styles, leaders can increase the productivity of the group.

END-OF-CHAPTER EXERCISES

Review Questions

1. What are five ways that understanding various learning styles can help leaders?
2. How do smart people find a way to succeed?
3. What three types of intelligence are grouped with the cognitive learning domain?
4. List 15 examples of vocations that would suit people with linguistic abilities.
5. List 11 examples of vocations for people with logical-mathematical intelligence.
6. List 11 examples of vocations for people with spatial intelligence.
7. What two types of intelligence are grouped with the affective learning domain?
8. List 13 vocations for people with good interpersonal abilities.
9. List eight vocations for people with intrapersonal intelligence.
10. What two types of intelligence are grouped with the psychomotor learning domain?
11. Why is it hard to place musical ability within one of the three domains of learning?
12. What is meant by teaching the total person?
13. What are the two kinds of perceiving and the two kinds of processing?
14. List seven advantages and nine disadvantages of divergers.
15. List four advantages and eight disadvantages of assimilators.
16. List five advantages and seven disadvantages of convergers.
17. List 7 advantages and 10 disadvantages of accommodators.
18. Why do leaders need to be aware of learning styles?
19. What are nine leadership traits of divergers?
20. What are 10 leadership traits of assimilators?
21. What are 10 leadership traits of convergers?
22. What are six leadership traits of accommodators?

Fill-in-the-Blank

1. Only the ____________________ learning domain is typically recognized in schools and universities.
2. ____________________ is the ability to respond successfully to new situations and the capacity to learn from past experiences.
3. Your ____________________ can be damaged because of a lack of understanding of intelligence.
4. Although intelligence tests consistently predict school success, they ____________________ to indicate how students will do after they graduate.
5. The ____________________ learning domain deals with attitudes and human relations skills; the ____________________ learning domain deals with manual dexterity skills.
6. ____________________ is a group of people working together, but their total output as a team is greater than their output if they worked individually.

7. It is generally agreed that intelligence is a consequence of both ________________ and the ________________.
8. There are two major differences in how we learn: how we ________________ and how we ________________.
9. In certain situations, some people sense and feel their way, while others ________________.
10. In processing experience and information, some of us are observers while some of us are ________________.

Matching

Terms may be used more than once.

_____ 1. Phlegmatic personality tendencies
_____ 2. Sanguine personality tendencies
_____ 3. Melancholy personality tendencies
_____ 4. Choleric personality tendencies
_____ 5. Democratic leadership tendencies
_____ 6. Authoritarian leadership tendencies
_____ 7. Vocational tendencies toward social work
_____ 8. Vocational tendencies toward mathematics
_____ 9. Vocational tendencies toward production supervisor and management jobs
_____ 10. Vocational tendencies toward marketing or entertainment

A. diverger
B. assimilator
C. converger
D. accommodator

Activities

1. Name a person in your class who best exhibits a cognitive learning style, one with an affective learning style, and one with a psychomotor learning style. Justify your choices.
2. Select four people in your class who show the four learning styles: diverger, assimilator, converger, and accommodator. Justify your choices.
3. Write a one-page essay on how people can be smart yet unintelligent at the same time.
4. Study the sections on cognitive, affective, and psychomotor learning. Determine which of the three learning styles most closely describes you. Write a short essay and read it to the class to see if they agree with your self-assessment.
5. Within the cognitive, affective, and psychomotor learning areas are seven areas of intelligence: linguistic, logical-mathematical, spatial, interpersonal, intrapersonal, bodily kinesthetic, and musical. Rank each of these according to your perception of your own intelligence areas or strength. Share these with class members and see if they agree.
6. Write four one-page case studies of four people you know, other than classmates, who each exhibit one of the following learning styles: accommodator, converger, diverger, and assimilator.
7. **Take it to the Net.**
   Explore learning styles on the Internet. Go to Metacrawler. (www.metacrawler.com). Enter "learning

styles test" into the search field. Browse some of the Web sites for learning styles tests. Take as many of the tests as you want. Print the results from two tests and explain whether or not you agree with the results and why. If you would like additional information type some of the search items listed below in the search engine.

Web sites

If you are having trouble finding a learning styles test, try some of the following sites:

<http://www.pepperdine.edu/~dmreymun/test.htm>

<http://webster.commnet.edu/HP/pages/lac/styles/styles.htm>

<http://www2.ncsu.edu/unity/lockers/users/f/felder/public/ILSdir/ilsweb.html>

Search Terms

learning styles

learning styles test

learning styles inventories

Notes

1. O. Kroeger and J. M. Thueson, *Type Talk at Work* (New York: Delacorte Press, 1992), p. 9.
2. Ibid. p. 11.
3. Ibid., p. 12.
4. R. J. Sternberg, "Styles of the Mind," quoted in *Learning Styles: Putting Research and Common Sense into Practice* (Arlington, VA: American Association of School Administrators, 1991), p. 22.
5. Ibid.
6. Ibid.
7. T. Armstrong, *Seven Kinds of Smart* (New York: Plume, 1993), p. 8.
8. Ibid.
9. H. Gardner and T. Hatch, "Styles of the Mind" quoted in *Learning Styles: Putting Research and Common Sense into Practice* p. 23.
10. J. Thorpe, *Learning Style Inventory Manual* (Vinemont, AL: Route l, Box 2685, 1978).
11. C. U. Tobias, *The Way We Learn* (Colorado Springs, CO: Focus on the Family Publishing, 1994).
12. Ibid., p. 132.
13. Ibid., p. 133.
14. *Learning Styles: Putting Research and Common Sense into Practice*, p. 16.
15. Tobias, *The Way We Learn*, p. 135.
16. Ibid., p. 136.
17. Ibid., p. 135.
18. Armstrong, *Seven Kinds of Smart*, p. 10.
19. J. Miner, *Organizational Behavior: Performance and Productivity* (New York: Random House, 1988), p. 130.
20. B. McCarthy, *The 4 Mat System: Teaching to Learning Styles with Right/Left Mode Techniques* (Barrington, IL: Excel, Inc., 1987), p. 3.
21. Ibid.
22. Ibid., p. 9.
23. Tobias, *The Way We Learn*, p. 16.
24. McCarthy, *The 4 Mat System*, p. 10.
25. Tobias, *The Way We Learn*, p. 16.
26. Ibid., p. 18.
27. R. N. Lussier, *Human Relations in Organizations: Applications and Skill Building* (Columbus, OH: McGraw-Hill, 1998).

28. Tobias, *The Way We Learn*, p. 24.
29. Lussier, *Human Relations in Organizations*, p. 42.
30. McCarthy, *The 4 Mat System*, p. 39.
31. Ibid.
32. Tobias, *The Way We Learn*, p. 24.
33. McCarthy, *The 4 Mat System*, p. 41.
34. Tobias, *The Way We Learn*, p. 35.
35. Ibid., p. 68.
36. H. Gardner, *Frames of Mind: The Theory of Multiple Intelligence* (New York: Basic Books, 1993).
37. Lussier, *Human Relations in Organizations*, p. 41.
38. Ibid., p. 42.
39. Ibid.
40. The terms diverger, assimilator, converger, and accommodator were used by David A. Kolb, et al., *Organizational Psychology: An Experiential Approach, 3rd ed.*, Englewood Cliffs, NJ: Prentice-Hall, 1979.

CHAPTER

Developing Leaders

Objectives

After completing this chapter, the student should be able to:

- Discuss the importance of personal leadership development
- Explain how to attain group acceptance
- Describe the type of individual who emerges as a group's leader
- Compare the relationship between ability, experience, and opportunity
- Discuss the improvement of leadership in the workplace
- Explain the types of leadership traits, abilities, and skills
- Select qualities of successful leaders
- Describe human relations leadership qualities and skills
- Describe technical human relations leadership qualities and skills
- Describe technical leadership qualities and skills
- Describe conceptual technical leadership qualities and skills
- Describe conceptual leadership qualities and skills

Terms to Know

initiative
traits
abilities
attributes
qualities
conceptual leadership skills
technical leadership skills
human relations skills
innate
value system
portfolio
opportunists

Leadership involves an observable, learnable set of practices. It is not something mystical that cannot be understood by ordinary people. Given the opportunity for feedback and practice, those with the desire and persistence to lead can substantially improve their abilities to do so.

Anyone can learn to lead. An FFA officer, a team facilitator, the middle manager, the account executive, the athletic team captain, the mail clerk, or just about anyone else who has a good reason to learn how to lead. To an enormous degree, their leadership skills will determine how much success they achieve and how happy they will be. Families, charity groups, sports teams, civic associations, and social clubs all need dynamic leadership.

There is no one correct way to lead, and successful leaders come in many personality types, as discussed in Chapter 2. Leaders are loud or quiet, funny or sincere, tough or gentle, boisterous or shy. They can be any age, any race, and either sex. The leadership techniques that work best for you are the ones you should nurture.

Leadership helps others arrive at a better understanding of themselves, others, and the issues at hand and then use this understanding to accomplish the goals that brought the members of the group together. Whether leadership is planned or unplanned, it always has a purpose and a goal. It is a process of human interaction.

There can be no leadership without someone to lead. The relationship is successful only as long as the co-workers wish to follow the leader. A leaderless society does not exist; whenever two or more people come together, there is no such thing as uncontrolled, unrestricted, or uninfluenced behavior.

Leadership is a group effort. The existence of any group is evidence of the members' willingness to work together rather than alone toward a goal. Working together is a give-and-take business, and the leader is the catalyst of the process. He or she is successful when the members find the group accomplishments greater than those that could have been achieved by individuals.

Some people have greater natural tendencies for leadership than others, but there is sufficient experimental evidence to prove that leadership can also be created, trained, and developed in people of normal intellectual ability and emotional stability who are willing to learn. A leader is a person who, on the whole, best lives up to the standards or values of the group.

Leader/Co-worker Relationship Quotes

When a genuine leader has done his work, his co-workers will say, "We have done it ourselves," and feel that they can do great things without leaders.
Eric Hoffer
Longshoreman-philosopher

You take people as far as they will go, not as far as you would like them to go.
Jeannette Rankin
Congresswoman

In the world people must deal with situations according to what they are, and not what they ought to be; and the great art of life is to find out what they are, and act with them accordingly.
Charles F. Greville
English political writer

I make progress by having people around me who are smarter than I am—and listening to them. And I assume that everyone is smarter about something than I am.
Henry Kaiser
Industrialist

No man will make a great leader who wants to do it all himself or to get all the credit for doing it.
Andrew Carnegie
Industrialist

Figure 4–1 The leader and co-worker constitute a team. These quotes show how leaders depend on co-workers.

Remember, leadership is an action, something you do, not merely a list of personality traits. The leader helps others develop their skills and share their knowledge with others. Psychologists say co-workers set the limits for what they want their leaders to be and where they want to be led.[1] Figure 4–1 provides some excellent quotes on leader/co-worker relationships.

Importance of Personal Leadership Development

Leadership is needed in all occupational areas. John, a state FFA Creed winner and two-year state finalist in public speaking, was loading freight with the United Parcel Service (UPS) to support his college expenses. At a group meeting, his supervisor said that UPS was looking for someone to speak to the corporate executives at their convention on improvements and efficiency in the freight moving system. John accepted the assignment because of his confidence in public speaking. His speech caught the attention of the UPS executives and he was seen as having future leadership potential in the company.

It is a challenge to learn leadership skills, and these skills develop respect from others. Leadership helps you understand others as well as yourself. Leadership helps individuals mature and develop their self-concepts. When we like ourselves, we like others. We are more secure; therefore, we are happier. The end result is that living conditions are improved.

Personal Leadership, Direction, and Success

Personal leadership is the ability to crystallize your thinking to establish an exact direction for your own life, commit yourself to moving in that direction, and then take determined action to acquire, accomplish, or become whatever that goal demands.[2] Take this attitude: "If it's going to be, it's up to me."

Success is controlled by our own actions. Our attitude determines our altitude. Obviously, it is harder for some to become successful because of financial situations, innate intelligence levels, or nonmotivated peers. However, we must find our strengths, deal with our weaknesses, and have the persistence and determination to accomplish our goals. Winners find a way. Losers find an excuse.

Except for physical limitations in certain areas, such as professional sports, people can accomplish just about anything they want to in life if they are willing to make sacrifices.

If you try to please everyone except yourself, you are destined for failure. Success is the attainment of our personal goals, not other people's goals. In this day of peer pressure, trends, and fads, we need to realize and accept that each person has been custom-made. Each of us has unique and personal traits. We are to be ourselves and not copy other people, except to learn positive traits from them. People are born originals, but most die copies.

Becoming a Successful Leader

The best way to learn leadership is to place yourself in situations requiring leadership action. It means you must not dominate but rather have a desire to serve, achieve goals, and leave things better than they were when you found them. Practice what you learn daily. To become a good leader:

1. Study the qualities of recognized good leaders and learn from their mistakes. Listen, but do not imitate.
2. Analyze yourself, picking out your weak and strong points. Set goals for improvement.
3. Learn how to take directions. If you cannot take directions, you may not be able to give directions.
4. Learn as much as you can about groups in general and how they function. Identify the types of people in a group.
5. Make and follow a plan to develop personal leadership skills.

Steps of Your Personal Leadership Plan

Step 1

Develop a vision and focus your thinking.
Successful leaders, whether they are leaders of others or simply effective leaders of their own lives, all have one thing in common: vision. Vision is the ability to have a clear picture of what you want to attain. Here are some examples of individuals who had a strong vision.

- Lee Iacocca had a vision of turning an almost bankrupt Chrysler Corporation into a profitable company and saved hundreds of thousands of jobs.
- Abraham Lincoln had a vision of freedom for all Americans that fulfilled the real needs of the people and gave them unity.
- Thomas Edison had a vision of providing light by harnessing the power of electricity.
- Ben Jordan, Herman Anderson, and many others, as well as myself, have a vision of producing fuel (hydrogen) from water economically, creating a potential free fuel which would add more economic stability to the country and eliminate future wars over oil.

Without commitment to a vision, we cannot lead ourselves in a definite direction. Vision separates the average from the great. Vision provides

- Direction
- A worthwhile destination
- Motivation
- Enthusiasm
- A sense of achievement
- Fulfillment of one's purpose in life.[3]

Step 2

Set goals.
Once you have your vision, set goals. Goal setting is the key to accomplishment. Develop a plan of action and pursue it. (Goal setting is discussed fully in Chapter 14.) What do you want to be like 5, 10, and 20 years from now? Remember, few people plan to fail, they just fail to plan. Set goals in the following areas: physical, family, social, financial, educational, ethical, and career.

Step 3

Develop initiative.
Be energetic and persistent to accomplish your goals. Take the **initiative**. There are three types of people in this area: those who make things happen, those who ask what's happening, and those who ask what happened. Don't let life pass you by. Go after it. Be positive. Look at the ways things can be done rather than the ways they cannot. Focus on the positive, and avoid negative people. George Bernard Shaw said, "Some men see things as they are and say why? Some dream things that never were and say why not?"

Step 4

Develop self-confidence.
The only way we gain self-confidence is to expose ourselves to risk. By risking ourselves, we gain the experience. Don't quit. Edison said he learned 999 ways the light bulb did not work before he found the one way it did. In the words of a modern-day inventor, there are a thousand ways and reasons that something will not work; we have to find the one way that will. As Winston Churchill said, "Ninety percent of all failure comes from quitting. If most failure comes from quitting, then to eliminate most failure in life, don't quit."

Step 5

Develop personal responsibility.
We must take personal responsibility for our own thoughts, actions, and feelings. A basic problem with society today is that most people don't want to take responsibility or be accountable for their

own actions. Leaders understand this and take control of their own destinies. If we blame others for our mistakes or claim for way we feel is not our fault, then we are not in charge of our lives. If we are not in charge, we will never exhibit true personal leadership. If we make mistakes, we should admit them. We must also be willing to take correction without being defensive and vengeful. Defensiveness is a sign of insecurity.

Step 6

Develop a healthy self-image.

Gary R. Hickingbotham said, "Self-image is our mental picture of the person we think we are." In life there are a number of laws that govern our results. The law of gravity works every time, whether we like it or not. Similarly, the self-image law works just as surely: We never rise above the picture we create for ourselves. People cannot exceed their self-imposed bonds of limitation.

In May 1954, part-time English athlete, Roger Bannister, pictured himself breaking the four-minute barrier for the mile race. Athletes before him could not run such a time because they saw it as impossible. Bannister did it because he saw himself doing it. It was only a matter of time before the Australian athlete, John Landy, also broke the barrier. Why could he do it? Because somebody else had done it before, so he then accepted that it could be done. Within a year, more than 10 other runners had run the four-minute mile.

Let your self-image run free. Let it be creative. Let it be innovative. Let it transport you to the pinnacles of success.[4] If you can dream it, you can achieve it.

Step 7

Develop self-organization.

One of the best traits of personal leadership is that of self-organization. You must know what to do next for the achievement of your goals. Consider the following points in self-organization.

- *Write down everything.* An old Chinese proverb says, "The shortest pencil is better than the longest memory."
- *Use a good diary system.* Use the style that works best for you. A favorite of many teachers is a pocket calendar that fits the front shirt pocket. This can be used in the summer when they don't wear a coat or jacket every day.
- *Make a daily to-do list.* Each night, list the tasks to be done the following day. Once this is done, arrange them in order of priority.
- *Use time management skills.* If you want something done, give it to a busy person. Why? They know how to work on priorities and overcome the habit of procrastination. Establish a weekly planner allocating specific tasks to specific time blocks in the week. Do exactly what you plan to do. Have short- and long-range goals, and you will find it easy to decide what to do on a daily basis.[5]

Step 8

Eliminate procrastination.

Procrastination is the tendency to put off tasks rather than take action on them now. To overcome procrastination, develop constructive habits to replace the destructive ones.

Gary Hickingbotham suggests four ways to overcome procrastination.

- Admit that you put things off. To admit this is a commitment to fix it.
- Learn about procrastination. Understand why you do it. The major reasons people put things off are

  insecurity (subconscious fears of achievement)
  confusion (indecision about what to do next)
  lack of priorities (poor planning)
  forgetfulness (poor memory)
  not accepting personal responsibility
  fear of change (staying in the "comfort zone")
  worry (fear of a future outcome)

monotony (when they find things boring, they procrastinate)

psychological fatigue (doing too much; not giving the brain sufficient time to rest)

- Choose a definite plan of action. List the steps you intend to carry out to achieve the desired response.
- Carry out your plan of action. Become a "do-it-now" person.

Step 9

Study.
People who wish to be good at a task need to study. If you want to be a good student, you must study. If you want to be a good doctor, you study medicine. It you want to be a good parent, you must study the art of parenting. If you want to be a good leader of yourself and others, you must study leadership. This means talking to successful people and asking them how they achieve results. It also means reading books on leadership and success. Throughout history, the remarkable achievers in life have been students who are eager to learn more and want to know how to fine-tune themselves.

Step 10

Magnify your strengths.
The first step toward success is identifying your own leadership strengths. Ask yourself what personal qualities or strengths you possess that can be turned into the qualities of leadership. Whatever talents you have (and we all have them, whether we have discovered them or not), use them to your advantage. Take advantage of your personality type, which is discussed in Chapter 2. Consider the following:

- For Robert L. Crandell, of American Airlines, his strength was an ability to anticipate change.
- For Mary Lou Retton, Olympic gymnast, it was her enthusiasm.
- For Hugh Downs, the veteran newscaster, it was his humility.
- For Ronald Reagan, it was his communication ability.

Dale Carnegie said that whatever strengths you have—a dogged persistence, a steel-trap mind, a great imagination, a positive attitude, or a strong sense of values—let them blossom into leadership. Remember, actions are far more powerful than words.[6]

Attaining Group Acceptance

We can have personal success, but most of us like to be accepted. It is a basic human need. Leadership effectiveness is likely to depend on whether the group accepts you. You are likely to attain group acceptance if you do the following.

- *Be knowledgeable about the purpose of the group.* Group members are more willing to follow when the leader appears to be well informed. The more knowledgeable you are, the more credibility you will attain.
- *Work harder than anyone else in the group.* Leadership is often a question of setting an example. When others in the group see a person who is willing to do more than his or her fair share for the good of the group, they are likely to support the person. The person seeking to lead must be willing to make personal sacrifices.
- *Be committed to the purpose of the group.* If group members detect lack of commitment in you, they may lose confidence in you.
- *Be willing to be decisive.* When leaders are unsure of themselves or unwilling to make decisions, their groups may ramble, become frustrated, or become short-tempered. Sometimes, leaders must make decisions that will be resented or cause conflict. Nevertheless, people who are unwilling or unable to be decisive are not going to maintain leadership for long. An old Arab proverb says, "An army of

sheep led by a lion would defeat an army of lions led by a sheep."

- *Interact freely with others in the group.* Participate fully in group discussions, but do not dominate the discussions. Share your ideas, feelings, and insights. Participate fully in the early stages of group work to find out whether you are able to influence others before you try to gain leadership. This is why political candidates announce that they are considering running for an office: They want to test their influence before making a commitment. The group in Figure 4–2 is freely interacting.
- *Exhibit human relations skills.* Effective leaders make others in their group feel good, contribute to group cohesiveness, and give credit where it is due. Although a group may have both a task leader and a human relations leader, the primary leader is often the one who shows human relations skills.[7]

Figure 4–2 Freely interacting with others is one of many ways to gain group acceptance. Whether at FFA camp or a leadership conference, bonding can occur to further enhance group acceptance. (Courtesy of *FFA New Horizons Magazine.*)

Types of Individuals Who Emerge as Group Leaders

Once you become accepted in a group, you will be considered for group leadership. For any group to be effective and achieve its goals, successful leadership must occur.

Within a group, there are three types of individuals who emerge as the group's leader: member-oriented, goal-oriented, and self-oriented. The member-oriented individual emerges when a conflict occurs within the group and the individual can solve the difficulty, mediate, or referee the conflict. When the group's goals and rewards look good, a goal-oriented member emerges as the leader. This type of leader thrives on task accomplishment. In the event that the leadership attempt fails, this leader will probably not attempt to lead again. Self-oriented leaders emerge when leading a group is prestigious or provides a direct reward. This type of leader enjoys the recognition a position of group leadership entails.

Leaders usually exhibit all three types to some extent. Group leaders are willing, motivated, energetic, and goal-oriented. They have a tendency to be more intelligent than the group and exhibit originality and initiative.[8]

Ability, Experience, and the Opportunity to Lead

To realize maximum leadership potential, all three elements—ability, experience, and opportunity—must occur together. Given the proper experience and the opportunity, anyone can emerge a leader should he or she so desire.

1. *Ability.* Whether a person's ability is due to a learned skill through formal or informal leadership training or due to a contributing heredity trait, such as personality type, competence is needed for leadership. In societies that have a class system, the association between leadership and the elite (high social class) leads many people to feel that leadership is an inherited trait. Eventually, those supporting the hereditary theories of leadership give way to other theories as people of all classes become leaders. To date, geneticists have not discovered a gene they could label the "leader gene."[9]
2. *Experience.* Although we cannot control our heredity, we can give our abilities some experience. Experience involves the development of leadership abilities into skills by participating in various activities. Learning and developing these abilities into leadership skills requires training and practice at school, camps, and seminars, as well as reading and learning on one's own.
3. *Opportunity.* Even though you have abilities and experience, you cannot lead until you have the opportunity to lead. Sometimes this means being in the right place at the right time. More often than not, it means being prepared to lead so you are ready when the opportunity arises.

Relationship of the Three Elements

An experienced or skilled person may never have the opportunity to lead. This may simply be due to never being selected as the leader or, as stated earlier, being in the right place at the right time. For example, many students would make excellent officers if only someone would nominate or select them. On the other hand, the opportunity to lead may occur with an inexperienced or unskilled person. If a person assumes a leadership position but has no prior experience or skill in leading a group, this individual will probably not achieve a high level of accomplishment.

IMPROVEMENT OF LEADERSHIP IN THE WORKPLACE

Business and industry use three basic approaches to improve leadership within the organization: selection, situational hiring or engineering, and training.

1. *Selection.* Selection is the quickest approach to improving leadership in the workplace. The company simply selects the person with the skills most needed by the group. To be effective, the business performs an analysis of the company's needs and then assesses each candidate's leadership record based on past performance. Finally, it selects a leader. For example, a company needs to improve its meetings. The company should select someone with experience in parliamentary procedure. Therefore, your ability, experience, and opportunity can lead to your selection as a leader.
2. *Situational hiring or engineering.* The second approach matches the situation with a leader. Manipulation (or engineering) is sometimes done before facing a situation to ensure the leader's success. Companies avoid placing individuals in situations in which failure is likely. For example, the increase or decrease of the leader's authority could depend on the situation he or she is about to experience. If the leader is capable, the company can leave more of the decisions to the selected leader.
3. *Training.* The training approach is a widely used method of improving leadership in the workplace. Usually companies work with the individuals they have rather than select someone new. Organizations send employees and members to seminars, workshops, and camps to improve their leadership skills.[10]

Types of Leadership Traits, Abilities, and Skills

In Chapter 1, we discuss seven categories of leadership: trait, power and influence, behavioral, situational, tradition, popularity, and combinational leadership. Actually, when most people think about leadership, they are referring to leadership **traits**, **abilities**, and skills. These are also referred to as leadership **attributes** or **qualities**.

Using these abilities and skills effectively, leaders get extraordinary things done in organizations, whether it is the FFA or the workplace. There are three major types of leadership abilities and skills: conceptual, technical, and human relations skills.

1. *Conceptual (thinking) skills.* **Conceptual leadership skills** are needed to analyze situations and generate ideas. Conceptual skill development involves placing individuals in situations they might face as leaders. These are the thinking skills, which are harder to develop than technical skills. However, the use of "think games" to solve problems, forecast events, and make decisions makes them easier to learn. Refer to Figure 4–3 for examples of conceptual skills.
2. *Technical (how-to-do-it) skills.* **Technical leadership skills** involve "doing." Improvement in technical skills may occur with practice in public speaking, time management, and using parliamentary procedure. Technical skills are the most easily obtained and retained of the three types of leadership skills. Refer to Figure 4–3 for examples of technical skills.
3. *Human relations (people) skills.* **Human relations skills** are needed to understand and work with people. Human relations skills relate to the ability to work with others ("people" skills). Leaders should receive training in working effectively with others within groups. People are the foundation of all groups; leaders must be able to deal effectively with members. Refer to Figure 4–3 for examples of human relations skills.

Technical skills are usually the easiest leadership skills to develop and retain. Conceptual skills are more difficult to learn than technical skills. Human relations (people skills) are a requirement for all leaders. They are not difficult to learn, but there is a tendency for leaders to forget or set aside human relations skills when under pressure to complete an activity.

Each type of skill is important for all tasks. Some leaders are better at one type of skill than another. Leaders must be able to recognize the talents and abilities of themselves and members to maximize the productivity of the group. Accurate assessment of members allows leaders to match members' skills with the skills needed to conduct the task. Refer to Figure 4–4 to see the relationship between types of leadership abilities (skills) and levels of leadership. Notice that human relations skills are equally important for top-, mid-, and lower-level (frontline) leaders.

Selecting Qualities of Successful Leaders

Many qualities, abilities, and skills are important in leading others. Together, they present a profile of a leader who has compassion and consideration for his or her co-workers while at the same time holding them strictly accountable for results. Some psychologists call this "tough love." The profile of a leader includes desired personal traits, abilities, and skills, such as honesty, integrity, and sensitivity, in addition to intangible factors, such as determination and courage.

Leadership Is Not an Exact Science

There is no simple formula for leadership success. The study of leadership is not an exact science, like biology, chemistry, or physics. The social science

Types of Leadership Traits, Abilities, and Skills

Type of Skill	Examples	Skills That You Might Possess
Conceptual (Thinking)	Analyzing a situation Thinking logically Combining concepts and ideas into a workable relationship Generating ideas Helping solve a group or individual problem Anticipating change Recognizing opportunities and potential problems	Good imagination Dedication Combining concepts and ideas into a workable solution Good problem-solving skills Creativity Logical thinking Good decision-making skills Anticipating problems Ability to think independently Foreseeing change Open-mindedness Welcoming new opportunities Persistent
Conceptual-Technical		
Technical (How-to)	Knowing a variety of ways to do things Understanding how to get things done Ability to establish and follow procedures Techniques for conducting a special activity Knowledge gained from study and experience	Communication skills Prepared speaking skills Extemporaneous speaking skills Parliamentary procedure Group organization, group dynamics, and leading discussions Goal setting and program of activities Time management Financial management Conducting successful meetings Organizational skills
Technical-Human Relations		
Human Relations (People)	Human behavior Interpersonal communication—how to get along with others Feelings about yourself and others The variety of attitudes and values people have Motives that others may have Good self-concept and self-esteem	Honesty Capacity for hard work Social skills Listening Cooperativeness Strong self-concept Enjoy working with people Sensitivity Positive attitude

Figure 4–3 Leadership skills and abilities are traditionally discussed by category—conceptual, technical, and human relations—but these are not always clearly separated. Thus, the conceptual-technical and technical-human relations categories were added. For example, we can be taught how to think logically and how to improve self-esteem.

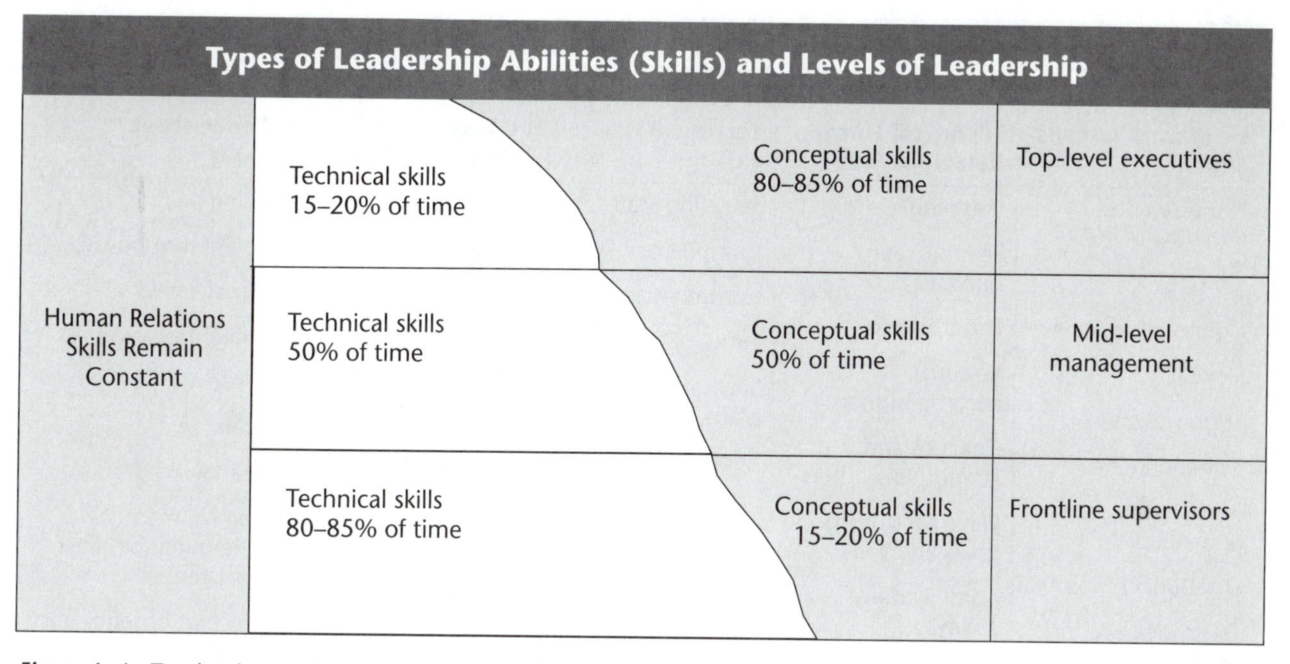

Figure 4–4 Top-level executives use conceptual (thinking) leadership skills 80 to 85 percent of the time. Frontline supervisors use technical (how-to-do-it) leadership skills 80 to 85 percent of the time. Human relations (people) skills are used equally for top-, mid-, and lower-level (frontline) leaders. (Courtesy of John M. Grogan Company.)

world isn't nearly as orderly as the physical world. It is not susceptible to rules. People are anything but uniform, and often they are hard to predict.

In Pursuit of Leadership Perfection

It is not possible to reach perfection in each leadership skill, but the capacity to lead gradually builds over time. Leadership development can be accelerated, however, by constantly visualizing the ideal leadership qualities and modeling your behavior after them. Figure 4–5 lists more than 50 leadership attributes. Several books were reviewed to arrive at this list, including books written by authors such as Shaw, Bennis, Nanus, Kouzes and Posner, Bethel, Manske, Drake, Covey, and Maxwell. Each author lists different qualities he or she feels strongly about. Many leadership abilities and skills were referred to by similar names. For example, *inspiring* was also referred to as *motivator*, *people-mover*, *enthusiastic*, *charismatic*, and *enlists others*. Representative names were selected and combined to arrive at the selected qualities of leadership. By studying these qualities, you will be exposed to numerous ideas that will help you improve your strengths and eliminate your weaknesses. You can also learn to

- Equip yourself with skills, tools, and techniques to maximize your natural leadership talents.
- Expose yourself to abilities and skills that will build the confidence you need to make leadership decisions.
- Sharpen your curiosity and reinforce your commitment to demonstrate leadership.
- Challenge yourself with questions that will help you discover how to be the best leader you can be.

Leadership is not something you learn once and for all. It is an ever-evolving pattern of skills,

Qualities of Successful Leaders				
Human Relations Skills	**Technical-Human Relations Skills**	**Technical Skills**	**Conceptual-Technical Skills**	**Conceptual Skills**
Honesty and integrity	Listening	Selecting staff	Intelligence	Vision
Sensitivity	Team (group) building	Competence	Problem-solving ability	Ability to motivate
Cooperation	Earns loyalty	Communication	Decision making	Decisiveness
Flexibility	Rewards accomplishments	Time management	Cooperative genius approach (synergy)	Straightforwardness
Supportiveness	Coaches and strengthens others	Sets goals and plans	Master of delegation	Creativity
Self-confidence	Plans short-term goals and wins	Achievement	Opportunist	Courage
Responsibility and dependability	Administrative ability	Responds to failure	Insists on excellence	Risk taking
Emotional maturity		Leads individuals and groups	Holds co-workers accountable	Commitment, determination, and persistence
Takes pleasure in work			Broad interests and abilities	Sense of urgency and initiative
Servanthood				Follow-through
Sets an example				Master of change
Availability and visibility				Wisdom
Shows confidence in people				Desire to lead
Uses power wisely				

Figure 4–5 The qualities of successful leaders are divided into five categories.

abilities, and qualities that grow and change as you do. To become a leader, you must first have clear definitions of leadership qualities, a mission, and an honest desire to make yourself better. Remember, we lead by example. Everything we say or do sends a message, sets a tone, or teaches people what to do or what not to do.

Five Leadership Quality Categories

An attempt was made in Figure 4–5 and the remainder of this chapter to place the leadership abilities into five categories: conceptual skills, conceptual-technical, technical, technical-human relations, and human relations. The proper category, however, is open to discussion. For example, innate intelligence (conceptual) is useless unless the mind is educated and cultivated (technical); and learning is enhanced with a good attitude (human relations). It is also interesting to note that a person may possess 49 qualities of leadership, but without the fiftieth, his or her leadership is null and void. For example, what good are the other qualities if you have no honesty and integrity?

Human Relations Skills

The following human relations skills tend to reflect a person's innate personality type or value system.

Honesty and Integrity

Successful leaders live by the highest standards of honesty and integrity. Recent surveys asked business and government leaders what quality they thought was most important to their success as leaders. Their unanimous answer was *integrity*. It is one thing to talk about integrity, and it is quite another to live that way. Honesty and integrity are best taught by example. More than anything else, co-workers want to believe their leaders are ethical and honest.[11]

> *A good name is seldom regained. When character is gone, one of the richest jewels of life is lost forever.*
>
> —*J. Hanes*

> *Integrity is not a given in everyone's life. It is the result of self-discipline, inner trust and a decision to be relentlessly honest in all situations in our lives.*
>
> —*Workman 1987 Page-A-Day Calendar*

> *Without integrity you do not have trust, and without trust you have nothing.*
>
> —*F. A. Manske, Jr.*

Sensitivity

A sensitive leader inspires loyalty. An effective leader is a flexible leader—someone who can change his or her style of leadership to suit the needs of co-workers and the situation. Sensitive leadership is not leadership that lacks strength or courage; it is not being soft or having less power. A sensitive leader has a heightened awareness of the issues, values, and people in a changing society. Sensitive leaders focus on the daily world in which they operate, particularly on the people they lead.[12]

Cooperation

A leader must have a mastery of interdependence. Americans have always believed strongly in self-reliance and competitiveness. However, in our complex, highly interdependent world, cooperation is much more important than competition.[13] Smart people realize they are successful when others succeed and are therefore oriented toward aiding each other to perform effectively. They encourage each other because they understand the other's priorities and want to help them be successful. Compatible goals promote trust.

President John F. Kennedy was so good at fostering cooperation that he was able to convert many in the Washington Press Corps, who were traditionally critical of U.S. Presidents, into "friends" who treated his administration kindly and with respect even while reporting on its problems.

Nothing is so energizing as a leader who is able to forge talented individuals into a team, for which each is willing to do his or her very best for the good of the group.

Flexibility

In Chapter 1, we discuss the situational leader. Situational leaders are flexible. They adapt to the occasion, being authoritative if followers are unwilling or democratic if the group is competent.

A flexible leader can adapt the organization to meet changing needs with minimal opposition. There is nothing more distressing or destructive to a creative group than having a leader reject an idea simply because the company has never done it before.[14]

Supportiveness

Leaders support their co-workers. When a co-worker knows the leader will be supportive, there is a healthy relationship. If workers or employees are afraid to make a mistake, they will not be creative or innovative. With supportive leaders, when mistakes are made in an effort to find a better way, encouragement is given rather than correction.

This does not mean leaders support everything, especially violation of policy, laziness, or insubordination. It would be foolhardy to give support to a person whom you know to be wrong. However, any leader at any level will sooner or later be in a situation in which they either need support themselves or find it necessary to support another. If workers have received support from the leader, the leader will receive support from workers when the occasion arises.

Self-Confidence

Self-confidence is belief in yourself. If we don't believe in ourselves, we cannot lead. If we have no self-confidence, we will not set goals or have a vision. We will not plan. Why should we? We don't have enough confidence in ourselves to make it happen.

Self-confident leaders display confidence in the correctness of their positions and in their capabilities. They project this confidence to their co-workers. They know they will find a way to reach their goals and complete their task.

Responsibility and Dependability

You must be able to follow directions before you can become a good leader. Workers must be responsible and dependable, therefore, leaders must be, too. Dependability involves getting to work on time; completing assignments; following through; giving an honest day's work; not missing school, work, or meetings without notification; and completing tasks on your own. The willingness to accept responsibility is equally important. A group likes to know that the leader is getting the job done. Responsibility means being reliable and accountable for your actions and meeting obligations.

Emotional Maturity

Some people just never grow up. Their feelings are hurt easily and they get angry if they don't get their way. They may pout, get into verbal battles, slam doors, refuse to talk, and exhibit self-pity. They don't ask for fear of rejection. They are afraid to confront someone when an issue needs discussion.

Maturity is the balance between courage and consideration. If people can express their feelings and convictions with courage, balanced with consideration for the feelings and convictions of others, they are mature, particularly if the issue is very important to both parties.[15] It is not enough to be nice, you have to be assertive; you must be empathetic as well as confident, considerate and sensitive, as well as brave.

Taking Pleasure in Work

If a person does not like the job he has, he should get one he likes. If a person does not like children, she should not teach. If an airplane pilot is afraid of flying, she should not fly. Similarly, if you are not comfortable leading, do not lead.

Successful leaders receive pleasure from their work. They enjoy leading, directing, and organizing. They set goals and get pleasure from achieving them. Once these goals are reached, others are set.[16] If there are no more goals to attain, they become bored or dissatisfied. Bill Walsh, Bill Parcells, and Joe Gibbs retired after multiple Super Bowl wins only to take other coaching positions with new teams. Why? They had accomplished their goals and did not enjoy their work any longer. Changing jobs renewed their goals, thus renewing their opportunities to get pleasure from work.

Servanthood

Leaders often serve their co-workers. Elected officials are referred to as public servants. Leaders have to speak at public meetings and maintain a schedule dictated by others.

Servant leaders are sensitive to the needs and feelings of their people. They are supportive of employees, helpful to them, and concerned for their well-being. Such leaders put the needs of others above their own interests. A great paradox of life is that the more you give of yourself, the more you receive. Morale and productivity are highest in work groups when the leader shows a high degree of consideration, servanthood, and concern. Getting a fair deal is the most important concern to employees at all levels.

The time is always right to do what is right.
—Martin Luther King

Sets Examples for Others to Follow

What you do as a leader exerts far more influence on employees or colleagues than what you say. Around your co-workers, you are constantly on stage. Everything you do is interpreted as acceptable behavior for them to follow. People are not changed by force or intimidation but by example. Great leaders never set themselves above their co-workers except in carrying responsibilities. Figure 4–6 lists some ways for leaders to set examples.

Availability and Visibility

Leaders should be available and visible to their organization. One of the most important maxims of leadership is this: Be there with your personnel. The great lesson of leadership from the military is that troops will put up with a difficult life, often with intolerable conditions, if their leaders do the same. Like radar, the good leader picks up obvious and not-so-obvious signals about conditions in an operation. Successful leaders always find time to make the rounds, chat with workers, show an interest in them, and listen to their concerns and ideas. History is full of examples of the impact of the commander's presence on the front lines. General Robert E. Lee visited the camp sites of his troops the night before each major battle. Often he would do this at the expense of getting no sleep himself. Opposing generals believed that Napoleon's presence on the battlefield was worth thousands of soldiers.[17]

Shows Confidence in People

One of the biggest mistakes leaders make is not having faith in people. You must trust people even if it involves some risk. Trusting people

Leadership: Set the Example for Others to Follow

1. Be totally committed to doing the best job possible.
2. Support yourself with the highest-quality people.
3. Maintain a cheerful, pleasant attitude.
4. Take the heat for your own errors—don't blame others.
5. Be willing to say, "I made a mistake" or "I don't know."

   There are no mistakes so great as that of being always right.
   —Samuel Butler
6. Avoid indiscreet, negative criticism of co-workers, press, and superiors.
7. Work hard and smart.
8. Stand up for the principles you believe in.

   Keep true, never be ashamed of doing right; decide on what you think is right, and stick to it.
   —George Eliot
9. Keep an open mind.
10. Be diplomatic.
11. Maintain a positive mental attitude.

    Nothing can stop the man with the right mental attitude from achieving his goal; nothing on earth can help the man with the wrong mental attitude.
    —W. W. Ziege
12. Develop a professional, energetic image.
13. Be a team player.
14. Be enthusiastic.
15. Treat your co-workers as you would like to be treated—with respect and dignity.

    Example is not the main thing in influencing others. It is the only.
    —Albert Schweitzer

Figure 4–6 As leaders, our walk is more important than our talk. We lead by example. Each of these attributes helps make a leader more successful by setting the example. (Source: F. A. Manske, Jr., *Secrets of Effective Leadership.* Columbia, TN: Leadership Education and Development, Inc., 1990.)

builds their confidence and stimulates them to do their best. Booker T. Washington said, ". . . few things help an individual more than to place responsibility upon him, and to let him know you trust him." Trust is the glue that keeps workers together.[18]

Uses Power Wisely

Leaders need to be strong enough to tackle tough problems and gentle enough to keep the solutions humane. They must be demanding enough to challenge others not to settle for easy answers and patient enough to know that progress takes time.

Power well used is like strong glue. It holds the other leadership qualities together. It energizes people and resources. When you use power wisely, you show people you can be trusted. They know you are sensitive and democratic and will not take advantage of them. When your co-workers trust you, you gain their loyalty and respect.[19]

TECHNICAL-HUMAN RELATIONS SKILLS

The following human relations skills can be learned rather easily. The previous group of human relations skills tended to reflect more of our **innate** personality type or **value system**.

Listening

Did you ever wonder why we were created with two ears and one mouth? Listening attentively and empathetically is one of the best ways to show respect for a co-worker. It demonstrates that you believe the individual has worthwhile thoughts and therefore is a valued member of the team. When people feel needed, they tend to take more interest in what they do. When co-workers no longer believe their leader listens to them, they start looking around for someone who will.

Leaders who make it a habit to bring out the thoughts, ideas, and suggestions of their co-workers and who are receptive even to disappointing news will be properly informed. Having all the appropriate information is very important to a leader's ability to make good decisions. (Listening is discussed in detail in Chapter 5.)

Team (Group) Builder

A leader builds group cohesiveness and pride. The thing that runs all successful organizations is the pride and sense of accomplishment that comes from striving for and achieving high levels of performance. Employers want to be a part of a winning team. Successful teamwork is exciting and offers personal rewards.[20] Deep within all of us is the desire to be a winner—to be somebody. Well-led co-workers feel needed, important, and part of a worthwhile enterprise.

Successful leaders do two key things to build team togetherness and pride.

They establish goals and visions of what could be for their group or organization.

They praise and recognize individual accomplishments.

> ***People are not motivated by failure; they are motivated by achievements and recognition.***
>
> ***—E. F. Fournies***

Earns Loyalty

The loyalty of your co-workers is priceless. It cannot be bought or secured by favors. It is not won overnight nor is it everlasting. Rather, loyalty is given by the group as long as it thinks its leader is worthy of it. Effective leaders make their co-workers feel good about themselves and their work. People thrive on the appreciation you show them—smiles, thanks, gestures of kindness. Elementary school teacher Ann Mabry, a friend of mine, says that students have to learn to like you before they will like to learn.

Co-workers will never be totally loyal until they feel you are sincerely devoted to the good of all and are doing everything in your power to help them succeed both as individuals and as a

Ways to Earn Loyalty

1. Develop warm, person-to-person relationships with your co-workers.
2. Be aggressive in serving your co-workers.
3. Confide in the people you trust.
4. Provide special rewards for your outstanding performers.
5. Recognize and show appreciation for other co-workers, too.
6. Give your co-workers the benefit of the doubt.
7. Never say you will do something for an employee, then change your mind.

Figure 4–7 Earning loyalty is a key characteristic of successful leaders. When the going gets tough for a leader, if he or she has earned loyalty, co-workers will be supportive. (Source: F. A. Manske, Jr., *Secrets of Effective Leadership.* Columbia, TN: Leadership Education and Development, Inc., 1990.)

group. Refer to Figure 4–7 for ways to earn the loyalty of your co-workers.

Rewards Accomplishments

Many believe that money is the greatest motivator, but studies have shown this is not true. The greatest morale booster of any group is simply being appreciated.

Appreciation provides encouragement. The word *encouragement* has its root in the Latin word *cor*, meaning "heart." When leaders encourage others through recognition, they inspire or motivate them with courage—with heart. When we encourage others, we give them heart. When we give heart to others, we give love.

James Kouzes and Barry Posner, in their book, *The Leadership Challenge*, stated that ". . . of all the things that sustain a leader over time, love is the most lasting. And it led us to suspect that just possibly the best-kept secret of successful leaders is love: being in love with leading, with the people who do the work and with what their organizations produce. Leadership is an affair of the heart, not of the head."[21]

Coaches and Strengthens Others

Leaders coach to improve performance. The most important responsibility of a leader is to develop personnel. Your success as a leader depends on how well your people perform. There is no greater monument at the end of one's career than the performance of individuals who have revealed their potential because of your coaching and leadership. Leaders create the kind of climate in their organization in which personal growth is expected, recognized, and rewarded. There are five common reasons why coaching to improve performance is necessary:

- Workers do not know what they are supposed to do.
- Workers do not know how to do it.
- Workers do not know why they should do it.
- There are obstacles beyond the workers' control.
- Workers have an attitude problem.[22]

Encourage your co-workers to explore new options instead of settling for the obvious. If they aren't making an occasional mistake or two, it's a sure sign they're playing it too safe. The role of the coach is to teach players to stand on their own two feet.

Good coaches know that real learning does not occur unless people are challenged to do their own thinking. Coaching for the playoffs is great, but the payoff is even greater for leaders who coach winners by asking the proper questions, encouraging and reinforcing positive behavioral changes, and setting an example of excellence. The coach in Figure 4–8 is motivating and challenging his players to do their best "and then some."

Figure 4–8 Not only is Boots Donnelly a winning football coach, he coaches and strengthens his players by instilling character, persistence, commitment, and moral responsibility. (Courtesy of MTSU Photographic Services.)

Good leadership consists in showing average people how to do the work of superior people.

—John D. Rockefeller

Plans Short-Term Goals and Wins

"The journey of a thousand miles begins with the first step." Good leaders convince co-workers that the impossible is possible. Small wins breed success and motivate co-workers. It moves co-workers toward greater heights. Successful leaders pursue a strategy of getting first downs or hitting singles rather than relying solely on the home run.[23]

Successful leaders work hard at finding ways to make it easy to succeed. They notice and celebrate movement in the right direction. When I was working with my students, researching how to run engines with hydrogen from water, our first goal was to run a small lawnmower engine for 10 seconds. The students accomplished this goal, and the win was tremendous. They had several other short-term goals and wins, including running a large tractor with hydrogen, running a car with hydrogen, and making hydrogen from water on board a vehicle. The big win was four years later when they set the world's land speed record for a hydrogen-fueled vehicle at the Great Salt Flats in Wendover, Utah, at the World Finals. Figure 4–9 shows the hydrogen car, which provides an example of short-term goals and wins. Each year the goal is to increase the previous year's world-record speed by a hydrogen vehicle by 10 miles per hour.

Administrative Ability

Leaders must have their priorities in order. They need to give attention to detail, especially with finances and personnel. Leaders must have the ability to move the followers and the organization forward. They have to be aware of the 80-20 rule: 20 percent of the people in any group do 80 percent of the work. Four critical administrative abilities follow.

1. *Identify the real problem.* Think about the symptoms and causes of the problem and gather information to help you identify the real problem. Discuss the problem situation, then write your key findings and make a conclusion.
2. *Manage time effectively and shift priorities as necessary.* Apply the 80-20 rule to your time,

Figure 4–9 This vehicle has the world land speed record for a hydrogen-fueled vehicle. By setting and accomplishing many short-term goals over a period of five years, the ultimate long-term goal was attained. (Courtesy of MTSU Photographic Services.)

assigning priorities to activities that will return the greatest benefit to you. Make a to-do list, and assign priorities.

3. *Explain work.* Explain in detail what you expect of your co-workers. Keep in mind expected results, delegated authority, and progress reports.
4. *Listen actively.* Develop a genuine interest in your employees and their activities. When they come to you with suggestions, listen.

Technical Skills

Selecting Staff

Leaders need to surround themselves with excellent people. Every person hired should be good enough to replace the leader. It takes a secure leader to do this, however. The opposite approach is to hire weak people to make the leader look good. Obviously, in this situation, everyone loses.

Hollie Sharpe, an executive, told the author about his job interview with Ohio Casualty Insurance. The vice-president who was interviewing Sharpe asked him where he wanted to be in five years. After thinking for a moment, Sharpe said, "Sitting in your chair." The vice-president was briefly stunned with the answer, then replied, "I like that." The company knew it was getting a good, aggressive person; he was hired.

No matter what leadership training you have had or what kind of a leadership coach you are, poor staff that are lazy or have bad attitudes will make your life as a leader miserable.

Competence

At the very least, leaders should know what they are doing. Education, either on the job by coming up through the ranks or at a trade or technical school, college, or university, is necessary for successful leaders. We just cannot get there if we do not know where we are going. Become competent before you take on a leadership position.

Communication

Verbal fluency is the number one predictor of promotion within an organization. It is also a major predictor of success in life. If you speak poorly, you are seen as unintelligent. If your communication ability is excellent, you are perceived as very intelligent. "Perception is reality"; you are what people think you are.

Chapter 6 discusses communication in detail. Chapters 7, 8, and 9 discuss giving speeches on the FFA Creed, public speaking, and extemporaneous speaking. Chapters 10 and 11 discuss parliamentary procedure. After studying these chapters and performing the activities, your communication ability should be greatly enhanced.

Time Management

Make every minute count. Visualize the future and prepare for it. Time is your most valuable personal resource. Use it wisely because it can't be replaced. Time management skills can be developed, leading to substantial improvement in your productivity. One aspect of becoming wise is knowing when to say no.

Sets Goals and Plans

We have already mentioned the importance of goals. Chapter 14 further pursues this leadership quality. Lack of goal attainment can cause sadness and depression. Truly successful leaders develop and pursue challenging goals; their goals are their driving force.

Achieves

Leaders are winners. They are achievers. This is why employers want to see a resume when hiring: Resumes provide a list of achievements. Many employers also want to see a **portfolio**. Portfolios are evidence of a person's performance, such as articles or artwork.

Achieving is habit-forming. Leaders that achieve once will achieve again. Achievers find a way to win (Figure 4–10).

Figure 4–10 Achieving doesn't just happen. Nonachievers ask what's happening. Achievers make things happen. This student is studying to achieve. (Courtesy of the National FFA Organization.)

Responds to Failure

For many, the word *failure* carries with it an end, a sense of discouragement. For the successful leader, however, failure is a beginning. Responding to failure is perhaps the most impressive and memorable quality of leaders. Successful leaders put all their energies into their task. They simply do not think about failure. They use words such as *mistake, bungle, false start, mess, setback,* and *error.* They never use the word *failure*. They believe a mistake is another opportunity to do something better.

Leads Individuals and Groups

Leading individuals and groups is one of the basic steps in becoming a leader. As a matter of fact, it is so basic that it is taken for granted. One way to learn to lead groups is by learning and practicing parliamentary procedure. Once parliamentary procedure is learned, you can conduct successful meetings. Learning group dynamics is also important for future leaders. Leading individuals and groups is covered in Chapters 5 and 12.

CONCEPTUAL-TECHNICAL SKILLS

The following leadership qualities fall somewhere between conceptual and technical skills. They are thinking skills, but they can be taught, so I call them conceptual-technical skills.

Intelligence

Less has been written about intelligence and leadership than about any other leadership quality. We really do not know what intelligence is or at least, what its relationship to leadership is. We are not saying that you do not have to be intelligent; however, there are so many different types of intelligence that any one type does not guarantee the ability to lead.

Consider the following dilemma: Keith has much wisdom. He just seems to know what to do, how to solve problems, and how to make excellent decisions. However, Keith did not complete high school due to family problems. Kim was an honor student and scored very well on the ACT, but she does not seem to know how to get things done. Sandra is an excellent speaker with a winning personality, but she barely graduated from high school. Randy is a near mechanical genius, but his test college entrance scores were too low to be admitted to college. Who should get the leadership role?

Any of these people could get the leadership role—they are all intelligent in different ways. Part of being intelligent is knowing your strengths and weaknesses. By surrounding yourself with excellent people who excel in areas in which you are weak, you can build a winning team.

The former President of the University of Tennessee, Dr. Andy Holt, once said, "Know a little about everything, but know much [be an authority] about something." Maximize your intelligent qualities and surround yourself with people who are strong in areas in which you are weak.

Problem-Solving Ability

Problem solving is a methodical process of finding solutions and making decisions. Xerox Learning Systems® identified problem solving and decision making as two of the top six critical skills for success of supervisory leaders. Problem solving is the second most time-consuming task of a supervisor.

All problems require certain steps to reach solutions. To solve a problem, follow the steps of the scientific method: Define the problem, gather data, determine and evaluate alternatives, select a workable solution, implement the solution, evaluate the solution, and implement the decision. Chapter 13 discusses these steps in detail.

Decision Making

Decision making is the process of choosing between alternatives to solve a problem. It is a basic ingredient of leadership. Decisions must be made when you are faced with a problem. The first decision is whether or not to take action to solve the problem.

Making poor decisions can get leaders removed. Decisions leaders make can affect the health, safety, and well-being of employees and the general public. Ethical decision making is becoming more important as new technologies become available. As with most leadership qualities, decision-making skills can be developed. Chapter 13 discusses the decision-making process.

Cooperative Genius Approach (Synergy)

A technique I use when working with the alternative fuels project is a process I call the *cooperative genius approach*. If 10 people focus on the same goal and each is one-tenth as smart as Thomas Edison or Albert Einstein, the group can accomplish great things. Therefore, I assign each person one-tenth of the puzzle.

This process can also be called synergy. Simply defined, it means the whole is greater than the sum of its parts. The relationship the parts have to each other is a part in and of itself. In *The 7 Habits of Highly Effective People*, Stephen Covey points out that synergy is everywhere in nature. "If you plant two plants close together, the roots co-mingle and improve the quality of the soil so both plants will grow better than if they are separated."[24] If you put two pieces of wood together, they will hold much more than the total of the weight held by each separately. The whole is greater than the sum of its parts. Synergy is teamwork at its best.

Master Delegator

The leader who attempts to "do it all" is heading for disaster. Successful leaders know they cannot know everything; as humans, they can only know and do so much. No matter how dedicated a leader may be, each individual's time and energies are limited. The successful leader delegates authority along with responsibility. Nothing is so frustrating as to have responsibility with only implied authority. A leader must feel secure before committing to delegation. Resistance to delegation comes from fear of losing control and the habit of holding on to work. Consider the following as you delegate.

- Planning is the first step in delegation; it largely determines the success and efficiency of the action.
- A key to effective delegation is delegating the right task to the right person.
- The delegator should specify the *what*, *when*, and *why* in delegating—but not the *how*.
- To delegate effectively, authority and responsibility must be given to the co-worker; however, the ultimate responsibility remains with the delegator.
- Delegated tasks should be evaluated by results rather than the means used to accomplish them.[25]

Opportunist

Opportunists take advantage of unplanned situations and turn them into positives. Opportunists don't see problems, they see opportunities. Risk and opportunity go hand in hand. "Reasons first, answers second" is the first rule of an opportunist. If you have enough reasons to seize the opportunity, you will find answers.

Effective leaders are not problem oriented, they're opportunity oriented. They feed opportunities and starve problems. They think preventively. Every problem is an opportunity.

When leaders see their co-workers' problems as opportunities to build a relationship instead of as negative, burdensome irritations, the nature of their interaction changes. Leaders become more willing, even excited, about understanding and helping their co-workers.

Insists on Excellence

Breaking through the "good enough" barrier to enter the zone of excellence requires total preoccupation with quality from everyone in the organization. It starts with the leader's attitude: Great leaders are never satisfied with current levels of performance. They constantly strive for higher levels of excellence.[26] If you insist on excellence, more than likely you will get it. This is called a self-fulfilling prophecy.

An effective leader brings out the best in people by stimulating them to achieve what they thought was impossible. In a classroom experiment with grade-school children, one teacher was told she had a group of high academic achievers. The truth was that they were average. Another leader was told she had a group of slow learners. Actually, they were also average. The result of the experiment was that the so-called high achievers progressed much more rapidly than the so-called slow group.

The intelligence level of both groups was the same. The difference in performance came from how the teachers related to their students. The "high achievers" were challenged, and the "slow learners" became bored with the slow, repetitious teaching style. The lesson here is that if you lead for high expectations, you will get them; lead for low expectations and you will get them.

Holds Co-workers Accountable

An effective leader is tough-minded when it comes to obtaining results from co-workers. Excuses are not tolerated when achievable goals are not met. What counts are the results, not that the person really tried. In the business world, if an employee does not measure up after retraining and repeated warnings, the leader must have the fortitude to take appropriate disciplinary action. Effective leaders seldom have to use this power, but the employees know they will do so if necessary. Effective leaders can handle occasionally being unpopular or even disliked by some people. Permissiveness is neglect of duty.

Charles Knight, chairman of Emerson Electric Company, said anyone who accepts mediocrity—in school, on the job, or in life—is a person who compromises; and when the leader compromises, the whole organization compromises.[27] Unfortunately, in every organization, there are some people (sometimes as much as 20 percent of the work force) who do not respond to anything you do: They just don't care. In this case, a leader must realize that eventually the welfare of the entire group suffers if the rotten apple is allowed to remain in the barrel.

Broad Interests and Abilities

Leaders have broad interests and abilities in order to obtain respect from the group and to be the best possible communicator. Can you imagine a business luncheon without the leader having any interest in sports, politics, or international affairs? Leaders are well-read and up to date on the news.

Conceptual Skills

Vision

Vision is the first critical dimension of effective leadership. Without vision there is little or no sense of purpose in an organization; efforts drift aimlessly. However, visions and intellectual strategies alone are insufficient to motivate and energize followers. Substantial results come only if co-workers accept a vision. They must believe the leader's vision is their vision; only then will they accept responsibility for achieving it.

Outstanding leaders are visionaries. They have to dream about what could be and involve others in their dreams. Like small creeks that grow into mighty rivers, the dreams of these leader visionaries eventually shape the course of history.

Fred Manske tells several compelling stories of visionary leaders.[28] The student Frederick W. Smith had a vision in the early 1970s. In a term paper for an economics course at Yale University, Smith explained his vision of an air express system that would provide overnight, nationwide delivery of urgent packages. At the time that vision seemed impractical and far-fetched. In fact, Smith's economics professor gave him a *C* on the term paper. Undaunted, Smith used the ideas formulated in that *C* paper to create Federal Express. Today the Federal Express Corporation has nearly 50 percent of the air express market in the United States and is consistently ranked in the top 10 best places to work in America.

Another business leader who envisioned a product that significantly affected America was Allen H. Neuharth, chairperson of the Gannett Corporation, a major publishing firm. His vision was that of a national newspaper delivered to homes and newsstands early each morning. Neuharth believed there was a whole new generation of readers who wanted the news presented in concise format, with colorful charts and graphs to explain world and national news, lifestyles, sports, and finance. When the first edition of *USA Today* reached the streets on September 15, 1982, many people doubted it would succeed. As the annual losses grew, it appeared that the skeptics were right, but Neuharth refused to abandon his dream. In late 1987, *USA Today* became profitable, with more than five million readers—the largest daily readership in America.

Martin Luther King, who led the great crusade for civil rights from 1955 to 1968, epitomized a leader with great vision and the tenacity to move ahead at all costs. Despite being jailed several times, stabbed, and stoned, King persisted in his efforts to fulfill his dream of a world of racial equality and improved living conditions for the poor.

Ability to Motivate

Leaders must be able to motivate or inspire others. Some leaders motivate with enthusiasm or charismatic behavior; others inspire and move people through persuasion. Whatever the method, leaders must understand motivation and be aware of both psychological and physiological needs.

Motivation includes recognizing people, including them, encouraging them, and teaching them the required behavior. A leader must make followers feel loved, trusted, respected, and cared for. Above all, leaders must motivate people to excel for themselves and for the organization.

Decisiveness

Decisiveness means making decisions quickly, firmly, and forcefully. Some decisions must be made immediately. Some believe that sometimes a wrong decision is better than no decision. Maybe, maybe not. However, consider this. You are in the midst of a battle. The leader must decide to either advance or retreat. Otherwise, by not giving an answer to someone, it may keep someone from pursuing other alternatives or finding solutions.

Successful leaders have the courage to take action where others hesitate. The leader must do what he or she thinks is right, not what someone else thinks should be done. In today's work environment, however, it is imperative that coworkers be involved in the decision-making process. People carry out decisions they have participated in making much more enthusiastically than they carry out orders. It has been proved that involvement leads to commitment. Nevertheless, the decision-making burden ultimately belongs to the leader.

Straightforwardness

Somewhat related to being decisive is being straightforward. To be straightforward is to "tell it like it is." In *Principle-Centered Leadership*, Stephen Covey said, "when you are living in harmony with your core values and principles, you can be straightforward, honest and up-front. And nothing is more disturbing to people who are full of trickery and duplicity (deceit) than straightforward honesty—that's the one thing they cannot deal with."[29] The person in Figure 4–11 is being straightforward, as evidenced by his facial expression.

Figure 4–11 We have to lead according to the situation and our personality type. This leader is being straightforward. (Courtesy of the National FFA Organization.)

Creativity

Leaders who make a difference are creative. They know that seeing things others cannot see is not just a quality of leadership, it is a responsibility. Creativity is as much an attitude as it is an aptitude. Walking on the moon and heart transplants were somebody's creative ideas that became realities. Without creativity, we would make very little progress.

Creativity is an attitude that sees people, places, and things bigger and better than they see themselves. As a creative leader, you will find positives where others do not, you will seek opportunity where others find only problems, and you will see answers where others have not yet recognized the questions!

Creativity and imagination are abilities distinct from the capacity to acquire knowledge. Albert Einstein said that imagination is "something more important than knowledge." To enhance creativity and productive innovation, leaders must let their imaginations flow.

Courage

Courage is the intangible leadership quality of which greatness is made. General George C. Patton said courage is "fear holding on another minute." It is demonstrated when a person exhibits perseverance or determination when faced with adversity or an unusual challenge. Courageous leaders do not suffer from the crippling need to be loved by everyone. For example, they are not afraid to say "no" to unreasonable requests and demands placed on them and to take positions on issues of importance. Though they may be disliked and ridiculed from time to time, the courageous leader earns respect in the end. On the other hand, those who try to be the "good guy" by pleasing everyone lose the respect of the entire organization. Success is never final and failure never fatal. It is courage that counts. Leaders with courage also face conflict openly and head-on, follow a difficult path in the face of danger, and stand up boldly for their beliefs and values.[30]

Risk Taking

Risk taking is an indispensable part of leadership. When we look at leaders who are making a difference, we see that they take the risk to begin, while others wait for a better time, a safer situation, or assured results. They are willing to take risks because they know that being overcautious and indecisive stifles opportunity. It is ironic that we are always in search of novelty, yet we can be deeply uncomfortable with taking risks.[31]

Many decisions would never be made if the leader would not take a risk. If we are to stay strong and healthy into the twenty-first century, leaders must use their risk-taking abilities to create an environment that supports and encourages innovation and creativity. When the leader acts as a role model for risk taking, everyone benefits.

Commitment, Determination, and Persistence

It is hard to discuss these three qualities of leadership separately. They are all traits of a leader who finishes the race, completes battles, and attains goals. They show the toughness of a bulldog, which has a nose that slants inward so it can breathe while it holds on.

Commitment is the glue to your success as a leader. It is the binding force that holds the other leadership qualities together and makes them work. Commitment is that intangible ingredient you reach for, deep inside, to help you through the tough times. It is the inner strength that keeps you going when everyone else gives up. Faithfulness and persistence are two key ingredients of commitment.

Good leaders show exceptional determination in pursuing their objectives. They never give up until they succeed. Timing is also important. Knowing when to push for changes and improvements often determines the success of obtaining them. People are not always psychologically ready to accept innovative ideas. With determination and timing, you cannot lose.

When 90 leaders were asked about the personal qualities they needed to run their organizations, "they never mentioned charisma, or dressing for success, or time management. Instead, they talked about persistence and self-knowledge; about willingness to take risks and accept losses, about commitment, consistency, challenge and learning."[31]

Sense of Urgency and Initiative

Without urgency, progress is slow. The attitude has to be, "Let's do it, and let's do it now!" Leaders seize the moment. However, it is a difficult challenge of leadership to maintain a high sense of urgency to accomplish difficult goals year after year. Many believe it is easier to build a winning team from scratch than to keep it continuously on top. Here are three ways for leaders to keep everyone energized and inspired:

1. Periodically change or rotate co-workers' responsibilities.
2. Take unusual or unexpected action.
3. Set the example of urgency for others to follow.

Initiative is recognizing our responsibility to make things happen. Taking initiative does not mean being pushy, obnoxious, or aggressive. Our basic nature is to act, not to be acted on. Many people wait for something to happen or someone to take care of them. People who end up with the good jobs and eventually become successful leaders are the active people who seek solutions to problems. They seize the initiative to do whatever is necessary, consistent with correct ethics, to get the job done.

Follow-Through

Some leaders are great at starting projects, but their enthusiasm fades quickly. Once the new idea or project does not run as smoothly or work as well as first planned, they let it die a slow death. If ideas or projects are worth doing, they should be given a chance to work. The natural timetable for a new idea or project to work may be slower than the expectations of the leader. If a project does not mature, the leader should follow through with a formal closure, not just let it die without proper evaluation from the group.

Master of Change

Leaders must realize that, basically, people do not like change. They like things the way they are. Psychologists tell us our fear of change comes from the loss of personal identity. When a major change occurs, our internal dialogue sounds something like this: "I won't know how to act," or "I won't fit," or "after this change, I won't be me anymore." Fear of change also comes from the fear of the unknown, fear of being different, or lack of trust in the change.[32]

Leaders need to be aware that all people do not have the same tolerance for change. People and their adaptability to change follow a bell-shaped curve (Figure 4–12). For example, early adopters of change comprise about 13 percent of the population, whereas those resistant to change comprise another 16 percent of the population.

Consider the introduction of the round hay baler. Very few farmers approved of the round hay baler at first, but after a few years went by, they adapted; the round hay balers followed the chart pattern exactly. Now, hardly anyone uses the traditional small square balers except for special purposes.

Three steps you can take to foster receptivity and openness to change are these.

1. Provide new information to expand your co-workers' thinking.
2. Provide new ideas to spark their creativity and broaden their horizons.
3. Provide new experiences to build a desire for and a belief in the value of change.

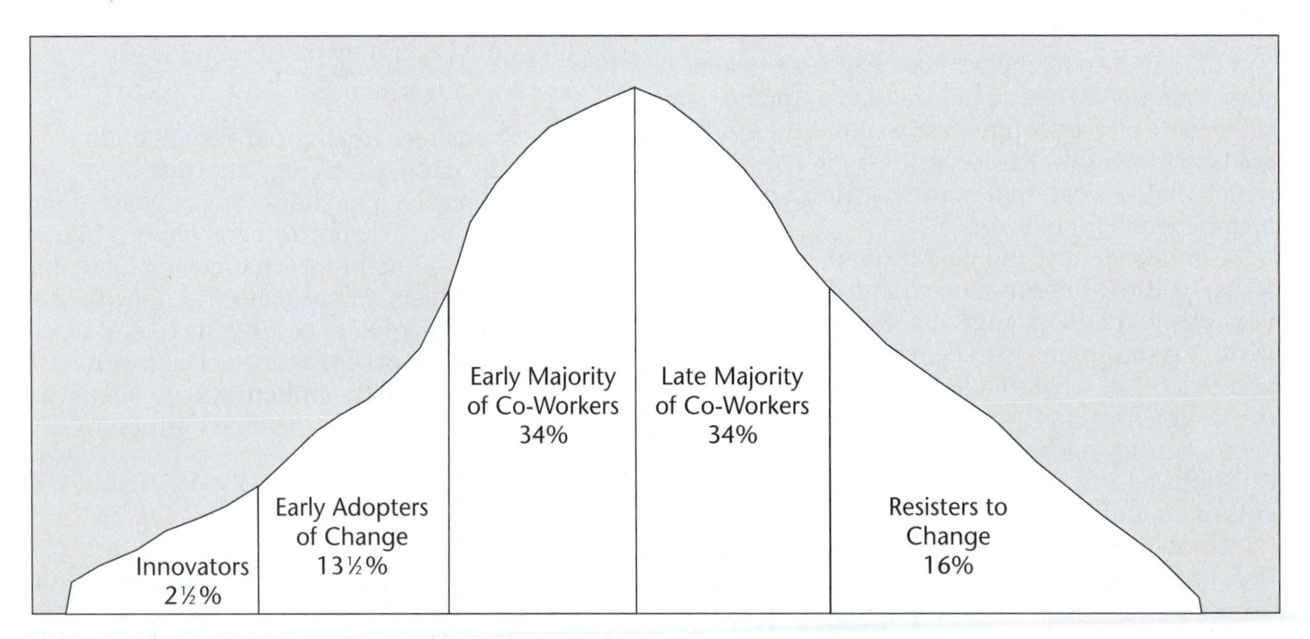

Figure 4–12 Change is a slow process. It takes 7 to 14 years from the inception of a new idea to its adoption and use. Creative thinkers create the changes. Early users are the first group of people to try the new idea, invention, or concept. The early co-workers watch the early adopters, who tend to be leaders, before they make the change. The late co-workers are open-minded, but they take much convincing to change. Die-hards change only because they have no other choice; either no other options are available or "everyone else is doing it."

Wisdom

Wisdom is knowing how and why human beings function and progress. It is having insight. To have wisdom is to have the ability to size up a situation quickly. Wisdom also includes awareness and timing. A leader with wisdom can differentiate between the authentic and the fake, between things worthwhile and things unworthy, between things lasting and those things that will fade quickly.

Knowledge for the sake of knowledge is not important; it *is* important that leaders have the wisdom to apply what they learn. The pursuit of wisdom builds character; it gives you the inner tools to lead others.

Desire to Lead

Regardless of your potential as a leader, it is of no value if you have no desire to lead. With leadership comes responsibility. Many people desire a simpler, worry-free lifestyle. Someone may be offered a leadership position but, due to family, economical, or physical reasons, decides to stay where he or she is and be quite happy with the decision.

However, each of us has leadership responsibilities beyond the workplace. Many of these leadership responsibilities are unavoidable, such as leadership within the family, and some are necessary because of public or community service. When you are faced with leadership responsibilities, be assertive, study, review the leadership qualities presented here, and seek help from peers and role models. Refer to Figure 4–13 for categories of organizations in which leadership skills may be used.

Categories of Organizations in Which Leadership Skills May Be Used

Organization	Examples
Work	Supervisor, firefighter, union officer
School	FFA, FCCLA, SKILLSUSA-VTCA, DECA, FBLA, BPA, student council
Civic/service clubs	Kiwanis, Lions, Rotary, Exchange, Jaycees, Masons, Shriners, Elks, Eastern Star
Economic organizations	Cooperatives, Cattleman's Association, chamber of commerce
Professional organizations	American Vocational Association, National Education Association, Business and Professional Women
Charitable organizations	Red Cross, United Givers Fund
Advisory committees	School advisory committees, community advisory committees
Special interest groups	County fair boards, American Legion, ecology groups
Political organizations	Republicans, Democrats
Support groups	FFA alumni, Parent-Teacher Association, Band Boosters
Age-oriented groups	Boy Scouts, Girl Scouts, senior citizen groups
Athletic groups	Little League, riding clubs, ski clubs

Figure 4–13 Leadership skills are used in the workplace as well as in a variety of organizations. (Adapted from Curriculum Instructional Materials Center, Stillwater, OK: Oklahoma Department of Education.)

CONCLUSION

No matter what your level of leadership, no matter what your leadership accomplishes, it is exciting to know you are in the process of becoming. You are an ever-changing composite of the things you say, the books you read, the thoughts you think, the company you keep, and the dreams you dream.

What you think is what you become. Your potential is endless. Your ability to make a difference has no boundaries. It is said that one person with belief is equal to 99 with only interest. If you think you can, you can. If you think you cannot, you cannot.

SUMMARY

Leadership is an observable, learnable set of priorities. Leaders come in many personality types. Leadership has a purpose and goal. Leadership is a group effort.

Personal leadership development is important. It is a challenge to learn leadership skills, and these skills create respect from others. Leadership helps you understand others as well as yourself. Leadership helps individuals mature and develop their self-concept.

The best way to learn leadership is to place yourself in situations requiring leadership action. Five steps to becoming a good leader are studying, analyzing yourself, developing, learning, and following a definite plan.

Steps of your personal leadership plan include developing a vision; setting goals; developing initiative, self-confidence, personal responsibility, a healthy self-image, and self-organization; eliminating procrastination; studying; and magnifying your strengths.

Most of us like to be accepted. You are likely to attain group acceptance if you become knowledgeable about the purpose of the group, work hard, become committed to the group, are willing to be decisive, interact freely with others, and exhibit human relations skills. Three types of individuals emerge as leaders: member-oriented, goal-oriented, and self-oriented.

To realize maximum leadership potential, three elements must occur: ability, experience, and opportunity. Business and industry use three basic approaches to improve leadership within the organization: selecting, situational hiring or engineering, and training.

There are three major types of leadership traits, abilities, and skills: conceptual (thinking) skills, technical (how-to) skills, and human relations (people) skills.

Human relations skills include honesty and integrity; sensitivity; cooperation; flexibility; support, self-confidence, responsibility, dependability, and emotional maturity; deriving pleasure from work; servanthood; setting an example; being available and visible; showing confidence in people; and using power wisely.

Technical-human relations skills include listening, team (group) building, earning loyalty, rewarding accomplishments, coaching and strengthening others, planning short-term goals and wins, and administrative ability.

Technical skills include selecting staff, being competent, being a communicator, managing your time, setting goals and plans, achieving, responding to failure, and leading individuals and groups.

Conceptual-technical skills include intelligence, problem-solving ability, decision making, cooperative genius approach (synergy), mastery of delegating, being an opportunist, insisting on excellence, holding subordinates accountable, and exhibiting broad interest and abilities.

Conceptual skills include vision, being a motivator, being decisive, being straight-forward, creativity, courageousness, being a risk-taker, commitment, determination, and persistence, sense of urgency and initiative, following through, being a master of change, wisdom, and the desire to lead.

End-of-Chapter Exercises

Review Questions

1. Define the Terms to Know.
2. What are five steps to becoming a good leader?
3. List 10 steps of your personal leadership plan.
4. Discuss the importance of a healthy self-image.
5. List four techniques to help develop self-organization.
6. List four steps to help overcome procrastination.
7. What are six ways to attain group acceptance?
8. What are the three types of individuals who emerge as a group's leader?
9. Explain the relationship between ability, experience, and opportunity to lead.
10. What are three ways to improve leadership in the workplace?
11. What are four things you can learn by studying the qualities of successful leaders?
12. Which of the three types of leadership qualities (conceptual, technical, and human relations) remain constant no matter what level of leadership? Explain.

Fill-in-the-Blank

1. Leadership is an observable, ___________________ set of priorities.
2. The leader is a person who, on the whole, best lives up to the standards or values of the ___________________ .
3. "If it's going to be, it's up to ___________________ ."
4. Success is controlled by our ___________________ actions.
5. Winners find a way. Losers find an ___________________ .
6. People can accomplish just about anything they want if they are willing to ___________________ .
7. If you try to please everyone but yourself you are destined for ___________________ .
8. Some men see things as they are and say why? Some dream dreams that never were and say ___________________ ___________________ .
9. Winners never quit, and quitters never ___________________ .
10. "In an argument, the first one to get mad ___________________ ."
11. If you can dream it, you can ___________________ it.

Matching

Terms may be used more than once.

_____ 1. People skills
_____ 2. Thinking skills
_____ 3. How-to-do-it skills
_____ 4. People skills easily taught
_____ 5. Thinking skills easily taught
_____ 6. Honesty and integrity
_____ 7. Listening
_____ 8. Communication
_____ 9. Problem solving
_____ 10. Vision

A. conceptual skills
B. technical skills
C. human relations skills
D. conceptual-technical skills
E. technical-human relations skills

Activities

1. Set three goals in each of the following categories: physical, family, social, financial, educational, ethical, career.
2. What do you want to be 5, 10, and 20 years from now? (Remember, "few people plan to fail, they fail to plan.") Write three goals each for where you want to be 5, 10, and 20 years from now. List one sacrifice or obstacle that you anticipate for each goal. For example, five-year goal—graduate from college; obstacle—studying and self-discipline.
3. Write five facts about modern-day leadership that you can use for the future. Share these with your classmates and get their reactions.
4. Select three conceptual skills you believe to be your strongest qualities. Write a paragraph on each defending your reasoning.
5. Select three conceptual-technical leadership skills, and repeat the process in question 4.
6. Select three technical leadership skills, and repeat the process in question 4.
7. Select three technical-human relations skills, and repeat the process in question 4.
8. Select three human relations leadership skills, and repeat the process in question 4.
9. Every student's name in your class will be placed in a container. Draw a name. Select five leadership qualities that person best exhibits. Write a paragraph on each and present it to the class.
10. Identify your five weakest leadership qualities. Write a short plan describing how you plan to improve on each of these.
11. Select one of the 50 leadership qualities and write a paper or lesson on how you would teach this to others.
12. **Take it to the Net.**

    Explore leadership on the Internet. Some Web sites that contain information on leadership have been listed below. Descriptions of each Web site are also provided. Pick one Web site and do as the description says. If you are having problems with the sites or want more information, you can use some of the terms below in the search field.

Web sites

<http://www.cdaconsulting.com/tips&tools.htm>

Provides tips and tools to enhance your leadership skills. Browse the tips and tools and pick five that you find most beneficial. Print all five and write a brief summary telling why you find them beneficial.

<http://www.successoptions.com/strategylibrary.htm>

Contains strategies, reports, articles, and book reviews dealing with leadership. Read some of the articles. Print one article and write a short summary of it.

Search Terms

leadership

leadership skills

effective leadership skills

Notes

1. Bob Patton, et al. *FFA: Learn, Grow, Become* (Stillwater, OK: Department of Vocational and Technical Education—Curriculum and Instructional Materials Center, 1984).
2. G. R. Hickingbotham, "Creative Leadership," in *Creative Innovators*, ed. Dorothy M. Walters (Glendra, CA: Royal Publishing Inc., 1988), p. 61.
3. Ibid., p. 65.
4. Ibid., p. 69.
5. Ibid., p. 70.
6. S. R. Levine and M. A. Crom, *The Leader in You* (New York: Simon & Schuster, 1993), p. 27.
7. *Plans for Training Professionals in Community Development* (Alexandria, VA: The National FFA Organization, 1980).
8. *Personal Skill Development in Agriculture* (College Station, TX: Instructional Materials Service, Texas A & M University, 1989), 8738-A, p. 2.
9. Ibid., 8738-A, p.1.
10. Ibid.
11. F. A. Manske, *Secrets of Effective Leadership* (Columbia, TN: Leadership Education and Development, Inc., 1990).
12. S. M. Bethel, *Making a Difference: Twelve Qualities That Make You a Leader* (New York: G. P. Putnam's Sons, 1990).
13. B. Nanus, *The Leader's Edge* (Chicago: Contemporary Books, Inc., 1989).
14. R. D. Collons, "Spotlight on Leadership" in *Leadership*, ed. A. Dale Timpe (New York: Facts on File Publications, 1987).
15. W. Bennis and B. Nanus, *Study Guide for Leaders: The Strategies of Taking Charge* (Newark, CA: Sybervisions Systems, Inc., 1986).
16. Ibid.
17. Manske, *Secrets of Effective Leadership.*
18. Bennis and Nanus, *Study Guide for Leaders: The Strategies of Taking Charge.*
19. Bethel, *Making a Difference: Twelve Qualities That Make You a Leader.*

20. Manske, *Secrets of Effective Leadership.*
21. J. M. Kouzes and B. Z. Posner, *The Leadership Challenge: How to Get Extraordinary Things Done in Organizations* (San Francisco: Jossey-Bass, 1987).
22. Manske, *Secrets of Effective Leadership.*
23. Kouzes and Posner, *The Leadership Challenge: How to Get Extraordinary Things Done in Organizations.*
24. S. Covey, *The 7 Habits of Highly Effective People* (New York: Simon & Schuster, 1990), p. 263.
25. J. Townsend, (Notes from "Leadership" class) Texas A & M University, College Station, TX.
26. Manske, *Secrets of Effective Leadership.*
27. Ibid.
28. Ibid., p. 4.
29. S. Covey, *Principle-Centered Leadership* (New York: Simon & Schuster, 1991).
30. Manske, *Secrets of Effective Leadership.*
31. Bennis and Nanus, *Study Guide for Leaders.*
32. Bethel, *Making a Difference: Twelve Qualities That Make You a Leader.*

CHAPTER

Leading Teams and Groups

Objectives

After completing this chapter, the student should be able to:

- Discuss the importance of democratic group leadership
- Explain the importance of groups
- Explain the importance of leading teams
- Describe the three types of groups and their subgroups
- Explain how to organize groups
- Explain group dynamics
- Analyze the five stages of group development
- Describe various types and forms of group discussion
- Demonstrate how to lead a group discussion

Terms to Know

participative management
facilitator
synergy
confronting
functional groups
task groups
rigorous
affinity
group dynamics
norms
tasks
cohesiveness
intragroup
intergroup
status
task roles
maintenance roles

Most of you either are leaders, would like to become leaders, or have been leaders. At least, you would like to understand how you are being led. You need a good idea of how groups function. As you develop personal and leadership skills, it is natural to apply them in group situations. Examples of groups are your FFA chapter, your church group, your group at work, or your group of friends (Figure 5–1).

The fundamental belief of American democracy is in the dignity of man. As a result of such a belief, we have devised the group process as a means of governing our affairs. In every phase of living, we find people working in groups to solve their problems.[1] Groups, when they have access to the facts, can usually make better decisions than one person.

DEMOCRATIC GROUP LEADERSHIP

In Chapter 1, we concluded that in most situations, democratic leadership is preferred. It is not only the American way, it is the most productive way. Therefore, the formal democratic group is neither laissez-faire nor autocratic. The laissez-faire group is characterized by its lack of organization. The autocratic group is under the domination of an individual or a "power clique." Congress is a democratic group. The Capitol (Figure 5–2), is a symbol of democracy.

Why Democratic Group and Team Leadership?

When compared to democratic groups, autocratic groups have greater motivation to work, more member satisfaction, and greater productivity. There is less discontent among members and less evidence of frustration and aggression. There is more friendliness, cooperation, and group-centered spirit when democracy prevails. Also, more individual initiative is displayed. There is evidence that a leadership style that emphasizes democratic team building is positively associated with high productivity and profitability.[2] Teamwork ensures not only that a job gets done but also that it gets done efficiently and harmoniously.

Figure 5–1 As human beings, we are group oriented. Membership in a group meets the basic psychological needs we all have. Groups have a synergy effect that allows us to accomplish things we could not do otherwise.

Figure 5–2 The Capitol is a symbol of democracy. Democratic groups have greater motivation to work, more member satisfaction, and greater productivity than other types of groups.

Why Not Autocratic Group Leadership?

The autocratic group cannot compete with the democratic group in all-around productivity. Within these groups are found excessive irritability, hostility, and aggression, often directed at fellow members as well as the autocratic leader. The members of such a group are apt to be apathetic in their general attitude even when secretly discontented. Group members are much more dependent and show little creativity. When the leader is absent, little or no action takes place. There is also evidence that the lack of team building can have a negative influence on the physical and psychological well-being of leadership personnel. Team building is not just a good idea, it is a necessity of biological life.[3]

Democratic Group Leadership and Attitudes

In industry, in the classroom, and even in the military, it has been demonstrated that the involvement of people in groups and the reaching of decisions in a democratic atmosphere leads to more favorable attitudes toward the decision. One study of industry revealed that participation in decision making by all members of the group resulted in greater productivity, less resistance to change, and a lower turnover rate than did either committee decisions or careful explanation of decisions made by others.[4]

Democratic Group Leadership and Peer Pressure

When group members commit themselves to act in a certain way, their decision is strengthened by an awareness that others are similarly committed. They do not wish to lose status by failing to follow through on a decision witnessed by their peers. One of the strongest motivating forces for an individual is to be respected and to have status in the eyes of the members of groups to which he or she considers it important to belong. This principle is one of the most important in establishing the superiority of group (democratic) action over individual (autocratic) action.

Importance of Group Leadership Skills

It is probable that without leadership no group could produce worthwhile action in the direction of its goals.[5] The democratic leader has the ability to perceive the direction in which the group is moving and to move in that direction more rapidly than the group as a whole. An individual is a leader when his or her ideas and actions influence others in the group.

IMPORTANCE OF GROUPS

People like to belong. A sense of belonging is a basic need that everybody has.

Why People Belong to Groups

In our quest to satisfy our needs and wants, we find that many of our needs are best satisfied through group affiliation and action. I hope you want to be a part of the FFA. Some of you want to be on the football, baseball, softball, or basketball team. Others want to be in the band. You like to choose with whom you socialize. Joining groups helps you meet your psychological needs. Other reasons people belong to groups are affiliation, proximity, attraction, activity, assistance, and tradition.[6]

Affiliation We all need to interact with others. We spend time around people we like. People who like the people they work with tend to have high levels of job satisfaction. Studies have shown that today's young workers value friendships at school and on the job more highly than either their parents or grandparents did. As a matter of fact, they work with the expectation of meeting new friends and joining a group or team. The individuals in Figure 5–3 are co-workers as well as

Figure 5–3 These co-workers are also friends. They are from various universities, high schools, and the state department of education, but they are unified as a group because of their common objective of improving agricultural education.

friends, which increases the productivity of the group.

Proximity We tend to form groups with the people we see often. People who have classes together, go to church together, work at the same place, and live in the same dorm tend to associate with each other.

Attraction We tend to be attracted to the people who have attitudes, personalities, social class, income, and interests similar to ours. We also like to associate with people we find physically attractive.

Activity We join a group because of the activity it performs. For instance, FFA is an integral part of agricultural education. Agricultural education is a part of the agriculture community, including production agriculture, agribusiness, and agriscience. The love for agriscience and the type of activities it involves brings people together (Figure 5–4).

Assistance We often join groups for the assistance they provide. Assistance with leadership development, travel, and awards may be a reason for joining the FFA. Once you graduate, you may wish to join a civic club such as the Kiwanis, Lions, or Rotary Club to develop business contacts and for the services they provide to a community.

Figure 5–4 People belong to groups because of affiliation, proximity, attraction, activity, assistance, and tradition. Interest in agriscience has brought the people in this picture together. (Courtesy of USDA.)

Tradition You may join a group purely from tradition, (your mother or father was an active member and it was expected that you would also be a member). Various communities have a rich tradition in certain groups: football, band, chorus, FFA, and so on. If you have the opportunity, you join in because everybody in the community expects you to be a part of the group or team.

Why People Don't Belong to Groups

There are many reasons that people do not belong to groups.

Unaware of the Group People may not know that a group focused on their interests exists. If they do know about a potential group they might like, they may not join because of the way it does things.

Insecurity and Feelings of Inferiority Students may feel insecure about joining a group, worry about whether or not they will be accepted by the group. Some people fear they lack the human relations skills to get along with other group members. They may feel inferior to them for reasons such as status, educational background, or even clothing.

Cannot Meet Expectations People may hesitate to join a group because they are not sure of the group's expectations of its members. They may feel that the other group members know much more than they do.[7]

Why Group Leadership Skills Are Important

Besides being necessary for athletic groups, high school groups, and civic groups, leadership skills are needed in the workplace.

Leader Evaluation Traditionally, managers or leaders had to concentrate on supervising individual employees rather than a work team. Supervisors' ratings or performances are evaluated on the results of the department's or group's performance as a whole, rather than on the results of each employee.

Big Part of Job Job supervisors report spending 50 to 90 percent of their time in some form of group activity. The current trend is to have employees or co-workers more involved in decision making. Industry calls this **participative management**. Workers at the beginning levels of work provide input to improve the company, which consumes much of the leader's time as a group leader or **facilitator**. However, this time is well spent if productivity is increased.[8] The supervisor in Figure 5–5 is taking time to listen to his employee.

Figure 5–5 Listening is part of being a good group leader. The supervisor above appears to adhere to participative management. (Courtesy of USDA.)

Better Group Performance Technology is useless if people do not feel they are a part of the team. Input from groups is needed. Group participation in decision making results in better decisions, with more commitment to their implementation.

LEADING TEAMS

Some say that TEAM could be an acronym for "Together Each Accomplishes More." Teamwork will be the norm for the twenty-first century. Tomorrow's leaders will have to become coaches. If you have ever played in or watched an athletic contest, you can understand the importance of teamwork. If the team loses, every individual loses. No chain is stronger than its weakest link.

Another positive outcome of teamwork is an increase in synergy. **Synergy** is the interaction of two or more parts to produce a greater result than the sum of the parts taken individually.[9]

Teams in the workplace are allowed to be **confronting**, questioning, policy-making, and, above all, participative. Teams are constantly evaluating conditions that keep it from functioning effectively. This involves both the leader (supervisor) and co-workers (employees). All must work together to set objectives and plan how to achieve them. The leader uses teamwork to solve problems. The bottom line is this: Organizations and departments that work together as a team generally outperform those that act as individuals, just like an athletic team. The leader of the future must be able to lead a group—or someone will be found who can. Consider the following as you build your team.

- Create a shared sense of purpose. People working together can accomplish tremendous things. What gives a team that special boost is the unified vision the individual members share.
- Make the goals team goals. Unless the whole team wins, each individual within the team loses if a particular goal is not achieved. However, if our personal goal is to do the best we can do and we achieve that goal, we are still winners by our own standards even if others do not agree.
- Treat people like the individuals they are. Individuals have different personalities, skills, hopes, and fears. A talented leader recognizes those differences, appreciates them, and uses them to the advantage of the team.
- Make each member responsible for the team product. People need to feel their contributions are important. Otherwise, they devote less than complete attention to their tasks.
- Share the glory, and accept the blame. When the team does well and is recognized, it is the leader's responsibility to spread the benefits. For criticism, a smart leader accepts whatever complaints arrive publicly but speaks privately with team members about improvements.
- Be involved; stay involved. Know what is happening. Empathize with what is happening around you. Get an intuitive feel for the team.
- Be a mentor. Team players are the leaders of tomorrow. It is the leader's job to develop the talents and strengthen the people on the team, help team members achieve goals, and reinforce the confidence you have in their abilities.[10]

Research on Teams

Many organizations are implementing teams that involve workers in tasks previously done by leader management. Research has shown that the use of teams causes an increase in productivity, improved quality of work, more positive attitudes, less member absenteeism, and a desire by members to stay on team.[11] The movement toward work teams appears to be a major trend in organizations.

The use of teams in this country was spurred by the American interpretation of Japanese management style. The Japanese continue to dominate many sectors of the global economy and give much of the credit for that success to their participative, team-based decision-making and management style.[12] It stands to reason that other nations would want to adopt successful leadership practices.

Purpose of Teams

A *work team* is a group of individuals with a common purpose. The existence of the team implies that members work together cooperatively to achieve a common goal.[13] It takes teamwork to make the team work.

Characteristics of Effective Teams

Team members must be carefully selected for compatibility and to ensure that each person is bringing something unique to the team. Teams need to be given a charge, mission, or goal. Teams are difficult to start and must be managed, or led, in order to start quickly. Consider the following five characteristics of teams.

- Team members must be fully committed to a common goal and approach. Members must agree that the goal is worthwhile and decide on a general approach.
- To succeed as a team, members must be accountable to one another and to the organization for the outcome of their work.
- Teams must develop a team culture based on trust and collaboration. Team members must be willing to compromise, cooperate, and collaborate.
- Teams must develop shared leadership among all members.[14] Figure 5–6 lists the characteristics of an enthusiastic team.
- The team needs enough power and authority to accomplish its tasks and implement its ideas.

Eleven Commandments for an Enthusiastic Team

1. Help one another be right—not wrong.
2. Look for ways to make new ideas work, not for reasons they will not.
3. If in doubt, check it out! Do not make negative assumptions about one another.
4. Help one another win, and take pride in one another's victories.
5. Speak positively about one another and about your organization at every opportunity.
6. Maintain a positive mental attitude no matter what the circumstances.
7. Act with initiative and courage, as if it all depends on you.
8. Do everything with enthusiasm; it is contagious.
9. Whatever you want, give away.
10. Do not lose faith; never give up.
11. Have fun!

Figure 5–6 Eleven commandments for an enthusiastic team.

Establishing an effective team is a challenging process that requires both interpersonal skills and extensive technical support. The development of trust, a common vision, and the ability to work well together depends on appropriate interpersonal skills. Once trust and goals are established, tackling complex tasks requires training. Many of these interpersonal and training functions are the leader's responsibility.

The Leader's Role in a Team

The role of a leader changes in a team environment to that of facilitator and coach. Leaders are also caretakers of their teams, helping them to achieve their goals by giving them instructions, encouragement, and resources. Leader-facilitators still fulfill many of the functions of traditional leaders, but they do so to a lesser extent and only when asked. They assist the team by obtaining the resources needed to solve problems and implement solutions, intervening only when needed. The leader's central activities are assessing the team's abilities and skills and helping it develop the necessary skills, which often includes getting the right type of training.[15]

Another role for the team leader is to make the team aware of its boundaries. Many teams fail because they take on too much or ignore organizational realities and constraints. For example, a team of schoolteachers assigned the role of revising the social studies curriculum may propose changes that affect other parts of the school curriculum and then be disappointed when its recommendations are not fully implemented. In this case, the role of the team leader is to keep the team focused on its specific task or integrate the team with others who may help implement its wider recommendations.[16]

Potential Problems with Teams

The team concept does not always work in the United States and other countries. Much of the team management concept originated in Japan, which has a more team-based culture. United States culture tends to be more individualistic and often conflicts with team-based approaches.

Australians have come up with a new concept: *collaborative individualism.*[17] Collaborative individuals are not limited by the boundaries of the group. They cooperate with their team while maintaining their internal motivation and conflict-tolerant skills. Such an approach may be more suitable to the United States than the Japanese search for consensus and conformity in a team.[18]

TYPES OF GROUPS

There are three types of groups. **Functional groups** and **task groups** are formal groups. The third type of group is informal groups.[19]

Functional Groups

Functional groups are formal groups that exist indefinitely. The group's objectives and its members may change over time, but the basic functions remain the same. Basically, these are social groups, teams, and clubs with which you are familiar: FFA, 4-H, Farm Bureau, various civic clubs, and many others.

Functional groups also exist within the work force. Most agribusiness companies have marketing, finance, production, and personnel or human resources departments. Each department makes up a functional work group. The supervisors of each department provide a link to managers, who link to the vice-president, who links to the president. Colleges and universities have the same type of organizational system. Many companies are moving away from this traditional pyramid structure because of the barriers in communication that it creates.

It might come as a surprise to some people, but the pyramids are tumbling down.[20] The traditional rigid structure is loosening to let people do their creative best, to fully develop the talent that has been lying dormant for years. Innovative, well-led organizations are using the team approach. Here are some examples.

- *Problem-solving teams.* The most widely known type of problem-solving group is the *quality circle,* initiated by the Japanese, in which employees focus on ways to improve quality in the production process. These problem-solving groups consist of hourly employees who meet to discuss ways of improving product quality, operational efficiency, and the work environment. These recommendations are then proposed to management for approval.[21]
- *Self-managing teams.* These multiskilled employees rotate jobs to produce an entire product or service.
- *Special-purpose teams.* These teams are often involved in developing new work procedures or products, devising work reforms, or introducing new technology. They provide a link among separate functions, such as production, distribution, finance, and customer service.[22]

Task Groups

Task groups are formal groups because they are sanctioned by the organization to perform a specific function. The task group is more commonly called a committee. Task groups have proved to be an effective problem-solving tool. Unlike work groups, they can be composed of members from different departments. Members usually report to their work groups when not meeting with the task force. Commonly used task groups include standing committees, ad hoc committees, and boards.[23]

Standing committees are permanent groups that exist to deal with year-to-year issues. Members usually serve a one-year term on the committee. When their terms are up, others may take their place. To keep things running smoothly, committees usually replace members on a rotating basis. Examples of standing committees in the FFA are the Finance Committee, Scholarship Committee, and Recreation Committee.

Ad hoc committees are temporary task forces formed for a specific purpose. They are temporary because they exist only until the task is accomplished. In an FFA meeting, someone may make a motion to buy a vending machine for the shop. Someone may refer this motion to an ad hoc committee. Why? There are many questions, such as principal approval, projected income and expenses, and management. The ad hoc committee would be appointed and asked to report back at the next FFA meeting. It would disband once the report is given.

Corporations have boards (a type of standing committee) of directors or trustees whose task is to approve the management of the organization's assets, set or approve policies and objectives, and review progress toward the objectives. The national FFA organization has a board of directors, as do most of the state FFA associations. FFA alumni groups, which serve the active FFA members, also have a board of directors.

Informal Groups

Functional and work groups are intentionally created by the organization; informal groups are not intentionally created. Informal groups are created when members join together voluntarily because of similar interests. Informal groups are sometimes referred to as cliques. These groups may help or hurt the organization as it pursues its objectives.

In the workplace, people from different departments may get together for breaks, lunch, and after work. Membership in these informal groups tends to be open and changes at a faster rate than formal groups. Informal groups start by not having an agenda, but if they share common concerns, they could establish goals and objectives. They exist primarily to meet the psychological need of belonging.[24]

ORGANIZING GROUPS

Levels of Group Membership

Figure 5–7 illustrates group membership. The outer line forming the circle is the boundary that indicates membership: Those meeting membership requirements are inside the circle, and those not meeting the requirements are outside the circle.

The group defines the boundary for membership. In some groups, the requirements might be an interest in joining; others might require dues. Graduates of your high school become alumni of the school. Those who did not graduate from that

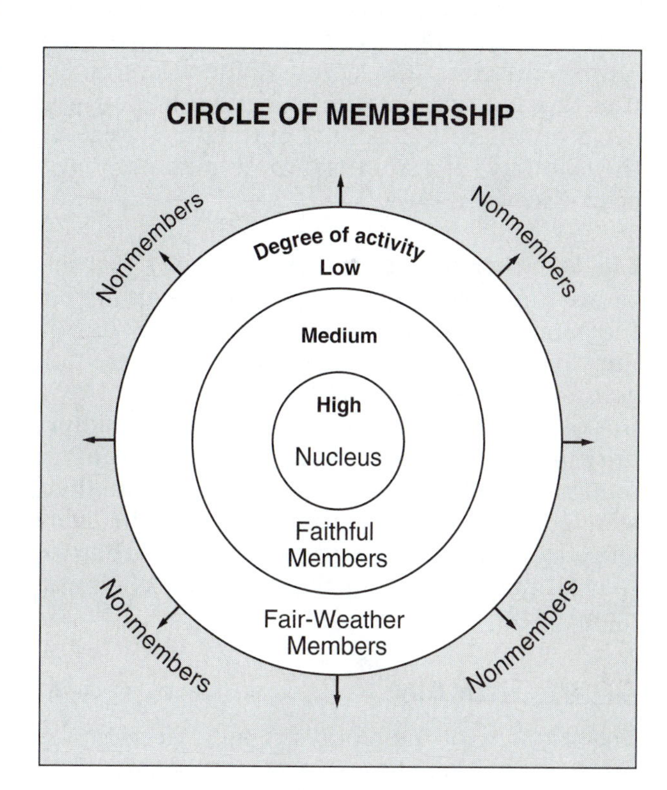

Figure 5–7 There are three levels of group membership: nucleus, faithful members, and fair-weather members. Due to your priorities, time commitments, and interests, you may be in the nucleus of one group and a fair-weather member of another.

school cannot become alumni and therefore cannot be a part of the group.[25]

Nucleus The degree of activity of members varies within a group. A circle is a visual representation of the activity level of members. Some members are highly active, perhaps officers or committee chairs. They are at the center of activity, the nucleus of the group. Much of the group's activities revolve around its motivation and leadership. Groups are good illustrations of the 80-20 rule, in which 20 percent of the group membership does 80 percent of the work. This is true of many civic clubs and high school groups.

Faithful Members The second level of membership includes individuals who are faithful and loyal to the group and attend all the meetings. Either they do not want to be or do not have the necessary abilities to be part of the inner circle. If they do have the abilities, these may not have been discovered yet.

Fair-Weather Members The third or outer circle are fringe or marginal members. They jump on the bandwagon when things are going great, but they are the first to criticize when things are not going well for the group. Most athletic fans are in this outer circle. If the team is going for a championship, they say, "Look at what *our* team is doing." If the team is having a losing season, they say, "Look at what *your* team is doing." The challenge to group leaders is to structure the group so that all members are in the inner two circles of membership.

Group Structure

The structure of the group strongly affects group productivity. People use groups to satisfy their needs for love, security, recognition, sense of accomplishment, and power through interaction with other group members. People seek out groups that allow them to meet their needs. Group members who belong by choice are most likely to satisfy these needs.

Strong groups develop by identifying and gratifying people's needs. It is important for group leaders to recognize that members join groups for social reasons rather than for its activities, even though the activities are the reason for the group's existence.[26]

Maintaining Strong Groups

Several factors help maintain strong groups. The major factor is that people go where the action is. Time is too valuable to be wasted, and if groups and meetings are not productive, people drop out.

People join, attend meetings, and help the group remain strong if the group satisfies their personal needs. Groups are strengthened when the group accomplishes its objectives. Groups are strengthened when the members have clearly defined goals. In defining these goals, each member must be able to recognize his or her role in achieving these goals.

The FFA Program of Activities is an excellent example of setting goals and objectives and setting the ways and means to achieve them (refer to the *Program of Activities Handbook*). A fully functioning FFA chapter assigns a task to every member. Pride in accomplishment and involvement is a great motivator.

Teamwork is an important factor in maintaining a strong group. Competition among members impairs teamwork and can result when individual roles are not clearly defined. Defining clear-cut roles for each member enhances teamwork by reducing competition and allowing members to work together toward a common goal.

Commitment of time and energy by group members is vital in maintaining a strong group. Members who invest time and energy in the group will likely value the group and feel strongly attached to it. **Rigorous** initiation requirements that require commitment also increase group **affinity**. A group requiring an investment increases its attractiveness to its members.

All participants in group activities need recognition for their contributions to the group. The feeling of being needed is an important part of maintaining the attractiveness of a group. Successful groups provide opportunities for members to make significant contributions.[27] Strong groups promote group interaction and provide a climate in which everyone has equal opportunity to earn higher status.[28]

Group Goals

Why does a group exist? It must have objectives, whether they are explicitly stated or only implied. Sometimes groups seem not to know why they exist, what they are trying to accomplish, or why they do what they do. They exist largely because they always have and people continue to hold meetings.[29]

Organized groups have purposes, goals, and objectives. They are well defined and integrated. Their objectives are both long range and short range. Goals are understandable, measurable, and chosen by the group. To accomplish goals, members must understand their roles in achieving the goals. There are several ways to clarify the goals of the group.

- Clearly define goals and have members summarize their meaning. Check with the group to see if others understand the intent of the goals.
- Describe goals in concise, simple, and measurable terms.
- Review past experiences, both successes and failures, in relation to goals. Make adjustments in goals to reflect current and future changes.
- Establish periodic progress checks as the group works toward achieving the goals.
- Use goals as road maps to guide the group to completion of its activities. Make sure the group stays on course.[30]

Establishing clear, concise goals is critical in building a strong group. All members should participate in developing and establishing goals because this enhances the likelihood that each member will understand and work toward these goals. This participation is likely to strengthen teamwork and group success. Again, the FFA Program of Activities format is an excellent way to clarify goals.

GROUP DYNAMICS

Group dynamics involves the study of groups, including the patterns of interactions that emerge as groups develop. As a leader, you need to understand how group dynamics affect performance. Groups consist of several processes or dynamics. Twelve of the most important are discussed here (Figure 5–8).

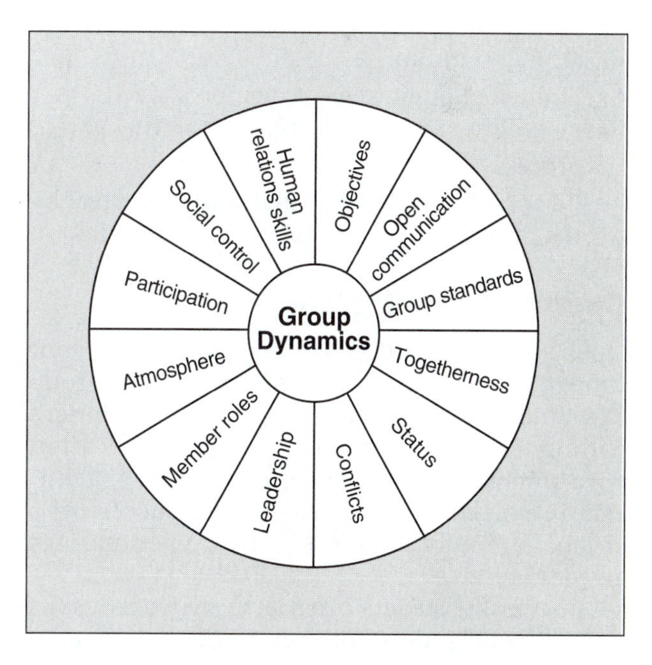

Figure 5–8 Group dynamics is the pattern of interactions or processes within a group. Twelve ingredients of group dynamics are shown here.

Objectives

There needs to be agreement on, and commitment to, clear objectives. To be effective, groups must agree on clear objectives and be committed to their achievement. Although agreement and commitment are separate issues, they are interactive. As group members participate in defining objectives, they tend to become committed to their attainment. The leader should allow the group to have input in setting objectives.[31] Acceptance of specific objectives improves performance. All objectives should be

- Dynamic and likely to promote action.
- Stated in terms that identify people and groups along with types of behavior or behavior changes expected.
- Compatible with general aims of the group.
- Achievable with the maturity and level of resources of the group.
- Able to lead the group to constantly higher levels of achievement.
- Varied enough to meet the needs of individuals in the group.
- Limited enough to meet members' needs.
- Evaluated to show evidence of the actual process.
- Cooperatively determined and accepted by the group.[32]

Open Communication

Open communication—freely expressing your thoughts—is an essential characteristic of group dynamics. Self-disclosure is a group member's willingness to engage in open communication with one or more group members. Feedback is also important for the growth and success of a group. For feedback to be effective, a person must give and receive it.

Successful leaders often achieve their results by paying attention not only to the members or individuals, but to group relationships, interactions, and communication. When formal communication is suppressed or ignored, informal lines of communication usually appear. A leader should answer certain questions about communications within his or her group.

- Does your group really work at ensuring open communication within the group?
- Are there definite means of open communication for sharing knowledge, plans, and decisions?
- Is there really two-way communication or just one-way communication?
- Have weaknesses of the formal communication system encouraged the development of cliques?[33]

Group Standards

Groups tend to form their own written or unwritten rules about how things are done. Group standards, which are sometimes referred to as **norms**, are the group's shared expectations of member behavior. Standards determine what should or must be done for the group to maintain consistent and desirable behavior. Group standards can be formal or informal, depending on the group.

Formal Groups Group standards or norms provide structure and order for the group. They help coordinate and focus the members' efforts on the goals and **tasks** of the group. Group members develop specific roles that function to fulfill the expectations established by the group. The leader may establish group norms or the nature of the group may imply them.[34]

Standards must be realistic and attainable. They should be understood by all group members. The group should be involved in setting standards, and group standards should be evaluated and upgraded periodically as necessary.

Informal Groups Group norms within informal groups develop spontaneously as group members interact daily. Each group member has cultural values and past experiences. Their beliefs, attitudes, and knowledge influence the types of norms developed. If a group member does not follow the norm, the other members usually enforce compliance, verbally or nonverbally, directly or indirectly.

An interesting thing about group norms is that one member's higher standards can be as negative to the group as lower standards. I took a job while in college with a freight trucking company loading and unloading trucks. Since I was raised on a dairy farm, my norm was to keep working until the immediate job or task was done. On my first day on the job, when break time came, I kept working because the truck wasn't loaded. I found out quickly that the union norm was that everybody stops working at break time.

Togetherness

When a group is working together, it is showing group **cohesiveness**. It is the attractiveness and closeness group members have for themselves and the groups. During the Revolutionary War, Ben Franklin told fellow patriots: "We must all hang together or we will all hang separately." This was a call for group cohesiveness. Groups stick together better when they provide services their members want and need. Togetherness and cohesiveness develop when groups deal with problems members think are important, recognize members' contributions, promote friendly interaction among members, and encourage cooperation. It is important for members to feel they belong, that they are making a contribution, and that their contribution makes a difference.[35]

Togetherness can be defined as group solidarity, moral, or *esprit de corps*. Group members feel a common concern and a stake in what happens to other members of the group and the group as a whole. Individual members feel they belong, are a part of, and have a common concern with the group; the word *we* is the natural expression. Other factors influencing togetherness and cohesiveness follow.

- *Objectives.* The stronger the agreement and commitment made to achieving of the group's objectives, the greater the cohesiveness of the group.
- *Size.* Generally, the smaller the group, the greater the togetherness. The larger the group, the more difficulty there is in gaining consensus on objectives and norms. Three to nine members seems to be a good group size for cohesiveness.
- *Homogeneity.* Generally, the more similar the group members are, the greater the togetherness. People tend to be attracted to people who are similar to themselves.
- *Participation.* Generally, the more equal the level of participation among group members, the greater the group's togetherness. Groups dominated by one or a few members tend to be less cohesive.
- *Competition.* The focus of members' competitive instincts affects togetherness. If the group focuses on **intragroup** competition and everyone tries to outdo one another, there will be low cohesiveness. If the group's competitive focus is **intergroup**, the members tend to pull together as a team to beat the rivals. It is surprising how much a group can accomplish when no one cares who gets the credit.
- *Success.* The more successful a group is at achieving its objects, the more together it tends to become. Success tends to breed togetherness, which in turn breeds more success. People want to be on a winning team. Notice that losing teams tend to argue more than winning teams.[36]

Leaders should strive to develop cohesive groups that accept a high level of productivity. Competition should be stressed against outside groups but not within the group. Leaders should strive to develop a cohesive, winning team that motivates the group to higher levels of success.

Status

Status is the perceived ranking of one member relative to other members of the group. Some group members may exhibit more power or status than other members. Behaviors toward a specific group member may influence that person's expression of status. Members displaying a higher level of status receive greater respect than those with lower levels of status.

Status Development Status is based on several factors, including job title, wage, seniority, knowledge or expertise, interpersonal skills, appearance, education, and occasionally race, age, and sex. Members who conform to the group's standards tend to have a higher status than members who do not.

Status Leader A group is more willing to overlook a high-status member breaking a standard. High-status members have more influence on the development of the group's norms. Lower-level members tend to copy high-status members' behavior and standards, therefore, high-status members have a major impact on the group's performance. To be effective, the leader needs to have high status in the group.

Conflict Conflict is another normal feature of groups. Conflict is helpful to a group in that it leads to growth, change, and cohesiveness. Members often express their hostility and conflict toward a group by confrontation. Confrontation effectively removes communication barriers and leads to understanding and acceptance.[37]

Leadership

The behavior of the group's leader influences the effectiveness of the group. Leadership, along with the group's standards, assists in directing the group toward the accomplishment of its goals and objectives.

Both leaders and members are needed for successful groups. Dynamic, creative leadership can do little if the group is apathetic. Ineffective leadership can dampen enthusiasm, but effective leadership ignites the spark that motivates the group to achieve its goals and objectives.

A democratic atmosphere and approach promotes dynamic interaction between leaders and members. Democratic leadership provides the most effective way of arriving at decisions in groups with voluntary membership. Democratic leadership is also superior in building group cohesiveness and cooperation. Autocratic and *laissez-faire* groups do not cooperate and interact with a purpose. Democratic leaders make everyone a part of the team.[38]

Atmosphere

Group atmosphere is the mood, tone, or feeling that permeates the group. A good atmosphere is essential for successful operation of a group. Effective leadership is needed to create a positive atmosphere.

Collective-Whole Atmosphere When individuals meet and work together, they respond as a whole to the prevailing group atmosphere. In groups with a warm, permissive, democratic atmosphere, there seems to be good motivation, member satisfaction, and productivity. Less discontent, frustration, and aggression is evident in these groups compared to other types. There is more friendliness, cordiality, and cooperation. Collective groups seem to foster individual thinking and creativity.

A group member's behavior is determined to a considerable extent by the group's reaction to him or her. Individuals who feel secure and perceive themselves as having adequate group skills more often take the lead in group activities. The total resources of the group can be tapped more fully when all individuals feel free to contribute and question. Motivation and morale reach high levels in a democratic, permissive atmosphere in which both the leaders and the members participate actively.

Authoritarian Atmosphere In an authoritarian atmosphere, the responsibility lies with the leader, and no one may participate or initiate action except at the dictates of the authoritarian leader. It is presumed that the leader knows best what the group should believe and do. Group member behavior is directed toward the leader's predetermined ends.

Democratic Atmosphere In a democratic atmosphere, leadership is shared by all, and individuals strive to recognize and play the roles needed for group productivity. The formal leader and other group members are responsible for creating conditions—including group atmosphere—under which group members can work together to accomplish chosen ends. The group members in Figure 5–9 are working in a democratic, productive atmosphere.[39]

Participation

Participation includes involvement through speaking and entering into discussions. *Breadth* of participation means how many members take part. Intensity of participation is a measure of how often members take part or how emotionally involved they become.

Patterns and Examples We can think of participation patterns as how people respond to each other. When a person enters the discussion, is he or she usually followed by certain others? Do a few people monopolize the discussion or is there opportunity for all to participate? Do we help everyone participate? Is the participation pattern leader centered or distributed throughout the group? Participation can also include attending meetings, being on committees, serving as an officer, helping with finances, being on work groups, washing dishes, or writing publicity.

Figure 5–9 The responsibility of a formal group leader is to create a positive group atmosphere. This leader appears to be accomplishing this objective. (Courtesy of the National FFA Organization.)

Why Participation? Research seems to indicate that individual and group productivity is related to the opportunities provided for member participation. These may include setting goals, deciding on means of attaining goals, and other aspects of discussion and decision making. Even when their ideas do not match the final group decision, people are happier and more productive when they have had an opportunity to participate and express themselves in the decision-making process.[40]

Social Control

The process of ensuring conformity to the expectations of group members is termed social control. It may take the form of rewards to group members for meeting group standards. Such rewards may include recognition, such as a perfect-attendance pin. Other rewards are less tangible and may take the form of being accepted or recognized by the group, a smile, a word, or a pat on the back.

Every group has standards and enforces them by varying degrees of social control. Some groups rely mostly on incentives or rewards for control, others more on fear or punishment. If groups are to be productive, members need to know the group standards and the means used to enforce those standards—the methods of control.[41]

Human Relations Skills

Human relations skills are the skills involved in working with people and getting along with people. Too often it is assumed that since we have lived all our lives with people, we must be proficient in human relations skills. Most of us, for example, have the ability to disagree with another without creating open hostility.

Leaders who understand and facilitate good human relations in their groups are the most successful. Some studies suggest that it is more important for leaders to understand and be skillful in human relations, individual motivation, and group process than to be highly proficient in the subject matter.[42]

Member Roles

Member role is the general expectation of the group member within the group. It is not to be confused with the role of group members in a discussion.

Organizations that specify the roles of group members are likely to have greater goal achievement than organizations that do not. If roles are not clearly defined, if there are overlapping roles, or if the defined roles leave responsibility for important functions unspecified, fewer goals are achieved.

A member's understanding of his or her role, how the role relates to other roles in the group, and the importance of the role all determine the member's sense of responsibility to the group and motivate the member to contribute.[43]

GROUP DEVELOPMENT

Each group is unique, and its dynamics change over time. All groups go through the same stages as they become a smoothly operating and effective team, chapter, club, organization, or department. However, the time a group spends in each of the developmental stages is different. The stages of group development are

Orientation (forming)
Dissatisfaction (storming)
Resolution (norming)
Production (performing)
Termination (adjourning)

Orientation (Forming)

When people form a group, they tend to bring a moderate-to-high commitment to the group. Because they have not yet worked together, however, they do not have the competence to achieve the task. When first interacting, members tend to be anxious about how they will fit in, what will be required of them, what the group will be like, and the purpose of the group, among other things.

The leader must create an open atmosphere within the group. The leader must promote interaction among group members and orient them toward the goals and objectives of the group. An open atmosphere is essential during the forming stage to promote the sharing of ideas. Group members must not be judgmental of others' ideas. The leader must assist others in being receptive to all ideas brought to the group.[44]

Dissatisfaction (Storming)

As members work together for some time, they tend to become dissatisfied with the group. They start to ask: Why am I a member? Is the group going to accomplish anything? Why don't other group members do what is expected? Often the task is more complex and difficult than anticipated; members become frustrated and have feelings of incompetence. They may even separate themselves from the group.[45]

Conflict is a normal process in group development. A leader should keep the communication channels open during this stage and maintain conflict at a controllable level. After a while, the members realize the self-defeating nature of conflict. They see the need to make decisions and move on to the next stage of development.

Resolution (Norming)

With time, members often resolve the differences between their initial expectations and the realities of the group's goals. As members develop competence, they often become more satisfied with the group. Relationships develop that satisfy group members' togetherness needs. They learn to work together as they develop a group structure with acceptable standards and cohesiveness. Commitment can vary from time to time as the group interacts.[46] As team-building efforts progress, group morale increases.

Production (Performing)

At the production stage, commitment and competence do not fluctuate much. The group works as a team with high levels of satisfaction of members' affiliation needs. The group maintains a positive group structure. Members are productive, which leads to positive feelings. The group dynamics may change with time, but the issues are resolved quickly and easily; members are open with each other.

Termination (Adjourning)

During termination, members experience feelings about leaving the group. In groups that have progressed through all four stages of group development, the members usually feel sad that the group is ending. (For groups that did not progress through the stages of development, a feeling of relief is often experienced.)[47] The group makes decisions concerning the roles of members and the procedures followed in accomplishing the group's goals and objectives.

Understanding the dynamics of groups and the stages of development provides managers, consultants, and other leaders with realistic expectations of group behavior. Recognizing group behavior traits enables one to use groups effectively and to promote group growth. Figure 5–10 provides an overview of the six stages of group development.

Group Discussion

Many people have tried to describe what goes on in a group meeting. Some of us have tried to review a meeting to determine why it was a success or failure. The personality types of individual members and the roles members play control a group discussion.

Regardless of your personality type, you can learn to perform any of the following roles. Some will come naturally. For other roles, we will have to work harder to perform. Familiarity with the roles will enable group leaders and members to analyze, more or less automatically, the roles being played by group members. Also, the leader can better lead discussions when he or she recognizes the members' roles. The group member is in a position to play the roles needed for group productivity, to encourage others, and to discourage roles that are not contributing to the group's work.

For a group to be productive, each member must contribute. Group members contribute to a group through the roles or jobs they perform. No one performs the same role throughout the course of the group. In addition, a person may perform an entirely different role in another group. Three categories of group member roles commonly observed in groups are task roles, maintenance roles, and individual roles. An understanding of these roles will increase group effectiveness.[48]

Task Roles

Task roles of the discussion group members are to define and solve problems. These roles include seeking information, presenting ideas and information, and summarizing. The following members' roles are necessary for accomplishing group tasks.

- *Initiator-contributors* propose new ideas or change ways of regarding group problems or goals.
- *Information-seekers* ask for clarification of suggestions, authoritative information, and facts pertinent to the problem being discussed.
- *Opinion-seekers* ask for a clarification of the values pertinent to what the group is undertaking.
- *Information-givers* provide authoritative facts or generalizations. In providing information,

Stages of Group Development			
Stage	**Task**	**Relationship**	**Leader**
Orientation (forming)	Set boundaries Clarify task Define task High commitment and low competence	High anxiety Size up one another Establish turf Protect turf	Helps set boundaries Be prepared Establishes open atmosphere
Dissatisfaction (storming)	Identify alternatives Look at "big" picture Task more complex than anticipated	Frustration Differences emerge Conflicts develop Confrontation	Mediates conflicts Keeps communication open
Resolution (norming)	Make decisions Select alternatives	Become more satisfied Listen to understand Compromise Share information	Decision making Strengthens relations Increases morale Builds team
Production (performing)	Productive Conduct task Follow through	Commitment Competence Work as team Positive group structure	Delegates information gathering Keeps things moving
Termination (adjourning)	Make decisions Closure of task Evaluate results	Evaluate process and roles Experience feeling of leaving	Interprets results for future action

Figure 5–10 Most groups go through these stages as they grow to become smoothly operating, effective teams.

they may relate personal experiences to the group problem.

- *Opinion-givers* express their own beliefs or opinions. They may provide these opinions as suggestions or alternatives to the group problem.
- *Elaborators* spell out suggestions in terms of examples or developed meanings, offer a rationale for suggestions previously made, and try to project how an idea or suggestion would work out if adopted by the group.
- *Summarizers* pull together ideas and suggestions under discussion to help determine where the group is in its thinking or action process.
- *Coordinator-integrators* clarify the relationships between ideas and suggestions, extract key pertinent ideas from member contributions, and integrate them into a meaningful whole.
- *Orienters* define the position of the group with respect to its goals, point to departures from

agreed-on directions or goals, or raise questions about the direction being taken by the group discussion.

- *Disagreers* take a different point of view, argue against, and imply error in fact or reasoning. They may disagree with opinions, values, sentiments, discussions, or procedure.
- *Evaluator-critics* subject the accomplishment of the group to some set of standards of group functioning in the context of the group task.
- *Energizers* prod the group to action or decision, attempting to stimulate the group to greater or higher-quality activities or discussions.
- *Procedural technicians* perform tasks such as distributing materials, arranging the seating, or operating equipment for the group.
- *Recorders* write down suggestions, ideas, and group decisions. The recorder in essence is the secretary, whether formally or informally.[49]

Maintenance Roles

Task roles are concerned with the content of discussion, and **maintenance roles** help the group work together smoothly as a unit. People who fill maintenance roles support one another, relieve tensions, control conflict, and give everyone a chance to talk.[50]

- *Encouragers* praise, agree with, and accept the contributions of others. They show warmth and solidarity in their attitude toward other group members, offer praise, and in various ways indicate understanding and acceptance of other points of view, ideas, and suggestions.
- *Harmonizers* bring the group together. They help reduce tensions and straighten out misunderstandings, disagreements, and conflicts. They recognize when the group is stagnating or tiring. They have a sixth sense for when to tell a joke and get the group to loosen up a little before returning to a task.
- *Compromisers* evaluate their position or opinion, admit any errors or unfairness, and agree to settle tensions by meeting halfway.
- *Gatekeepers* help keep communication channels open. Gatekeepers ensure balanced participation, keeping those who tend to dominate in check and encouraging those who tend to be shy. Examples: "We haven't got the ideas of Mr. Hall yet," or "Let's limit the length of our comments so everyone will have a chance to speak."
- *Standard-setters* express standards for the group to aim for in its functioning or apply standards in evaluating the quality of group discussion.
- *Group observers* keep records of various aspects of the group process and feed such data, with proposed interpretations, into the group's evaluation of its own procedures.
- *Followers* go along with the group. They basically accept the ideas of others and serve as an audience for the group discussion and decision making.[51]

Individual Roles

Group members attempt to satisfy their own psychological needs, which are irrelevant to the group task or team building. This reduces the effectiveness and productivity of the group.

- *Aggressors* may work in a variety of ways, such as deflating the status of others or expressing disapproval of member's values, acts, or feelings. They attack the group or the problem it is working on and show envy of another's contribution by trying to take credit for it.
- *Dominators* try to enforce authority and superiority within the group. This is done by manipulating, flattering, assuming they are superior to others, and acting authoritative. They may downgrade the contributions of others.
- *Blockers* tend to show a negative, stubbornly resistant, disagreeing attitude. They do not like change. Blockers appear to oppose others' ideas for no reason. They attempt to maintain or reopen an idea after the group has rejected it.

- *Anecdoters* relate present situations, stories, and tasks to their own experiences. The experiences of the anecdoter may have positive or negative affects on the attitudes of other group members.
- *Recognition-seekers* like to hear themselves talk. They seek any type of acknowledgment by the group. They boast, report on personal achievements, and make comments to prevent being placed in an inferior position.
- *Help-seekers* strive for sympathy from other group members. This may involve expression of insecurities, personal confusion, and self-pity.
- *Playboys* display a lack of involvement in the team effort. This takes the form of inattention, frequent wisecracks, horseplay, and other distracting behavior.
- *Special-interest pleaders* speak for whatever fits their individual needs.[52]

Leading a Group Discussion

Now that we know how to cope with individual members, we can proceed with leading a discussion. There are several factors to consider when planning for and carrying out a group discussion. Group discussion is rapidly becoming one of the most widely used forms of communication. With team building being an integral part of the workplace, leaders may spend up to 50 percent of their time in meetings. Consider the following factors when leading a group discussion.

Planning A successful discussion should be planned. Planning includes choosing a topic and types of questions, preparing an outline, and studying the relationships between the background knowledge, experience, and ideas of the participating team.

Beginning the Discussion A group discussion should start and end on time. Group members should be introduced to each other to build group cohesiveness.

Introducing the Topic Stating the topic at the start of the discussion prevents uncertainty about the subject to be discussed.[53]

Questioning Do not call on a group member to answer a question before asking the question. Other group members do not concentrate on a question directed to someone else. Provide about 10 seconds of thought and reflection time and then call on a group member to respond. This procedure keeps the group's attention directed to the question.[54]

Regulating Communication During the discussion, several leadership tasks must be addressed. All members should feel free to speak their minds fully and frankly. The leaders must not only invite but encourage members to contribute to the discussion. The idea that each individual's ideals are valuable should be stressed, but participation needs to be balanced. A leader must be prepared to step in when members begin to argue. Leaders are also expected to keep the discussion on track. Nevertheless, group leaders should also help members express their ideas clearly.

Concluding the Discussion There are two major concluding functions of group and team leaders. First, when leaders feel the group has adequately covered the discussion question or a preset time limit has almost been reached, leaders should summarize the major ideas and outcomes of the discussion. They must be careful not to overload the summary with their own ideas. Second, leaders should save enough time for all members to disagree with the summary or to express minority opinions if they wish.[55]

Conflict Resolution

Due to personality conflicts, there will always be problems within groups. Since those with choleric personalities believe they are always right, when two cholerics have a difference of opinion, sparks will fly. When we learn our personality weaknesses, we can better cope with conflict. If we truly are convinced the other person is wrong, we enter into conflict with one of two options: fight or flight. Epictetus said, "What concerns me is not the way things are but rather the way people think things are."[56]

Communication Communication is the key element affecting conflict: both its cause and remedy. Open communication is the means by which disagreement can be prevented, managed, or resolved. The lack of open communication can drive conflict underground and create a downward spiral of misunderstanding and hostility.[57]

Problem (Conflict)-Solving Resolution Plan Once open communication is attained, resolve conflicts by using the problem-solving process, which is discussed fully in Chapter 13. Most groups have found value in having a plan to follow when resolving conflicts. A step-by-step plan helps everyone stay focused on solving the problem and preserves the self-esteem of all involved. The problem-solving plan has six steps.

1. Define the problem.
2. Collect facts and opinions.
3. Consider all solutions proposed.
4. Define the expected results.
5. Select the solution(s).
6. Implement the solution(s).[58]

Conflict-Resolution Behaviors Good conflict-resolution behaviors help you resolve conflict in almost any situation you encounter. They allow you to benefit from positive disagreement without having those disagreements escalate into out-of-control personality conflicts that damage the morale and productivity of the organization. The five basic behaviors follow.[59]

1. *Openness.* State your feelings and thoughts openly, directly, and honestly without trying to hide or disguise the real object of your disagreement.
2. *Empathy.* Listen with empathy. Try to understand and feel what the other person is feeling and to see the situation from the other person's point of view.
3. *Supportiveness.* Describe the behaviors you have difficulty with rather than evaluating them. Be willing to support the other person's position if it makes sense to do so.
4. *Positiveness.* Try to identify areas of agreement and emphasize those. Be positive about the other person and your relationship.
5. *Equality.* Treat the other person and his or her ideas and opinions as equal. Give the person the time and space to completely express his or her ideas, and evaluate all ideas and positions logically and without regard to ownership.

Discussion Techniques

Several techniques can be used to secure group action. Techniques have been likened to the vehicle that helps move a group toward its goals. These meeting or session techniques are designed to either bring information and understanding to a group or move a group to action.

- *Small group discussion* is face-to-face mutual interchange of ideas and opinions between members of a relatively small group (usually 5 to 20). Refer to Figure 5–11 for room arrangement for a small group discussion.
- *The huddle/discussion 66 method* is derived from the idea of six people discussing a subject for six minutes. Figure 5–12 shows the room arrangement for this discussion technique.
- *Buzz groups* are an alternate method of breaking a large group into small segments to facilitate discussion. Often, only two people are in a buzz group (Figure 5–13).
- *Symposiums* are a group of talks, lectures, or speeches presented by several individuals on the various phases of a single subject.
- *Panel discussions* take place in front of an audience by a selected group of people (usually three to six) under a moderator.

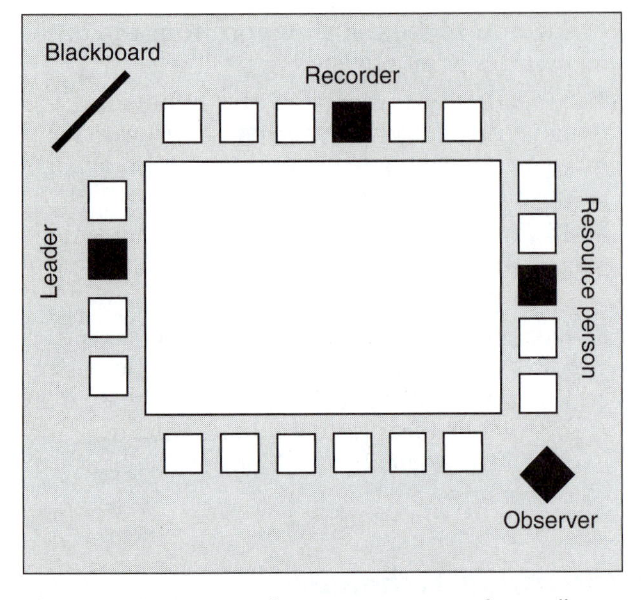

Figure 5–11 Suggested room arrangement for small group discussion.

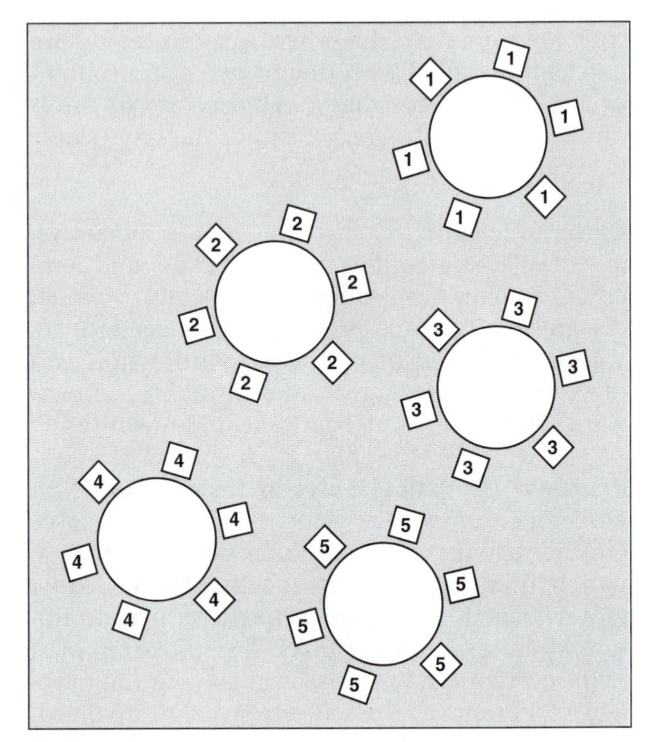

Figure 5–12 Seating arrangement for huddle/discussion 66.

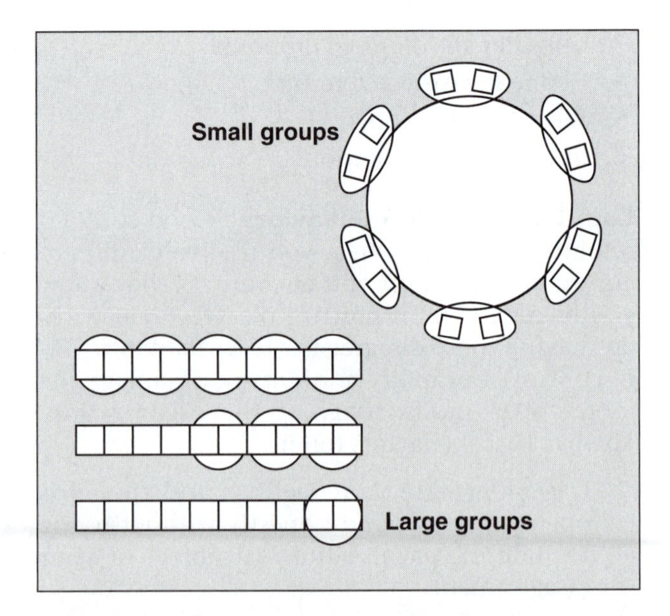

Figure 5–13 Buzz group arrangement for both small and large groups.

- *Interrogator panels* are an interrogation-discussion interchange between a small group of knowledgeable individuals (the panel) and one or more interrogators often under the direction of a moderator.
- *Committee hearings* involve several people questioning an individual.
- *Dialogues* are discussions carried on in front of a group by two knowledgeable people.
- *Interviews* are the questioning of an expert on a given subject by an interviewer who represents the group.
- *Brainstorming* is a type of small group interaction designed to encourage the free introduction of ideas on a restricted basis and without any limitations as to feasibility.
- *Role playing* is the dramatization of a problem or situation in the general area of human relations.[60]
- *Nominal technique* is an alternative form of group communication designed to reduce two problems: (1) the amount of time group discussions take and (2) members who try to

dominate a group discussion. A group using the nominal technique begins by asking each member to write down a list of possible solutions to the group's problem. Each member is then asked to state one idea from the list. Each idea stated is written down so that all may see the entire list. The third step consists of a brief discussion of ideas on the board. This discussion takes place primarily to clarify but may also involve some evaluation of ideas. Next, a secret vote is conducted, with each member ranking the ideas in order of personal preference. Finally, the rankings are tabulated and the solution with the highest ranking becomes the group's solution.[61]

SUMMARY

Understanding the group process makes for a better group leader. Leaders should know why people join groups, why people do not join groups, why group leadership skills are important, and the importance of teamwork.

There are three types of groups: functional groups, task groups, and informal groups. Some groups are formal and some are informal. Regardless of the type of group, group members appreciate the democratic process and respond with greater productivity. Status within any type of group has to be earned.

All group members are not equally committed to groups. There are nucleus members, faithful followers, and fair-weather members. Group membership will probably remain strong if the group is cohesive, accomplishes objectives, works as a team, and provides recognition.

Group dynamics is the pattern of interactions or processes in a group. The dynamics include objectives, open communication, standards or norms, togetherness, status, conflict, leadership, atmosphere, participation, social control, human relations skills, and member roles. Groups go through five stages: forming, storming, norming, performing, and adjourning.

The leader must be able to lead group discussions. Members assume three roles during a discussion: task, maintenance, and individual roles. Task roles have the mission of selecting, defining, and achieving goals. These roles include the initiator-contributor, information-seeker, opinion-seeker, information-giver, opinion-giver, elaborator, orienter, disagreer, energizer, procedural technician, recorder, evaluator, critic, coordinator-integrator, and summarizer.

The maintenance role focuses on the cohesiveness of the group. These roles include the harmonizer, compromiser, gatekeeper, standard-setter, group observer, and follower.

Individual roles are performed to meet the member's own psychological needs. These roles include the aggressor, blocker, recognition-seeker, help-seeker, playboy, and special interest pleader.

Various techniques can be used to secure group action. These include small group discussion, huddle groups, buzz groups, symposiums, panel discussions, interrogation panels, committee hearings, dialogues, interviews, brainstorming, role playing, and the nominal technique.

END-OF-CHAPTER EXERCISES

Review Questions

1. Define the Terms to Know.
2. List six reasons people are attracted to groups.
3. Why not use autocratic group leadership?
4. List three reasons people don't belong to groups.

5. List five considerations as you build your teams.
6. List five things that research says about teams.
7. What is the purpose of teams?
8. What is the role of the leader in using teams?
9. What are the potential problems with using teams?
10. What are the three types of groups?
11. What are three types of task groups?
12. Name 12 dynamics or processes typical of most groups.
13. In what ways can conflict benefit a group?
14. Name the five stages of group development.
15. What are the three levels of membership?
16. What five things need to exist to maintain strong groups?
17. Name three categories of group member roles.
18. List 11 discussion techniques.
19. What are five ways to clarify group goals?
20. List 11 factors that influence togetherness and cohesiveness.
21. What are six steps a leader should follow when leading a group discussion?
22. What are seven things to consider as you build your team?

Fill-in-the-Blank

1. ____________________ involves the study of groups, including the patterns of instruction that emerge as groups develop.
2. Written or unwritten rules about how things are done are group ____________________ .
3. When groups are working together, it is group ____________________ .
4. ____________________ is the perceived ranking of one member relative to another member.
5. The behavior of the group's leader ____________________ the effectiveness of a group.
6. ____________________ is working with people and getting along with people.
7. ____________________ effectively removes communication barriers and leads to understanding and acceptance.
8. When a group secures conformity to the expectations of its group members, it is termed ____________________ .
9. Group ____________________ is the prevailing mood, tone, or feeling that permeates the group.
10. ____________________ ____________________ can be brought about by communication and problem solving, along with openness, empathy, supportiveness, positiveness, and equality.

Matching

_____ 1. Maintains a negative attitude
_____ 2. Keeps group on task
_____ 3. Develops suggestions further
_____ 4. Wants sympathy from others
_____ 5. Draws group ideas together
_____ 6. Provides authoritative facts
_____ 7. Defines the group's position
_____ 8. Group mediator
_____ 9. Lacks group involvement
_____ 10. Wants others' opinions clarified

A. information-giver
B. elaborator
C. opinion-seeker
D. harmonizer
E. blocker
F. summarizer
G. gatekeeper
H. help seeker
I. playboy
J. orienter

Activities

1. Remember a time when you participated in a group. Recall any group processes and development stages that become apparent in the group.
2. Observe a group and make a checklist of the various roles performed by members.
3. Before attending a future meeting, list what you expect to gain from the meeting. Afterward, think about whether your expectations were met. If not, explain why.
4. Observe someone leading a group discussion. Make a list of the leader's strengths and weaknesses.
5. Consider a group meeting you have attended. What was the atmosphere during the meeting? Were you comfortable or uncomfortable? Explain your feelings and what could be done to make you feel better about the group.
6. Your teacher will divide you into groups of four to seven members. Give your group a name, or use your FFA chapter as the group. Each member of the group will answer the following questions.
   a. What is the purpose of our group?
   b. Why do individuals join our group?
   c. What problems have we experienced in our group?

   Next, select a recorder and a person to report to the total group by sharing your ideas on each of the three questions.
7. Write a one-page case study or scenario describing a situation in which the leader should involve the use of a team or group.
8. **Take it to the Net.**

   Listed below are some Web sites that contain information on teams and groups. Some of the sites contain articles and some contain activities. Browse the sites and find an article or activity you think is interesting. Print the article or activity and describe how it relates to this chapter. Some search terms are also listed if you want additional information.

Web sites

<http://www.timedoctor.com/teambuild.htm>
<http://www.successcoach.com>
<http://www.teambonding.com/home.htm>

Search Terms

group development
group work
leading groups
group dynamics
teambuilding

Notes

1. G. M. Beal et al., *Leadership and Dynamic Group Action* (Ames, IA: Iowa State University Press, 1962), p. 23.
2. B. L. Reece and R. Brandt, *Effective Human Relations*, 5th edition (Boston: Houghton Mifflin Company, 1995), p. 323.
3. G. Lippitt et al., "Cutting Edge Trends in Organization Development, *Training and Development Journal*, July 1984, p. 62.
4. R. N. Lussier, *Human Relations in Organizations: Applications and Skill Building* (Columbus, OH: McGraw-Hill Higher Education, 1998).
5. Beal, *Leadership and Dynamic Group Action*, p. 31.
6. Ibid., p. 74.
7. Ibid.
8. Lussier, *Human Relations in Organizations.*
9. Reece and Brandt, *Effective Human Relations*, p. 323.
10. S. R. Levine and M. A. Crom, *The Leader in You* (New York: Simon & Schuster, 1993), pp. 103–109.
11. J. P. Howell and D. L. Costley, *Understanding Behaviors for Effective Leadership* (Upper Saddle River, NJ: Prentice-Hall, 2001), pp. 270–274.
12. A. Nahavandi, *The Art and Science of Leadership*, 2nd edition (Upper Saddle River, NJ: Prentice-Hall, 2000), p. 168.
13. Howell and Costley, *Understanding Behaviors for Effective Leadership*, p. 271.
14. Nahavandi, *The Art and Science of Leadership*, p. 164.
15. Ibid., p. 167.
16. Ibid.
17. Ibid., p. 168.
18. Ibid.
19. L. H. Lamberton and L. Minor, *Human Relations: Strategies for Success* (Homewood, IL: Irwin, 1995), pp. 131–135.
20. Levine and Crom, *The Leader in You*, p. 98.
21. Daniel F. Jennings, *Effective Supervision: Frontline Management for the '90s* (Minneapolis/St. Paul: West Publishing Company, 1993), p. 231.
22. Reece and Brandt, *Effective Human Relations*, p. 324.
23. Lussier, *Human Relations in Organizations.*
24. Ibid., pp. 275–276.
25. *Personal Development in Agriculture* (College Station, TX: Instructional Materials Service, Texas A & M University, 1989), 8742-A, p. 1.

26. Ibid., 8742-A, p. 2.
27. Ibid., 8742-A, p. 3.
28. R. F. Verderber, *Communicate!* (Belmont, CA: Wadsworth, 2001).
29. Beal, *Leadership and Dynamic Group Action*, p. 43.
30. *Personal Development in Agriculture*, 2742-A, p. 2.
31. Lussier, *Human Relations in Organizations*.
32. Beal, *Leadership and Dynamic Group Action*, p. 143.
33. Ibid., p. 88.
34. *Personal Development in Agriculture*, 8741-G, p. 1.
35. B. R. Stewart, and M. L. Amberson, *Leadership for Agricultural Industry* (New York: McGraw-Hill, 1978), pp. 65–66.
36. Lussier, *Human Relations in Organizations*.
37. *Personal Development in Agriculture*, 8741-G, p. 2.
38. Stewart and Amberson, *Leadership for Agricultural Industry*, pp. 71–72.
39. Beal, *Leadership and Group Action*, pp. 82–83.
40. Ibid., pp. 88–89.
41. Ibid., pp. 93–94.
42. Ibid., p. 112.
43. Ibid., p. 100.
44. *Personal Development in Agriculture*, 8741-G, p. 3.
45. Lussier, *Human Relations in Organizations*.
46. Ibid.
47. Ibid.
48. *Personal Development in Agriculture*, 8741-H, p. 1.
49. Beal, *Leadership and Group Action*, pp. 103-105.
50. Verderber, *Communicate!*, p. 261.
51. Beal, pp. 106–107.
52. Ibid., pp. 107–109.
53. J. O'Connor, *Speech: Exploring Communication* (Upper Saddle River, NJ: Prentice-Hall, Inc., 1998).
54. *Personal Development in Agriculture*, 8741-H, p. 4.
55. O'Connor, *Speech: Exploring Communication*, pp. 116–118.
56. T. Alessandra and P. Hunsaker, *Communicating at Work* (New York: Simon & Schuster, 1993), p. 91.
57. Ibid., p. 92.
58. Reece and Brandt, *Effective Human Relations*, p. 365.
59. Alessandra and Hunsker, *Communicating at Work*, pp. 106–107.
60. Beal, *Leadership and Dynamic Group Action*.
61. Lussier, *Human Relations in Organizations*.

SECTION

Communication and Speaking to Groups

CHAPTER

6

Communication Skills

Objectives

After completing this chapter, the student should be able to:

- Define communication
- Explain the relationship between communication and leadership
- Explain the purpose of communication
- List the forms communication takes
- Explain the communication process
- Recognize and overcome communication barriers
- Describe and use techniques to improve your listening, reading, writing, and speaking
- Discuss nonverbal communication skills
- Develop skills using feedback
- Discuss the importance of self-communication and interpersonal communication

Terms to Know

communication
output-based communication
input-based communication
nonverbal communication
message
sender
encoded
channel
receiver
decoding
perception
feedback
interference
barriers
hearing
listening
body language
kinesics
proxemics
doodling
self-communication
interpersonal communication barriers

Words are a small part of **communication**. When people speak, the message communicated is only 7 percent attributed to the words they use. Voice expression accounts for 38 percent and body gestures account for 55 percent. Therefore, we can communicate about anything we want to anyone if we know how to say it. The FFA members in Figure 6–1 are communicating with elementary students as they make a "Food for America" presentation.

In this chapter, you will see why communication is important, what purposes it serves, and what forms it takes. You will learn that communication is a complex process involving distinct elements that work together to convey accurate messages. You will examine barriers to communication and learn ways to overcome them. You will develop an understanding of the communication styles people use, and learn how you can adapt to each style. Finally, you will be given suggestions to improve your listening, reading, writing, and nonverbal communication skills as well as feedback techniques and interpersonal communications.

Figure 6–1 There are many ways to practice and enhance your communication skills through agricultural education and the FFA. These students were selected to make a "Food for America" presentation. (Courtesy of FFA New Horizons Magazine.)

COMMUNICATION AND LEADERSHIP

Communication may be defined as the process of sending and receiving messages in which two or more people achieve understanding.[1] The message itself may come from an oral or written source, or from a gesture. It is important to realize that if the person sending the message and the person receiving it do not agree on the meaning of the message, then true communication has not taken place.[2] This may be the cause of many hurt feelings, conflicts, and problems. Miscommunication causes misinterpretation, which can create confrontation.

All forms of communication are important as you work with others. The American democratic system depends on skill in group communication. Parliamentary procedure is used in local, state, and national government. Public speaking is a necessity for people in leadership positions. Group communication skills and conducting meetings are skills needed by all leaders. Agricultural education and the FFA help develop communication skills. Members recite the Creed (Chapter 7), and are involved in public speaking (Chapter 8), extemporaneous speaking (Chapter 9), parliamentary procedure (Chapters 10 and 11), leading group discussions (Chapter 5), and conducting meetings (Chapter 12).

1. Seventy-five percent of each workday is consumed by talking and listening. Seventy-five percent of what we hear we hear imprecisely, and 75 percent of what we hear accurately we forget within three weeks. Communications, the skill we need the most at work, is the skill we most lack.[3]

2. Managers spend 70 percent of their time communicating in some way. The time communicating can be broken down as follows: 9 percent reading, 16 percent writing, 30 percent talking, and 45 percent listening.[4]
3. The majority of problems are caused by poor communication—by people who are unable or unwilling to communicate.[5]

The ability to communicate can be learned. It must be learned by people who want interesting and rewarding careers or who want to take part in community affairs. Positions of responsibility, pay increases, and promotions go to those who can speak effectively. These skills are essential for advancement in business, politics, or community service.

The development of improved communication skills is crucial to your success as a student, as an employee, and as a social person. The coworkers in Figure 6–2 are communicating as they perform their daily tasks.

Figure 6–2 Effective communication in the workplace is an essential life skill. (Courtesy of USDA.)

THE PURPOSES OF COMMUNICATION

Communication is an important aspect of life. We are expected to communicate with our family, friends, teachers, and coworkers daily. How well we communicate helps determine how successful we are in school, in the workplace, and in building relationships with others. There are three primary purposes or goals for communication.

1. *To inform.* Communication may simply seek to give information to another person. This information may be necessary to solve problems, make decisions, increase understanding, or build social relationships. An agricultural education teacher giving a lecture on genetic engineering is an example of communication used to inform.
2. *To influence.* Communication may also seek to influence another person's behavior. Examples of communication used to influence others are persuasive speeches, commercials, and giving instructions or directions. Remember, influence is leadership.
3. *To express feelings.* Communication may be used to tell how a person feels in a situation. For example, is the person happy, angry, or discouraged?[6]

It should be noted that although these purposes for communication have been separated into groups, any message could have one, two, or all three of these purposes. For example, an angry employer might reprimand an employee who is late for work and explain that company policy states that employees who consistently arrive late can be fired. In this communication, each of the three purposes is present: information is given concerning company policy, influence is present in the employer wanting employees to arrive on time, and feelings are expressed in the employer's reprimand.

FORMS OF COMMUNICATION

Communication can be output based (i.e., I communicate) the person produces the communication; input based (i.e., I listen, or receive), or nonverbal (i.e., behavior and the physical environment convey messages). Forms of **output-based communication** are speaking (formally and casually) and writing (all forms). Forms of **input-based communication** involve listening and reading.[7] Forms of **nonverbal communication** are physical behavior, such as gestures and body language. The physical environment, such as arrangement of furniture, use of space, and building design also can be forms of nonverbal communication.[8]

Figure 6–3 Asking for a date is a difficult communication process for many young people. This student is showing a bit of nervousness as he makes his call.

THE COMMUNICATION PROCESS

The process by which a message is sent from one person to another seems relatively simple at first glance; however, a number of elements may help or hinder the understanding of the intended message. The elements that exist in any form of communication are (1) the situation, (2) the message, (3) the sender, (4) the channel, (5) the receiver, (6) feedback, and (7) interference.[9] A situation demonstrating the communication process follows. After the setting, each element of the communication process is defined and identified.

Setting

John is doing his homework one evening. He begins thinking about the big ball game coming up next weekend. John wants badly to ask Jennifer to go to the game, but he is unsure she would want to go with him. After thinking about it, John decides to call Jennifer and ask. When the call is placed, John begins by talking about how important the big game is and how many of their friends will be there. John finds, however, that he has to repeat himself often because there seems to be a bad connection. John also finds he is so nervous that he cannot seem to come right out and ask Jennifer to go to the game. Finally, he comments that he would like to go to the game if someone would go with him. Jennifer understands John's hint and says she would be glad to go to the game with him. The nervousness can be seen on John's face as he talks to Jennifer in Figure 6–3.

Situation

The situation is when and where communication takes place.[10] In this example, the communication takes place one evening in John and Jennifer's houses. In any communication, it is important to decide if the occasion and the physical setting are appropriate for the communication. If John had tried to ask Jennifer to the game during a math test, the situation obviously would not have been appropriate.

Message

The **message** is whatever is intended to be communicated by one person to another. In the example, the message is John's desire to see if Jennifer would like to attend the ball game. It is important to remember to make communication clear and precise. In our example, John's message could have been misunderstood because he was not precise and definite when asking Jennifer to go to the game.

Messages are more effective if they are of reasonable length; if they are correct, concise, and interesting; and if they are delivered in a timely and orderly manner.[11] Remember that not only do words carry the message, but so do voice, gestures, and facial expressions. Being aware of all the ways in which your message is sent helps you be sure the message you want to communicate is the one the other person actually receives.

Sender

The **sender** is the person who wishes to communicate (send a message) with someone else. In the example, the sender is John. John must first formulate the message he wishes to communicate. The message is then **encoded**, or put into a form that the sender believes the receiver will understand. Encoded messages may be in the form of words, gestures, sounds, or numbers.[12] John's message was to ask Jennifer to go to the game. He encoded his message in words when he made the phone call. The effectiveness of the sender can be influenced by several factors.

- *Purpose and motive.* The purpose and motive (reason) of the sender are important. Generally speaking, the better the motive, the better the message is received.
- *Status or position.* The status or position of the sender may affect how well the message is received. We are more likely, for example, to listen more closely to what an employer tells us to do than to what a co-worker tells us.
- *Personality.* Personality factors, such as attitude, credibility, reputation, perceptions, and abilities, affect how others view the sender and accept the message sent. Obviously, the better the personality characteristics of the sender, the better the chance of the sender and the message being accepted.[13]

Channel

The **channel** is the means by which the sender chooses to communicate the message. The channel John used to send his message was the telephone. The appropriateness, efficiency, and dependability of the channel determines its effectiveness. The channel used may need to be changed based on the time available for communication (a phone call is faster than a letter), the importance of the communication (an emergency requires quick action and therefore the fastest form of communication available), and the need for response (if a response is needed immediately, a phone call is much quicker than getting in the car and going to see the person).[14]

Receiver

The **receiver** is the one for whom the message is intended. The receiver takes the message and converts it into a form that can be understood—this is **decoding** the message. Jennifer decoded all of the talk about the game and realized John wanted her to go to the game. It is essential to realize that the receiver has her own **perception**, or way of understanding, the message, based on her own beliefs, knowledge, and ways of organizing information.[15] Thus the saying "Perception is reality." Whether we are right or wrong, the way we perceive something is the way it is, at least in our own minds. The receiver may not understand a word, phrase, or thought in the same way you do. Efforts should be made to ensure that both the sender and receiver have the same message.

Feedback

Because each person has his own way of understanding your message, it is important that the sender get feedback. **Feedback** is the way the receiver responds to the message. It allows the sender to know if the message was clearly understood. If the receiver does not provide feedback, the sender may check for understanding by asking questions or interpreting body language, such as puzzled looks.[16] Jennifer telling John she would be glad to go to the game with him was a form of feedback. John was able to determine that the intent of his call was understood—even though he had difficulty in expressing his true message.

Because feedback helps the sender determine the effectiveness of the message, time and opportunities for feedback from both the sender and receiver are a must. Some organizations provide training in feedback techniques.

Interference

Interference is anything that hinders communication. Interference may come from sources outside of the receiver (such as noise in a classroom) or from the receiver (not paying attention, or doing another activity at the same time).[17] Interference came from a poor phone connection and quite possibly from John's nervousness. Obviously, to convey a message that the receiver will clearly understand requires minimizing interference.

Now the communication process is complete. What seems to be a simple process is actually made up of several complex choices and assumptions. A diagram of the communication process is provided in Figure 6–4.

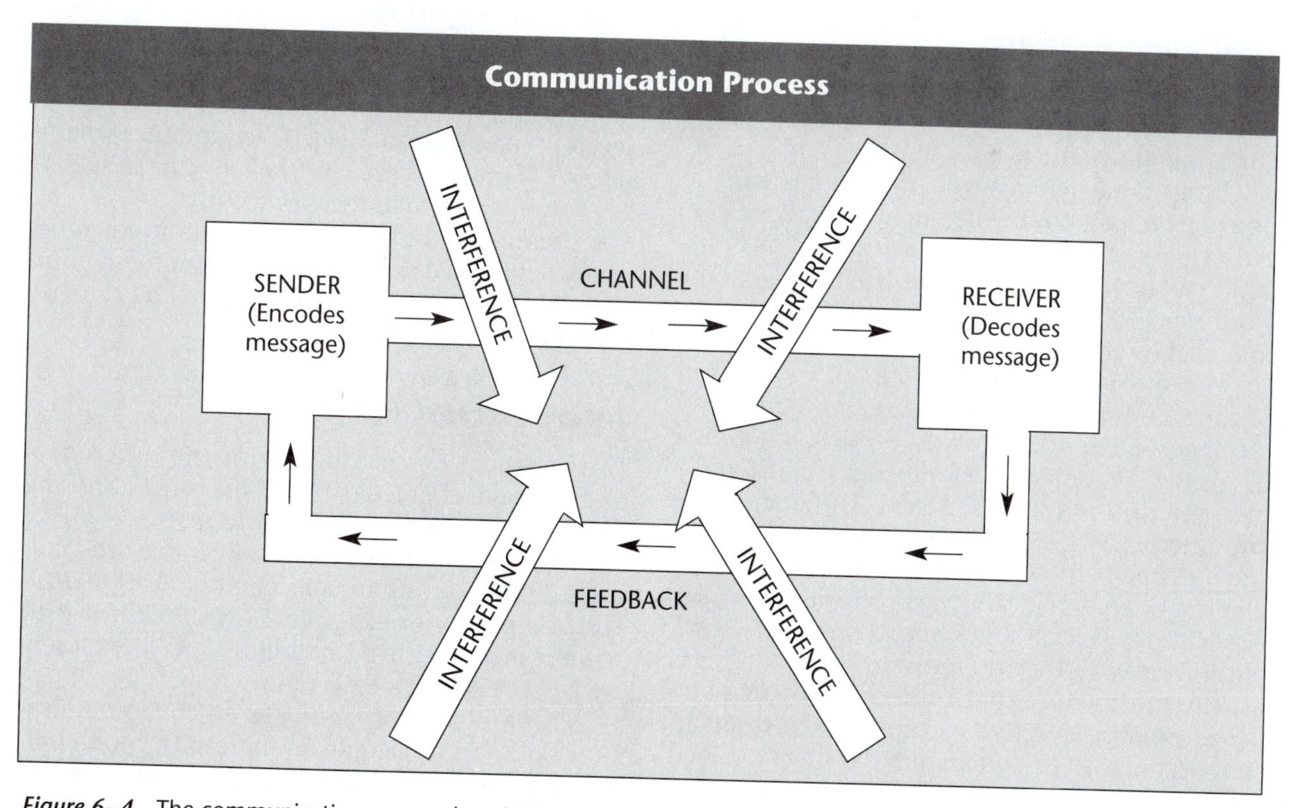

Figure 6–4 The communication process by which a message is sent from one person to another seems relatively simple at first glance; however, a number of elements are present that may help or hinder the understanding of the message.

BARRIERS TO EFFECTIVE COMMUNICATION

Much of what is communicated is misunderstood or understood only in part. Sometimes this may be simply a result of a poor phone connection or bad handwriting. More often, the reasons for ineffective communication are more complex. An understanding of the barriers to effective communication can help us avoid problems that hinder our own communication. It may also help us understand why others are unable, or seemingly unwilling, to communicate.

Four primary types of barriers hinder communication: (1) language barriers, (2) interpersonal (sender and receiver) barriers, (3) situational-timing barriers, and (4) organizational structure and procedural barriers.[18]

Language Barriers

Language barriers deal with the way our words are understood. Words and phrases are constantly changing in meaning.

What a particular word or phrase means to one person may not be the same to someone else. This is called semantics. As a leader, you must make sure receivers understand what you say. The word *slowly* may be twice as fast to the learner as intended by the instructor.[19]

When words are incorrectly used, are used out of context, are too specialized, are too vague, have too many different meanings, or have emotional overtones, it stands to reason that the intended message can be misunderstood. Understanding and interpretation can also be hindered if the entire message is too complex or is presented in an illogical or grammatically incorrect way.

Selective perception occurs when people block out information they do not want to hear. This is usually information that conflicts with their values or beliefs. Selective perception can result in stereotyping. A person may have preconceived ideas about something and therefore apply selective perception to confirm those ideas. For example, environmentalists might have trouble getting a community to support additional environmental controls on a local factory. This resistance could result from the community's fear that the idea might affect their jobs and the local economy. They hear what they want to hear.

Poor listening habits can be caused by a number of things. A person can have her mind on something else or have already thought about what she might say. Perhaps the receiver is just not interested in the subject or has a poor knowledge of the subject. Whatever the case, poor listening skills can reduce the effectiveness of the communication process.

The credibility of the sender affects how people receive a message. If the receiver perceives the sender as not living up to her word, communication barriers will exist. This can be caused by status differences between communicators or the receiver's negative feelings toward the sender. As a leader, you must build your credibility for people to trust and respect you.

Filtering is the manipulating of information so it is perceived as the receiver desires. This occurs often with organizations that have several levels of management. Each level can filter what it doesn't want the next level to know or change it to promote particular ideas or feelings.

The mind avoids detail and cannot remember many details in a short time. Consider these numbers: 10, 12, 6, 8, 0, 63, 4, 16. How many can you repeat in proper order?[20]

Interpersonal Barriers

Interpersonal barriers deal with the differences and personal characteristics of the sender and the receiver that hinder communication.

On a personal level, age, status, role, and cultural differences often make communication difficult. Consider, for example, how difficult it is to communicate with a small child, or how closely we pay attention to a person we respect.

People also have varying levels of communication skills. Some people are naturally good communicators. Others have taught themselves to communicate well. Confident people and those who speak well or have a persuasive manner tend to be good communicators.[21] Prejudiced and insincere people as well as those who have withdrawn, insecure, defensive, or apathetic (uncaring) personalities may not be good communicators because of these traits.[22]

Remember, people tend to view the message you send in light of their experiences, background, and personality (perception). People often hear only what they want to hear. Understanding this is essential for improving communication.

Past experiences with others also may affect our communication. If a person has been sarcastic, untrustworthy with confidential information, unwilling to cooperate, or resistant to new ideas and change, it is likely that future attempts at communication will be difficult.[23]

Understanding the personalities of others can make our communication easier or at least provide an explanation for why communication seems difficult. The effects of leadership style, personality type, and communication style are discussed later in the chapter.

Situational-Timing Barriers

Situational-timing **barriers** deal with the time and place communication takes place.

People in today's society are so hurried and so bombarded with information that they do not have time to process everything they hear or read. An individual may receive up to 1,500 messages during a single day.[24] There simply is not enough time to process so many messages.

The timing of communication, however, may be unrelated to hours and minutes. If a person is under a great deal of stress or is in a personal crisis, any effort to communicate with that person may be ineffective. Be sensitive to such people. You have probably learned that if your parents or teachers are in a bad mood, you don't ask for something if you are not sure they will say yes. Wait until the time is right (mood changes) and you have a better chance of getting a yes. A forced or rushed answer is often negative.

The place where communication occurs may make communication difficult. If the environment in which communication occurs is hostile, lacks privacy, or is uncomfortable, communication can be affected negatively. The amount of noise, or interference, in the environment also affects how well we understand and can be understood.

We need to match our communication style with the situation. Earlier in the book, we studied situational leadership; the same principles apply here. For example, when using the autocratic style, the sender initiates a closed, strong presentation, whereas the laissez-faire style uses an open, mild form. The mild-to-strong component can be used with any of the styles, but the strong messages tend to be used more often with the autocratic and democratic styles.[25]

Organizational Structure and Procedural Barriers

Organizational structure and procedural barriers deal with how and through what structure a message goes from the sender to the receiver. Barriers can be found in the workplace but can also be found in schools, churches, or any place where people have varying roles or levels of authority in an organization.

Communication may be hindered by something as simple as the fact that offices are spaced too far apart for people to communicate easily.

Organizations may lack policies that promote effective communication. There may be too many people who must pass on the sender's message to the person who acts on the message, due to the size of the organization.

The more levels there are in an organization and the farther the receiver is from the sender, the harder it is to communicate a message effectively. The potential for distortion exists when a message is transmitted through a series of individuals at different management levels in the organization. Parts of the message may be ignored, reinterpreted, omitted, or misunderstood. You may have played the game "Gossip" when you were a child. Remember how the message was completely different when it reached the tenth person in the circle? Communication in organizations is not much better if messages are not given in written form as well as verbally.

There may be an overreliance on impersonal forms of communication (such as memorandums) that make people feel isolated in the organization rather than an important part of the organization.[26]

OVERCOMING COMMUNICATION BARRIERS

Today, agriculture is a technology-oriented industry that includes production, agriscience, and agribusiness. Most people who work in the agricultural industry do not work on farms and ranches, but rather are employed in the feed, seed, farm machinery, fertilizer, chemical supply, food-processing, and related businesses. Fine-tuned communication skills are needed to be successful and as we have seen, there are numerous barriers to effective communication. We present some general guidelines for improving overall communication in this section. Later we look at specific ways communication can be improved in the areas of listening, reading, speaking, and writing.

Overcoming barriers can be accomplished in at least three primary ways: (1) improving perception, (2) improving the physical process of communicating, and (3) improving relationships and speaking ethically.[27]

Improving perception involves putting yourself in another person's place rather than focusing only on your own experiences and background. Essentially, this means you try to understand the point of view of those with whom you communicate. Perceptions can be influenced by age, status, personality, cultural background, and feelings. If you realize how the other person is seeing and experiencing the world around him, you may be able to put your own communications in a form he can understand. In other words, talk at the person's level. Remember also that you have your own perceptions; take time to examine them to be sure they are accurate and are not based on false information or prejudices.[28]

Improving the physical processes of communication involves paying attention to the elements of the communication cycle described earlier. Be sure the situation (time and place) is appropriate. As a sender, make an effort to encode messages clearly by selecting words and/or gestures carefully. Use an appropriate channel to send your message. Pay close attention to the feedback (verbal and nonverbal) the receiver gives so that you can be assured your message was understood. As a receiver, make an effort to decode messages accurately and provide feedback to the sender so that misinterpretations can be corrected.

Improving relationships involves building trust between yourself and those with whom you are communicating. Trust, honesty, and confidence are essential to good communication; therefore, follow through on promises you make, and maintain confidences. Once again, put yourself in the other person's place to understand how that person feels and thinks.

Speaking Ethically

Communication is a form of power and therefore carries ethical responsibilities. When the power of communication is abused, the results can be disastrous. Adolf Hitler was unquestionably a persuasive speaker. With his communication, he galvanized the German people into following one ideal and one leader, but his goals were horrifying. He remains the ultimate example of why the power of the spoken word needs to be guided by a strong sense of integrity. Make sure your goal is ethically sound. Even if your goal is worthy, you are not ethical if you use questionable methods to achieve it. In other words, the ends do not justify the means. Over the years, experts have set up four basic guidelines for ethical methods in communicating:

- Be well informed about your subject.
- Be honest in what you say.
- Use sound evidence.
- Employ valid reasoning.

IMPROVING COMMUNICATION SKILLS

Listening

Listening is an important skill for following directions, avoiding mistakes, getting along with others, and learning.

Hearing is simply receiving sound. Most of what we hear is ignored.[29] **Listening** is a conscious mental effort to understand what is heard. For example, when you truly listen to something, you receive the message, then you interpret it by putting it in a form you can understand. Next, you evaluate the message to see if you agree or disagree, and finally, you respond.[30]

Although statistics show that the average person can speak from 100 to 200 words per minute, the brain can actually process at least four times as many words.[31] Although this would seem to make listening easy, consider that studies indicate that listening consumes 45 percent of all the time we spend communicating, yet a listener hears only 70 percent of the information sent and remembers 50 percent or less.[32] In other words, listening takes up a major portion of our communication time, and yet it is a poorly developed skill.

Listening Barriers What barriers make it difficult to develop good listening skills and habits? Often, the environment is simply too noisy for good communication; or the words used by the speaker may be too difficult or unfamiliar for the listener to comprehend. Barriers also can be found in our own poor listening habits. In *The Art of Public Speaking*, Stephen Lucas discussed the following poor habits.[33]

- Lack of concentration is a common barrier to good listening skills. Something the speaker says may trigger an unrelated thought in the listener, who focuses on that thought while shutting out the speaker. The listener may have pressing problems or worries that are distracting. In either case, it is not that the listener is not paying attention on purpose, the mind has just wandered.
- Listeners may concentrate too hard. When listeners try to remember every single thing that is said, the main message may become lost in all the details.
- Listeners may jump to conclusions. They may anticipate what the speaker is going to say, or they may object strongly to the beginning of what the speaker says. When these things happen, listening may stop before the entire message is completed or understood.
- Focusing on the speaker's appearance or way of speaking may distract listeners. Grooming, clothes, speech defects, accents, or mannerisms may so irritate listeners that they do not pay attention to what is being said.
- Other poor habits that hinder good listening skills are completing the speaker's sentences, not making eye contact, preparing responses to what is being said rather than listening, interrupting, and doing other things while the speaker is talking.

Effective Listening Skills Overcoming listening barriers requires a conscious effort to remember, understand, and evaluate what is said. Guidelines follow for developing effective listening skills.[34]

- Eliminate noise and other distractions that may draw attention away from the speaker.
- Be quiet. You cannot listen effectively if you are talking.
- Put the speaker at ease by being friendly and attentive.
- Let the speaker know you are interested in what is being said. Make eye contact, and ask questions. Repeat what is said in your own words. In other words, provide feedback.
- Make notes if there is a great deal of information to remember.
- Listen for main ideas.
- Listen to the entire message, even if you think you object to what is being said. Be patient and attempt to see the speaker's point of view. Do not worry about how you will respond or become angry to the point of arguing or criticizing.
- Try to put aside your opinions of the speaker's appearance, mannerisms, or accent. Focus on the message, not the person.

Remember, being a good listener requires effort and practice and is essential to becoming a good communicator.

Reading

Reading is important for success in school, as much of the information you are taught comes from textbooks. The ability to read is necessary in most occupations. In the workplace, professionals keep current on new innovations or important changes in their fields by reading journals or training materials. Many people read some kind of newspaper or magazine on a regular basis, such as *FFA New Horizon*, *Progressive Farmer*, and *Successful Farming*. To be a well-informed consumer requires the ability to read information in advertisements and on labels to compare prices and determine important features of products. Can you imagine applying pesticides without being able to read directions? In short, how well you read can determine your success in school, in the workplace, and as a well-informed individual.[35]

Some of the same problems that hinder our ability to listen can also hinder our ability to read and understand what is read. Distractions from the surroundings and from other thoughts can keep us from comprehending what is read. Almost everyone has, at one time or another, been reading only to realize that other thoughts have kept the information from "sinking in."[36] Many people concentrate better if a fan, heater, or something that produces a noise is used while reading. This tends to keep noises from distracting and creates the same effect as the sounds from ocean waves, rain hitting a tin roof, or a crackling fireplace.

Will Rogers once said, "I love words but I don't like strange ones. You don't understand them, and they don't understand you. Old words are like old friends—you know 'em the minute you see 'em." This noted humorist expressed the views of many people. If the information read is written in professional or technical terms or uses unfamiliar words, our ability to understand may be hindered. When the reading is not interesting to us personally, it is difficult to concentrate on the meaning and relevance of the information.

Another common problem readers may have is believing that everything in print is true. This is especially true when figures are presented or when the author is well known or considered an authority. Readers may also tend to "read more into" a selection than the author has intended.[37]

Improving Reading Skills Reading requires receiving the words and processing them in your mind until you understand them. The following are suggestions for helping you become a more effective reader.

- Concentrate on what you are reading. Eliminate noise distractions, and try to focus your thoughts on the material being read.
- Begin by reading introductions, section headings, and summaries first to get an overall idea of the author's purpose.
- At the end of a section or heading area, stop to ask yourself if you have understood what the author has written. If not, reread.
- Look up unfamiliar words in a dictionary, or try to define them based on the context (the words around the new word).
- Become familiar with *jargon*, or words that are common to a specific subject or occupation.
- Read critically and without prejudice. Do not assume everything you read is true. Look for false logic or erroneous statements.[38]
- Make the reading mean something to you, your job, or your interests.
- Look at any graphs, charts, and other visual aids to clarify or simplify what is presented.

A conscious effort to improve your reading skills will benefit you in school, the workplace, and as a well-informed person. The researcher in Figure 6–5 reads constantly to do her best at her job.

Writing

Writing is a form of communication in which messages are put into words. The ability to write effectively is important to success.

Sometimes we are unclear about some of the things we write. Through words we try to touch each other's minds, and although words flow

Figure 6–5 Reading is an essential life skill. For horticulturalist Henrietta Chambers, extensive reading is part of her job as a researcher. (Courtesy of USDA.)

freely, it is curious how seldom and how fleetingly minds meet.

Poor writing usually occurs for one or more of the following reasons: The intent or purpose of the writing is unclear; the thinking of the writer appears confused; the ideas are not presented in an organized manner; sentence or paragraph structure and grammar are poor; the author is unaware of or writes inappropriately for the audience; or the style of writing is inappropriate for the occasion.[39]

Guidelines for Effective Writing Regardless of whether you are writing an essay, a term paper, a business letter, or a memorandum, the following guidelines will help you convey your message effectively.

- Know your audience. Know the level of understanding the reader has about your subject. Choose words and phrases the reader will understand.
- Know why you are writing. Should you write at all? What is the purpose of your communication? Are you giving information, asking for something, confirming information, or making a complaint?
- Be knowledgeable about your subject. You cannot write effectively unless you understand what you are writing.
- Present your ideas clearly, in a logical order. Be sure your ideas flow naturally and smoothly from point to point.
- Be precise. Say what you have to say as briefly as possible. Do not try to impress the reader with long sentences and big words.
- Stay on the topic. Everything you write should relate to your purpose for writing. Preparing an outline before writing helps you stay on the topic.
- Use correct grammar. Check with writing handbooks or textbooks if you are unsure.
- Use correct style. For example, business letters are written in a different format than friendly letters. Ask teachers or employers about the formats they prefer for written communication.
- Proofread your writing. Reading out loud or having someone else read your work can be helpful. Be sure you have made your point in a way your audience can understand. Check for errors in spelling and grammar.[40]

In the workplace, written communications, such as business letters or memos, provide a record of the message sent to a specific person on a particular date. People more readily act on written business communication than spoken conversations. In school, essay examinations, term papers, and theses all require good writing skills to relay how well you have understood a body of material.[41]

Remember, writing is one form of communication that can be kept and reviewed. Now that word processing software is common, with spell-checker and other technologies, your formal writing can be improved. Written communication reveals how well you develop, organize, and present your thoughts.

Speaking

Oral communication is a common and quick way to send a message. Oral communication may take place face to face or on the telephone; it may involve conducting a meeting, leading a group discussion, or making presentations.[42] Nearly every day you talk to teachers, friends, family, employers, and co-workers. You may be called on to speak formally in a business situation or informally in conversations with friends. In any of these situations, your success depends on how well you express yourself orally.

Problems with speech may arise from not speaking so the listener can understand, not paying attention to the person being spoken to, using a tone of voice that makes the listener defensive, using poor grammar or pronunciation, not giving the listener an opportunity to respond, and confusing the listener by drifting away from the subject.[43]

Practicing Effective Speaking Skills Preparing and giving speeches, conducting meetings, and leading discussions are covered in depth in later chapters. There are several suggestions to use in any speaking situation to increase the effectiveness of your communication.

- Speak clearly. Concentrate on correct pronunciation, talking loudly enough, and speaking at an appropriate pace (speed).
- Make eye contact with your audience. You can hold a person's attention by looking directly at them instead of away from them. If you are speaking to a large group, make an effort to speak to all parts of the room.
- Use a pleasant tone of voice. Your message will be better accepted, even if it is critical, if you sound positive and friendly. The way you say something may be as important as what you say in some cases.
- Avoid using slang in formal situations. Use good grammar and appropriate terminology. Using good grammar makes you sound educated and knowledgeable.
- Be sure your words are understood. Allow the listener to respond with questions or comments. Watch for nonverbal communication, such as puzzled looks or nods of affirmation.
- Keep to your subject. Avoid rambling.
- Be brief but thorough.[44]

Good speaking skills are essential for everyday communication with people at school or at work. The impression you make may well affect future relationships with friends, teachers, employers, and customers.

NONVERBAL COMMUNICATION

Nonverbal communication is sending a message with body movements, behaviors, and the physical environment rather than with words. Some have estimated that 55 percent of communication is nonverbal, with voice accounting for 38 percent and words, 7 percent. The *way* you send a message can be as important as the message itself.

Improving your nonverbal communication can help you relay messages to others more effectively and may help you understand how others feel and think about your message.

> ***WARNING: Nonverbal communication is difficult to understand accurately. Never use nonverbal cues alone in determining how a person feels or thinks. Nonverbal communication signs are usually used to support or refute a verbal message.[45] For example, a person may tell you he will do something for you but shake his head negatively at the same time. This may tell you the person will do what you have asked but is not happy about doing it. Also, some people have sleep disorders: "When they stop, they flop." Sleeping in class or meetings could be misinterpreted as disinterest. "Actions speak louder than words" may be true, but understanding those actions is difficult.***

There are many types of nonverbal communication. Although certain behaviors are thought to mean certain things, be aware that some behaviors may be habits rather than signs of deeper meaning. For example, sitting with your arms crossed in front of you is interpreted as shutting out incoming messages even though it may simply be your preferred way of sitting, as illustrated in Figure 6–6.

Although some nonverbal communication is difficult to understand accurately, there are some easily communicated nonverbal signals, such as excitement, joy, pain, boredom, anger, disappointment, sadness, enthusiasm, and interest.

Body Language

If you have ever seen a good pantomimist perform, you know how much can be communicated through **body language**, or **kinesics**. There are entertainers who can create a mood or tell a story simply through body movement, gesture, and facial expression. Most people similarly pantomime a bit in their everyday conversation.

Facial expressions are the best way to communicate emotions and feelings. A smile or a frown sends a message about your attitude. Tightly closed lips indicate disapproval. The eyes also provide clues: If a person's pupils enlarge, they are responding positively to you; smaller pupils indicate a negative feeling. A speaker's sincerity increases with eye contact: The more eye contact a speaker uses, the more trustworthy he or she seems to an audience. Eyes can convey love, hate, interest, indifference, or anger. Other facial expressions include wrinkling eyebrows and biting the lip.

Touching is a nonverbal cue that leaders or speakers use effectively in certain situations. However, touching is easily misunderstood or misinterpreted, so use it carefully. Some people feel comfortable with a pat on the back or a hug as a show of your excitement for their success. Others might take a hug as a sexual advance. Be alert to signs that indicate that the other person is misunderstanding your actions. If someone misinterprets your actions, set them straight immediately.[46]

Body movements can convey many messages to others. Do you form opinions about people based on the way they walk? Posture and even how people stand or sit can give signals to others.

Proxemics is the science of how spacing and placing tells others messages. Leaders or people with a great deal of respect command empty space or room around them. Political leaders, for example, often have more space around them than would a journalist or co-workers. Leaders often choose how close to approach others. How close is too close when you are talking to others? What is a comfortable distance for you to be from

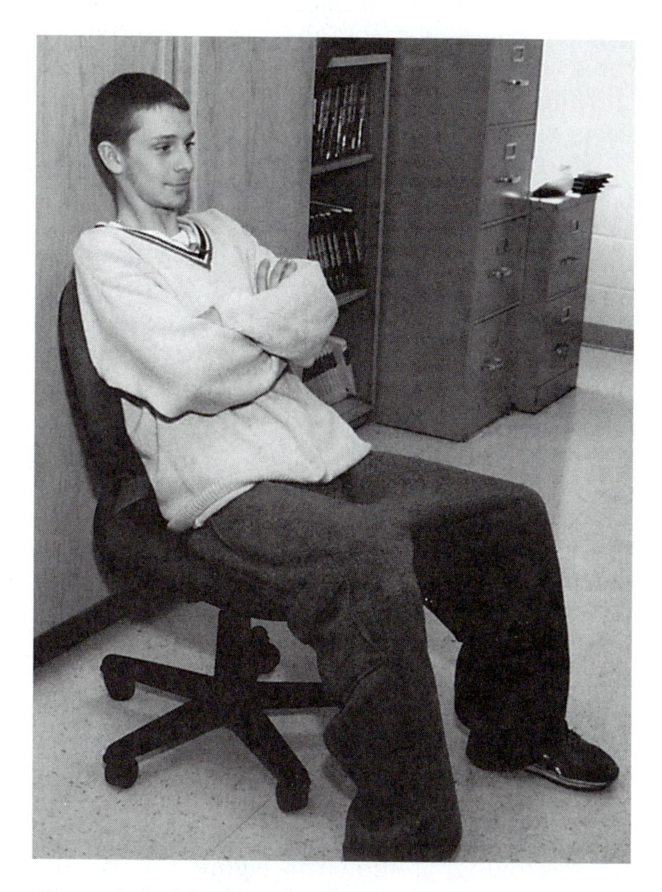

Figure 6–6 Even though this may be the student's preferred way of sitting, it nonverbally communicates that he is not interested in the program.

a speaker or leader? How does it make you feel if someone stands very close to you when talking to you?[47] Preferences vary with every individual. Regional and cultural diversity contribute to these differences.

The way a person walks may also convey nonverbal messages. A slow step may communicate a lack of confidence and lack of decision-making ability. A quick, lively walk generally communicates happiness.

Posture Standing or sitting erectly may communicate interest and alertness. Slumping while sitting or standing is thought to be a sign of tiredness or a lack of interest. Nervousness or boredom can be conveyed by swinging a leg or tapping fingers on a desk. Boredom may also be communicated by fidgeting, doodling, and continual shifting in a seat. Shrugging the shoulders is thought to be a sign of indifference.[48]

Putting a hand to the mouth or putting the head in the hands can be a way of showing objection or surprise at a message. Hands placed on the hips with elbows pointing outward may be a sign of anger. Arms folded across the chest is thought to mean that a person is defensive or closed to the message. Arms at the side may relay the idea of openness and relaxation. Pointing a finger while speaking may be used to communicate unhappiness or to show authority.

Crossing the legs is a gesture that says many things about the person. When someone crosses his/her legs and moves his or her foot in a slight kicking motion, the person is probably bored. Crossed legs often mean a narrow line of thinking, one not open to negotiation. Individuals who cross their legs seem to be the ones who give you the most competition and need the greatest amount of attention. If crossed legs are coupled with crossed arms, you really have an adversary.

Stroking the chin is a gesture of thinking and evaluating. It seems to say, "Let me consider." Chin stroking, as shown in Figure 6–7, is used when people go through a decision-making process. Another such gesture is that of pulling on the beard. A congruent facial expression to these gestures is a slight squinting, as if trying to see an answer to the problem in the distance.

There are some gestures that we have learned to accept as stall tactics. One such gesture is that of removing the glasses and putting the earpiece of the frame in the mouth. Taking the glasses off and throwing them on the table is a stall play. Pipe smokers are supreme stallers. The ritual of smoking takes time, usually away from conversation, if the smoker so desires.

Boredom is another emotion shown in different forms of body language. One that is the most

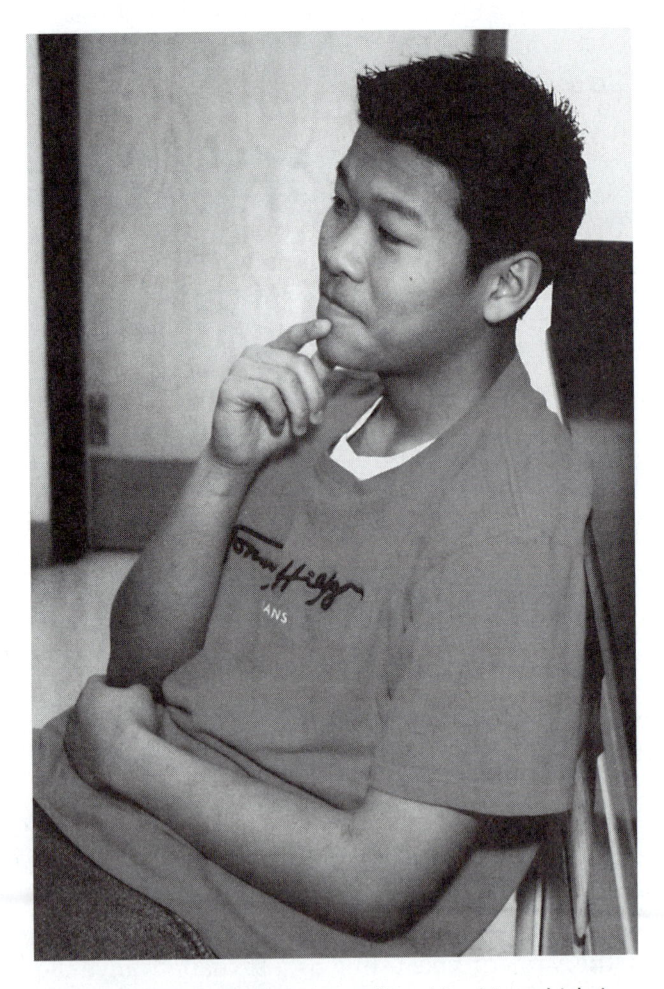

Figure 6–7 This student is stroking his chin, which is a nonverbal indication that he is thinking and evaluating.

recognized is drumming on the table or tapping the foot. Some psychiatrists believe that when we are impatient we try to duplicate a prior life experience when we felt safe and secure, such as when we were in the womb. The drumming is the same type of beat as the mother's heartbeat. Another motion associated with boredom is the head-in-the-hand signal. This coupled with drooping eyes is a gesture of regret. Another boredom sign is **doodling**. When a person doodles, interest is waning. For the most part, it is believed that a person who doodles lets that interfere with open communication; their listening ability is jeopardized.

Another emotion that has body language as its form of communication is nervousness. Clearing the throat is often associated with nervousness and anxiousness. When a person gets nervous, mucus forms in the throat. The natural thing to do is to clear it, making that familiar sound often heard from the speaker's podium. Sometimes, however, throat clearing is simply a habit. This habit, if it continues, is a signal that the speaker is uncertain and apprehensive.

Anxious people often make sounds that indicate their feelings. An example of a sound is the "whew" sound. It signals, "Whew, I'm glad that's over with." People usually use this sound when they want to communicate that some task or obstacle has been overcome. It is possible the person has caught his breath and the "whew" sound is literally a sigh of relief. Another sound is that of whistling. This sound is used to build courage or confidence.

Frustration is an emotion that has many features in body language. The first feature is one that sounds like short breaths or an angry bull. Highly emotional people take deep breaths and expel the air slowly, making long, sighing sounds. Other features are clenched hands and wringing hands. These are usually observed when someone is on the hot seat. People with either of these features are tense and very difficult to relate to. Try to put them at ease before you try to communicate with them. On the other hand, some people do not like to be bothered when they are in certain moods.

Physical Environment Large offices with expensive decor may convey that the person who occupies the office is important in the organization.[49] Generally speaking, the person seated behind a desk is thought to be in charge. Certain desk arrangements in classrooms, such as horseshoe designs or desks grouped in small circles, may indicate a desire for open dialogue, whereas rooms arranged with desks in straight rows may indicate the teacher's preference for lecturing.

In terms of improving your own nonverbal skills (recognizing and sending), consider the following ideas.

- Study the meanings of body language so you recognize them. Use appropriate gestures to drive home your meaning.
- Become aware of your own nonverbal communication by having others watch you.
- Work to convey open, friendly messages with your body language. Smiles, erect posture, and positive nods of the head are good ways to send such messages.
- Interpret nonverbal communication *only* as a way of confirming or refuting a verbal message.
- Realize that nonverbal communication is at times imprecise and should be interpreted with care. If you are unsure if a person is communicating something to you with body language, ask tactful questions that might help you understand.
- Be sensitive to the physical environment. Use appropriate seating arrangements to match the type of communication environment you want. Be aware that if you are behind a desk, the other person may feel like a subordinate (that you are his or her boss). This may make communication more difficult.

Being aware of the messages others send with body language can help you determine the effectiveness of your communication. Also, being aware of your own body language and using it to your advantage may help you communicate your ideas to others.

FEEDBACK

Feedback is the way the receiver responds to a message that helps the sender know if the message was understood. If the receiver gives no feedback, the sender may need to determine if the receiver understood the message by drawing out feedback. Getting feedback is a necessary step in the communication process; without it, the cycle is incomplete.

Some teachers have a special talent for reading the faces of their students as they teach. Of course, this is not limited to teachers. When looks of confusion, boredom, or lack of understanding are observed, they change tactics. The feedback they are receiving tells them that they have to do something quick because the communication is ineffective.

False Feedback

There are problems with feedback that may hinder its effectiveness. Not giving or not allowing time for feedback are obvious problems. If feedback is given by the receiver, it may be vague or untruthful. If feedback is sought from the sender, many times it is in this form: "Are there any questions?" When this question is asked, and no questions follow, the sender may believe that the message has been completely understood. The fact may be, however, that the receiver may feel awkward or ignorant if he or she asks a question, or the receiver may not understand enough of the message to even ask a question.[50] I can remember being a freshman in college. I was so lost in some classes I couldn't even think of a decent question to ask. I think the professor thought we understood; we were so lost we didn't know anything to ask.

Improving Feedback

To improve feedback, consider the following suggestions:

- As a sender, encourage feedback. Do this by allowing time for feedback. Be patient and encouraging while making others feel comfortable enough to respond to you.
- Be aware of nonverbal messages. Puzzled looks may tell you that your message is not understood. Also, be aware of your own nonverbal messages that may discourage communication, such as rolling your eyes or sighing when you are asked to repeat something you have said before.
- Make it your responsibility to be understood. As a sender, you can seek (or as a listener you can provide) feedback instead of simply waiting for it to happen.
- Ask specific questions about the details of your message to be sure you are understood. Avoid the "Do you have any questions?" trap.
- Use paraphrasing techniques. Use a statement such as, "Please explain to me in your own words what we need to do so I can be sure I have explained myself well." This type of question takes pressure off the listener because it sounds like you are checking your own performance, not doubting whether the listener has paid attention.[51]

Figure 6–8 Notice the intense listening skills of the students. (Courtesy of USDA.)

The students in Figure 6–8 are exhibiting good listening skills. Refer to Figure 6–9 for a summary of improving communication skills.

SELF-COMMUNICATION AND INTERPERSONAL COMMUNICATION

A person must be able to self-communicate if he or she is going to effectively communicate with other people (Figure 6–10). Remember the advice of Socrates, "Know thyself." When you know who you are and you understand your personality type, communication style, learning style, and their relationship, you are in touch with your feelings and have peace of mind about your life goals. We call this self-communication. The following drill may help you communicate better with yourself.

Self-Communication

Visualize a day in your life five years from now. What do you want to be? Ask yourself the following questions.

1. What am I doing now to accomplish this objective?
2. What is keeping me from doing it?
3. What resources am I not using?
4. What do my actions now have to do with this goal?

Next, evaluate the past five years. Ask yourself the following questions.

Improving Communication Techniques

Skills	Listening	Reading	Writing	Speaking	Nonverbal	Feedback
Techniques	Eliminate distractions Be quiet Show interest Make notes Be objective Observe nonverbal cues	Concentrate Find overall meaning Summarize periodically Think critically Avoid prejudice Use a dictionary Make relevant to yourself	Know audience Know purpose Know subject Be precise, clear, brief Stay on topic Use good grammar Use correct style Proofread	Speak clearly Make eye contact Use pleasant tone of voice Use good grammar Observe non-verbal cues Stay on topic Be brief, but thorough Be assertive	Understand meaning of non-verbal behaviors Know your own nonverbal behaviors Interpret body language Be open with your own non-verbal actions Use with care Be sensitive to physical environment	Encourage feedback Observe nonverbal cues Ask specific questions Use paraphrasing techniques Take responsibility for being understood

Figure 6–9 Good communication skills and techniques are an asset to any leader. Six areas of communication, with key things to remember for each, are summarized here.

1. What have I accomplished?
2. How did I do it?
3. Why did I succeed?
4. Why did I not succeed?
5. What was I doing right?
6. What was I doing wrong?
7. How can I alter my behavior?
8. What is happening to me in relationship to other people?

Figure 6–10 The ultimate goal of interpersonal and self-communication is happiness. These FFA members and their advisors appear to be excellent at interpersonal communication. (Courtesy of the National FFA Organization.)

These may seem to be simple questions, but it is difficult to answer all of them. It is hard to be honest with ourselves and admit we have done some things we may regret. Furthermore, it is hard to admit that we could have accomplished much more if we had communicated more effectively with ourselves and had set better goals. Also, if we had taken time to communicate with ourselves and had defined some timetable objectives in our life, we would now be happier with ourselves. We would also have a feeling of self-fulfillment. Once a person has achieved **self-communication**, a major barrier to communicating with other people has been broken.

Interpersonal Communication

The next step toward getting through to people is that of interpersonal communication: How can you get close to people? In other words, how can you make people "loosen up" and better understand their feelings? Don't let the following develop into **interpersonal communication barriers**.

Honesty John O. Stevens makes some interesting observations about getting close to people. Stevens says that honesty is crucial for interpersonal communication. Stevens says, "I have to first be honest with myself and get in touch with my experiencing, and take responsibility for it by expressing it as my experiencing."

Awareness We have to be aware of our own experiences and be able and willing to make others aware of these experiences. We have to send clear messages about our awareness, our experiences, our feelings, and our needs. On the other hand, we have to be aware of the messages that others send.[52]

Communicate Clearly To communicate clearly with another person it is helpful to do the following.

1. Send concise statements that are as clear as possible.

2. Pause periodically to allow the listener time to understand what you are saying. (It might be useful to think of yourself as talking in paragraph form, not whole chapters at a time.)
3. Be aware of your own feelings.
4. Be aware of your intentions in communicating with your listener.
5. Express yourself as honestly as possible.
6. Maintain contact with the people you are talking: Pay attention to their verbal and nonverbal responses, maintain frequent eye contact, and watch for cues to their understanding of your message (or lack of it).
7. Be aware of your own nonverbal messages (twisting your ring, leaning forward, laughing, hesitating, biting your lips, the quality of your voice) and notice if these "messages" are consistent with the verbal message you are sending. For example, are you saying that you are calm, but your voice is shaking and you have crossed and uncrossed your legs a dozen times?

Not What, But How It has often been said that it is not what you say but how you say it. The "how" of the message is complex and must be subdivided. Every message contains information about how it should be taken. Cues at this level will tell the receiver whether the message from the sender is to a question, a joke, a statement, or an opinion.

Relationship to the Receiver Every message contains information concerning how the sender perceives his relationship to the receiver, whether he considers them friends, enemies, or equals. Information about the relationship between sender and receiver is transmitted in many different ways, including (1) words, or verbal behaviors; (2) nonverbal behavior, namely, gestures, tone of voice, posture, and facial expression; and (3) the situation itself or the context in which the communication occurs.

Directness One aspect of being direct involves the sender's "owning" the thoughts, feelings, ideas, evaluation, or expectations they convey rather than attributing them to some other source. This kind of responsibility for oneself reflects the difference between saying, "You are obnoxious" and saying, "I really am angry at what you're doing." In this instance, I can be direct about my own feelings since I know them firsthand; I can send the second message legitimately. Sending the first message can lead to almost nothing but interpersonal difficulty for a number of reasons. One problem with the "You are obnoxious" maneuver is that it is usually the first move in the "blame game," a pattern of interaction that usually results in stalemate. It goes without saying that directness is not synonymous with bluntness, tactlessness, and accusativeness.

Nonjudgmental Another primary aspect of interpersonal communication is knowing how to present one's thoughts and feelings in a way that is nondemanding, nonabsolutist, and nonjudgmental. Bad examples follow.

"You shouldn't interrupt." (absolutist)
"It isn't right to interrupt." (judgmental)
"Shut up and let me talk." (demanding)

Respect the Other Person's Position Still another aspect of interpersonal communication is knowing how to respect and understand the position of the other person. This involves, as mentioned earlier, knowing how to listen nonjudgmentally to what the other person is thinking and feeling. It is primarily in this way that we show our love for others. Consider the following quote from Marshall Rosenberg.

> ***Love is a state of complete attention, without intruding thoughts and motivations. Contrary to general belief, love is not just a feeling or emotion. The opposite of love is not hate; the opposite of love is . . . judgmental thinking.***

Happiness A person who has mastered the art of interpersonal communication may get satisfaction and fulfillment out of life. After all, what is brilliance, wealth, or fame, if one cannot share happiness? It appears that the epitome of happiness is this closeness and friendship that we have with people. This can only be achieved when a person knows how to communicate interpersonally.

SUMMARY

Communication is the process of sending and receiving messages in which two or more people achieve understanding. Communication is important because so much of each day is spent in some form of communication and because many problems that occur in our relationships with others are due to miscommunication.

The purpose of communication is to inform or influence others and to express feelings. We communicate when we send messages (speaking, writing), receive messages (listening and reading), or use nonverbal messages (body language). The communication process involves a situation in which a sender provides a message through a channel to a receiver. The receiver responds with (or the sender seeks) feedback. The process may be broken down by interference from numerous distractions.

Factors that may hinder communication are language barriers, interpersonal barriers, situational-timing barriers, and organizational/procedural barriers. These factors may be overcome by improving perception, by improving the physical process of communication, and by improving relationships. To communicate with other people, we must first be able to communicate with ourselves. The ultimate communication for most of us is interpersonal communication.

Communication of all types can be improved by understanding the problems associated with each and by making a conscious effort to practice techniques designed to foster and improve communication skills.

END-OF-CHAPTER EXERCISES

Review Questions

1. Define the Terms to Know.
2. Define communication.
3. List three situations in which communication is important.
4. List and give examples of three primary purposes of communication.
5. List and give examples of three forms communication takes.
6. List and briefly describe seven elements of the communication process.
7. List four barriers to communication and give an example of each.
8. Explain three ways to overcome communication barriers. Give an example of each.
9. Name at least three suggestions for improving each of the following skills: listening, reading, writing, and speaking.
10. Explain how and when nonverbal communication should be used.
11. When using feedback, is simply asking "Are there any questions?" a good method for checking understanding? Why or why not?
12. Name five characteristics that must be involved for interpersonal communication to take place.

Fill-in-the-Blank

1. ________________ barriers deal with the way words are understood.
2. ________________ barriers deal with personal characteristics of the sender and receiver that hinder communication.
3. ________________ barriers deal with how and through what structure a message goes from the sender to the receiver. Barriers of this type are common in the workplace.
4. ________________ barriers are those that deal with the time and place a communication takes place.

Matching

_____ 1. The process of sending and receiving messages to achieve understanding.
_____ 2. When and where communication takes place.
_____ 3. The person who wishes to start communication with someone else.
_____ 4. The responses a receiver gives the sender so the sender knows his or her message was understood.
_____ 5. The means a sender uses to convey encoded messages.
_____ 6. A person's understanding of a message based on personal beliefs, knowledge, and ways of organizing information.
_____ 7. Person for whom a message is intended.
_____ 8. That which is intended to be communicated from one person to another.
_____ 9. Anything that hinders the sender from making the message understood.
_____ 10. Process by which the sender puts the message in a form the receiver will understand, such as words, numbers, gestures.

A. perception
B. communication
C. receiver
D. encoding
E. message
F. sender
G. channel
H. feedback
I. situation
J. interference

Activities

1. Write a one-page essay expressing your feelings about an issue in your life or something in the news. Have the class evaluate your ability to communicate your feelings.
2. Write an imaginary communication scene similar to John and Jennifer. Have a partner identify each element of the communication process.
3. Read an article from a favorite magazine. Write a brief summary of the article outlining the main points.
4. Read the summary you wrote for question 3 to a classmate. Have the classmate listen carefully and tell the main points of your summary. Check the classmate's listening skills by comparing the answers to your summary. If there are misunderstandings, reread the summary or rewrite if it is your writing that caused the misunderstanding.

5. Watch a classmate, teacher, or favorite actor. Write down the nonverbal forms of communication you see. Also note how the nonverbal communication matched (or did not match) the verbal communication used.
6. During a class discussion, jot down ways students provide feedback (verbally and nonverbally). Share your findings in small groups.
7. For 5 or 10 minutes, you and a partner develop a brief role-play situation in which you demonstrate one of the barriers to communication (language barrier, interpersonal barrier, situational-timing barrier, or organization structure and procedural barrier). Perform the role-play for the class. Have them guess which barrier you are demonstrating. Have a class discussion about how the communication could be improved.
8. Visualize a day in your life five years from now. What do you want to be? Write an answer to each of the following questions.
   a. What type of job, educational accomplishment, and personal goals do I want to have achieved five years from now?
   b. What am I doing now to accomplish these objectives?
   c. What is keeping me from doing this?
   d. What resources am I not using?
   e. What do my actions now have to do with this?
9. **Take it to the Net.**

   Explore communication skills on the Internet. Below, various Web sites are listed that contain information on communication skills. Under each Web site is a description of the site and directions. Browse as many of the sites as you want. Pick one and follow the directions provided under it. If you have problems with the Web sites listed below or just want more information, some search terms are also provided.

   Web sites

   <http://www.queendom.com/communic.html>

   Contains a test that is designed to evaluate your general level of communication skills. Take the test and print the results. Write whether you agree with the results or not.

   <http://www.selfgrowth.com/comm.html>

   Contains a lot communication skills information. It includes sponsor websites, articles, and additional websites, all containing information on communication skills. Browse through the articles and pick one you find interesting. Read the article and print it. Write a summary of the article.

   <http://www.acjournal.org/

   The site for the *American Communication Journal*, containing several articles on communication. Browse the articles. Pick one and read it. Print the article and write a summary of it.

   Search Terms

   communication skills

   communication

   effective communication

   importance of communication skills

Notes

1. J. O'Connor, *Exploring Communication* (Upper Saddle River, NJ: Prentice-Hall, 1984).
2. R. N. Lussier, *Human Relations in Organizations: Applications and Skill Building* (Columbus, OH: McGraw-Hill, 1998).
3. R. Maidment cited in R. N. Lussier, *Human Relations in Organizations.*
4. C. Caskey cited in R. N. Lussier, *Human Relations in Organizations.*
5. M. R. Hansen cited in R. N. Lussier, *Human Relations in Organizations.*
6. Lussier, *Human Relations in Organizations.*
7. W. D. St. John, *A Guide to Effective Communication* (Nashville, TN: Dr. Walter D. St. John Enterprises, Inc., 1970.)
8. D. Jennings, *Effective Supervision: Frontline Management for the '90s* (Minneapolis/St. Paul: West Publishing Company, 1993).
9. Ibid.
10. S. Lucas, *The Art of Public Speaking* (New York: Random House, 1986).
11. St. John, *A Guide to Effective Communication.*
12. Jennings, *Effective Supervision.*
13. St. John, *A Guide to Effective Communication.*
14. Ibid.
15. Lucas, *The Art of Public Speaking.*
16. Jennings, *Effective Supervision.*
17. Lucas, *The Art of Public Speaking.*
18. St. John, *A Guide to Effective Communication.*
19. *Personal Development Skills in Agriculture* (College Station, TX: Instructional Materials Service, Texas A & M University, 1990), 8741B, p. 2.
20. Ibid.
21. Jennings, *Effective Supervision.*
22. St. John, *A Guide to Effective Communication.*
23. Ibid.
24. John Noe, *People Power* (Nashville, TN: Thomas Nelson Publishing, 1986).
25. Lussier, *Human Relations in Organizations.*
26. St. John, *A Guide to Effective Communication.*
27. *Communication Barriers* (Nashville, TN: Tennessee Department of Education, 1993).
28. Ibid.
29. O'Connor, *Exploring Communication.*
30. *Communication Barriers.*
31. Jennings, *Effective Supervision.*
32. *Communication Barriers.*
33. Ibid.
34. Jennings, *Effective Supervision.*
35. *Improving Written Communication Skills* (Skills for Success #51), (Nashville, TN: Tennessee Department of Education, 1993).
36. Ibid.
37. St. John, *A Guide to Effective Communication.*
38. Ibid.
39. Ibid.
40. Ibid.

41. *Improving Written Communication Skills.*
42. Lussier, *Human Relations in Organizations.*
43. *Improving Oral Communication Skills* (Skills for Success #50), (Nashville, TN: Tennessee Department of Education, 1993).
44. Ibid.
45. *Improving Nonverbal Communication Skills* (Skills for Success #52), (Nashville, TN: Tennessee Department of Education, 1993).
46. *Personal Development Skills*, 8741-E, pp. 1–2.
47. Ibid., 8741-E, p. 1.
48. Lussier, *Human Relations in Organizations.*
49. Jennings, *Effective Supervision.*
50. Lussier, *Human Relations in Organizations.*
51. Ibid.
52. J. O. Stevens, *Awareness: Exploring, Experimenting, Experiencing* (Moab, UT: Real People Press, [no date available]).

CHAPTER

7

Reciting (FFA Creed)

Objectives

After completing this chapter, the student should be able to:

- Recite the FFA Creed
- Explain why you should participate
- Demonstrate basic principles of speaking
- Demonstrate practice methods
- Reduce stage fright and nervousness
- Establish rapport with the audience
- Analyze practical and fun ways to say the Creed

Terms to Know

creed (FFA)
agricultural education
curriculum
attitude
Greenhand degree
delivery
convey
winner
self-confidence
self-deprecating
insecure
monotone
chapter
rapport
lectern
spontaneity
tone
inflections
articulation
pronunciation
force
intangible
clone

Reciting the FFA **Creed** is the first of many organized speaking activities in **agricultural education**. Although the FFA Creed is an excellent document to memorize and study, our purpose for this chapter is to use it as a tool to begin public speaking.

Here is an excellent plan for students to follow to master their speaking skills.

1. Recite the Creed—9th or 10th grade
2. Parliamentary procedure—10th grade
3. Prepared public speaking—11th grade
4. Extemporaneous speaking—12th grade

Due to the class arrangement of many schools, this plan of progress may not be appropriate. However, speaking can begin at any grade level. All the speaking examples listed could be learned in the same class, or in other parts of the **curriculum**. Speaking is fun. With an open mind and a proper **attitude**, speaking can be as natural and enjoyable as playing sports. The FFA member in Figure 7–1 is enjoying herself as she says the FFA Creed.

Figure 7–1 Speaking can be fun if you develop the proper attitude. Loosen up, enjoy yourself, and remember that if you mess up, your classmates are laughing with you, not at you.

LEARNING THE FFA CREED

Memorizing the Creed is an example of the symbolic leadership mentioned in Chapter 1. Memorizing the Creed is a requirement of the first degree attained in the FFA, the **Greenhand degree**. Memorizing is tough and sometimes boring, but for attaining leadership, "when the going gets tough, the tough get going."

Since the Creed script is already written, it allows you more time to work on **delivery** when speaking. Part of effective delivery is emphasizing important words. The Creed has many expressive words that should be said with feeling.

Key Words in the FFA Creed

There are no dull subjects, but there are dull people. Select key words in each paragraph of the Creed. Some students find a picture in an agricultural magazine or some other source that illustrates a key word, sentence, or a paragraph. Selecting pictures that **convey** meaning to you may help you memorize the Creed and say certain words with sincere feeling and expression.

Every sentence within the FFA Creed tends to revolve around one or more key words. In the first paragraph, the following are key words: *believe, future, agriculture, faith, words, deeds, achievements, present, past, generations, promise, better, enjoy,* and *struggles*.

The Creed can be said effectively if certain words are emphasized. Say words with the feeling of the meaning they convey. For example, say *better* with positive facial and vocal expressions. Say *struggles* with negative facial and vocal expressions.

As you memorize the Creed, you may wish to develop a paragraph similar to the one in Figure 7–2. This accomplishes two things: First, it helps you memorize the Creed; and second, it helps you emphasize the key words when you recite the Creed.

Consider the non-italicized words in the FFA Creed, reprinted below. Write each paragraph with

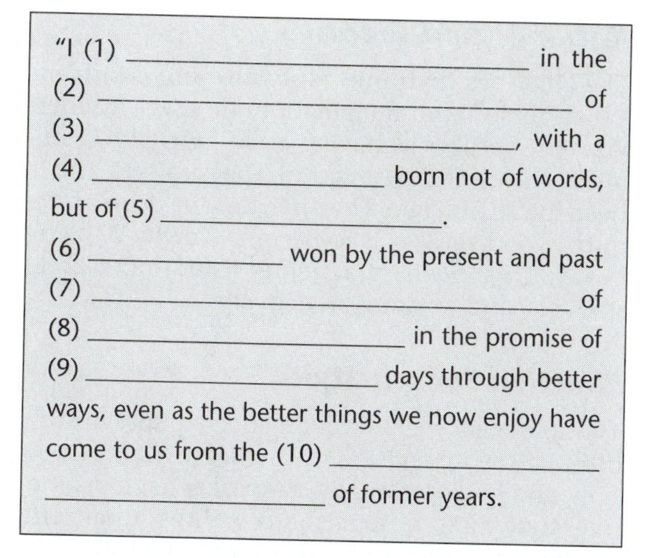
"I (1) ____________________ in the
(2) ____________________ of
(3) ____________________, with a
(4) ____________________ born not of words,
but of (5) ____________________.
(6) ____________________ won by the present and past
(7) ____________________ of
(8) ____________________ in the promise of
(9) ____________________ days through better
ways, even as the better things we now enjoy have
come to us from the (10) ____________________
____________________ of former years.

Figure 7–2 As you memorize the Creed, you may wish to develop a paragraph similar to the one above. This accomplishes two things: (1) it helps you memorize the Creed, and (2) it helps you emphasize key words when you recite the Creed.

blanks for the key words (see Figure 7–2). You may wish to change which key words are selected.

> *I* believe *in the* future *of* agriculture, *with a* faith *born not of words, but of* deeds. Achievements *won by the present and past* generations *of* agriculturists *in the promise of* better *days through better ways, even as the better things we now enjoy have come to us from the* struggles *of former years.*
>
> *I* believe *that to live and* work *on a* good *farm, or to be* engaged *in other agricultural pursuits, is* pleasant *as well as* challenging*; for I know the* joys *and* discomforts *of agricultural life and hold an inborn fondness for those* associations *which, even in hours of* discouragement, *I cannot* deny.
>
> *I* believe *in leadership from ourselves and* respect *from others. I believe in my own ability to* work efficiently *and* think clearly, *with such* knowledge *and skill as I can* secure, *and in the ability of progressive agriculturalists to serve our own and the public interest in* producing *and* marketing *the product of our* toil.
>
> *I* believe *in less dependence on* begging *and more* power *in* bargaining*; in the life* abundant *and enough* honest wealth *to help make it so*—for others *as well as* myself*; in less need for* charity *and more of it when needed; in being* happy *myself and playing* square *with those whose* happiness *depends upon me.*
>
> *I* believe *that American agriculture* can *and* will hold true *to the best* traditions *of our national life and that I can* exert *an* influence *in my* home *and* community *which will stand* solid *for my part in that* inspiring task.[1]

REASONS TO PARTICIPATE

If you are like most high school students, some of you are dragging your feet, feeling that this is not for you. Maybe you feel it is not worth the effort. You are probably thinking, "I can't win with speaking, so why try?"

You Win by Participating

It is common for students to feel that the **winner** will always be someone else, but this is not true. You can be a winner. In fact, the only way you cannot be a winner is if you refuse to participate. David Jameson in his *Leadership Handbook* tells us why. Most people have made the mistake of equating being "number one" with being the only winner. We focus our attention on the athletes who are declared the champions and consider them the winners and all their competition losers. However, we may notice two or three years later several star college players who were members of the so-called losing teams. In fact, many stars come from last-place teams; but as individuals, they were winners. They became winners because they participated and learned athletic skills. So we need to look beyond today when we think about who the winners really are.

You Can Win in the Game of Life

Many occupations require speaking skills. In educational terminology, we call these affective skills. Lawyers, salespeople, preachers, politicians, auctioneers, and many other occupations need speaking skills.

The winner of a public speaking contest may end up in an occupation in which very little public speaking is required, whereas students who do not place first may have what it takes to enter the field of politics or assume a position of responsibility in the community. The skills learned by participation in speaking help them become winners.[2]

When you say, "I can't win, why should I try," you are being very shortsighted. The truth is that by not participating, you may be establishing a lack of success in life.

By Participating, You Cannot Lose

Opportunities for learning many essential leadership skills arise only occasionally. Therefore, everyone who wants to be a long-term winner should be anxious to participate when the opportunity arises, recognizing that the real winning may be several years in the future. Because of your involvement now, however, you will be ready for the day when the bigger contests come.[3] By participating, you cannot lose. You only lose when you quit or do not even try.

BASIC PRINCIPLES OF SPEAKING[4]

Reasoning and Feeling

Communication involves both reasoning and feeling. The able speaker both thinks and cares. The correct attitude is one of friendly good will, and a desire to help, to inform, or to inspire. If a speaker doesn't believe in the message, lacks **self-confidence**, is apologetic, is immature, or is **self-deprecating**, listeners will lack confidence in what is said.

Rational and Creative

The able speaker thinks rationally and creatively. When reason and imagination are used, listeners view the speaker as having good judgment, common sense, competence in the subject, and a healthy self-image. The first task in developing into a good speaker is to become worthy of being heard. This means practicing, listening to good speakers, and working on delivery.

Effective Speech Style

The able speaker achieves an effective speech style through the integration of language, vocal expression, and body action. Every speaker has a distinctive style that reveals many things about the person. An **insecure** speaker uses impersonal language and speaks in a **monotone**. Because of fear of the audience, the style becomes a shell in which the speaker can withdraw from the audience. Most of the time when an audience disagrees with the speaker, it is not because of the statements made but because of an irritating voice or manner. The speaker must adapt the style and delivery to the audience.

Observe the Audience

The able speaker observes the audience as the speech is given, watching facial expressions and body movements that indicate interest and attention. The able speaker can distinguish between the frown that indicates "You're making me think hard" and the frown that indicates "You're all wrong."

Persuade the Audience

To persuade an audience, the speaker must understand how the listeners feel, what they want, and what they need. A speech must begin with the listeners; otherwise, it is a waste of time. The greatest artists of persuasion are those who understand people because they genuinely like them. No technique can substitute for a sincere concern and respect for other people.

David Jameson, a past National FFA officer, suggests that you can easily master what he calls the "Big Four" in speaking skills. These "Big Four" common-sense skills are:

- *Volume*—Always speak loud enough to be heard easily in the back row of the room.
- *Tempo*—Speak slowly enough so the audience can grasp and reflect on what you are saying.
- *Diction*—Speak clearly enough that every word is distinct, so the audience hears every word.
- *Enthusiasm*—Speak with enthusiasm, so people will know you mean what you are saying, find it easier to pay attention, and remember more of what was said.

Here is what we are accomplishing for you in the minds of the audience when we help you master these four basic skills. First, you impress an audience because they can easily hear what you are saying. People are also impressed when you speak slowly and clearly enough for them to understand and digest what you are saying. Last, and most important, enthusiasm makes your speech seem shorter and easier to remember.[5]

Figure 7–3 Videotaping your Creed presentation and evaluating it is an excellent way to improve delivery. (Courtesy of the National FFA Organization.)

Practicing

The old saying is "practice makes perfect." Is this really true? Consider most people's handwriting and you will agree that practice does not make perfect. In reality, perfect practice makes perfect. Practice as well as you can. You will pursue perfection through self-evaluation and evaluation by others. Practice with the intention of doing a better job each time.

There are a number of ways to practice the Creed. Use a tape recorder to make sure you are speaking loud and clear enough and that you have expression in your voice. Use a videotape recorder to rehearse your gestures. Evaluate how you emphasize with your hands and facial expressions. The person in Figure 7–3 is taping presentations of the FFA Creed. Once several members are taped, the teacher and the students can evaluate each speaker.

Figure 7–4 Saying the FFA Creed in front of a mirror is a good way to help overcome stage fright and improve your facial expressions and hand gestures. (Courtesy of the National FFA Organization.)

It may come as a surprise to you, but one of the most effective ways to practice is in front of a mirror (Figure 7–4). You feel silly at first. You will be embarrassed even though no one may be around. You will even have trouble concentrating at first. Eventually, though, you will gain self-confidence and get more comfortable speaking before others.

Practice the Creed in front of your family, before members of your **chapter**, your chapter advisor, or other adults. There is no substitute for frequent and regular practice with the intention to get better each time. More is said about practicing toward the end of the chapter and in Chapter 7.

STAGE FRIGHT AND NERVOUSNESS

Frequently, when students are asked to speak in front of a group, they say, "Why, I can't get up in front of a group and speak. My knees would be knocking and I would be so nervous that I would make a fool of myself. I'd like to do it, but I just can't."

Adults as well as students fear getting up in front of an audience. Your classmates are probably just as scared as you are. Their knees will be weak and their voices quivering, so you are not alone. Even professional speakers admit they often become quite nervous before a speech or when making a sales presentation to a group.

The ability to communicate to groups of people is a skill that can make a critical difference in your career and in your ability to share information, ideas, experience, and enthusiasm with others. A study conducted by AT&T and Stanford University revealed that the top predictor of professional success and upward mobility is how much you enjoy public speaking and how effective you are at it.[6] The speaker in Figure 7–5, in addition to his many other leadership characteristics, has attained a high position because of his excellent verbal ability.

Figure 7–5 MTSU President Dr. James Walker is an excellent leader. Besides his leadership ability, one of his greatest attributes is his speaking ability. (Courtesy of MTSU Photographic Services.)

A survey conducted on fear asked more than 2,500 Americans to list their greatest fears. The greatest fear (41 percent) was speaking before a group. Ironically, only 18 percent included death as one of their greatest fears. Other studies rank speech-making near the top as one of their greatest fears.[7]

The #1 Fear . . . The #1 Success Predictor

Nervousness Is Normal

Nervousness is a very normal reaction. Sir Winston Churchill compared his prespeech fear to the sense that a nine-inch block of ice was sitting in the pit of his stomach.[8] Even powerful speakers like Abraham Lincoln and Franklin D. Roosevelt were nervous before speaking. Actually, most people tend to be anxious about doing anything important in public. Actors are nervous before a play; politicians are nervous before a campaign speech; athletes are nervous before a big game. The ones who succeed have learned to use their nervousness to their advantage.

Train the butterflies to fly in formation.
—Alessandra and Hunsaker

Although fear never goes away entirely, professionals know you can make the butterflies "fly in formation." You can learn to manage your fear. In other words, it is perfectly normal—even desirable—to be nervous at the start of a speech. The question is, how can you control your nervousness and make it work *for* you rather than *against* you? What happens is this: Your brain and body muscles become supercharged. Your body is carrying out its natural function of preparing you to meet a special situation. Good speakers can channel this energy properly.

The only thing we have to fear is fear itself.
—Franklin Roosevelt

Controlling Your Nervousness

We have now learned that some stage fright will always be with us. However, this is good news. Without some stage fright, we probably would not be at our best. The key is to control your nervousness and not get rid of it completely. The following steps show how.

Step 1

Prepare thoroughly to control stage fright.
Since most stage fright comes from a fear of not succeeding, thorough preparation can guarantee that about 90 percent of your Creed delivery will go smoothly.

Thorough preparation also gives you the right level of confidence once you have started speaking. If you ever miss a line or make a mistake, don't call attention to it. Correcting yourself does not help, and the mistake may go undetected.

Step 2

Relax before you speak to prevent physical tension.
Some speakers feel their neck, stomach, or facial muscles tighten. Remember, these are symptoms rather than causes of stage fright. J. Regis O'Connor suggests the following relaxation techniques to help reduce physical symptoms of stage fright.[9]

Force yourself to yawn widely several times. Fill your lungs with air each time by breathing deeply.

Let your head hang down as far as possible on your chest for several moments. Then slowly rotate it in a full circle, at the same time allowing your eyelids to droop lazily. Let your mouth and lower jaw hang open loosely. Repeat this rolling motion five or six times, very slowly.

Sit in a slumped position in a chair as if you were a rag doll (Figure 7–6). Allow your arms to dangle beside the chair, your head to slump on your chest, and your mouth to hang open. Then tighten all muscles one at a time, starting with your toes and working up your body to your neck. Next, gradually relax each set of muscles, starting at the top and working back down to your toes. Repeat this process several times.

Figure 7–6 The rag-doll technique is a way to practice relaxing before speaking.

Step 3

Realize that audiences tend to be sympathetic to the problem of stage fright.
Most listeners realize what you are going through and have sincere sympathetic understanding. They expect to feel the same way when they are speaking. If you don't do as well as you wish, listeners usually react in a friendly and encouraging fashion.

Step 4

Develop the right attitude.
If you look on stage fright as something positive, you can learn to harness the nervous energy. Controlled stage fright then aids you in becoming a successful speaker. Just before the delivery of the Creed, talk to yourself as follows:

> *What I am feeling are symptoms of stage fright. It is great that my body prepares me for such events. This tense feeling is just what I need to sharpen my thinking and help me reach my full potential once I start speaking.*

Step 5

Concentrate on the Creed.
Quit thinking about yourself and concentrate on the Creed. You have a chance to persuade, entertain, and develop leadership. Don't think "I *have to* give a speech today," but think "I *have a* speech to give today!"

Step 6

Concentrate on your audience.
When you are delivering the Creed, search the faces of your audience, whether it is your classmates or some other group. Look at the interactive listeners while speaking. Their obvious appreciation for your speech will be a great confidence builder for you. Avoid negative listeners. If you concentrate on trying to please them, you will likely disinterest the other 90 percent of your audience.[10] By concentrating on your audience, you will have little time to think about yourself, and your stage fright will be controlled and your confidence will increase.

Step 7

Exhibit a self-confident and pleasant smile.
Smiling relaxes you and the audience more than anything else. For presentations other than the Creed, inject a little humor. Humor has long been used to reduce the tension between the speaker and the audience. Getting a laugh builds confidence rapidly. It assures the speaker that there is little to fear from the listeners.

Step 8

Move around.
While delivering a speech, moving will help you reduce nervous tension and other symptoms of stage fright. Movements that help communicate your message nonverbally are the only kind you should use. Such movements release nervous tension just as effectively as distracting ones, such as twirling a pencil, pacing, and tossing your hair.

Step 9

Handle specific symptoms of nervousness.
Symptoms of stage fright may bother some more than others. Successfully handling symptoms that you find particularly annoying will help build your confidence.

Step 10

Speak as often as you can.
Confidence in speaking is built more by experience than anything else. Practice, practice, practice! The more often you speak publicly, the better your delivery will become.

Establishing Rapport with the Audience

Rapport is the feeling of warmth and harmony between speaker and listener. Before the audience can be persuaded or stimulated, it must have sympathy for the speaker. Whatever happens during the speech, you must make happen. Although rapport is important throughout the speech, it is most important at the beginning.

The following are ways[11] to establish rapport.

- Walk to the platform or **lectern** with quiet assurance. This projects confidence and puts the audience at ease.
- Stand quietly before beginning. A nervous speaker tends to begin without a moment's hesitation and talk too rapidly. A pause before beginning gives the audience a chance to get quiet and the speaker a chance to adjust to the situation. It also draws attention and makes the audience feel something important is about to happen.
- Look directly at the audience. Failure to look directly at the audience or taking only brief glances in that direction leaves the impression that you are remote and detached. It may be necessary to glance at notes occasionally, but when beginning to talk, you should look directly at the audience.
- Appear friendly and pleasant. If you are friendly, the audience tends to respond with equal warmth. The audience reflects the speaker. The speaker who has a genuine interest in people has an advantage. Friendly casualness and **spontaneity** in the right amounts are disarming and inviting.
- Dress appropriately. We live in a casual age, but casualness need not deteriorate into sloppiness. A company president said that he has found that people who feel themselves to be above good grooming are also above the patience and attention that the work requires.

For an official FFA Creed contest, wear the FFA official dress. Female members are to wear a black skirt, white blouse with official FFA blue scarf, black shoes, and official jacket zipped to the top. Official dress of male members is black slacks, white shirt, official FFA tie, black shoes, black socks, and official jacket zipped to the top.[12] The FFA members in Figure 7–7 are in official dress.

Figure 7–7 These FFA members are in official dress.

Don't forget the proper uses of the FFA jacket that relate to official dress at speaking contests.

- The jacket should be worn on official occasions with the zipper fastened to the top. The collar should be turned down and the cuffs buttoned.
- School letters and insignia of other organizations should not be attached to or worn on the jacket.
- No more than three medals should be worn on the jacket. These should represent the highest degree earned, the highest office held, and the highest award earned by the member.[13]

Delivery of the Creed

The delivery of the Creed in a contest is based on a 100-point scale. I realize that most of you are not saying the Creed to enter a contest, but presenting the material in this section from a contest perspective helps us focus on what top-quality speakers strive to master. Figure 7–8 shows an

Creed Speaking CDE Score Card

Judge's Name: ____________________ Section: ____________

Flight: ____________________ Room: ____________

	Possible Points	One	Two	Three	Four	Five	Six	Seven	Eight
Voice									
• Quality	40								
• Pitch	40								
• Force	40								
• Articulation	40								
• Pronunciation	40								
Voice Total:	200								
Stage Presence									
• Personal Appearance	50								
• Poise & Posture	40								
• Attitude	35								
• Confidence	35								
• Personality & ease before audience	40								
Stage Presence Total:	200								
Power of Expression									
• Fluency	30								
• Emphasis	35								
• Directness	35								
• Sincerity	35								
• Communicative Ability	35								
• Conveyance of thought and meaning	30								
Power of Exp. Total:	200								
General Effect									
• Extent to which speech was understandable, convincing, pleasing and held attention.	100								
General Effect Total:	100								
Response to Questions*									
• Ability to answer satisfactorily the questions asked by the judges indicating familiarity with the subject.	300								
Response Total:	300								
Gross Total Points:	1000								
Less Time Deduction**:									
Less Accuracy Deduction***:									
Net Total Points									
Contestant Ranking:									

* Judges will select three questions per round. Questions will be selected from a data bank of questions that were developed by the Creed Speaking Career Development Event Committee.

** – 1 point per second over, determined by the timekeepers

*** – 20 points per word, determined from by the accuracy judges.

Figure 7–8 The National FFA Creed Speaking Career Development Event Score Card. The score card includes five areas; except for accuracy, the other parts are basically the same as the public speaking contest. (Courtesy of the National FFA Organization.)

example of the National FFA Creed Speaking Career Development Event Score Card.

Voice

The voice is the sound effect of personality. A good personality without an effective, pleasant voice can be compared to a good movie without sound. Just as we are judged by appearance, so are we judged by voice.

Quality Quality refers to tone of voice. Voice is often considered the mirror of the mind. A friendly tone of voice makes conversation a pleasure. **Tone** refers to the modulations of the voice. No two people have the same tone, and there are about four octaves difference between male and female voices.

Tone of voice greatly affects the importance a listener attaches to what is said. In fact, tone of voice can cause the listener to place very little importance on an important statement. The speaker projects mental and emotional attitudes to the listener by the tone of voice. Good posture is important to the tone of voice because voice tone and control depend on a constant air flow through the lungs. Obviously, the tone of voice should be varied during a speech or conversation.

Pitch Pitch is the height or depth of the tone. Most people speak with a medium or low pitch. Pitch is a determining factor in voice quality. Listeners become uncomfortable when the pitch is extremely low or high.

Changes in pitch are known as **inflections**. They give your voice luster, warmth, and vitality. It is the inflection of your voice that reveals whether you are asking a question or making a statement, whether you are being sincere or sarcastic. Your inflections can also make you sound happy or sad, angry or pleased, dynamic or listless, tense or relaxed, interested or bored.

Some people fall into a monotone unconsciously. You can guard against this problem by recording your speeches as you practice them. If all your sentences end on the same inflection—either upward or downward—work on varying your pitch patterns to fit the meaning of your words. As with breaking any other habit, this may seem awkward at first; but it is guaranteed to make you a better speaker.[14]

Articulation **Articulation** refers to the way the tongue, teeth, palate, and lips are moved and used to produce the crisp, clear sounds of good speech. Most people are capable of producing vowel and consonant sounds clearly, but they fall into lazy habits. They become unwilling to exert that extra bit of effort needed to produce clear speech, especially during conversation. Unfortunately, when such bad habits are carried over into public speaking, an audience may show little respect for the speaker who sounds sloppy.[15]

Many speakers suffer from minor articulation problems such as adding a sound to a word (ath*a*lete for athlete), leaving out a sound (li*b*ary for li*br*ary), transposing sounds (re*va*lent for rel*e*vant), and distorting sounds (tru*f* for tru*th*). Although some people have consistent articulation problems that require speech therapy (such as consistently substituting *th* for *s*), most of us are guilty of carelessness that is easily corrected. When practicing the Creed, therefore, concentrate on moving your tongue, lips, and lower jaw vigorously enough to produce crisp, clear sounds. Be especially careful of the consonants that can be slurred or dragged. Consider the following examples.

Word(s)	Misarticulation
going to	*gonna*
did you	*didja*
specific	*pacific*
ought to	*otta*
will you	*wilya*

Practice the exercises at the end of the chapter. They are the ultimate articulation testers.

Pronunciation **Pronunciation** is the form and accent a speaker gives to the syllables of a word. Pronunciation can also play a part in determining the degree of respect given to a speaker.

What makes a certain pronunciation correct and another incorrect is usage. Once enough people agree to pronounce a word in a certain way, that becomes the correct way. If you are unsure of how to pronounce a word you are going to use, look it up in a dictionary.[16]

Stephen Lucas observed that there are, of course, regional and ethnic dialects that affect pronunciation.

> ***In New York people may get "idears" about "dee-ah" friends. In Alabama mothers tell their children to clean up their rooms "rat now." . . . Such dialects are fine as long as your audience does not find them objectionable or confusing. Otherwise, you should try to avoid them, for they may cause listeners to make negative judgments about your personality, intelligence, competence, and integrity.***[17]

Because constant mispronunciation suggests a person is ignorant or careless (or both) you will want to try to correct mistakes you make. Consider the following.

Word	Common Error	Correct Pronunciation
genuine	*gen-u-wine*	*gen-u-win*
theater	*thee-ate-er*	*thee-a-ter*
athlete	*ath-a-lete*	*ath-lete*
family	*fam-ly*	*fam-a-ly*
particular	*par-tik-ler*	*par-tik-yu-ler*

Force **Force** refers to volume and the variety in volume for the purpose of emphasis. A speaker who is constantly loud makes everything sound important, and eventually, nothing sounds important. One who always speaks softly is difficult to hear. The audience eventually gives up straining to hear.

Important or key ideas should be spoken with sufficient force so those in the back row can hear them easily. Transitions to new sections of a speech, the start of the conclusion, and the parts in which you wish to be dramatic with a kind of "stage whisper" may be spoken more softly, but should be audible to listeners in the rear of the room.

Stage Presence

Stage presence includes personal appearance, poise and body posture, attitude, confidence, personality, and ease before an audience.

Personal appearance Personal appearance is a combination of grooming and dress. Wear official FFA dress, as discussed earlier. Grooming includes a professional image. Styles vary from year to year but whatever the style, there are those that tend to rebel against the norm. Make sure your grooming is acceptable to your advisor and the majority of the chapter members.

Personal appearance plays an important role in speech making. Listeners always see you before they hear you. Just as you adapt your language to the audience and the occasion, so should you dress and groom appropriately. No matter what the speaking situation, try to project favorable first impressions, which are likely to make listeners more receptive to what you say. The members in Figure 7–9 would make favorable impressions as they delivered a speech. The student in Figure 7–10 would not make as favorable an impression.

Figure 7–9 These FFA members have delivered many speeches. They always make favorable impressions, partly due to their appearance.

Figure 7–10 This young man would not make a positive impression with his appearance.

Poise and Body Posture Poise and body posture show self-confidence. When one thinks of poise, one thinks of an NFL quarterback standing in the pocket, not nervous as he delivers his pass on target as six big 300-pound linemen are charging toward him. A speaker should exhibit the same self-confident poise as he or she stands to deliver a speech.

What should I do with my arms and hands? No simple answer can be given. However, basic hand and arm positions must be both natural to you and suited to the audience and the total speaking situation. Consider the following positions.

One or both arms hanging naturally at your side

One or both hands resting on (not grasping) the speaker's stand

One or both hands held in front of your stomach

Fingertips touching with hands held in front of your stomach

These positions look quite natural for some speakers but not for others. Learn from the reactions of your teacher and classmates which work best for you.

Attitude Attitude of the speaker can affect the attitude of the audience. If you exhibit a positive attitude, your audience will react the same. Act as if there is not another place that you would rather be or anything else that you would rather be doing at this moment. Positiveness is contagious.

Confidence Confidence is attained by knowledge and preparation. You nonverbally exhibit confidence by good eye contact, a relaxed appearance, good body posture, and a facial expression that commands respect and attention.

Personality Personality involves being enthusiastic, lively, expressive, cheerful, inspiring, charming, and energetic. When you present your speech, be enthusiastic about what you have to say. Enthusiasm is by far the most important element of effective speaking. A speaker who looks and sounds enthusiastic will be listened to, and that speaker's ideas will be remembered.

Ease Before Audience Ease before the audience assures the audience that you are calm and everything is under control. Stephen Lucas provides a good description of ease before the audience.[18]

> ***As you rise to speak, try to appear calm, poised, and confident, despite the butterflies in your stomach. When you reach the lectern, don't lean on it, and don't rush into your speech. Give***

yourself time to get set. Arrange your notes just the way you want them. Stand quietly as you wait to make sure the audience is paying attention. Establish eye contact with your listeners. Then—and only then—should you start to talk.

Power of Expression

Power of expression includes fluency, emphasis, directness, sincerity, communicative ability, and the conveyance of thought and meaning.

Fluency Fluency is the smooth and easy flow of the speech. Do not be too fast or too slow. End a sentence, pause, then start the next sentence. Although varieties of loudness and speech are needed for emphasis, a smooth and easy-flowing speech can still convey meaning.

Emphasis Earlier, key words in the FFA Creed were discussed. These key words should receive emphasis. Say words to express the feeling they convey.

Directness Directness means using language that adapts to the needs, interests, knowledge, and attitudes of the audience. In most situations, the more direct you can make your language, the more appropriate it will be. Although you cannot do this when delivering the Creed, to increase directness, you can use personal pronouns, ask questions phrased to stimulate a mental rather than verbal response (theoretical questions), share common experiences, and create stories to illustrate a point (hypothetically).

Sincerity Sincerity is being genuine, honest, and straightforward as you present your speech. Your speech or delivery of the Creed must be genuine to make a lasting positive impression. You must believe what you say. Politicians are often doubted because the audience cannot figure out if they are sincere or only trying to get elected. The slightest hint of insincerity turns many listeners against a speaker instantly.

Communicative Ability Communicative ability is conveying your thoughts and feelings to the audience so they can be understood. Eye contact lets us communicate nonverbally, conveying truthfulness, intelligence, attitude, and feelings.

The quickest way to establish a communicative bond with your listeners is to look them in the eye, personally and pleasantly. Avoiding eye contact is one of the surest ways to lose them. At best, speakers who refuse to establish eye contact are perceived as ill at ease. At worst, they are perceived as insincere or dishonest. You should look at the audience 80 to 90 percent of the time you are talking.[19]

Conveyance of Thought and Meaning Conveyance of thought and meaning is a transfer of your feelings and perceptions to the audience. Perception is reality. When you speak, it is not what you think, it is what you convey that the audience judges.

Thoughts and meaning are expressed largely by facial expression. Your facial muscles can be arranged in an almost infinite variety of positions to express a wide range of emotions. They can show the degree of interest you have in your speech and give certain information about your personality.

Audiences respond positively to honest and sincere expressions that reflect your thoughts and feelings. Think actively about what you are saying, and your face probably will respond accordingly. If not, practice in front of a mirror to improve your facial expressions.

General Effect

General effect includes the extent the speech is interesting, understandable, convincing, pleasing, and holds attention. If you combine the preceding delivery characteristics according to your preferences, your speech will be interesting, understandable, convincing, pleasing, and hold attention. It will also be specific, logical, organized, motivating, credible, and persuasive.

There is also an **intangible** trait that occasionally contributes to general effect. It may be a quality that only you have that can be used to your advantage. If it is good or it works, use it.

Practical Steps and Fun Techniques for Saying the FFA Creed

A proper attitude toward speaking can be developed if you make it fun. Following are some fun techniques you might want to try. They may help you develop your confidence in reciting.

Rolled-Up Pants Legs

When the author starts teaching the Creed, he rolls up his pants leg nearly to his knees. This amuses the students and makes them relax. It sets the tone that they are going to have fun while learning. It begins to lessen stage fright because they know they cannot look more foolish than their teacher when they get in front of the class.

Why rolled-up pants legs? You have to be around a barnyard during a spring thaw to fully appreciate it. For those that have not, it gets real deep. It also gets real deep when we start speaking. We have to roll up our pants, wade in, and have fun (Figure 7–11).

Figure 7–11 This person is using the rolled-up pants legs technique to relax the students and make speaking fun.

Eggshell Analogy

Students tend to be bound up in a shell when they start speaking: They are tight, nervous, and restricted. Visualize breaking out of a shell as if you were a newborn chick. Move your arms freely and stretch. When your teacher permits, take turns saying, "I believe" in a fun and dramatic fashion.

Toothpaste Tube

Have you ever thought that you were out of toothpaste and you set the empty tube aside? You forget to get a new tube so you have to try to squeeze some more out of the old tube. To your surprise, you get more toothpaste to come out. The next day, you again forget to get a new tube. You go back again to the old tube and you squeeze a little harder. Again, you get toothpaste. This can go on for a week. Finally, you tear open the tube and scrape the toothbrush on the inside of the tube to get that last bit. Don't forget the top.

You are that way with your intelligence and your speaking ability. There is more within you. Just when you have done the best you think you can do, you can do better. Every one of you can learn to speak better than you ever thought possible if you keep squeezing. Just when you think it is all gone, there is still more "toothpaste" or speaking ability within you.

Football Field/Gymnasium Practice

Some of us speak too softly and shyly when we first start speaking. To solve this, go to the football field or gym (Figure 7–12). Stand beneath one football goalpost or basketball goal while the teacher and the rest of the class stand beneath the other. Say one or more paragraphs of the Creed so they can clearly hear you.

Public speaking is not a shouting match. However, once you return to the classroom, you will be amazed at how easy it is to speak louder. It is then easy to tone down and clean up the rough edges.

Figure 7–13 This advisor and student are practicing the face-to-face imitation technique.

Model Speaker

Sometimes learning is best done by following a model, in this case, a model speaker. Maybe your teacher could show you videotapes of some excellent Creed speakers. The National FFA Supply Service sells videos of the National Public Speaking Contest. Learn from the best.

Face-to-Face Imitation

Another good technique for learning how to deliver the Creed is to stand opposite your teacher or someone else with a desk or table between you.

Figure 7–12 Practicing your delivery on the football field helps you learn to speak louder.

Put your fingers on the table, relax, and look the person straight in the eye (Figure 7–13). Your teacher or a more advanced speaker says a line (sentence) using good voice, stage presence, power of expression, and general effect. You immediately follow, imitating him or her. Do not move to the next line until you have sufficiently practiced the first line.

The idea is not to be a **clone** of your teacher but to develop a speaking style. Disc jockeys, ringmasters, play-by-play athletic announcers, and newscasters each have their own styles. Creed speakers and FFA public speakers also have a certain style.

Note: The next steps are for those who are trying to win the "big one." However, you may wish to practice just for fun.

Hayloft

Most of you do not have a hayloft, but you can create your own "hayloft" out of a park, field, warehouse, or whatever is available. Get in the hayloft and have fun speaking. Use exaggerated gestures, body positions, voice variations, and facial expressions. After being silly, get serious and pretend that you are speaking to 22,000 people at the National FFA Convention.

Balcony

Practicing voice projection is best done from a high place, such as a balcony or a second-story window. It gives you a sense of power and command of the situation. Practice the Creed several times. It is both fun and challenging, but remember safety. The only drawback is the funny looks you may get when cars pass by.

Turkey Test

The author used to have turkeys that offered an evaluation on voice pitch, tone, and variation. (Really!) The male turkeys (gobblers) would gobble when they heard a loud or high-pitched noise. To pass the turkey test, the Creed candidate would have to emphasize certain words loudly enough to make the turkeys gobble. The louder and more expressive the candidate, the more the turkeys would strut and gobble. There are modern electronic devices that produce a sound or move a gauge to measure voice force, but they are not as much fun as turkeys.

It is the same with entertainers. The better they perform, the better the response or applause. If you are really good, the audience will give you a standing ovation.

Tears

You know you have mastered the Creed delivery when a judge pulls out a handkerchief and wipes away the tears. The perfect delivery of the Creed may project so much feeling that it brings tears from the audience.

National FFA Creed Speaking CDE

The following are the rules for the National FFA Creed Speaking Career Development Event (CDE). Most states follow the same rules as the national contest.

Event Rules

- The National FFA Creed Speaking CDE is limited to one participant per state, who must qualify in grades 7, 8, or 9 and must compete at the next national convention following the state qualifying round.
- It is highly recommended that participants be in official FFA dress in each event.
- The National FFA Creed Speaking CDE follows the general rules and policies for all National FFA CDEs.
- The National FFA Officers and National Board of Directors are in charge of the event.
- Three to six competent and impartial persons are selected to judge the event. At least one judge should have an agricultural/FFA background. Each state with a speaker provides a judge for the national event.

Event Format

- The event includes oral presentation as well as answering questions directly related to the FFA Creed. Each contestant is asked three questions per round, with a five-minute limit. The questions used change as the contestant progresses to semifinal and final rounds of competition. The questions are formulated annually by the Creed Speaking CDE committee and do not include two-part questions. Sample questions are not available prior to the event.
- Members present the FFA Creed from the current year's official FFA manual.
- The event is a timed activity, with four minutes for presentation. After four minutes, one point is deducted for every second over the set time.
- The national event is conducted in three rounds: preliminary (consisting of six to eight speakers per section), semifinals (two sections of eight speakers each), and finals (four participants). No ranking is given, except for the final four.

- Event officials randomly determine the speaking order. The program chairperson introduces each participant by contestant number and in order of the drawing. No props are to be used. Applause is withheld until all participants have spoken.
- Each contestant must recite the FFA Creed from memory. Each contestant begins the presentation by stating "The FFA Creed by E. M. Tiffany." Each contestant ends the presentation with the statement ". . . that inspiring task. Thank you."
- Contestants are held in isolation until their presentation. Contestants are not allowed to have contact with any outside persons.
- At the time of the event, the judges sit in different sections of the room in which the event is held. They score each participant on the delivery of the Creed, using the score sheet provided.
- Two timekeepers record the time each participant takes to deliver his or her speech. Timekeepers are seated together.
- The content accuracy judges record the number of recitation errors during delivery. The accuracy judges are seated together.
- When all participants have finished speaking, each judge totals the score of each speaker. The timekeepers' and accuracy judges' records are used in computing the final score for each participant. The judges' score sheets are then submitted to event officials to determine final ratings of participants.
- Participants shall be ranked in numerical order on the basis of the final score to be determined by each judge without consultation. The judges' ranking of each participant are then added, and the winner is the participant whose total ranking is the lowest.

Tiebreakers

Ties are broken based on the greatest number of low ranks. The participant with the greatest number of low ranks is declared the winner. If a tie still exists, the event superintendent ranks the participant's response to questions. The participant with the greatest number of low ranks from the response to questions is declared the winner. If a tie still exists, the participant's raw scores are totaled. The participant with the greatest total of raw points is declared the winner (refer to Figure 7–8).

SUMMARY

Learning and reciting the Creed can be fun. Reciting the Creed is one of the first things done in an agricultural education program in the FFA in pursuit of leadership. It is hard to lead if you can't communicate, which includes speaking before groups. Principles learned as you present the Creed also apply to public speaking, extemporaneous speaking, and parliamentary procedure.

Basic principles of speaking include using reason and feeling, using an effective speaking style, observing the audience, and persuading the audience. Every speaker should master delivery by using the correct volume, tempo, diction, and enthusiasm.

Polish your skills by using a tape recorder, video recorder, practicing in front of a mirror, and practicing in front of friends. Practice helps in overcoming stage fright and nervousness. You can also practice and learn to establish rapport with your audience. As you practice your delivery, work on your voice, stage presence, power of expression, and general effect.

To control your nervousness, prepare thoroughly, relax before you speak, realize audiences tend to be sympathetic, develop the right attitude, concentrate on the Creed, concentrate on your audience, exhibit a self-confident smile, move naturally, and speak as often as you can.

Establish rapport with the audience. The following are ways to establish rapport: Walk to the platform or lectern with quiet assurance; stand quietly before beginning; look directly at the audience; appear friendly and pleasant; and dress appropriately.

Remember that voice includes quality, pitch, articulation, pronunciation, and force. Stage presence includes personal appearance, poise and body posture, attitude, confidence, personality, and ease before an audience. Power of expression includes fluency, directness, sincerity, communicative ability, and conveyance of thought and meaning. General effect includes the extent the speech is interesting, understandable, convincing, pleasing, and holds attention.

Some practical and fun ways for saying the Creed include rolling up your pants legs; practicing on a football field, gym, or hayloft; watching a model speaker; face-to-face imitation; and passing the "turkey test." Finally, you know that you have mastered Creed delivery if you say it with such feeling and emotion that you bring tears from the audience.

END-OF-CHAPTER EXERCISES

Review Questions

1. Define the Terms to Know.
2. What are the advantages of participating in reciting the FFA Creed?
3. What are five basic principles of speaking?
4. The old saying, "Practice makes perfect," is often used. Do you agree? Why or why not?
5. What did studies by AT&T and Stanford University reveal as the top predictor of professional success and upward mobility (job promotions)?
6. Explain the statement, "Train the butterflies to fly in formation."
7. List 10 ways to help control your nervousness.
8. What are five ways to establish rapport with the audience?
9. What is the official FFA dress for males? Females?
10. What is meant by *general effect*?
11. What are 11 practical and fun things to do to make saying the Creed more effective and enjoyable?

Fill-in-the-Blank

1. By participating, you cannot ________________ . You can only ________________ when you quit or don't even try.
2. The number one fear of most adults (even above death) is ________________ .
3. No more than ________________ medals should be worn on the FFA official jacket.
4. ________________ ________________ is a combination of grooming and dress.
5. ________________ is attained by knowledge and preparation.
6. ________________ is the smooth and easy flow of speech.
7. ________________ means using language that adapts to the needs, interest, knowledge, and attitudes of the audience.
8. ________________ is being genuine, real, honest, and straightforward.

Matching

_____ 1. Refers to the tone of voice
_____ 2. Height or depth of voice
_____ 3. Shaping of speech sounds into recognizable oral symbols
_____ 4. Refers to volume for the purpose of emphasis
_____ 5. Self-confident and calm
_____ 6. Enthusiastic, lively, expressive, cheerful, enthusiastic
_____ 7. Form and accent a speaker gives to the syllables of a word

A. force
B. pitch
C. pronunciation
D. quality
E. poise
F. personality
G. articulation

Activities

1. Write the first paragraph of the FFA Creed. Cut out two or three pictures that convey the meaning, and tape or paste them beneath the paragraph. Write the other four paragraphs following the same procedure.
2. Rewrite Figure 7–2 in your notebook. Fill in the key words.
3. Write paragraphs two through five of the Creed and underline the key words.
4. Practice saying the Creed in front of a mirror. Write a short essay about what you felt and learned as you practiced.
5. Demonstrate three relaxation techniques to help reduce physical symptoms of stage fright.
6. Stand behind a lectern. Just before delivery of the Creed, talk to the audience in a loud and dramatic fashion. Say:

   What I am feeling are symptoms of stage fright. It is great that my body prepares me for such events. This tense feeling is just what I need to sharpen my thinking and help me reach my full potential once I start speaking.

7. Practice articulating the following phrases:
   The big black bug bit the big black bear, and the big black bear bled.
   Peter Piper picked a peck of pickled peppers. A peck of peppers, Peter Piper picked.
   I would if I could, couldn't, how could I; you couldn't without your could, could you? I couldn't, could you?
   Sid said to tell him that Benny hid the penny many years ago.
   Fetch me the finest French-fried freshest fish that Finney fries.
   Three gray geese in the green grazing; gray were the geese and green was the grazing.
   Shy Sarah saw six Swiss wristwatches.
   One year we had a Christmas brunch with Merry Christmas mush to munch. But I don't think you'd care for such. We didn't like to munch mush much.
8. Compile a list of positive delivery characteristics exhibited by other members of the class.
9. Practice the Creed on the football field or in the gym. Half the class can stand between one goalpost (basketball goal) and the other half of the class can stand beneath the other goalpost. There are several ways to practice. To add variety, you may wish to alternate paragraphs between those under each goalpost.

10. Practice saying the Creed by following any of the other practical and fun techniques described in this chapter. Write a report on your experiences.
11. Recite at least one paragraph of the FFA Creed in front of the class in official dress. Use as many of the techniques learned in this chapter as you can.
12. **Take it to the Net.**

    Go to the National FFA Web site. Browse the Web site and all the things going on in the FFA. Try to find the FFA Creed and tips on reciting it.

    Web site

    <http://www.ffa.org>

Notes

1. The Creed was written by E. M. Tiffany, and adopted at the 3rd National Convention of the FFA. It was revised at the 38th Convention and the 63rd Convention.
2. D. B. Jameson, *Leadership Handbook* (New Castle, PA: LEAD, 1978), p. 210.
3. Ibid., p. 211.
4. Adopted from Dr. Joe Townsend's "Leadership" class, Texas A & M University, College Station, TX.
5. Jameson, *Leadership Handbook*, pg. 53.
6. T. Alessandra and P. Hunsaker, *Communicating at Work* (New York: Simon & Schuster, 1993), p. 169.
7. "What Are Americans Afraid of?" *The Braskin Report*, July, 1973, p. 53.
8. Alessandra and Hunsaker, *Communicating at Work*, p. 170.
9. J. R. O' Connor, *Speech: Exploring Communication* (Upper Saddle River, NJ: Prentice-Hall, Inc., 1998), p. 135.
10. Ibid., p. 137.
11. *Personal Development Skills in Agriculture* (College Station, TX: Instructional Materials Service, Texas A & M University, 1989).
12. *Official FFA Manual* (Alexandria, VA: National FFA Supply Service, 2001).
13. Ibid.
14. S. Lucas, *The Art of Public Speaking*, Second Edition (New York: Random House, 1986), p. 231.
15. O'Connor, *Speech: Exploring Communication*, p. 212.
16. Ibid., p. 12.
17. Lucas, *The Art of Public Speaking*, p. 234.
18. Ibid., p. 239.
19. Ibid.

Prepared Speaking (FFA Public Speaking)

CHAPTER

Objectives

After completing this chapter, the student should be able to:

- Plan a speech
- Analyze the audience
- Select a topic for the speech
- Gather information for the speech
- Record the ideas in the speech
- Prepare an outline for the speech
- Write a speech
- Practice the speech
- Present the speech
- Answer questions about the speech
- Evaluate speeches
- Speak to special groups

Terms to Know

empathize
brainstorming
quotations
statistics
simile
metaphor
personification
hyperbole
irony
ethical
defamation of character
introduction
body
pangs
ardently
salutation
deliberate

Speeches are given to inform, to persuade, or to integrate the members of the audience (as in pep talks, speeches of welcome, and introductions). People also listen for the same reasons.

Through persistence you can achieve your speaking goal. The member in Figure 8–1 is practicing speaking before his friends. They will make suggestions, and he will practice again.

Speaking skills increase a person's effectiveness and also influence the decisions of others. An effective communicator should provide people with information and should influence others, changing their attitudes or behaviors. Speaking in public is an art form nearly as old as humanity itself. Speaking skills are a major factor in selecting the president of the United States, chairperson of the board, and other leaders. More often than not, a speech gives a clear statement of the speaker's beliefs and actions.[1] Effective public speaking is influence. Influence is leadership.

PLANNING A SPEECH

As a speech is planned, consider the purpose, the audience, and the occasion. Think like the audience. If you can **empathize** with the audience, you will be able to plan a better speech. The speaker in Figure 8–2 is good at empathizing with the audience.

Analyze the Audience

Do you select the topic first or do you analyze the audience? It depends on the situation. More often than not, however, the audience should be analyzed before the topic is selected. Can you imagine the featured speaker at the Young Republicans Banquet speaking on the virtues of being a Democrat, or vice versa?

The audience is rather obvious and predictable for an FFA Public Speaking Contest. It consists of agricultural education students, agricultural education teachers, and those interested

Figure 8–1 Practice does not make perfect, but perfect practice does make perfect. Practicing before friends with evaluations aids in the pursuit of perfection. Classroom instruction prepares students for theory and after school work based learning opportunities as important parts of Supervised Agricultural Experience Programs.

Figure 8–2 Zig Ziglar is showing how to get in touch with the audience as he delivers a speech at the National FFA Convention. (Courtesy of *FFA New Horizons* Magazine.)

in the field of production agriculture, agribusiness, or agriscience. From this analysis, a topic can then be selected.

The rules of the FFA Prepared Public Speaking Contest state, "Contestants may choose any current subject for their speeches which is of an agricultural character (nature), and which is of general interest to the public." Specifically, "Official judges of any FFA Public Speaking Contest shall disqualify a contestant if he/she speaks on a nonagricultural subject." As you can see, analysis of the audience is predetermined for the FFA Public Speaking Contest.

Analyze the audience by finding out as much information as possible concerning the audience. In some instances, this may be impossible until minutes before the actual presentation. It may be helpful to know the number of people in the group, their ages, their interests, their knowledge of your topic, whether it is a formal or informal group, their occupations, their educational levels, sensitive subjects, the setting, the time, the place on the program, room size, seating arrangement, whether speakers stand, microphone availability, and special occasions. For example, if you are speaking at a banquet to a group of agricultural education teachers, you might want to allude to how their jobs have an effect on the food that was consumed during the meal.

Be Audience Centered

Good speakers are audience centered. They know the primary purpose of speech preparation is not to display their knowledge, demonstrate their superiority, or express their anger. The primary purpose should be to gain a desired response or please the listener. You need not waiver from your beliefs to get a favorable response from your audience. You can remain honest to yourself while adapting your speech to the needs of a particular audience. When analyzing your audience, keep three questions in mind:

- To whom am I speaking?
- What do I want them to know, believe, or do as a result of my speech?
- What is the most effective way of composing or presenting my speech to accomplish that aim?[2]

The answers to these questions influence every decision you make along the way: selecting a topic, determining a specific purpose, settling on your main body and supporting materials, organizing the speech, and, finally, delivering the speech.

SELECT A TOPIC

First, choose a topic that interests you. Next, choose a topic in which you are knowledgeable or have an interest in becoming knowledgeable. Third, pick a topic of interest to your audience. If possible, choose an area in which you are experienced, have convictions, and are accomplished. If you feel strongly about your topic, this feeling will be conveyed in your delivery. If you have knowledge and experience, you will do a better job in responding to questions.

When searching for a topic for an FFA speech, you may wish to start with five general agricultural areas: agriscience and technology, agribusiness, agrimarketing, international agricultural relations, and agricultural communications. List topics that interest you in one area. Quickly jot down every word or phrase you know relating to that topic. Spend no more than one or two minutes on each topic. From the compiled list, select the most interesting topic. If necessary, compile a new list for this topic if it still seems too general. This process is called **brainstorming**. Brainstorming is an uncritical, nonvaluative process of generating ideas, much like the word-association process. For example, one student, while brainstorming, wrote down Willie Nelson.

Willie Nelson made her think of Waylon Jennings. Jennings reminded her of cowboys. Cowboys suggest the Wild West. The Wild West suggests horses. Horses reminded her of the Pony Express. The Pony Express was an early form of mail service. Suddenly, this student remembered a magazine article she had read comparing the U.S. postal service to mail service around the world. The idea clicked

Figure 8–3 Selecting a speech topic can be a challenge for some students. These students are brainstorming and associating words to come up with a topic.

in her mind. After considerable research, she developed an excellent speech entitled "The U.S. Postal Service: Not Perfect, but Still First Class."[3]

That is a long way from Willie Nelson. If you started out free-associating from Willie Nelson, you would probably end up somewhere completely different. This is what brainstorming is all about. The students in Figure 8–3 are brainstorming and associating words to come up with speech topics.

Figure 8–4 lists ideas for topics for speeches in five areas: agriscience and technology, agribusiness, agrimarketing, international agricultural relations, and agricultural communications. Whatever topic you choose, nearly every speech will require some additional research before you are

Agriscience and Technology	Agribusiness	Agrimarketing	International Agricultural Relations	Agricultural Communications
Genetic Engineering	What Is Agribusiness?	The Role of the Internet in Agrimarketing	Free Trade versus Protectionism	Food Safety in Agriculture
Cloning	How Big Is Agribusiness?	The Role of Government in Marketing Agricultural Products	The Value of Exporting to the Agricultural Economy	Importance of Positive Media in the Agriculture Industry
Biotechnology in Agriculture	The Impact of Vertical Integration in the Agricultural Industry	Market Alternatives for the Production Agriculturalist	NAFTA—Friend or Foe of the American Agricultural Industry?	Why Does the Production Agri-culturalist Need to Communicate?
Use of Global Positioning Satellites	Individual versus Cooperative Marketing	Creating a Market for Your Agricultural Product	GATT—Friend or Foe of the American Agriculture Industry?	What Does the General Public Need to Know about Agriculture?
Alternative Fuels	Why Are Agribusinesses Increasing While the Number of Production Agriculturalists Are Decreasing?	Marketing in a Global Economy	The Impact of Weather in Importing Countries on U.S. Agricultural Prices	Agriculture: Friend or Foe of the Environment?

Figure 8–4 Selecting a topic for an FFA speech can be challenging. The topics above may help you think of other ideas for a speech.

ready to deliver it. Give yourself plenty of time in advance of your speaking date to allow for this additional preparation.

GATHER INFORMATION

Benjamin Franklin once said, "An empty bag cannot stand upright." Gathering information is of the utmost importance. Without solid material, your speech will fold like Franklin's empty bag. Start your research by checking your personal books and magazines. *Farm Journal* and *Progressive Farmer* are excellent sources of general information. Breed associations and trade and business magazines also are excellent sources.

Consult not only articles and books but organizations and experts. Most research for talks is done best in a good library, with the help of good librarians. Library research should prove effective for you even if you know very little about using a library card catalog or reference works like *the Reader's Guide to Periodical Literature*. Librarians can show you where to look for what you need. They can also be very helpful even if you happen to know a great deal about using libraries.[4]

You need not limit yourself to library sources. You can also do research by requesting information from organizations active in the field about which you are talking. Men and women who are experts in their fields are often glad to furnish comments or opinions when such information is tactfully requested by letter or by phone.

Government officials and their aides often provide information on issues of public importance. An example is the U.S. Department of Agriculture (USDA). They have experiment stations throughout the United States. USDA produces a monthly magazine of related research and publish hundreds of bulletins, which are available at little or no cost. A list of these can be obtained from the U.S. Government Printing Office, Washington, DC 20402. Information is on file at the National Agriculture Library in Beltsville, Maryland. With the appropriate computer equipment, such as a modem and communication software, you can do a library search from your school library or classroom. Your state extension office is also an excellent source of materials for your speech. Local extension offices may also have the information you need. If you are in doubt about which government agency is responsible for a particular subject, try calling or writing your congressional representative. His or her staff can point you in the right direction.

If the subject is controversial, you can be pretty sure there is an organization for it or against it—and probably both. If the issue is exploring for oil in the national parks, it is obvious that the oil companies are for it while environmental groups like the Sierra Club are against it. What about tobacco products? The American Tobacco Institute is for the use of tobacco; the American Cancer Society opposes its use. The *Encyclopedia of Associations*, available in most libraries, lists special interest groups, along with their mailing addresses. Most of these groups offer free pamphlets and literature.

Speakers can find quotations to support their ideas in sources such as *Bartlett's Familiar Quotations*, *Brewer's Dictionary of Phrase and Fable*, *Granger's Index to Poetry*, and the *Oxford Dictionary of Quotations*. Don't forget the obvious. Most libraries have a recent edition of *Encyclopedia Britannica*, *Encyclopedia Americana*, or *World Book Encyclopedia*. Many topics are already outlined.

There are several other sources of information. Some of them follow:

- Livestock breed associations
  Examples: American Quarter Horse Association
  American Hereford Association
  Holstein Fresian Association
- Interviews
  Examples: President of a civic club
  Farmer or rancher
  Agribusiness person
  Agricultural scientist
- Education facilities
  Examples: State universities

Local colleges
Private research groups

- Newspapers
  Examples: *The New York Times*
  Wall Street Journal
  Local newspapers
- Almanacs and atlases
  Examples: *World Almanac*
  Rand McNally Atlas
- Surveying
  Examples: Public opinion
  Suggestions for improvement
- Biographical sources
  Examples: *Who's Who in America*
  Dictionary of National Biography
  Dictionary of American Biography
- Microfilm indexes
  Examples: *COM (Computer Output Microfilm)*
  Magazine Index
  Business Index
- Computer databases
  Examples: ERIC (Educational Resources Information Center)
  Agridata Network
  AG STAT
- World Wide Web

Gathering materials for a speech is like gathering information for any project. For most topics, you will uncover far more sources than you can use. Skim or rapidly go through the material. If you are evaluating a magazine article, spend a minute or two finding out what it covers. If it is a book, read the table of contents carefully, look at the index, and skim pertinent chapters. You can then decide which sources should be read in full, which should be read in part, and which should be abandoned. Minutes spent in such an evaluation may save you hours of reading. In Figure 8–5, an FFA member is going straight to the source as he gathers information on his speech by interviewing a farm family. Interviews are sources and should be referenced just as books are referenced.

Figure 8–5 Interviews are an excellent source of information. This student is getting information from a farm family.

RECORD YOUR IDEAS

When gathering material, write each item of information on a note card with the name of the source, page number, and author. This method allows easy movement of the information from one place to another. Use **quotations** and **statistics** in your speech only when they are needed to make a point. The audience wants to hear your opinions, not another author's opinions.

Make sure the source and date of information are accurate. It is common courtesy to cite the author's name. This also helps support your information. Keep the true meaning of the quotations and statistics. Use quotations and statistics in the original context.

Information from Books

On each note card, write the note, the author, title, page number, and a heading indicating the subject of the note (Figure 8–6). The subject heading is particularly important. It simplifies the task of arranging your notes when you start to organize the speech.

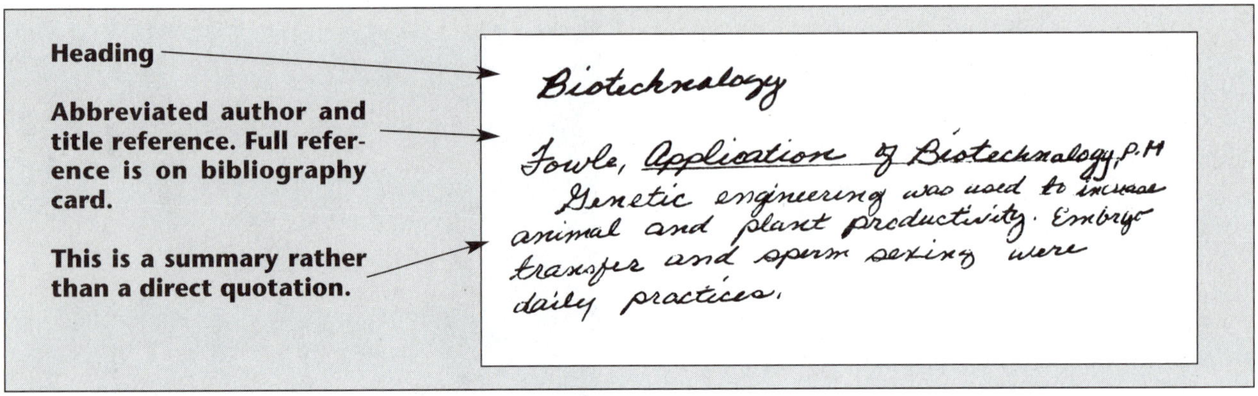

Heading

Abbreviated author and title reference. Full reference is on bibliography card.

This is a summary rather than a direct quotation.

Biotechnology

Fowle, Application of Biotechnology, p.14
Genetic engineering was used to increase animal and plant productivity. Embryo transfer and sperm sexing were daily practices.

Figure 8–6 Book note card example.

Information from Magazines

For magazine articles, write the note, topic, name of article, name of the author (if one is given), magazine title, year and date, and the page number on each note card (Figure 8–7). Use a separate card for each note. Take plenty of notes. Even though you are not certain you will use a particular piece of evidence in your speech, record it and the source in which you discovered it.

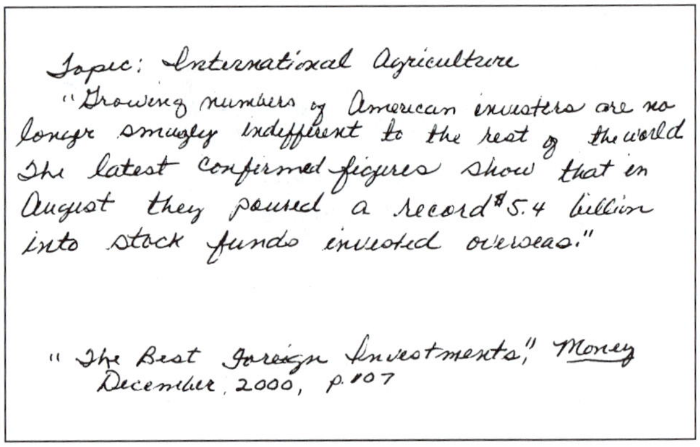

Topic: International Agriculture
"Growing numbers of American investors are no longer smugly indifferent to the rest of the world. The latest confirmed figures show that in August they poured a record $5.4 billion into stock funds invested overseas."

"The Best Foreign Investments," Money December, 2000, p.107

Figure 8–7 Magazine note card example.

Source Citations

In a written report, ideas taken from other sources are designated as footnotes. However, in a speech, these notations must be included within the context of your manuscript. This will not only help the audience evaluate the context, but it will also add to your credibility. Figure 8–8 gives examples of several source citations.

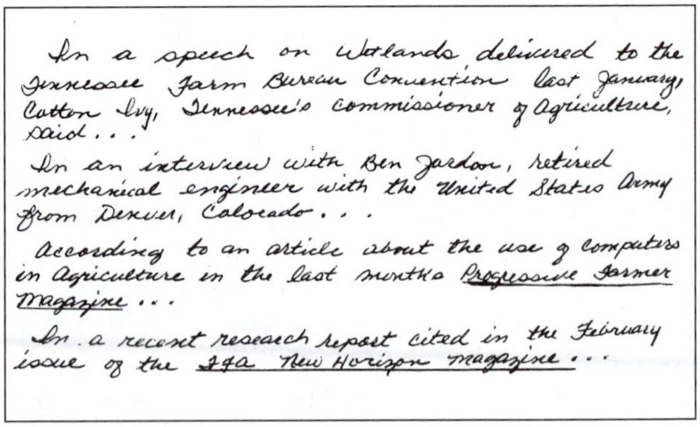

In a speech on Wetlands delivered to the Tennessee Farm Bureau Convention last January, Cotton Ivy, Tennessee's commissioner of Agriculture, said . . .

In an interview with Ben Jardon, retired mechanical engineer with the United States Army from Denver, Colorado . . .

According to an article about the use of computers in Agriculture in the last month's Progressive Farmer Magazine . . .

In a recent research report cited in the February issue of the FFA New Horizon magazine . . .

Figure 8–8 Examples of source citations.

MAKE AN OUTLINE

An outline helps you prepare the speech. An outline is the beginning of actually putting your speech together. There are three main reasons for using outlines.

1. To help you recognize the speech's strengths and weaknesses

2. To help you organize and develop your ideas
3. To help you save time when writing the speech

While preparing the outline, you decide what you will say in the introduction, how you will organize the main points and supporting materials in the body of the speech, and what you will say in the conclusion.

Format

A standard set of symbols should be used. Main points are indicated by Roman numerals, major subdivisions by capital letters, minor subheadings by Arabic numerals, and further subdivisions by lowercase letters. Although further breakdown of ideas can be shown, a speech outline is rarely subdivided beyond the level shown in Figure 8–9.

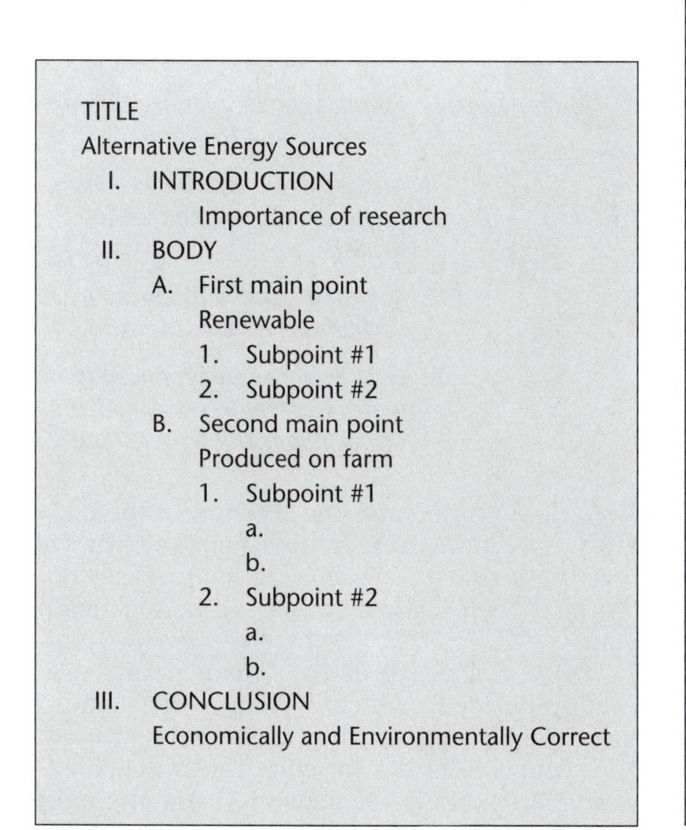

TITLE
Alternative Energy Sources
I. INTRODUCTION
Importance of research
II. BODY
A. First main point
Renewable
1. Subpoint #1
2. Subpoint #2
B. Second main point
Produced on farm
1. Subpoint #1
a.
b.
2. Subpoint #2
a.
b.
III. CONCLUSION
Economically and Environmentally Correct

Figure 8–9 Topical outline style.

Topical Outline The topical outline style uses key words or phrases to express the order of the thoughts to be presented in a speech. This topical outline can be used for a speaker's notes.

Sentence Outline The sentence outline style uses complete sentences to express the order of the thoughts to be presented in a speech (Figure 8–10). The sentence outline is more complete and therefore usually a better outline for beginning speakers

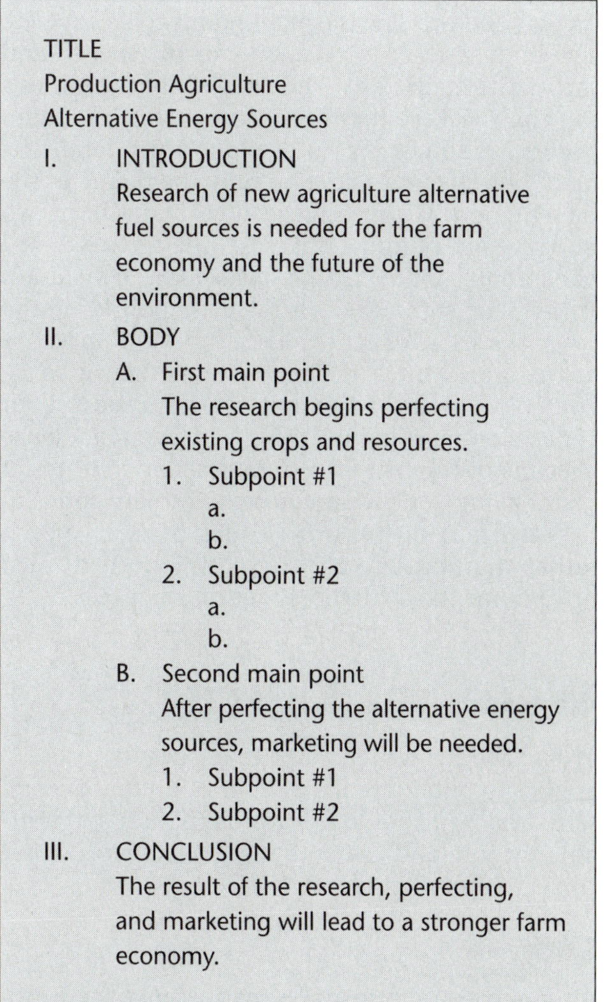

TITLE
Production Agriculture
Alternative Energy Sources
I. INTRODUCTION
Research of new agriculture alternative fuel sources is needed for the farm economy and the future of the environment.
II. BODY
A. First main point
The research begins perfecting existing crops and resources.
1. Subpoint #1
a.
b.
2. Subpoint #2
a.
b.
B. Second main point
After perfecting the alternative energy sources, marketing will be needed.
1. Subpoint #1
2. Subpoint #2
III. CONCLUSION
The result of the research, perfecting, and marketing will lead to a stronger farm economy.

Figure 8–10 Sentence outline style.

to use. Using complete sentences enables you to see (1) whether each main point actually enhances the theme of your speech goal and (2) whether the wording makes the point you want to make.

Parts of an Outline

An outline consists of the introduction, body, and conclusion. Since the introduction is the beginning of the speech, it should motivate the audience to listen to your speech. The introduction should get the audience's attention, state the purpose, relate the importance of the topic to the audience, and preview the main points to be covered. The body is the central theme of the speech and carries the main ideas. The body of the speech carries the speaker's message. It consists of the main points, subdivisions, and supporting detail. Its main objective is to develop fully the theme of the presentation. The speech usually contains from two to five main points. Otherwise, the audience will have trouble following the speaker. The conclusion reviews or summarizes the main points of the speech or calls for action. The end of the speech is just as important as the beginning and must be accorded its proper allotment of preparation. Your conclusion must do two things: give your audience a feeling of completeness or satisfaction in terms of your having done what you set out to do, and center attention on the whole idea of your speech rather than on any part of it.[5] In short, it must drive home the central message of the speech.

WRITE THE SPEECH

The adage for writers, "Write the way you talk," applies to public speaking. Your talk should represent you. You should use words that are comfortable for you and the audience. Select words that convey your exact message.

Body

Although you may assume that because the introduction is the first part of the speech, that is where you should begin. How are you going to write the introduction until you have considered the material that you will be introducing? Therefore, start with the body. Write the main points; arrange or order the main points in a logical sequence; and select effective language, examples, and quotations.

Main points are the key building blocks of your speech. They are the ideas that you want your audience to remember. The main points should be specific, vivid, and parallel. They should be of equal importance. The wording should explain exactly what the speaker means and arouse interest. The main points of a speech about alternative fuels for agriculture follow.

Specific purpose: To inform my audience that a production agriculturalist can produce fuel in times of national crisis.

Central idea: Production agriculturalists can produce their own ethanol, soy-diesel, methane, and hydrogen.

Main points:

I. Ethanol can be produced from corn raised on the farm.

II. Methane can be produced from manure in dairy and feed lot facilities.

III. Soy-diesel can be produced from soybeans raised on the farm.

IV. Hydrogen can be produced from water with solar, wind, hydro, or nuclear as the power source.

These four points form the skeleton of the body of the speech. If there are four major alternative fuel types that production agriculturalists can produce, then logically there can be four main points.

Once you establish your main points, you need to decide in which order to write them. A speech can be organized in many different ways. Your goal is to find a structure that will achieve your goal and help the audience make the most sense of your materials. These organizational patterns can help an audience follow your speech:

- *Time or chronological order*—follows a chronological sequence of ideas or events.
- *Space order*—tells the audience there is a special significance to the positioning of the information, such as top to bottom.
- *Topical order*—emphasizes categories or divisions of a subject.
- *Causal order*—emphasizes the causal relationship between the main points and the subject of the speech resulting from specific conditions.
- *Reasons order*—emphasizes why you believe an audience should believe in a statement or behave in a particular way.
- *Problem solution order*—the main points are written to show:

  There is a problem that requires a change in attitude or behavior.

  The solution you are presenting will solve the problem.

  The solution is the best way to solve that problem.

Supporting materials back up ideas for your main point. When selecting supporting materials, make sure they are directly relevant to the main points they are supposed to support. This material comes from the material gathered earlier, when planning the speech. Each item of information listed under a key element heading may contain further subpoints. These subpoints show the relationship between each idea. Use illustrations, examples, and anecdotes for each subpoint. List enough material to have both quantity and quality. Remember, it is always easier to edit than to search for new material when time is short.

Choosing effective language for a speech is like choosing the right clothes for a special occasion.

> ***Words, are the garments with which speakers clothe their ideas.***
>
> *—J. Regis O'Connor*

Wise speakers choose their words very carefully to display their ideas effectively. This involves using simple words, using more concrete and specific words, and restating certain words. Choosing effective language includes clarity, creating levels, emphasis, and using figures of speech, such as **simile**, **metaphor**, **personification**, **hyperbole**, and **irony**.

When writing the speech, be aware of your legal and **ethical** responsibilities. A speaker must refrain from statements that may harm another's reputation (**defamation of character**). Concerning ethical issues, a speaker must remember he or she is personally accountable for their speech. Do not make statements without first checking the facts. A speaker must allow free choice by the audience (hear both sides of the issue). Ethical speaking builds respect for the speaker and others.[6]

Introduction

Once the body of the speech is ready for practice, begin working on the **introduction**. (Prepare two or three different introductions to determine which will be most effective.) The introduction may be one sentence or many, depending on the length of the speech. A good introduction should grab the audience's attention, create a feeling of credibility and good will, and lead directly into the body of the speech. The audience may be a "captive audience," but physical presence does not guarantee mental presence. The speaker's first goal is to create an opening that will earn the audience's undivided attention. Do something to gain the attention of the audience.[7]

- Tell a joke.
- Pound the speaker's stand.
- Make a loud noise.
- Ask a question.
- Tell a story.
- Use a quotation.
- Use a personal reference.
- Create suspense.
- Give a compliment.

All these methods grab attention, but attention may wane if the remaining introduction does not lead directly into the speech. With most speeches, the audience experiences a feeling of good will if they perceive the speaker as sincere and concerned for them.

The introduction must also focus the audience on the goal of the speech. To focus attention, simply tell the audience what you are going to tell them in the speech. This is the last statement of the introduction and leads into the **body**. This is usually the main purpose statement.[8]

Conclusion

All's well that ends well.

—Shakespeare

Write the conclusion last. This offers the speaker one last opportunity to remind the audience of the speech content. It reinforces the audience's understanding of, or commitment to, the central idea. Do not startle the audience by ending too abruptly, nor ramble so long that they become exhausted. The most common types of conclusions are these.

- Summarize the main points
- Use a story
- Be humorous
- Appeal
- Make an emotional impact

As with an introduction, write two or three different conclusions to find the most effective one.[9]

PRACTICE THE SPEECH

It is important to remember the following considerations when practicing the speech.

Practice Time Limits

When should you begin preparing your speech? Many students of public speaking do not allow themselves enough time for planning, research, speech writing, and practice. Consequently, they do themselves an injustice by not leaving sufficient time to do the job they are capable of doing. It is not unusual for a student to attempt to write, memorize, practice, and deliver a speech for a contest within a week. Even an accomplished speaker would be heavily taxed to do a job of this magnitude in that short a time span.

Think in terms of the minimum time that will enable you to do your best. You should start working on the speech at least two months before the delivery date. It should be completed within a month. This allows a month to do the necessary research and writing and another month to memorize, practice, and otherwise prepare yourself for the speech date. Two weeks is the minimum time needed. If you can devote more time to it, it will benefit you to do so.

To succeed in public speaking, you must be willing to give much time to practice. Continue to practice even beyond the point when you think sufficient time has been devoted. Revise sentences so they are easier for you to say; change words if they are difficult for you to pronounce; find substitutes.

Memorize the speech as quickly as possible so that more time can be given to practice. Too frequently high school students become convinced all too early that they are fully prepared, and as a result, slow down on their practicing. This can cause overconfidence, which can cause you to forget your speech and thus make a dismal showing.

Practice Methods

Demosthenes, the famous Greek orator, developed his speaking ability by going off into the woods alone and standing on a rock while speaking to an imaginary audience. There are several resources available to help you practice your speech.

1. *School classes and teachers.* Most teachers will be happy to have you appear before their classes for practice. Ask them to listen to your speech privately, as well as in class, and to offer constructive criticism.
2. *Home and mirror.* Practice at home as much as possible. Times when you would otherwise be

doing nothing, practice that speech! Stand in front of a mirror and observe your facial expressions and gestures. This will help you see yourself as others see you. Do you like what you see in the mirror? If not, try to do something about it.

3. *Auditorium.* When opportunity permits, go into the auditorium for practice (Figure 8–11). This gives you an opportunity to become accustomed to different acoustic conditions. It also gives you the additional advantage of learning to look at greater space as you talk.
4. *Civic organizations.* Arrange to talk to every civic organization and other group outside the school that you possibly can. This will give you a decided advantage in becoming accustomed to facing an audience and an opportunity to get over those **pangs** of stage fright that haunt the average beginning speaker.
5. *The video camera.* Use of the video camera is perhaps one of the finest modern methods of speech rehearsal. With this device, you can record your speech and then watch and listen to the tape. This gives you an excellent opportunity to detect your mistakes and correct them. Record your speech, watch and listen to it, then rewind and record it again. When satisfied that you have it about right, ask your speech teacher, as well as others, to listen to it. Solicit their suggestions and criticisms, then practice again and again.

Figure 8–11 This student is practicing in the auditorium. It gives her an opportunity to become accustomed to different acoustic conditions.

Refer to Chapter 7 for other fun and practical ways to practice.

Things That Need Practice

When you are practicing your speech, be aware of the following suggestions.

Your Smile Loosen up, and smile a little for your audience! You will know when to smile at the right time in your speech; you will know the passages that require a look of seriousness. Practice these facial expressions just as **ardently** as you practice the words of your speech.

Gestures As soon as you have your speech memorized, work on the finer points of delivery. Practice your gestures and get used to making them gracefully. The problem of when and how much to gesture with hands is a difficult one, but it is safe to say that the amateur public speaker should leave most of the hand gestures to the experienced speaker. Gestures must come naturally and may often be used to add emphasis or give expression.

Many experienced speakers greatly overdo the use of gestures. This is usually due to lack of early training and practice when they were beginners. It is much better to use no gestures than to use too many. If you use gestures too much, you will find that the interest of the audience is focused on your movements rather than the message in the subject matter.

Head and Eyes One of the best ways of giving that added touch of emphasis without distracting the audience is with the head and eyes. A gentle nod, shake, or backward movement of the head may have a far greater effect than you might expect.

You often hear the remark, "She talks with her eyes." When you are talking to an audience, you are talking to individuals. Do not make the horrible mistake of looking at the floor, the ceiling, or out a window while talking. Look those individuals in the eyes, and they will feel that you are talking to them. You may find this distracting when you first try it, but with a little practice, this feeling will disappear and you will find that you really are talking to them and enjoying the feeling.

Sincerity Practice being sincere in what you say. When people are willing to spend their time listening to a speaker, they deserve to hear a speech that is the result of much thought and one in which the speaker shows sincerity. The speaker who delivers a speech in a disinterested way should not be insulted if members of the audience walk out.[10]

Be sincere as you write the speech, as you practice it, and especially as you deliver it.

PRESENT THE SPEECH

Chapter 7 gives a detailed description of delivery or presentation. The score card and divisions are addressed specifically. This section reviews presentation techniques that are more suited to a prepared speech than FFA Creed delivery. Refer to Chapter 7 also for overcoming stage fright and nervousness.

The actual delivery of the speech is much easier if you have done a good job of preparation. A few things should be considered: the **salutation**, being **deliberate**, using the hands, using the body, humor, dress and physical appearance, where to stand, notes, and special considerations.

Salutation The salutation should be pleasant and inclusive. The exact wording depends on the particular audience. Some speakers begin simply with "Ladies and Gentlemen," and then launch into the speech. In many instances, a better approach might be: "Mr. Chairperson, fellow members, distinguished guests, and friends, I shall speak to you on the subject . . ."

Being Deliberate Try to be deliberate from the time you rise to your feet until the time to sit down again. One of the most common mistakes is to appear to be in a hurry. This gives the impression that you are nervous. A deliberate manner of speaking helps you enunciate your words more clearly. When talking rapidly, we tend to run the words together. This can be a serious handicap to a public speaker.

Using the Hands What to do with your hands will probably be one of your problems—suddenly you have no place to put them. This need not be true. No one else is apt to notice your hands until you start using them in an unaccustomed manner.

Let your hands appear natural. If they want to clasp behind you, let them do that; if one wants to stray to a pocket, let it stray; if one wants to hold onto a lapel, let it do so. When you start hunting for places to put them, you are asking for trouble.

Using the Body One of the cardinal rules of public speaking is not to sway back and forth. This is a very common error of the beginning speaker. However, slight movements of the body coordinated with movements of the head or limbs are occasionally permissible to add emphasis. A slight step forward to indicate emphasis or change of pace in the speech, or perhaps a slight backward step in certain instances, may be used to your advantage. However, it is well to avoid walking about the stage or too much slight movement.

Humor Nothing puts an audience more at ease at the beginning of a speech than a good joke.

This, of course, is not common practice in contest speeches and is not recommended as such. However, at times it might be used to advantage even in a contest speech. It is almost never advisable for the beginning speaker to try to incorporate humor into his speech.

When occasion calls for humor in a speech, keep it clean. It is not necessary, so do not take the risk of making yourself unpopular with your audience.

You may not be a good joke teller. If not, do not try to tell jokes. You may improve your speech by including anecdotes, examples, and illustrations to help people understand the point you are discussing. Examples are word pictures, and people understand pictures better than abstract ideas. The experienced speaker in Figure 8–12 appears natural as he injects humor into his speech.

Figure 8–12 You may improve your speech, as this FFA member has done, by injecting anecdotes, examples, illustrations, and humor, if appropriate.

Dress and Physical Appearance The importance of physical appearance should not be underestimated. The general appearance that you present can have a definite impact on your listeners. For any official FFA speech, you should be in official FFA dress.

Let neatness be the watchword in this area. Clothes that are clean and neatly pressed, gleaming teeth, well-groomed hair, and, for men, a clean-shaven face are seemingly small items but they merge to help create a pleasant picture.

Where to Stand Where should you stand while delivering the speech? Usually this decision is left entirely up to you. Sometimes, especially in contests that are conducted in large auditoriums, the speakers are encouraged to stand behind the lectern so their voices will carry well over the public address system.

A good point to remember is that you should stand where the audience can hear you best. If the acoustics are poor, stand near the edge of the stage. If there is a lectern and you can adjust to it, stand behind it. This is especially helpful as a place to put your hands. If there is a table, stand either at the left or right of it.

Notes You should know your speech so well that notes are not necessary. However, if you have to use notes, write your notes large enough to be seen at a glance. Use a system to keep your note cards in order, such as an outline or numbers. Don't hold your notes in your hands; place them on the lectern so your hands will be free.

Special Considerations As a speaker, you must know how to deal with a number of special situations: the lectern, microphones, distractions, and following a written speech.

- *Lectern.* A speaker's stand is intended to provide a convenient place to lay notes or a manuscript. When one is not available, notes must be held inconspicuously in the hands.
- *Microphone.* A microphone can be very helpful when speaking to a large audience. To get the maximum use from a microphone,

however, you should practice with it, test it shortly before the speech, and keep your mouth a constant 10 to 12 inches from the head of the microphone at all times.

- *Distractions.* If distractions or interruptions occur during a speech, they must be handled calmly and with poise. If the interruption is minor, it is usually wise to ignore it; if it creates a major disturbance, you must handle it decisively and with wisdom.[11]
- *Following a written speech.* For contest purposes, should one follow the written speech exactly? It is up to the judge and the organization. In real life, nobody does or should follow their speech exactly. Daily news, daily events, introductions, previous speakers, and other things on the program may necessitate last-minute changes to make things more meaningful. Minor word changes in a delivery should never be a factor in the outcome of a contest.

ANSWER QUESTIONS

The rules for many public speaking contests allow up to five minutes for the judges to ask the speaker questions on his subject matter after the speech. Therefore, you must be well acquainted with the subject since it is supposed to be studied and researched well. Be sure you can answer all questions completely; continue reference reading up to the contest date. Ask your teacher and others to prepare lists of questions that might be asked on the subject matter; prepare answers for them.

During the questioning period, follow these guidelines.

- Be deliberate. Take time to think through the answer and organize your reply.
- Be complete. There is no scarcity of words, so use enough of them to render a complete answer. If you answer with several points, count them off on your fingers as you answer.
- Answer with confidence. A hesitating answer convinces the judges that you do not know much about the question.
- If you do not know the answer, say so without hesitation, but do not show embarrassment. Do not try to run a bluff; if you have some knowledge of the answer, admit that your information is limited on the question, then go ahead with what you know.
- If you did not hear or did not understand the question, ask the person to repeat or rephrase it. Some speakers, while stalling, say, "I'm glad you asked that question." Obviously, if everyone used that phrase, the judges would know what you were doing. Be creative rather than leaving dead space while thinking.

LISTEN AND EVALUATE

General Evaluation

Evaluating speeches not only provides the speaker with an analysis of where the speech went right and where it went wrong, but it also gives you insight into techniques to use as you prepare your own speech. If a speech has good content, is well organized, uses language well, and is presented well, it is more likely to achieve its goal.[12]

FFA Speech Evaluation

The FFA Prepared Public Speaking Contest uses the score sheet shown in Figure 8–13. Scoring includes two parts. The first part includes content and composition of the manuscript. You may wish to ask your English teacher to review your speech for needed corrections. The content could be reviewed by friends, professionals, family, teachers, and others who may provide input. Before you enter a contest, there is no excuse for not scoring near perfect on the first part if you are sufficiently prepared.

The second part scores the delivery of the speech, including voice, stage presence, power of expression, response to questions, and general effect. Figure 8–13 provides a further breakdown and explanation of each of these areas. Refer to Chapter 7 in which the FFA Creed score card is

Prepared Public Speaking CDE Score Card

	Max. Points	One	Two	Three	Four	Five	Six	Seven	Eight
Content of Manuscript	200								
• Importance and appropriateness of the subject • Suitability of the material used • Accuracy of the statements • Evidence of purpose • Completeness and accuracy of bibliography									
Manuscript Composition	100								
• Organization of contents • Unity of thought • Logical development • Language used • Sentence structure • Accomplishment of purpose conclusion									
Voice	100								
• Quality pitch • Articulation • Pronunciation • Force									
Stage Presence	100								
• Personal appearance • Poise and body posture • Attitude confidence and personality • Ease before an audience									
Power of Expression	100								
• Communicative ability including: fluency, emphasis, directness, sincerity • Conveyance of thought and meaning									
Response to Questions*	300								
• Ability to answer the questions on the speech which are asked by the judges, indicating originality, familiarity with subject and ability to think quickly.									
General Effect	100								
• Extent to which the speech was interesting, understandable, convincing, pleasing and held attention									
Gross Total Points:	1000								
Less Time Deduction**:									
Net Total Points									
Contestant Ranking:									

* Judges should meet prior to the event to prepare and clarify the types of questions to be asked.
** – 1 point per second over, determined by the timekeepers

Section: ________ Flight: ________ Room: ________ Judge Signature: ____________________

Figure 8–13 The National FFA Prepared Public Speaking Career Development Event Score Card.

discussed, for a comprehensive explanation of the delivery section.

The points attributed to each item are included in the Judge's Score Sheet in Figure 8–13. Note that points are deducted for time if your speech is too long or too short.

National FFA Prepared Public Speaking CDE

The National FFA Prepared Public Speaking CDE is designed to develop agricultural leadership by providing for member participation in agricultural public speaking activities and stimulating interest in leadership and citizenship. The national event is held in conjunction with the national FFA convention.

Event Rules

- The National FFA Prepared Public Speaking CDE is limited to one participant from each state association.
- The National FFA Prepared Public Speaking CDE follows the general rules and policies for all National FFA CDEs. A participant cannot serve on the national officer nominating committee.
- Each participant's manuscript must be the result of his or her own efforts. It is expected that the participant will take advantage of all available training facilities at school in developing speaking ability. Facts and working data may be secured from any source but must be appropriately documented.
- Participants report to the event superintendent for instructions at the time and place shown in the current year's program for National FFA events.
- It is highly recommended that participants be in official FFA dress, as defined in the current official *FFA Manual*.
- **Each state with a speaker shall provide a competent individual to judge the national event.** Judges for the final round are representatives from agricultural education-related organizations. Three to six competent and impartial persons are selected to judge the event.
- The Board of National Officers and National Board of Directors of the FFA are in charge of this event.

Event Format

The speaker must provide the following.

- Fifteen double-spaced typewritten copies of the speech on $8\frac{1}{2} \times 11$-inch white bond paper, with a cover page that gives the speech title, participant's name, state, and date. The body of the manuscript must have 1-inch margins, and 10–12 characters per inch. Follow the American Psychological Association style manual for developing references and bibliography. Do not bind, but place a staple in upper left corner. Manuscripts not meeting these guidelines will be penalized.
- A signed statement of originality on the certification form provided through your state FFA association.
- A complete and accurate bibliography used in writing the speech. All participants in the National FFA Prepared Public Speaking CDE should give credit for any direct quotes, phrases, or special dates used in the manuscript in order not to be guilty of plagiarism. The Boards of National Officers and Directors of FFA at the October 1960 meeting in Kansas City, Missouri, adopted the following resolution:

  A bibliography MUST be included as part of the public speaker's manuscript, and direct quotes from any source of information must be marked in "quotes" on the manuscript and be identified in the bibliography. Failure to do so will automatically disqualify a participant. This applies to all events above the local level.

Time Limit

Each speech shall be a minimum of six minutes in length and a maximum of eight minutes. Each participant is allowed five minutes additional time in which he or she is asked questions relating to his or her speech. Participants are to be penalized one point per second on each judge's score sheet for being under six minutes or over eight minutes. No time warnings will be given.

Judging

- Event officials will randomly determine the speaking order. The program chairman introduces each participant by name in order of the drawing.
- A participant is permitted to use notes while speaking, but deductions in scoring may be made for this practice if it detracts from the effectiveness of the presentation. No props are to be used. Applause shall be withheld until all participants have spoken.
- A timekeeper(s) records the time used by each participant in delivering his or her speech, noting undertime or overtime, if any, for which deductions are made. Timekeepers should be sitting together.
- Prior to the event the content and composition of all manuscripts are judged and scored by qualified individuals using the manuscript score sheet. Manuscript scores are averaged and supplied to the presentation judges after they have scored the oral presentation. Manuscript comment cards are completed by manuscript judges and presented to the contestants at the awards function. Presentation judges are furnished with typewritten copies of the participants' manuscripts, which they use to formulate questions.
- At the time of the event, the judges are seated in different sections of the room in which the event is held. They score each participant on delivery, using the score sheet provided. They also complete a judge's comment card, which is presented to the participants at the award function.
- Each judge formulates and asks questions. Questions shall pertain directly to the speaker's subject. Questions containing two or more parts should be avoided. Judges score each participant on the ability to answer all questions asked by all judges. The full five minutes for questions should be used.
- When all participants have finished speaking, each judge totals the score on composition and delivery for each participant. The timekeeper(s) record is used in computing the final score for each participant. The judges' score sheets are then submitted to event officials to determine final rankings of participants.
- Participants shall be ranked on the basis of the final score to be determined by each judge without consultation. The judges' ranking of each participant shall then be added, and the winner is the participant whose total ranking is the lowest. Other placings are determined in the same manner.

Tiebreakers

Ties are broken based on the greatest number of low ranks. Participants' low rankings are counted, and the participant with the greatest number of low rankings is declared the winner. If a tie still exists, the event superintendent ranks the participant's response to questions. The participant with the lowest ranking from the response to question is declared the winner. If a tie still exists, the participant's raw scores are totaled. The participant with the greatest total of raw points is declared the winner.

SPEAKING TO GROUPS AND ORGANIZATIONS

Speaking at contests and in classes is formal. Informally, we must be poised and act naturally. Here are suggestions to follow when you speak to groups and organizations.

- **Acknowledge your introduction.** Graciously acknowledge the introduction. If the chairperson has been funny at the expense of the speaker, the speaker should laugh good-naturedly. If the chairperson has expressed real admiration for the speaker's accomplishments, a joke should not be made at the chairperson's expense. The response to an introduction should not be overdone, nor should it be completely ignored.

- **Show appreciation.** Express pleasure for the opportunity to speak. This is done directly, sincerely, and briefly. Direct your comments to the program committee or whomever invited the speaker.
- **Relate to local interest.** Refer to matters of local interest. If there is a particular event or person for which a town is famous, the speaker can establish rapport by mentioning it. If speaking before a civic club or charitable organization, the speaker should mention some accomplishments of the organization. When a speaker tactfully displays an awareness of the audience or community, this is a compliment for the audience or community.
- **Respond to the mood.** Respond to the mood of the audience. The speaker should attempt to pick up on the group feeling and capitalize on it. A statement such as, "As I listened to the reports, I sensed that this organization really wants to do something for the community," will enhance rapport. In addition, some circumstance that the audience is aware of should be included in a comment. Almost invariably, there is something related to the room, the occasion, the stage, the noise outside, or something of little consequence that lends itself to an amusing comment.[13]

The groups and organizations in Figure 8–14 are possible places to deliver a speech. Be sure that you know something about the organization before you address them. This builds rapport quickly.

POINTS TO REMEMBER

The speaker's immediate purpose is to communicate ideas and feelings to the listeners. Communication is the transfer of meaning. The speaker is aware of the audience's response and is able to read the many clues given by the audience.

The speaker's basic purpose is to relate to the listeners to achieve some goal. Communication is a means and not an end in itself. The speaker

Optimist Club	Lions Club	Secondary school principals
State Board of Education	Jaycees	Rotary Club
Kiwanis Club	Women's Club	Veterans of Foreign Wars
Chamber of Commerce	American Legion	Labor unions (AFL-CIO, etc.)
Sales and Marketing Executives	Century Club	Medical society
Local business or industry board meetings	House of Representatives and/or Senate	Youth organizations (school, vocational, special programs, Boy Scouts, etc.)
Vocational associations	Toastmasters	Better Business Bureau
League of Women Voters	City Council meetings	Agency to Reduce Crime
State Education Association	Knights of Columbus	Retail Merchants Association
Parent-Teacher Association	Local school board	Public school assemblies

Figure 8–14 These organizations are places where you could deliver speeches while in high school or after you graduate. Some of these groups are also good places to practice your speech if you are entering a speaking contest.

communicates to the listener to get something done. It may be to share information, solve mutual problems, relax tensions, or inspire improved performance. The purpose of a speech should be determined before it is given.

The speaker makes an accurate analysis of the speech situation and adapts the speech to it. No two speech situations are ever identical, and the able speaker makes a quick analysis of the situation. The following questions should be answered in the analysis: Am I prepared? Am I enthusiastic about the subject? What will be the audience's reaction to the subject? How do factors such as time, length of program, and meeting room influence the situation? Does the audience have an interest in the subject, or does interest have to be created? Can I improve?

Here are a few points that might well be remembered.

- Open the speech with a sentence that secures the attention of the audience.
- End the speech in a forceful manner.
- Take appropriate pauses and do not allow yourself to run out of breath.
- Maintain good posture while speaking.
- Keep the voice well modulated; use variety of tone and pitch.
- Strive for correct pronunciation and enunciate clearly.
- Do not eat a heavy meal immediately before speaking.
- Invite constructive criticism.
- Cultivate a sincere interest in people.
- Constantly strive to increase your vocabulary.

The power of speech is too great a weapon to let it fall solely into the hands of radicals, demagogues, and "ne'er-do-wells." Orators who are people of vision, such as Webster, Clay, and Lincoln, will always be needed in shaping the destinies of our country and of the world. People who possess the inherent qualities that could be developed to meet these needs have an obligation to humankind.

The field of public speaking has never been crowded with good speakers. As in all other fields, there is always plenty of space for honest people of high principles and high ideals and people who are willing to sacrifice. To be a good public speaker, you must be willing to sacrifice, willing to give of yourself, and remain conscious of the fact that you are serving humanity.

You are challenged to work hard, study earnestly, practice unceasingly, and make yourself a good speaker. The richness of the rewards will far exceed your fondest dreams!

SUMMARY

Speeches are given to inform the audience, to persuade, or to bring the members of the audience together. Through persistence you can achieve your speaking goal. Speaking skills increase a person's effectiveness and influence the decisions of others. Effective public speaking is influence. Influence is leadership.

As a speech is planned, consider the purpose, the audience, and the occasion. Good speakers are audience centered. They know the primary purpose of speech preparation is not to display their knowledge, demonstrate their superiority, or express their anger. The primary purpose should be a desired response or to please the listener.

When selecting a topic, choose one that interests you. Next, choose a topic in which you are knowledgeable or have an interest in becoming knowledgeable. Third, pick a topic of interest to your audience. When searching for a topic for an FFA speech, you may wish to start with three general agricultural areas: production agriculture, agribusiness, and agriscience.

Start gathering information for your speech by researching your own books and magazines. Breed associations and trade and business magazines are excellent sources. Consult not only articles and books but organizations and experts. Most research for speeches is done in a good library, with the help of good librarians.

When recording your ideas, write each item of information on a note card with the name of the source, the page number, and the author. Use quotations and statistics in your speech only when they are needed to make a point. The audience wants to hear your opinions, not another author's opinions.

An outline helps you prepare the speech. It is the beginning of actually putting your speech together. An outline helps you recognize the speech's strengths and weaknesses, organize and develop your ideas, and save time when writing the speech. The two types of outlines are topical and sentence outlines. An outline includes the introduction, body, and conclusion.

In writing the speech, write the way you talk; use words that are comfortable for you and the audience. Select words that convey your exact message. Start by writing the body of the speech and list the main points. Once the body is completed, begin working on the introduction. Write the conclusion last.

Practice the speech. There are several resources available to help you practice your speech, including school classes and teachers, the home and mirror, auditorium, civic organizations, and the video camera. When you are practicing your speech, be aware of your smile, gestures, and head and eyes. Be sincere as you write the speech, as you practice it, and especially as you deliver it.

The actual delivery of the speech is much easier if you have done a good job of preparation. As you deliver your speech, consider the following: salutation, being deliberate, using the hands, using the body, humor, dress and physical appearance, where to stand, notes, use of lectern and microphones, and distractions.

The rules of many public speaking contests allow up to five minutes for the judges to ask the speaker questions on his or her subject matter after the speech. You must be well acquainted with the subject since your speech is supposed to be well studied and researched.

If a speech has good content, is well organized, well written, and presented well, it is more likely to achieve its goal. The FFA Prepared Public Speaking Contest evaluates the content and composition of the manuscript. The delivery of the speech is also evaluated, which includes voice, stage presence, power of expression, response to questions, and general effect.

When speaking to groups and organizations, graciously acknowledge the introduction. Express pleasure for the opportunity to speak. Refer to matters of local interest. Last, respond to the mood of the audience.

END-OF-CHAPTER EXERCISES

Review Questions

1. Define the Terms to Know.
2. How is public speaking related to leadership?
3. Should you select the speech topic first or analyze the audience before selecting the speech topic?
4. Briefly explain how to gather (review) materials for a speech.
5. What are nine sources of information for preparing your speech?
6. What are three reasons for using outlines?
7. What are the three major parts of an outline?
8. Explain the speaker's ethical responsibility when writing a speech.
9. List six organizational patterns within the body of the speech.
10. What are nine things that can be done in an introduction to gain attention?

11. Which do you believe would be the best practice method for you? Explain.
12. Discuss the speaker's use of notes when speaking.

Fill-in-the-Blank

1. ________________ is an uncritical, nonevaluative process of generating ideas, much like the word-association game.
2. When information is recorded on note cards from books, write the note, the ________________, ________________, and the page number.
3. The most common types of conclusions are ________________, ________________, ________________, ________________, and ________________.
4. Using a ________________ ________________ is perhaps one of the finest modern methods of speech rehearsal.
5. A ________________, ________________, ________________, and ________________ are four things that need practice when rehearsing a speech.
6. The scoring for content and composition of a speech is ________________ points and ________________ points are allowed for the delivery.

Matching

____ 1. Entering fully, through imagination, into another's feelings
____ 2. A device to connect internally or externally to a computer and telephone to get information electronically through the telephone wires into the computer
____ 3. A figure of speech that consists of exaggeration used to emphasize a point
____ 4. A figure of speech containing a direct comparison that omits the words *like* or *as*
____ 5. A type of humor in which the literal meaning expressed is the opposite of the meaning intended
____ 6. A figure of speech that gives human qualities to inanimate objects, ideas, or nonhuman creatures
____ 7. A figure of speech that presents a brief comparison of two basically unlike items using the words *like* or *as*
____ 8. The greeting in a speech
____ 9. Being slow and careful in deciding what to do

A. simile
B. personfication
C. irony
D. hyperbole
E. modem
F. metaphor
G. empathize
H. deliberate
I. saluation

Activities

1. Divide a sheet of paper into three columns. Label column 1 with your major or vocation, such as "art history"; label column 2 with a hobby or activity, such as "chess"; and label column 3 with a

concern or issue, such as "water pollution." Working on one column at a time, brainstorm a list of at least 20 words or phrases for each column. Then, check three of the words or phrases in each column that are most compelling, of special meaning to you, or of potential interest to your audience. From this exercise, select a topic for your speech.

2. Your library probably has the *Encyclopedia of Associations* in its reference section. Using this book, find the names and addresses of three organizations you might write to for information about one of your speech topics.
3. Using the card catalog, locate three books on the subject of your next speech. Following the format explained in this chapter, prepare a preliminary bibliography card for each book.
4. In the *Reader's Guide to Periodical Literature,* find three articles in magazines on the subject of your next speech. Prepare a preliminary bibliography card for each article, and track down the articles in the stacks or in the periodicals section of the library. Follow the format explained in this chapter.
5. **Take it to the Net.**

   Research your speech topic on the Internet. Go to a search engine. A useful search engine is Metacrawler <http://www.metacrawler.com>. Type one of the five potential topics (agriscience and technology, agribusiness, agrimarketing, international agricultural relations, or agricultural communications) into the search field. Browse the Web sites. Pick five that you think will be useful in preparing your speech and print them out.
6. List three other sources that would provide information for your next speech topic.
7. Select a speech topic and gather information for a prepared speech.
   a. *Select a speech topic.* It would probably be best to select a topic that you already know something about and that interests you. Choose from the areas of agriscience and technology, agribusiness, agrimarketing, international agricultural relations, and agricultural communications.
   b. *List references.* List possible references that may help you find background information to develop your speech topic.
   c. *Record information.* Read the following guidelines for recording information.

      Recording key information from references:
      - Record information on 3 × 5 cards.
      - If you are in doubt about whether or not to take a note, take it. It is easier to discard a note you do not need than to track down information again.
      - Summarize information in your own words without distorting the original thought.
      - Do not copy information word for word from the original unless your source concisely sums up a great deal of information and you want to incorporate that quotation into your speech.
      - If you copy a quotation, make sure you enclose it in quotation marks so you will know later that the note was quoted verbatim.
      - Cite the source and date of the information you have recorded on each card. Use the format presented in this chapter.[14]
8. Write the outline for your prepared speech. Use either the topical outline (Figure 8–9) or sentence outline (Figure 8–10).
9. Write the body of your speech. Use the outline you prepared in activity 8. Read through your speech and make modifications as needed.
10. Write an introduction to your speech using the outline you prepared in activity 8.

11. Write a conclusion to your speech using the outline you prepared in activity 8.
12. Practice your speech using one of the methods discussed in this chapter. Write a paragraph explaining which practice method works best for you.
13. Take part in a class discussion on the topic "What I find most difficult about speech delivery." Notice during this discussion if many of your classmates have the same problems you experience.
14. Collect pictures from magazines to show different gestures. Display these pictures in the classroom, and ask students to identify the ideas being communicated by each picture.
15. Watch TV commercials for one night. Take notes on commercials, and report to the class on which commercials make the best use of gestures to convey their messages.
16. Prepare the final draft of your speech. Make sure the introduction, body, and conclusion flow together. Review the speech closely. Reword any parts that need improving.

Notes

1. Appreciation is extended to Dr. Joe Townsend, Professor, Agricultural Education, Texas A & M University, for sharing notes from his "Leadership" class about speech writing.
2. S. E. Lucas, *The Art of Public Speaking*, Second Edition (New York: Random House, 1986), p. 67.
3. Ibid.
4. E. Ehrlich and G. R. Hawes, *Speak for Success* (New York: Bantam Books, 1984), p. 148.
5. *Effective Communication Skills* (Stillwater, OK: Oklahoma Department of Education—Vocation and Technical Education, [no date available]).
6. J. Townsend, "Leadership" class notes.
7. Ibid.
8. Ibid.
9. Ibid.
10. E. Johnson, *Public Speaking* (Nashville, TN: Tennessee Department of Education, Division of Vocational-Technical Education, [no date available]).
11. J. R. O'Connor, *Speech: Exploring Communication* (Upper Saddle River, NJ: Prentice-Hall, 1998).
12. R. F. Verderber, *Communicate!* (Belmont, CA: Wadsworth Publishing Company, 2001).
13. *Understanding the Importance of Effective Communication Through the Spoken Word*, Publication No. 8369 (College Station, TX: Instructional Materials Service, 1989).
14. *Effective Communication Skills*.

CHAPTER

9

FFA Extemporaneous Public Speaking

Objectives

After completing this chapter, the student should be able to:

- Deliver an extemporaneous speech
- Discuss the advantages and disadvantages of extemporaneous speaking
- Prepare and research a speech
- Develop an extemporaneous speech
- Prepare an outline of a speech
- Implement various practice procedures
- Describe procedures to follow in the speech preparation room
- Discuss the use of note cards
- Explain the parts of an extemporaneous judge's score card
- Answer questions about your speech after delivery

Terms to Know

extemporaneous
impromptu
scantily
ambiguous
simultaneous
earnestness
slovenliness
verbose
exaggeration
mastery
credence
rudiments
animate
discreet
acronym
captivate
empathy

Speech delivery may be classified into four types: **extemporaneous**, **impromptu**, reading from manuscript, and memorized. Eventually, you should be able to deliver all four styles. In this chapter, we study extemporaneous speaking. Joe Clark, who inspired the movie *Lean on Me* is delivering an extemporaneous speech at the National FFA Alumni Convention (Figure 9–1). The most widely used and basic to the proper practice of the other styles, extemporaneous speaking develops the fluency needed for impromptu talks and the communicativeness needed for speaking either from manuscript or from memory. Mark Twain said, "It takes three weeks to prepare a good ad-lib speech."

Unlike the impromptu speech, which is speaking without preparation, an extemporaneous speech uses only a set of brief notes on a speaking outline after a limited amount of preparation time. The exact wording is chosen at the moment of delivery.[1]

The extemporaneous speech is carefully prepared. You know exactly how you will begin the speech, how you will develop each main point, and how you will conclude the speech. It is not **scantily** prepared. However, you do not write out the speech word for word, nor do you memorize it. A true extemporaneous speech, then, is a speech in which the ideas are firmly fixed in the mind, but the exact words are not memorized.[2]

Figure 9–1 Joe Clark delivering an extemporaneous speech at the National FFA Alumni Convention.

Keys to Effective Extemporaneous Speaking

The ability to speak *ex tempore* depends on nothing more than reading, studying, gaining information, and memorizing. The extemporaneous speech should develop from careful planning and thorough rehearsal. At the moment of presentation, you must recall only your organization, supporting materials, and planned mode of delivery.[3] The extemporaneous speaker's preparation is extensive and painstaking. It entails research, practice, delivery, and answering questions.

Adaptability of Extemporaneous Speaking

The final delivery of the speech can be adapted to the occasion because the speaker is not chained to a fixed sequence of words. Details can be offered, if necessary, during delivery. If the listeners look puzzled, the speaker can stop to restate an idea, add an illustration, make a comparison, or define an **ambiguous** phrase. Extemporaneous speakers who are sensitive to the immediate responses of the audience do not freeze their language prior to delivery. A speaker's style is given its final mold during the act of delivery.[4]

The circumstances of delivery, then, demonstrate that the speaker is in part creating the speech as it is being delivered. Sometimes, as in the impromptu mode, speakers build the entire speech as they talk. In fact, in every impromptu speech (as well as in most conversations), creation and presentation are **simultaneous**. No other artists, not even actors or musical performers, build their entire work in front of their audience. "Strictly construed, a speech that is built for delivery to a specific group of hearers is not realized—it does not live or exist—until it is spoken. Thus, the oft-repeated truism that a speech lives but once."[5]

ADVANTAGES OF EXTEMPORANEOUS SPEAKING

The first advantage of an extemporaneous speech is that you can adapt it to unforeseen situations; it is flexible. Suppose a group to which you were speaking included members of another group you spoke to previously. You would feel uncomfortable telling the same joke or using the same illustration. Therefore, you turn aside from your planned presentation and adapt.

New ideas flash into your mind through the inspiration of the audience. Therefore, you adapt your thoughts as you go along. New illustrations for ideas previously planned come to you at the last moment, and you include them as you speak. An extemporaneous speaker must always be ready to meet unforeseen speaking situations as they develop.

The second advantage of the extemporaneous speech is that it promotes a personal relationship between the speaker and the audience. The extemporaneous speaker is not glued to a manuscript, nor is she thinking about each memorized word. Since the speaker is prepared, even overflowing, with subject matter, she can talk about it personally with the audience. There should be no barriers between the speaker and the audience. Henry Ward Beecher said, "A written sermon is apt to reach out to people like a gloved hand; an unwritten sermon reaches out the warm and glowing palm, bared to the touch."[6]

Of course, you don't automatically speak better when extemporizing than when reciting a memorized speech. You will probably not speak nearly as well, but with experience, you can learn to speak better. The reasons are obvious. Neither your mind nor your eyes are on a manuscript. You are not trying to call up each memorized word. Instead, you face a live audience: as you talk, they respond to what you say. They show their response by facial expression, by nods of the head, and by close attention. This response arouses you to greater **earnestness**, greater sincerity, and greater power. You stimulate the audience; the audience stimulates you.[7]

DISADVANTAGES OF EXTEMPORANEOUS SPEAKING

A potential disadvantage of extemporaneous speaking is **slovenliness**. If you do not prepare carefully, you will ramble. You may also fall into slovenly or careless habits of using words such as "very necessary," or "most important." You may fall back on words that save you from really having to think: "I believe in *true Americanism* and in *justice* and *liberty*." The four italicized words, as they are used here, have no real meaning. Actually, you may mean that people have the right to free elections.

A careless extemporaneous speaker often adopts pet words and phrases that he repeats over and over. For example, each new idea may be introduced by *also*, and summarized by *don't you see?* This gets tiresome after a few minutes.[8]

The careless extemporaneous speaker can be **verbose**, for brevity requires exactness. It takes time to be brief. If you do not plan exactly, you will wander and digress, using many unnecessary words. Lord Lyons, eminent British diplomat, once began a letter with these words: "You will pardon the length of this letter, I have not the time to be brief."[9]

Extemporaneous speaking, unless watched carefully, may lead to **exaggeration** (Figure 9–2).

Figure 9–2 This speaker is beginning to exaggerate as he delivers his speech. With extemporaneous speech, spurred by the excitement of speaking, people sometimes expand or stretch the truth.

Spurred on by the excitement of speaking, a speaker may expand or stretch the truth. A few may become many, sometimes becomes always, and seldom becomes never. This tendency can develop into a habit, one that is hard to break and dangerous to keep.

PREPARATION AND RESEARCH

You may think that **mastery** is not a factor in the extemporaneous speech in which you must collect, organize, and present your thoughts almost simultaneously. Since your speech is not memorized, a great wealth of information is needed. You will find great comfort in a large storehouse of information and a broad background of reading and experience.[10] Some say the FFA Extemporaneous Speaking Contest is actually an information contest. This is partially true. Words without substance leave a speech wanting.

Preparation

Get started by investigating and analyzing your topic areas. Arrange your ideas, select material for the development of these ideas, prepare a suitable outline, and choose a method of attack. Even though you will not know your specific topic, you can still rehearse your delivery in the three topic areas. With preparation, appropriate words will come at the moment of delivery. The best extemporaneous speech develops from a solid foundation of rigorous preparation. Once sufficiently prepared, you will find the freedom, flexibility, and spontaneity needed to adapt to the shifting reactions of your listeners.[11]

Research

Notebook Start by preparing a notebook for the contest. Your notebook will become the key to your success. Topics will cover the themes of agriscience and technology, agrimarketing and international agricultural relations, food and fiber systems, and urban agriculture, so arrange the notebook into these four areas.

Read! Read! Read! There is no substitute for research as you prepare for the contest. Start by scanning material in your agricultural education classroom. As you see articles in the three topic areas, photocopy them and put the copied articles into your notebook. If you own the magazine, you may simply cut and tape the article into your notebook. Place tabs for subtopic areas. For example, international agricultural relations is one of the three sections in your notebook. Within this section, place tabs for your subtopics, such as General Agreement on Tariffs and Trade (GATT) and North American Fair Trade Agreement (NAFTA). The student in Figure 9–3 is preparing her notebook with sections and tabs.

Figure 9–3 Extemporaneous speaking in the FFA is really an information contest. This student is preparing a notebook and grouping information into tabbed sections.

Sources There are a variety of sources to use as you research your topics. The *Agriculture Yearbooks* are an excellent source. They are done yearly by topic areas. For agrimarketing, there is another yearbook called *Marketing U.S. Agriculture*. Another topic in the contest is international agriculture relations. The yearbook *U.S. Agriculture in a Global Economy* is an excellent source for this theme. For the other theme areas, agriscience and technology, articles can be found in the *Americans in Agriculture* yearbook. These books can be secured through your local congressperson.

Textbooks, magazines, research publications, and newspapers are all excellent sources for material. Many textbooks have the same titles or at least cover the same theme areas. Trade publications, such as a breed association magazine, address many relevant issues. *Progressive Farmer*, *Successful Farming*, and *Hoard's Dairyman* provide information in approachable language. They are useful in helping you understand an issue. Research publications can be technical, but many are written in readable language. One such publication is *Agricultural Research*, which is published monthly by the U.S. Department of Agriculture. Articles in this publication are reports of the latest research at various USDA research stations throughout the country.

Newspapers provide up-to-date information. Besides your community and city newspapers, *USA Today* has articles on many of the three major theme areas. *USA Today* also has a magazine with articles reprinted from the daily newspaper.

Many agricultural education programs have access to the AG/ED Network, which is a national online educational database. In the AG/ED Network, you will find many different agricultural lessons. Of course, there is an abundance of information on the Internet. By using search engine techniques, practically anything on the four subject areas of extemporaneous speaking can be found.

Use whatever resources you need to reach your objectives as long as they are within the rules. Don't forget the *Reader's Guide* and other library resources. Interviews are also appropriate for specialty information. The rules in the FFA Extemporaneous Speaking Contest limit sources to five items, four books and a notebook. The notebook may contain no more than 100 pages.

The more research you do, the more knowledgeable you become. Knowledge leads to confidence, which makes you more able to relate to the audience. The better you relate to the audience, the better speaker you will be. Finally, the more you research, the better you will be able to answer questions about your speech.

DEVELOPING YOUR SPEECH

You now have in your possession information and ideas to develop your speech. Remember, for an impromptu speech, you don't know what you will speak about, but for extemporaneous speaking, you at least have a general idea of your topic. Start practicing on developing your topic in the three topic areas.

Be sure that your information for your main points is accurate and sufficient. If necessary, provide additional facts, statistics, examples, and other materials that will help give **credence** to your main points. Draw on your own resources, but also verify your knowledge of a subject through additional observations, interviews, and reading.[12] As you develop your talk, prepare a method of attack that will work on any topic selected. Five possible methods of organizing and attacking a topic instantly are (1) six honest servants; (2) PREP; (3) past, present and future; (4) object or a visual aid; and (5) order of events.

Six Honest Servants

Rudyard Kipling, when asked how he became a prolific writer, replied, "I have six honest servants. They've taught me all I know. Their names are who, what, when, where, why, and how." In six words, you have the **rudiments** of all good speaking, writing, and reporting.[13] By using this formula, you should always be successful because you will be specific, organized, concise, and interesting. Consider the following.

Terry Young from Middle Tennessee State University set the World Land Speed Record for a hydrogen-fueled vehicle at the Bonneville Salt Flats in Wendover, Utah, in 1992. Young and his professor, Cliff Ricketts, wanted to demonstrate the viability of hydrogen from water as a fuel and show the speed and power of an environmentally clean vehicle. This was made possible by building an engine based on the physical principle of thermodynamics.

This paragraph includes all six honest servants.

PREP

PREP is an acronym for point, reason, example, and point.

Point: State your point.
Reason: Have a reason for your point.
Example: Give one example in the form of a statistic, comparison, incident or experience, illustration, or exhibit or demonstration. Any talk without a concrete example is weak.
Point: Restate or paraphrase your point or position.[14]

An example follows.

Point: Biotechnology can lead to great economic benefits in the area of ornamental horticulture.
Reason: Through plant genetics, multiple offspring can be created from one plant.
Example: Hundred of roses can be cloned by using tissue cultures from cells from a single rose petal.
Point: Because of biotechnology, less land, labor, and resources are needed to produce equal amounts of products, leading to tremendous economic gains.

You can use your time wisely by using the PREP formula. Take any theme from the three areas mentioned earlier and apply the PREP formula to develop an extemporaneous speech. You may amaze yourself with the results.

Past, Present, and Future

The past-present-future formula applies to every **animate** or inanimate thing you could mention. Again, the rules of extemporaneous speaking apply: Be specific, brief, and organized.[15] If you apply the past, present, and future formula to agricultural power technology, the flow will come easily. For example:

Past: In the early part of this century, fields were cultivated with horses and mules.
Present: Today, we cultivate fields with powerful four-wheel-drive tractors equipped with air conditioners, radios, CD players, and Global Positioning Units.
Future: What does the future hold? Already in the news we hear about no-till cultivators equipped with pesticide applicators to prevent insects, weeds, and diseases. The tractors can be remotely controlled by computers in your living room. Satellites from space can give you information about the condition of your crops.

Object or a Visual Aid

Using a visual aid is a modified version of show and tell.[16] For example, if you get a topic on international agriculture, your shoes could be used to support your speech. You could make the point that leather is produced by American cattle, cured and processed in Mexico, made into shoes in Thailand, and marketed in Europe by wholesalers to retailers in the United States. You could also use an item that you have in the pocket of your FFA jacket that illustrates a point you wish to make about agriscience technology or agrimarketing.

Don't overlook a chance to use something visual that's small and **discreet**. As a general rule, if you are not already using it or if it does not fit inconspicuously into your pocket, do not use it. Examples are bracelets or key chains with a quotation, pens, pencils, cards, a medicine bottle, credit cards, or money.

Some people use an adaptation of the object/visual aid method by using the **acronym** SPEECH, which stands for

Subject
Point
Enthusiasm
Exhibit
Concise (or ***C***lear)
Humor

Robert L. Montgomery suggests that all you really need is an object for the exhibit and the rest will come easily. You make a point about it concisely, clearly, and enthusiastically. It is clearer because you have a visual aid, something to show as you talk about it. The listeners can readily relate to it.

Order of Events

One other method of organizing your speech is to use a time sequence in which you discuss events in terms of the hour, day, month, or year, moving forward or backward from a certain time. A space order would take you from east to west, top to bottom, front to rear. Using causal order, you might discuss certain causes and then point out the results that follow. A special order is one of your own devising.[17] Whatever you use, make sure it flows smoothly and logically.

Outlining Your Speech

An outline organizes the results of your research as you develop your speech. It charts the main points and subpoints along the course that your speech will follow. Main points support or explain your topic. Subpoints support or explain either your main points or still other subpoints.

A fully developed outline is divided into three main sections: an introduction (which stimulates audience interest, paves the way for good relationships between the speaker and the audience, and discloses the speaker's subject and possibly the particular line of attack); the body or discussion (which is the major part of the outline and unfolds the steps in the development of the speaker's theme); and the conclusion, the capstone to the entire talk.[18] Refer to Chapter 8 for more detail on outlines.

Curtis Childers, former National Extemporaneous Speaker and National FFA President, offers a slightly different slant to an outline. Feel free to use a traditional outline or the outline prepared by Childers (Figure 9–4).

Attention-Getter

Make sure the attention-getter is as dynamic as possible. Tell a story, or use an analogy. Many speakers use a line from a popular song. When the song "The River" by Garth Brooks was a hit, many speakers alluded to it. Many stories can be told about historical figures or statements they have made. The primary purpose is to **captivate** the audience's attention so they think, "Hey, this is going to be interesting."

Childers's Extemporaneous Outline

I. Childers's attention-getter
II. Topic introduction
III. Main points
 A.
 B.
 C.
IV. Topic conclusion
V. Tie-in to the attention-getter

Figure 9–4 Curtis Childers uses this type of outline for his extemporaneous speeches. The attention-getter, which is usually short (15 to 30 seconds), comes before the topic is introduced. He goes back to the attention-getter story after the conclusion of the speech.

Topic Introduction

This is where most speeches begin. Introduce the extemporaneous speech just as you would a prepared speech. Introduce the topic and tell them what you are going to tell them. You basically want to say (1) why this topic is important and (2) why you need to know this, or how this information will benefit the audience personally, financially, or professionally.

State the Main Points

In Chapter 8, we stated that three to five main points are necessary. Three is usually preferred, but never use more than five. This is where you "tell them." Time is consumed quickly if you have subpoints for each of the main points. Use recognizable, everyday examples. If agrimarketing is the topic, the Associated Milk Producers slogan, "Milk does a body good" can be used as an example. "Pork, the other white meat" was used by the National Pork Council. Beef and poultry producers have new campaigns annually. Be up to date and relevant with your examples, illustrations, or stories.

Tie the main points together. To introduce the main points, use examples. For example, "Grow them, Know them, Show them."[19] As short and simple as it is, the phrase leaves a positive impression on the audience. Use this to illustrate the main points about educating the media about agriculture. This phrase is not the topic introduction nor is it an introduction to the main points. Here is an example of three main points illustrated in this phrase.

A. The media needs to be educated about the facts of production agriculture ("Grow them").

B. Use public relations and human relations skills to get to know members of the media better ("Know them").

C. Invite the media to the farm to spend a day experiencing production agriculture ("Show them"). This step is not necessary with every delivery. However, it certainly adds clarity and understanding if it can be worked into the speech.

Topic Conclusion

As you prepare your outline, simply remember (whether you write it down or not) to tie your main points together. Summarize the main points, or "tell them what you just told them."

Tie-in to the Attention-Getter

Go back to the story or whatever you used as an attention-getter and show a brief connection to your main points.

PRACTICING YOUR EXTEMPORANEOUS SPEECH

To quote Lloyd George, former Prime Minister of England: "trust in the inspiration of the moment—that is the fatal phase upon which many promising careers have been wrecked. The surest road to inspiration is preparation. I have seen many men and women of courage and capacity fail for lack of industry. Mastery in speech comes from mastery in one's subject."[20]

Practice Builds Success

Preparation and practice are the keys to self-confidence, enthusiasm, and competence. Mastery comes from study and practice or in one word: *preparation*. There is just one sure cure for all our negative speech habits, weaknesses, faults, and mannerisms. Get truly excited about practice and preparation, and 99 percent of your negatives will disappear. Because attitude is so important for success in any kind of speaking, especially extemporaneous speaking, adhere to the following advice from Oliver Wendell Holmes: "Success is the result of mental attitude, and the right mental attitude will bring success in everything you do."[21]

Figure 9–5 By speaking in front of teachers and other adults, you gain confidence in delivery and in topic area knowledge.

Public Practice

Set up formal practice times with your agricultural education teacher, your classmates, and the speech teacher in your school. Practice as if you were actually at the contest. Draw a topic, prepare for 30 minutes, use only the five references, and actually deliver an extemporaneous speech. Practice will improve the flow and transition of your speech. You will learn how to connect the attention-getter and introduction to the main points. The more you practice, the more you will be able to convert your thoughts into a speech. The FFA member in Figure 9–5 is practicing in front of the speech teacher and another adult.

Preplanned Introduction

It is wise to think up a few good sentences for an introduction to ensure against a fumbling start. Prepare introductions for each of the three topic areas. Remember, this is an extemporaneous speech and not impromptu speaking. Therefore, it is appropriate to prepare introductory thoughts even before you draw a topic.

Preplanned Conclusion

To prevent a total collapse of your speech, you may wish to have a few good sentences as assurance against an inconclusive ending. Obviously, these sentences must be open-ended and flexible since you don't know your specific topic. As in the introduction, thoughts can be focused for each of the three theme areas.

Practice Writing Main Points

When you draw a topic area during practice, quickly select three to five points. This is the time to prepare your thoughts. Practice finding data and statistics to support your points by using your four books and the notebook. Practice aloud to test your ideas. Fix the outline or main points in your mind. Develop an ear for the sound and swing of your talk. Keep thinking out your speech as you talk.

Practice Thinking Skills

When you first practice your outline, you are likely to find that you take too much time and that you are wordy and rambling. On this first practice, your statement of the main points may be confused, and even your telling of a simple story may be hard to follow. This is because your thinking is confused. Why does your choice of words get better after several trials? Because your thinking has improved. As you rehearse ideas extemporaneously, you literally think out loud; and as you word ideas differently, you are clarifying your thoughts. Although wordiness is the most common difficulty, searching for words is sometimes encountered. After stating your idea, in choosing analogies for a speech you may not be able to think of anything to say. What is the trouble? Again, it is your thinking. You have an idea but you have not thought it through; you have not developed it. So practice out loud. Try stating the thought in different ways. By talking about it, you clarify your thinking.

Practice Choosing the Right Words

As you practice, you must also choose words that will communicate your thinking to prospective listeners. The words that best clarify your own thinking may or may not be the right words for the audience. Therefore, you should practice with your prospective audience consciously in mind, varying your choice of words from practice to practice to suit your listeners.[22]

Specific Practice Sequence

- Organize three to five main points in your mind.
- Organize the specific and supporting material in your mind. Do not memorize it as you would a prepared speech. Picture or visualize it rather than memorizing.
- Read the outline silently and slowly.
- Read the outline aloud.
- Lay aside the outline and rehearse the speech aloud. If you forget any part, go on to the next part that you do remember. You are now trying to picture the whole thought pattern. Do not be tangled by details.
- Study your outline and note the places where you forgot, skipped parts, or got out of sequence. Patch these up in your mind. Then read the outline again aloud, slowly and thoughtfully.
- Lay aside the outline and rehearse the speech aloud from start to finish.
- Rehearse the speech formally 5 to 10 times (this does not include the previous silent rehearsals.) Start with thoughts on paper in the outline. Then, transfer them to thoughts in your mind. Now you get ready to create the speech while standing on your feet.
- Rehearse standing. Plan your posture and actions.
- Rehearse in a room about the size of the room in which you will speak. Get the feel of your voice filling the room. Make sure everyone can hear you.[23]

Learning to speak extemporaneously is like learning to read, learning to listen, or learning to use a typewriter. It takes practice. Practice builds knowledge. Knowledge builds confidence. Confidence builds success.

Preparation Room

Topic Selection

Thirty minutes before you are to speak, you will be asked to "draw" a topic. According to the rules of the FFA Extemporaneous Public Speaking Contest, the contestants "draw three specific topics relating to the industry of agriculture." After the contestant selects a topic, all three topics are returned to the original group before the next drawing. Obviously, select the topic for which you are best prepared.

Resources Needed

Make sure you have your notebook and four other references along with a pencil and note cards. To be more efficient with your time, wear a watch or have a stopwatch.

Get Comfortable and Organized

Take off your FFA jacket and shoes if it helps. The speaker in Figure 9–6 is getting relaxed as she prepares her speech in the preparation room. Once you start to prepare, eliminate everything not needed. For example, if the agrimarketing book is needed for your topic, eliminate or set aside the other three books. Then you will only have one book and your notebook in which to find information.

Select Three to Five Main Points

Quickly combine all relevant resources to be used for your topic. Brainstorm and write down words quickly as they come to mind. Organize the speech. Select three to five main points. Use your resources for specific information. Don't forget to use the table of contents or index to speed up your research. Some contestants mark the outer pages of their reference books with magic markers. These are color coded for selected topics. By doing this, they can get to information quickly. Make notes on the back of your outline note cards to get ready for possible questions that you anticipate.

Figure 9–6 This contestant has removed her jacket and shoes to relax as she prepares for her speech in the preparation room.

Supply Supporting Facts

As you add comments or supporting evidence, have some facts and figures. Do not just prepare a flowery presentation without any substance. Transfer what you know to your listeners. In other words, let people know what you know about the topic. Let your thoughts come through. Do not just read somebody else's thoughts—you could lose your credibility.[24]

Manage Your Time

Remember, you have only 30 minutes in the preparation room. Spend 10 to 15 minutes searching out materials and writing down facts in your outline. Spend the remaining time rehearsing your presentation. Go over in your mind at least two or three times what you are going to say. Go over and over the three to five main points. Make your presentation to the wall, a picture, or whatever is available in the preparation room.[25] Do not forget to go over your attention-getter, introduction, and conclusion, along with your main points.

Prepare Yourself Mentally

Other contestants will probably be in the room with you. Do not bother the others. Talk at a normal pace as you practice. If you cannot practice your speech aloud, go over your presentation mentally several times. Ready yourself mentally to perform. In athletics, we would say, "Psych yourself up."

NOTE CARDS

The contest rules state that "a contestant will be permitted to use notes while speaking, but deductions in scoring may be made for this practice if it detracts from the effectiveness of the presentation." It is only reasonable to assume that most beginning speakers will need note cards (Figure 9–7). Therefore, if you do use them, be discreet. A good way to be discreet is to fold the note card and place it in the palm of your hand. Refer to your notes when needed, but don't read them. Use your note cards as an aid, not a crutch. Try to limit looking at your note cards only as you go to your next point. Have an outline skeleton before you go to the preparation room. Once you began your preparation, fill in your information for the speech.

Preplanned Outline

TOPIC: New Ways of Marketing and/or Advertising Agriculture Products

I. ATTENTION GETTER

Tell Grocery Store Strike Story

II. TOPIC INTRODUCTION

Products are advertised for the 2000s to increase sales by stressing health, convenience, and packaging.

III. MAIN POINTS

A. Products are advertised and marketed to promote health:
1. Fat-Free
2. Low-Fat
3. Low Calories
4. Low Cholesterol

B. Products are marketed for the out-of-home working mother (and Father)
1. Pre-cooked
2. Microwave
3. Instant
4. Frozen

C. Packaging of Products
1. Bright Packaging
2. Small Quantities
3. Easy Access
4. California Raisins
5. Ziploc Packaging

IV. TOPIC CONCLUSION

As you can see, maybe the good old days are not as good after all.

V. TIE-IN ATTENTION GETTER

The Strike is Over and we are glad.

Figure 9–7 A sample note card with an outline skeleton that has been completed in the preparation room.

DELIVERY OF YOUR SPEECH

Chapter 7 provides extensive coverage on delivery: Refer to it for a discussion of power of expression, voice, stage presence, and general effect. Chapter 8 has some additional comments about delivery in prepared speaking. The comments in this section tend to relate more to extemporaneous speaking, although they could apply to any type of speaking. All of your preparations culminate either directly or indirectly in the delivery of your speech—in your voice, words, and body movements. Your goal is to talk *with* people, not *at* people. If you are convinced of the importance of your topic, you will be able to transfer these feelings to your audience.

When you approach the lectern to speak, take your time. Begin when you are ready. Your poise will help establish rapport with the audience.

Transferring Feelings

Since you don't follow a script with an extemporaneous delivery, you can be in more contact with the audience. Through your words, voice, and mannerisms, you can respond to the developing reactions of the listeners. Two words underline this concept: *empathy* and *rapport*.

Empathy is a process by which you involve yourself imaginatively and sympathetically with the thoughts and feelings of other people. You adopt their interests, try to see things as they see them and feel as they feel, even though you may not subscribe to their views.[26]

When both the speaker and listener project themselves into the minds and emotions of others, the interaction that follows is known as rapport. Rapport is the hallmark of good speaker-audience relationships. Without rapport, the communicative act is always incomplete and therefore imperfect; sometimes it is completely blocked.[27]

Conversational Norm

Former FFA Extemporaneous Public Speaker and National FFA President Curtis Childers suggests that you use a conversational norm as you speak. However, this is not true in all portions of the country. Charismatic enthusiasm is the norm. Your FFA advisor will know what your regional norm is.

Since lively conversations exemplify the attributes of communicative delivery, the conversational norm furnishes the best guide to the delivery of all speeches. Qualities of voice and action that characterize lively conversation—directness, animation, variety, and spontaneity—are appropriate to all speech, formal as well as informal. Many people have the mistaken notion that public speech calls for an oratorical manner that is mysteriously different from the natural mode of their private speech. The truth is that if you can talk easily, directly, and responsively in conversation happily, you may proceed with assurance in adopting the conversational norm.

Here are a few suggestions that will help strengthen your conversational style of delivery during your extemporaneous speech.[28]

Interest The first requisite of good delivery is an evident interest in what you are saying and in the people with whom you are talking. Given this interest in your subject and your listeners, many of the most common problems in delivery will vanish.

Think Think of what you are saying while you are saying it. Mental and emotional drifting can be spotted in a second; the speaker's manner and voice give him away. Speakers are not really communicating unless they are in touch with what they are saying as they speak.

Respond Respond to your audience. Think about what you are saying while you are saying it, but think about it in relation to your audience. Every audience sends out signals. Tune in on them to see if they carry messages of understanding, puzzlement, interest, boredom, weariness, or disapproval. Ask yourself, "Am I making myself heard? Would another example help?"

Captivate Hold onto your listeners by talking things over with them instead of just talking at them.

Voice Use your voice to carry meaning and feeling. The greatest value of the conversational norm is that it guides you in the use of your voice. Listen to the ordinary conversations around you. You will hear pleasant and unpleasant voices, clear and slovenly enunciation, good and poor diction, differences in dialect, mispronunciations, and grammatical errors. However, through it all usually comes remarkable meaning and feeling.

Body Language Speak with physical animation and directness. This can best be explained by the following.

> ***Speaking is action in which mind and body cooperate. Bodily action supplements and reinforces words; it energizes thought: it reveals you as a person who is self-confident rather than self-conscious. A listless person fails to sustain interest for very long; an anxious, distracted person communicates his distress; an overwrought person wears us out by trying too hard. A nice balance between relaxation and tension contributes to poise and directness.***[29]

Eye Contact Establish eye contact. People like personal attention. When you look your audience straight in the eye, you are taking notice of them. You are saying, "I invite you to share this information or observation." This simple gesture contributes to good human relationships. It also helps you adapt sensitively to the reactions of others.

Total Group Speak to the total group. If you are talking to a large group, you cannot focus your attention on everyone at once. Simply shift your attention smoothly from one point of the audience to another. Exclude no one. Do this easily and naturally without swinging your head back and forth. Usually there is no need to single out individuals unless you happen to be especially interested in their reactions. Even then, be careful not to make them feel uncomfortable.

FFA EXTEMPORANEOUS PUBLIC SPEAKING CDE

The national FFA Extemporaneous Public Speaking CDE is designed to develop the ability of all FFA members to express themselves on a given subject without having prepared or rehearsed its content in advance. This gives FFA members an opportunity to formulate their remarks for presentation in a very limited amount of time. The event is held in connection with the national FFA convention.

Event Rules

- The National FFA Extemporaneous Speaking CDE is limited to one participant from each state association.
- The National FFA Extemporaneous Speaking CDE is open only to students who were regularly enrolled in agricultural education during the current calendar year or who are still in high school but have completed all the agricultural education offered. When selected, participants must be active members of a chartered FFA chapter and the National FFA Organization. A member representing a state association may participate in the National FFA Extemporaneous Speaking CDE only once. A student can participate in only one of the speaking events at the national level in a given year.
- It is highly recommended that participants be in official FFA dress for each event.
- Copies of the rules and score sheet are supplied to participants in advance of the national event.
- The Boards of National Officers and Directors of the FFA are in charge of this event.
- Three to six competent and impartial persons are selected to judge the event. At least one judge should have an agricultural background. Each state with a speaker shall provide a judge for the national event.

Event Format

- Twelve topics shall be prepared by the event superintendent, including three each from the following categories: (a) agriscience and technology, (b) agrimarketing and international agricultural relations, (c) food and fiber systems, and (d) urban agriculture.
- Participants are admitted to the preparation room at 15-minute intervals and given exactly 30 minutes for topic selection and preparation.
- The officials in charge of the event will screen reference material on the following basis:

  Limited to five items.

  Must be printed material such as books or magazines and/or a compilation of collected materials. To be counted as on item, a notebook or folder of collected materials may contain no more than 10 single-sided pages, numbered consecutively (cannot be notes or speeches prepared by the participant or notes prepared by another person for the purpose of use for this event).
- Each speech shall be the result of the participant's own effort using approved reference material, which the participant may bring to the preparation room. No other assistance may be provided. Participants must use the uniform note cards provided. Any notes for speaking must be made during the 30-minute preparation period.
- A list of all possible topics is given to and reviewed by the judges prior to the beginning of the event.
- Event officials randomly draw speaking order. The program chairman introduces each participant by name and in order of the drawing. A participant is permitted to use notes while speaking, but deductions in scoring may be made for this practice if it detracts from the effectiveness of the presentation. Applause shall be withheld until all participants have spoken.
- The national contest is conducted in three rounds: preliminary, semifinals, and finals. No ranking is given except for the final four speakers.
- Two timekeepers record the time for each participant in delivering his or her speech, noting undertime or overtime, if any, for which deductions should be made. Timekeepers sit together.
- At the time of the event, the judges are seated in different sections of the room in which the event is held. They score each participant on the delivery of the production using the score sheet provided.
- Each judge shall formulate and ask questions. Questions shall pertain directly to the speaker's subject. Questions containing two or more parts should be avoided. Judges score each participant on the ability to answer all questions asked by all judges. The full five minutes should be used.
- When all participants have finished speaking, each judge totals the score on each participant. The timekeepers' record used in computing the final score for each participant. The judges' score sheets are then submitted to event officials to determine final ratings of participants.

Tiebreakers

Ties are broken based on the greatest number of low ranks. The participants' low rankings are counted, and the participant with the greatest number of low rankings is declared the winner. If a tie still exists, the event superintendent ranks the participant's response to questions. The participant with the lowest ranking from the response to questions is declared the winner. If a tie still exists, the participants' raw scores are totaled. The participant with the greatest total of raw points is declared the winner.

JUDGE'S SCORE SHEET

Be sure that you know how you are being evaluated. The following provides an explanation of the judge's score sheets (Figure 9–8).

Extemporaneous Public Speaking CDE Score Card

Judge's Name:____________________ Flight: ________________

Round: ____________________ Room No.: ________________

	Max. Points	One	Two	Three	Four	Five	Six	Seven	Eight
Content Related to Topic • Appropriateness of the total speech content to the topic selected • Extent to which the speaker addressed the topic selected • Suitability of the material used • Accuracy of the statements • Relationship to the content of agriculture	300								
Organization of Materials • Organization of contents • Unity of thought • Logical development • Language used • Sentence structure • Accomplishment of purpose — conclusion • Material related to sub-topic	100								
Power of Expression • Communicative ability including: fluency, emphasis, directness, sincerity • Conveyance of thought and meaning	100								
Voice • Quality, pitch • Articulation • Pronunciation • Force	100								
Stage Presence • Personal appearance • Poise and body posture • Attitude, confidence and personality • Ease before an audience	100								
General Effect • Extent to which speech was interesting, understandable, convincing, pleasing and held attention • Evidence of purpose	100								
Response to Questions* • Ability to answer the questions on the speech, which are asked by the judges, indicating originality, familiarity with subject and ability to think quickly.	200								
Gross Total Points:	1000								
Less Time Deduction**:									
Net Total Points									
Contestant Ranking:									

* Judges should meet prior to the event to prepare and clarify the types of questions to be asked.

** – 1 point per second over 6 minutes and – 1 point per second under 4 minutes, determined by the timekeepers

Figure 9–8 The National FFA Extemporaneous Public Speaking Contest Score Card with an explanation of the points. (Courtesy of the National FFA Organization.)

Time Limit

For contest purposes, each speech should not be less than four nor more than six minutes in length, with five minutes additional time allowed for related questions, which shall be asked by the judges. You will be penalized one point per second for being over six minutes or under four minutes.[30] The time begins when the speaker starts the speech. Speakers may use a watch to keep a record of the time. It may be advisable to secure a stopwatch. If it is discreetly placed on the lectern, it should not be a distraction.

Content Related to Topic

Preparation is the key to extemporaneous speaking. This does not mean having a "canned" speech for each of the three theme areas. Select your topic from the four choices. Make three to five points about the topic and add any supporting data or statistics for each point.

Knowledge of the Subject

Know your subject. Make sure your information is correct, important, and appropriate. Make sure the examples and stories you use to illustrate your main points are accurate and relevant. For example, you would not use artificial insemination as an illustration for genetic engineering.

Organization of Material

The key words are *smooth flow* and *transition*. Make sure the introduction, body, and conclusion have unity of thought. The main points should be presented in a logical order, such as time sequence or causal order. Use appropriate language and sentence structure.

Introduce the topic. Introduce the main points. Discuss the first main point and use illustrations to explain it. Discuss the second main point. Tie it in with the first point. Address the third main point. Show the connection with the first two points. Your conclusion should cover the main points. If you used an attention-getter, refer to it again as you conclude.

See Figure 9–8 for the judge's score card and an explanation of the score sheet points. Voice, stage presence, power of expression, and general effect are discussed in Chapter 7.

ANSWERING QUESTIONS

The only difference between answering questions for prepared public speaking (discussed in Chapter 8) and extemporaneous speaking is the unknown factor. With prepared speaking, you can somewhat anticipate your questions. Obviously, this is more difficult for extemporaneous speaking. First, you don't know your topic until the contest. Second, it is hard to predict or anticipate questions from the judges until you are in the preparation room. The true challenge for this contest is answering the questions. Preparation is vital to answering questions effectively. There is no substitute. Consider the following suggestions or techniques.

Ask for Questions

Once you conclude your speech, say something like, "I now would like to address a few questions." Many speakers just stand in silence. This statement provides a smooth transition from the speech to answering questions.

Positive Response

For questions that have two or more potential answers, you may want to introduce your answer with, "in my opinion," or "as I see it," or "my present thinking is this." The judge may not agree with your opinion, but the answer is not presented as a truth (only your belief or opinion). Technically, your opinion cannot be considered an incorrect response. In other words, you may not have given the judge the answer he or she wanted, but you gave a positive response.

Stall Tactfully

Stall tactfully and discreetly. Sometimes you have no idea how to respond. Rather than having 30 seconds of dead silence, which may seem like 30 minutes, say, "I'm glad you asked that question," or "Would you repeat the question please?" This could give you the needed time to regroup and continue without any noticeable break. However, don't get into the habit of using these phrases.

If You Do Not Know

What do you do if you don't know the answer? You have two choices. You can say, "I don't know," or you can say, "I didn't come across that information, but I did find that . . ." Another possible response would be, " I don't know, but if you had asked . . .".

Response Time

How much time should you spend answering a question? Remember, you have five minutes to answer questions. If you give short answers, you could get as many as six to eight questions. If you elaborate on your answers, you could get as few as two to four questions. Therefore, if you get a question in an area in which you are very knowledgeable, you may respond to it longer than usual, but do not overdo it. If you get a hard question, give a response and move on to the next question.

Data and Statistics

Use data and statistics when appropriate. Your credibility is greatly enhanced when you can quote data to support your answers. The more you read, the better you will be able to do this.

Conversational Norm

Don't forget that you are still performing. Speakers often forget the importance of voice, power of expression, stage presence, and general effect when they are answering questions. Use the same delivery style for answering questions as you do for your four-to-six-minute presentation. When the judge calls time, say softly, "Thank you." It seems awkward or disrespectful to sit down or leave the lectern without some closure or response. You and your advisor may think of other appropriate responses. What does the thank you mean? It is a short phrase that acknowledges their effort in permitting you to develop your speaking skills and abilities.

Summary

There is little point in extemporaneous public speaking if you follow a preconceived method of attack on your subject. Do not allow yourself to become panicky. Remember, some nervousness is a good sign of readiness. Your audience will expect nothing extraordinary from you because they too know you are speaking extemporaneously. Actually, they probably will be very encouraging. If you approach your speech with poise and determination, your chances of success are exceedingly good. Well-rounded knowledge obtained from a consistent reading program will assist you immeasurably. FFA Extemporaneous Public Speaking is a challenge. Take the challenge. This event will truly help you develop your speaking abilities.

Being able to speak extemporaneously and answer questions on an unknown topic with only 30 minutes of preparation is influence. Influence is leadership. Few people ever develop this ability. Although almost anyone can learn with practice, verbal ability, speaking, and thinking on your feet are skills that help get you promotions once you enter the workforce.

Preparation is the key. Read! Read! Read! There is no substitute for research. Read articles and assemble a notebook. Have a section for each of the theme areas: agriscience and technology, agrimarketing, and international agriculture relations. Select four other books that can be used later in the preparation room.

The advantages of extemporaneous speaking are that (1) you can adapt it to unforeseen situations, (2) it promotes a personal relationship between the speaker and the audience, and

(3) it leads to superior delivery. The disadvantages are that it can lead to (1) slovenliness, (2) verboseness, and (3) exaggeration. As you develop your speech, make sure your information is both accurate and sufficient. Draw on your own resources, but also verify your knowledge of a subject through additional observations, interviews, and reading. Develop a method of attack that will work on any topic selected. Five methods of organizing a topic are six honest servants; PREP formula; past, present, and future; an object or a visual aid; and order of events. An outline organizes the results of your research as you develop your speech. One method of outlining is to have an attention-getter, topic introduction, body (three to five main points), topic conclusion, and tie-in to the attention-getter.

Practice your speech. Preparation and practice are the keys to self-confidence, enthusiasm, and competence. Set up formal practice times with your agricultural education teacher (FFA advisor), classmates, and speech teacher. Preachers, other teachers, friends, and parents are also people who could assist you. Remember, you have only 30 minutes while you are in the preparation room. Spend 10 to 15 minutes on your outline and the remaining time rehearsing your presentation. Take your five references, eliminate everything not needed, and work on selecting three to five points.

Note cards should be used discreetly. A good way to be discreet is to fold the note card and place it in the palm of your hand. Try to look at the card only once—as you go to your next point.

Once you begin, approach the lectern, reflect for a moment, be natural, be confident, and look sharp. Deliver your speech conversationally. There may be exceptions for certain regions of the country. Express your feelings by exhibiting empathy and rapport. Be aware of the content of the judge's score sheet. Be aware of your five- to six-minute time limit. Your speech will be evaluated on content related to topic, knowledge of subject, organization of material, power of expression, voice, stage presence, general effect, and responses to questions.

Answering questions is the true challenge for the extemporaneous speaker. Give clear and thorough answers. Do not forget that you are still performing. If you do not know an answer or if you need more time to organize your thoughts, use some techniques to discreetly stall and make a transition statement. Then give a positive response.

END-OF-CHAPTER EXERCISES

Review Questions

1. Define the Terms to Know.
2. What is the difference between extemporaneous and impromptu speaking?
3. What are three advantages of extemporaneous speaking?
4. What are three disadvantages of extemporaneous speaking?
5. List five types of publications that could be used as resources for preparing an extemporaneous speech.
6. Is it possible or appropriate to have a preplanned introduction and preplanned conclusion before you speak on an unknown topic? Explain.
7. What are five things you should do in the preparation room once you have drawn a topic?
8. What are eight things that will help your conversational style of delivery?
9. Explain what you would do if you were asked a controversial question that may have two or more potential answers?
10. What can you do if you don't know an answer to a question?

Fill-in-the-Blank

1. For the FFA Extemporaneous Speaking Contest, ________________ minutes before you are to speak, you will be asked to draw your topic.

Insert the number of points after each category on the extemporaneous public speaking judge's score sheet.

2. Content related to subject: _____ points
3. Organization of material: _____ points
4. Power of expression: _____ points
5. Voice: _____ points
6. Stage presence: _____ points
7. General effect: _____ points
8. Response to questions: _____ points
9. Total points: _____ points

Matching

_____ 1. Start with a story or an analogy
_____ 2. "Tell them what you are going to tell them"
_____ 3. "Tell them"
_____ 4. "Tell them what you told them"
_____ 5. Go back to the story at the end of the speech
_____ 6. Speaking on the subject selected
_____ 7. Gain through research and preparation
_____ 8. Smooth flow and transition

A. topic conclusion
B. tie-in to the attention-getter
C. main points
D. attention-getter
E. topic introduction
F. knowledge of the subject
G. organization of material
H. content related to topic

Activities

1. Prepare a notebook for extemporaneous speaking. Arrange the notebook into three major sections. Title the three sections agriscience and technology, agrimarketing, and international agricultural relations. Read magazines, journal articles, newspapers, pamphlets, extension bulletins, and other materials. Copy (cut and tape when appropriate) articles that address the three theme areas. Place them in the notebook. Place tabs within each major section. Arrange subtopics together. For example, tab a section called "agricultural exports." Have at least five subtopics tabbed for each section of your notebook. Find at least one article for each subtopic.
2. **Take it to the Net.**
   Search the Internet for information on the five theme areas (agriscience and technology, agrimarketing, food and fiber systems, urban agriculture, and international agricultural relations). Go to a search engine. Try Metacrawler <http://www.metacrawler.com> as your search engine. Search each of the five theme areas separately by typing them into the search field. Find as many useful Web sites as you can and add them to the notebook you started in question 1.

3. Select a topic or make up one within one of the three theme areas mentioned in activity 1. Prepare a one- to two-minute talk using one of the following methods of organizing and attacking the topic: six honest servants, PREP formula, past, present, and future, object or a visual aid, order of events.
4. Refer to the outline section in this chapter. Select a topic from one of the three topic areas and prepare an outline using your notebook and reference books. Follow the Childers's outline style.
5. Select a person to listen to your speech from the outline prepared in activity 3. As you practice, get the person to list five areas in which you need to improve.
6. Deliver your speech to the class. Since you have begun preparing this extemporaneous speech as you completed the previous activities, you will have had more than 30 minutes to prepare. That is okay, we have to start somewhere. Note: At the appropriate time, your teacher will have you draw a topic, prepare for 30 minutes, and deliver your speech.
7. Answer questions about your extemporaneous speech.

Notes

1. S. E. Lucas, *The Art of Public Speaking* (New York: Random House, 1986), p. 229.
2. W. G. Hedde et al., *The New American Speech* (New York: J. B. Lippincott Company, 1963), p. 144.
3. W. W. Braden, *Public Speaking: The Essentials* (New York: Harper & Row, 1966), p. 78.
4. D. C. Bryant and K. R. Wallace, *Fundamentals of Public Speaking*, Fourth Edition (New York: Appleton-Century-Crofts, 1969), p. 20.
5. Ibid.
6. Hedde, *The New American Speech*, p. 144.
7. Ibid.
8. Ibid., p. 145.
9. Ibid.
10. Braden, *Public Speaking: The Essentials*, p. 76.
11. J. H. McBurney and E. J. Wrange, *Guide to Good Speech*, Fourth Edition (Englewood Cliffs, NJ: Prentice-Hall, 1975), p. 20.
12. R. L. Montgomery, *A Master Guide to Public Speaking* (New York: Harper & Row, 1979), p. 59.
13. Ibid.
14. Ibid., p. 60.
15. Ibid., p. 61.
16. Ibid.
17. C. S. Carlile, *Project Text for Public Speaking* (Boston, MA: Addison-Wesley, 1997).
18. McBurney and Wrange, *Guide to Good Speech*, p. 26.
19. "Competitive Public Speaking: The Extemporaneous Speech Contest," Videotape (Lubbock, TX: Creative Educational Video, 1993). Note: Curtis Childers narrates.
20. Montgomery, *A Master Guide to Public Speaking*, p. 67.
21. Ibid.
22. McBurney and Wrange, *Guide to Good Speech*, p. 27.
23. Hedde, *The New American Speech*, pp. 146–147.
24. "Competitive Public Speaking: The Extemporaneous Speech Contest."
25. Ibid.
26. McBurney and Wrange, *Guide to Good Speech*, p. 25.
27. Ibid.
28. Ibid., pp. 26–28.
29. Ibid., p. 29.
30. "National FFA Contest," Bulletin No. 4, 2000–2001, (Alexandria, VA: The National FFA Organization), pp. 135–138.

SECTION 3

Leading Individuals and Groups

CHAPTER

10

Basic Parliamentary Procedure

Objectives

After completing this chapter, the student should be able to:

- Discuss the characteristics of a presiding officer
- Demonstrate the procedure for handling a motion
- Describe the standard characteristics of a motion
- Describe the purpose and types of voting
- Demonstrate the following motions:
  - Main motion
  - Motion to amend
  - Previous question
  - Refer to a committee
  - Lay motion on the table
  - Take motion from the table
  - Motion to postpone definitely
  - Motion to postpone indefinitely
  - Suspend the rule
  - Rise to a point of order
  - Appeal from the decision of the chair
  - Division of the house (assembly)
  - Reconsider a motion
  - Motion to recess
  - Motion to adjourn
- Explain the proper order of business (agenda)
- Discuss common parliamentary errors and misconceptions

Terms to Know

parliamentary procedure
presiding officer
chairperson
out of order
poise
recognition from the chair
discuss
decorum
precedence
vote
majority vote
two-thirds vote
general (unanimous) consent
main motion
amendable
subsidiary motions
germane
primary amendment
secondary amendment
debatable
immediately pending motion
standing committees
special committees
adjourned meeting
order of the day
incidental motion
violation
parliamentarian
tie vote
vote by voice (viva voce)
privileged motion

Agricultural education and the FFA have been credited with developing leadership. In Chapter 1, we discussed the areas within the FFA that contribute to leadership development. One such area is **parliamentary procedure**. The real question is how does parliamentary procedure contribute to leadership development?

In Chapter 1, we discussed behavioral leadership as one of the seven categories of leadership. Behavioral leadership includes democratic, authoritarian, and situational leadership. Except in certain situations, Americans tend to favor democratic leadership. If democratic leadership is the best type of leadership, how do we learn to be democratic leaders? The answer lies within parliamentary procedure.

Parliamentary procedure rules guarantee that a meeting proceeds in a purely democratic fashion. Everyone has equal rights to discuss every proposal and everyone gets a vote. The process ensures that things done by the organization are supported by the majority of its members. Using correct parliamentary procedure in organization meetings is the most democratic way of doing things.[1]

The rules of parliamentary procedure are based on consideration for the rights of the majority, individual members, and absentees. Under the rules of parliamentary law, a group is free to do what it wants to do with the greatest measure of protection for itself and consideration for the rights of its members.[2] Besides local, state, and national elections, the most democratic experience in our country occurs when a person participates in a meeting according to the rules of parliamentary procedure. This is democratic leadership in action.

Parliamentary procedure has evolved over the years, starting from a few basic rules in the English Parliament. Other democratic groups and societies started using parliamentary procedure, and the rules gradually expanded and changed. Today, *Robert's Rules of Order, Newly Revised*, has become the standard code of parliamentary procedure.

This chapter introduces parliamentary procedure to the inexperienced student. After reading this chapter, the student should be able to run a meeting as a capable **presiding officer**, as well as handle a motion by going through all the sequential steps. The student should be able to handle a variety of privileged, subsidiary, and incidental motions.

After practicing and becoming competent with the parliamentary procedure abilities in this chapter, you should be a capable, thinking, and active **chairperson** or member of any organization that conducts its business according to parliamentary procedure. The information presented here covers most of the parliamentary procedure that a member would encounter in most business meetings. Chapter 11 goes a step further for those desiring more advanced and in-depth knowledge of parliamentary procedure.

CHARACTERISTICS OF THE PRESIDING OFFICER

The presiding officer of a meeting is ordinarily called the president or chairperson. In a group that is not permanently organized, this officer is known as the chairperson. In an organized group, the presiding officer's title is usually prescribed by the bylaws; *president* is the most common title. The term *chair* refers to the person who is actually presiding at the time, whether that person is the regular presiding officer or not.[3]

Addressing the Presiding Officer

The presiding officer should be addressed by an official title. If president, "Mr. or Madam President" is proper; if chairperson, "Mr. or Madam Chairperson." "Mr. or Madam Chairperson" is always appropriate and is recommended as a standard practice to avoid the common error of saying "Mr. or Madam President" when someone other than the president is the chair. The sex of that person is designated by the use of "Mr." or "Madam."[4]

The presiding officer speaks of himself or herself only in the third person. The presiding officer never uses the personal pronoun "I" to refer to himself or herself as long as he or she is the presiding officer. In actual parliamentary proceedings he or she always refers to himself or herself as "the chair." If the chair is obliged to rule that a motion is **out of order**, he or she should say, "The chair rules that the motion is out of order."[5]

Qualities of a Good Presiding Officer

The presiding officer should possess certain qualities that guarantee the rights of the majority, the minority, and individual members. The presiding officer should maintain order and be in complete control of the assembly at all times. The presiding officer should

- *Be fair.* Members on both sides of the question should be respected and heard. In fact, the chair should purposely alternate pro and con speakers on a motion as much as possible.
- *Possess good judgment.* Each situation should be handled thoughtfully and seriously. The chair should direct discussion to the topic under consideration.
- *Manifest* ***poise***. The chair should be erect in posture and speak with a strong, clear voice. The chair should also face the assembly and appear confident. The leadership of the chairperson should be apparent.
- *Have a working knowledge of parliamentary procedure.* It is recommended that the chairperson be well educated in parliamentary procedure. He or she should be able to make quick and accurate rulings. The chair should refer to the parliamentarian as little as possible.
- *Be deliberate and tactful.* The chair should not conduct business too hastily nor delay matters unnecessarily. The chair should also try to maintain group harmony in all situations.[6] The chairperson in Figure 10–1 appears to be a good presiding officer.

Figure 10–1 A good presiding officer should be fair, possess good judgment, manifest poise, be deliberate and tactful, and have a working knowledge of parliamentary procedure. (Courtesy of California's Fullerton FFA.)

PROCEDURE FOR PROPERLY HANDLING A MOTION

One of the basic elements of conducting parliamentary procedure effectively is receiving and disposing of motions in the correct manner. There are eight steps in receiving and disposing of a motion (Figure 10–2).

Step 1

A member rises and addresses the chair.

Step 2

The member gains **recognition from the chair**.

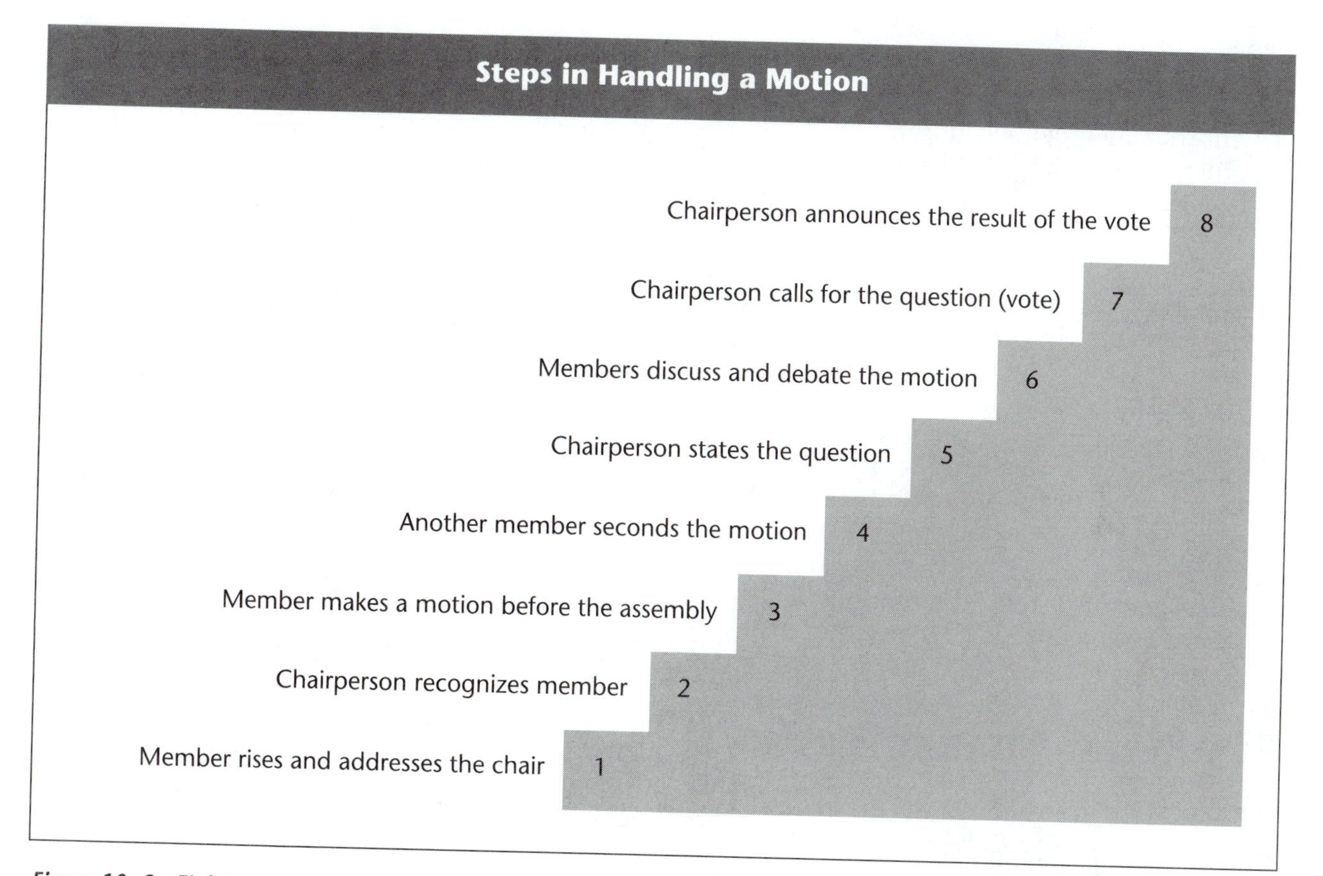

Figure 10–2 Eight steps are needed for a single motion to go through the proper procedure.

Step 3

The member makes a motion before the assembly.

Step 4

Another member seconds the motion.

Step 5

The chairperson restates the motion.

Step 6

The members **discuss** and debate the motion.

Step 7

The chairperson calls for the question (vote).

Step 8

The chairperson announces the result of the vote.[7]

It is vital for the chairperson to know and be able to follow these steps automatically when handling a motion.

Example

1. Member rises and addresses the chair. Member stands and says, "Mr. Chairperson."
2. Chairperson recognizes member. Chairperson says, "Mr. Speaker."

3. Member makes a motion. Member states, "I move that . . ."
4. Another member seconds the motion. Another member says, "I second the motion."
5. Chairperson states the question. Chairperson says, "It is moved and seconded that . . . (repeating the motion)."
6. Members discuss and debate the motion. Chairperson asks, "Is there any discussion?" Members debate the motion following rules of **decorum**.
7. Chairperson calls for the question (vote). After discussion has apparently ceased, the chairperson says, "If there is no further discussion, all in favor say 'Aye.' All opposed say 'Nay.' "
8. Chairperson announces the result of the vote. Chairperson states, "The motion passes," or "The motion fails."

STANDARD CHARACTERISTICS OF A MOTION

The five major classifications of motions are main, subsidiary, incidental, and privileged. When each of these is applied to a motion as a question, it helps the presiding officer keep his or her thoughts organized. When you know the characteristics of each motion, you will know how to classify them by asking yourself certain questions. Proper classification leads to correct parliamentary procedure during a meeting. They are as follows:

- Does this motion take **precedence** over any other motion? What motions can be pending without causing this motion to be out of order? What motions can be made and considered while this motion is pending?
- To what type of situation is this motion applicable? What motions, particularly subsidiary motions, are applicable to this motion, if any?
- Is this motion in order when another motion is on the floor?
- Does this motion require a second?
- Is this motion debatable?
- Is this motion amendable?
- What percentage rate is required for its adoption?
- Can this motion be reconsidered?[8]

Purpose of Seconding a Motion

The purpose of seconding a motion is to keep the assembly from wasting its time on a motion that only one member wishes to consider. A second shows that at least two people in the assembly are in favor of considering the motion; it does not necessarily mean they are in favor of passing the motion. By requiring a second on most motions, the assembly prevents one member from wasting time with useless, time-consuming motions.

If a motion that requires a second does not receive one, it dies from lack of support. The chairperson states, "The motion dies for lack of a second." The chairperson can then call for further business.

PURPOSE AND TYPES OF VOTING

The purpose of a **vote** on a motion is to see if the assembly is in favor of or opposed to that particular motion. Most of the motions requiring a vote require a **majority vote**. A majority vote is defined as more than half of the votes cast by those legally entitled to vote.

Two-Thirds Majority Vote

A **two-thirds vote** is required on motions that take away the rights of members, such as the motion to suspend the rules, previous question, rescind, or the motion to limit debate. In figuring a two-thirds vote, a chairperson must remember it is two-thirds of those voting who are legally entitled to vote.

An easy way to determine a two-thirds vote is to double the number of negative votes. If the number of negative votes is equal to or less than the number of affirmative votes, then there is a two-thirds majority vote. For example, for a motion requiring a two-thirds vote to pass, if the affirmative vote is six and the negative vote is three, it is a two-thirds majority vote. When the negative vote is doubled, it becomes six. Since six is equal to the number of affirmative votes, it is a two-thirds majority vote. On the other hand, if the negative vote, doubled, is more than the positive votes, it is not a two-thirds vote. For example, six positive and four negative votes are cast. Since the four is doubled to equal eight, and eight is more than six, the motion fails.

General Consent

Another type of vote is **general (unanimous) consent**. It is useful when there seems to be no opposition in routine business or on questions of little importance because it often saves much time. General consent is usually used to adopt the minutes. The chairperson says, "Are there any corrections to the minutes? If not, they stand approved as read." General consent can be used either to adopt a motion without going through the eight steps discussed previously or to take action without the formality of a motion.[9]

If it is obvious the motion is favored by all in the group, the chair may state, "If no one objects, we will adopt this motion (or, it passes)." If one member of the group objects, however, it is not unanimous, and the chair must conduct a regular vote, first calling for the affirmative and then negative votes.

Methods of Voting

There are five major methods of voting: voice, hand, standing, secret ballot, and roll call. For votes about which the chairperson is relatively sure or most people feel the same way, a voice vote is used. The chairperson often asks for a voice vote for an adjournment. Raising the right hand is the vote of choice for most meetings. For larger groups or more precise voting, a standing vote is often used. For local, state, and national government-elected officials, a roll-call vote is used. The votes are a matter of record. With new technology, many elected officials have buttons at their stations. When they press the "aye" or "nay" buttons, the vote is recorded on a display board. If you think your vote doesn't make a difference, consider the following.

- One vote gave America the English language instead of German.
- One vote brought Texas into the Union.
- One vote gave Rutherford B. Hayes the presidency.
- One vote gave Adolf Hitler leadership in the Nazi party.[10]

MAIN MOTION (ORIGINAL)

Rationale

An original **main motion** introduces new business before the assembly. A main motion is a formal suggestion of action to be taken on an item of business by means of the group discussing and debating it and ultimately voting on it for approval or rejection.

Explanation

Main motions are the foundation of parliamentary business. There can be only one main motion on the floor at a time. A main motion takes precedence over no other motion. Therefore, a main motion may be made only while no other business is pending on the floor. Since a main motion introduces new business before the assembly, a second is required to consider the motion. A main motion is debatable and **amendable** and requires a majority vote to pass. If a main motion receives the majority vote required to pass, it becomes the official statement of the action taken by the group. A motion should be worded in a clear, concise, and complete form appropriate for its intended purpose. All official motions should be introduced with the two words "I move" or three words "I move that." According to Sarah Corbin Robert in *Robert's Rules of Order, Newly Revised*, "no main motion is in order which conflicts with the bylaws (or constitution) or rules of the chapter, organization, or assembly. If such a motion is adopted, even by a unanimous vote, it is null and void."

Example

Member: *Mr. or Madam Chairperson.*

Chairperson: *Mr. or Madam Speaker.*

Member: *I move that our FFA chapter have a hayride.*

Chairperson: *Is there a second to the motion that our chapter have a hayride?*

Another Member: *I second the motion.*

Chairperson: *It is moved and seconded that our chapter have a hayride. Is there any discussion?* (There may or may not be discussion.)

Chairperson: *If there is no further discussion, all in favor say "Aye."* (Response.) *All opposed say "No."* (Response.) *The motion passes or fails.* (See Figure 10–3.)

Note: Either way, the chairperson asks for further business.

Figure 10–3 Parliamentary procedure is simply a process for majority rule or democracy in action. If the FFA chapter votes to have a hayride and school policies are not violated, the chapter will have a hayride. (Courtesy of Bruce Haley.)

MOTION TO AMEND

Rationale

The motion to amend is probably the most widely used of the **subsidiary motions**. The purpose of the motion to amend is to modify or change the wording and in some cases the meaning of the motion to which it is applied. A motion to amend may also be referred to as an amendment. There are three ways to amend a motion:

- insert (or add)
- strike out (or subtract)
- strike out and insert or substitute

Explanation

For an amendment to be considered, it must be **germane** to the motion to which it is being applied. "A germane amendment is one that is closely related to or having bearing on the subject of the motion to be amended. . . . The purpose of germaneness is so that no new subject can be introduced under the pretext of being an amendment."[11]

There are two types of amendments: primary and secondary. A **primary amendment** is one that is applied to any amendable motion, except the motion to amend. A **secondary amendment** is applied to a motion to amend. A secondary amendment is **debatable**; however, it is unamendable. Therefore, there may be only two motions to amend pending on a motion at a given time. For example, there is a main motion on the floor to have a hayride. There is also an amendment on the floor to add "with the FCCLA on October 14." There is also a motion to amend the amendment on the floor to strike the words *October 14* and insert *October 13*.

The motion to amend takes precedence over a main motion and the subsidiary motion to postpone indefinitely. The motion to amend requires a second and is debatable and amendable. A secondary amendment is also debatable, but it is not amendable. The motion to amend requires a majority vote to pass; if it is adopted, the motion it was applied to should be changed accordingly. If the motion to amend fails, the motion it was applied to remains as it was.

Note: A member's vote on an amendment does not obligate him or her to vote in a particular way on the motion to which the amendment applies.

Example

Member: *Madam Chairperson.*

Chairperson: *Madam Speaker.*

Member: *I move to amend the motion by adding the words* on October 25 at 7:00 PM *after the word* hayride.

Chairperson: *Is there a second to the motion that we amend the main motion by adding the words* on October 25 at 7:00 PM *after the word* hayride?

Another Member: *I second the motion.*

Chairperson: *It has been moved and seconded that the words* on October 25 at 7:00 PM *be added to the motion. Is there any discussion?*

Note: Chairperson takes a vote and if it passes, proceeds to attend to the main motion as amended. If the motion to amend fails, the assembly returns to discussion of the main motion.

Previous Question

Rationale

The purpose of the previous question motion is to end discussion and bring a motion to a vote. It is used when members are tired of long debates or when members feel the majority know how they want to vote without continued discussion. It is useful when the assembly is taking more time than is needed to consider a motion. The previous question not only immediately closes debate, but it also stops amendments or the **immediately pending motion** and other pending motions, as the motion may specify. It also prevents the making of any other subsidiary motions except the higher-ranking "lay on the table."

Explanation

Consider the following example as an explanation of the previous question motion. A main motion is pending that also has an amendment pending. During discussion of the motion to amend, a member moves the previous question. If the member specifies that the previous question is for all items of business on the floor and the previous question passes, the assembly would vote immediately on the motion to amend and then on the main motion. If the member does not specify the motions to which it applies, the chair should interpret that the previous question is applied only on the immediately pending motion, which in this example is the motion to amend.

The previous question is commonly used by legislators to bring a motion to a vote before the opposing side can gain enough votes for their position.

The previous question takes precedence over all debatable or amendable motions to which it is applied, and over the subsidiary motion to limit or extend limits of debate. It yields to the subsidiary motion to lay on the table and to all privileged motions. It requires a second, is not debatable or amendable, and requires a two-thirds vote to pass.

Example

There is a main motion pending that our FFA chapter send two delegates to the National FFA Convention and pay their expenses. As soon as the motion is open for discussion, a member moves the previous question. This is done as a strategic tactic, so the opposition does not have a chance to gain enough votes for their cause.

Member: *Madam Chairperson.*

Chairperson: *Madam Speaker.*

Member: *I move the previous question.*

Chairperson: *Is there a second?*

Another Member: *I second the motion.*

Chairperson: *This motion is nondebatable and requires a two-thirds vote. All those in favor raise your right hand.* (Response.) *All those opposed raise your right hand.* (Response.)

Note: If the motion passes, the next business is to vote on the motion. If it does not pass, the original motion is still open for debate.

Refer to a Committee

Rationale

The motion to refer to a committee is used on items of business that cannot or should not be disposed of immediately by the group.[12] To send business to a committee is especially helpful when the group needs more detailed study on a time-consuming or complicated matter. By referring a motion to a committee, the assembly can save time by entrusting action to those particularly qualified to act and avoiding long, unproductive discussion.[13]

A committee is a body of one or more persons elected or appointed by an assembly or society to consider, investigate, or take action on certain matters or subjects, or to do all of these things. A committee is not itself considered a form of assembly.[14]

Explanation

There are three major types of committees. Two of these are commonly used; the other is rarely used. **Standing committees** and **special committees** are commonplace in most organizations, but a committee of the whole is used only in special instances.

A standing committee has a continuing existence. Most organizations have several standing committees that carry out most of the committee work. The FFA Program of Activities consists of standing committees. Examples of standing committees are the Budget and Finance Committee, the Ethics Committee, and the Executive Committee. A special committee is created for a special purpose, and it goes out of existence as soon as it has completed a specified task.

An example of a special committee of an assembly is one that has been appointed to study the possibility of hosting a celebrity basketball game fund raiser. Since this item of business is obviously a one-time special purpose, a special committee is developed. After the committee is finished with its business, it is dissolved.

Some details may be included when the motion is made or amended during discussion:

- How many members are on the committee
- Who the members are to be on the committee
- When the committee should report back to the assembly
- Whether the committee has the power to make a decision (power to act)
- The duties the committee is to perform
- The type of committee: standing or special

If a motion to refer to a committee passes, any such details not given in the motion are then decided by the chairperson.

The motion to refer to a committee requires a second; it is debatable, amendable, and requires a majority vote to pass. The motion to refer to a committee takes precedence over the main motion, postpone indefinitely, and the motion to amend. The motion to refer to a committee is sometimes called the motion to refer or the motion to commit.

Example

During the discussion on the main motion that the FFA chapter sponsor a $500 scholarship for a graduating senior who will attend college, it is obvious that this matter needs some detailed study.

Member: *Madam Chairperson.*

Chairperson: *Mr. Speaker.*

Member: *I move to refer this motion to the Scholarship Committee; and they should report back at the next month's meeting and that this committee have the power to act.*

Chairperson: *Is there a second?*

Figure 10–4 This member has been properly recognized and is discussing a motion to refer to a committee.

Another Member: *I second the motion.*

Chairperson: *It has been moved and seconded that we refer the motion to the Scholarship Committee. Is there any discussion?*

Note: If the motion to refer to a committee passes, the matter goes to the standing scholarship committee. This committee has the power to act and is to report back at the next meeting. After the chair instructs the secretary to inform the committee members of their duties, the floor is open for further business.

If the motion to refer to a committee fails, the assembly returns to discussion of the main motion, "That our FFA chapter sponsor a $500 scholarship for a graduating senior who will attend college." The group of members in Figure 10–4 are discussing a motion to refer to a committee.

Lay Motion on the Table

Rationale

To lay a motion on the table when passed temporarily lays aside an item of business so the assembly may immediately handle a more urgent matter. The motion to lay on the table has no set time at which the question (motion) is to be resumed. Only a majority vote through the motion to take from the table can bring back the tabled motion before the assembly.

Explanation

The motion to lay on the table is not intended to kill the motion to which it is applied, but only to lay it aside in such a way that it may easily be resumed. However, in cases of organizations "holding regular business sessions at least within quarterly 'time' intervals, a question laid on the table remains there until taken from the table or until the close of the next regular session: if not taken up by that time, the question dies."[15]

Since a motion that has been laid on the table is still within control of the assembly, no other motion on the same subject is in order as long as the original motion is still on the table. It is also out of order to move to lay a motion on the table if there is evidently no other matter requiring immediate attention.

The motion to lay on the table takes precedence over the main motion, over all subsidiary motions, and over any incidental motions that are pending when it is made. The motion to lay on the table requires a second, is not debatable, is not amendable, and requires a majority vote to pass.

Example

During discussion of the main motion "that our FFA chapter participate in June Is Dairy Month," the assembly learns that the school secretary has just been admitted to the hospital. So the assembly may adopt a motion to send the secretary some flowers, but it must first lay aside the main motion concerning "June Is Dairy Month."

Member: *Mr. Chairperson.*

Chairperson: *Madam Speaker.*

Member: *I move to lay on the table the main motion that our FFA chapter participate in "June Is Dairy Month."*

Chairperson: *Is there a second?*

Another Member: *I second the motion.*

Chairperson: *It has been moved and seconded that we lay the main motion to participate in "June Is Dairy Month" on the table; this motion is nondebatable and unamendable and requires a majority vote.*

Note: A vote is taken by the chairperson; if the motion to lay on the table passes, it is set aside until a motion to take from the table is made or until a time limit terminates it. If the motion to lay on the table fails, the assembly returns to discussion of the main motion concerning "June Is Dairy Month."

TAKE MOTION FROM THE TABLE

Rationale

The motion to take from the table means bringing again before the assembly a motion or a series of adhering motions that were previously laid on the table. The motion to take from the table is not in order unless at least one item of business has been transacted since the motion was laid on the table.

A motion that has been laid on the table remains there and can be taken from the table "during the same meeting or the next meeting after it was laid on the table. If not taken from the table within these limits of time, the (tabled motion) dies; however, it can be reintroduced later as a new question."[16]

Explanation

The motion to take from the table is an unclassified motion and is in order only when no other question is pending. It requires a second, is not debatable, is not amendable, and requires a majority vote to pass. If the motion to take from the table passes, the question is back on the floor with any subsidiary motions that were pending at the time it was tabled. If it fails, the tabled motion remains tabled.

Example

A main motion was laid on the table earlier in this meeting, and at least one item of business has been transacted since that time. While no business is pending, the following business takes place.

Member: *Mr. Chairperson.*

Chairperson: *Mr. Speaker.*

Member: *I move to take from the table the main motion that our FFA chapter participate in "June Is Dairy Month."*

Chairperson: *Is there a second?*

Another Member: *I second the motion.*

Chairperson: *It has been moved and seconded that we take from the table the main motion that our FFA chapter participate in "June Is Dairy Month"; this motion is neither debatable nor amendable and requires a majority vote to pass. We will now proceed to vote.*

Note: If the motion passes, discussion returns to the main motion, which was tabled, that our chapter participate in "June Is Dairy Month." If the motion fails, the floor is open for further business.

MOTION TO POSTPONE DEFINITELY

Rationale

The motion to postpone definitely puts off action on a pending motion and fixes a definite time for its future consideration. A main motion can be postponed only to the next scheduled meeting or to a later time in the current meeting. If a matter of business that was postponed needs to be discussed before the next regularly scheduled meeting, which had a specified time, it is necessary to provide an additional **adjourned meeting**, and then postpone the motion to a special meeting. This special meeting time could be set in an amendment to a motion to adjourn (see Chapter 11).

Explanation

If the motion to postpone definitely passes, the question to which it is applied is taken up at the time that was specified in the motion to postpone. The motion becomes the **order of the day** (on the agenda) for the time to which it is postponed.

The motion to postpone definitely requires a second, is debatable and amendable, and requires a majority vote to pass. It takes precedence over the main motion and the subsidiary motions to postpone indefinitely, amend, and refer to a committee. The motion to postpone definitely is sometimes referred to as the motion to postpone to a certain time.

Example

Members of an FFA chapter are discussing the main motion "that our FFA chapter volunteer as group leaders at summer camp." Because it is early in the winter and most students have not made plans for the summer, the following transactions take place.

Member: *Mr. Chairperson.*

Chairperson: *Madam Speaker.*

Member: *I move to postpone the main motion until our next meeting in March.*

Chairperson: *Is there a second?*

Another Member: *I second the motion.*

Chairperson: *It has been moved and seconded that we postpone the main motion that our FFA chapter volunteer as group leaders at summer camp until our next meeting in March. Is there any discussion?*

Note: If the motion passes, it is treated as unfinished business in the March meeting. If it fails, the assembly returns to the discussion of the main motion.

Motion to Postpone Indefinitely

Rationale

The purpose of a motion to postpone indefinitely is to stop the passage of a main motion without letting it come to a vote by the assembly.[17] The motion to postpone indefinitely is used when it is in the best interest of the assembly not to take a position on a particular main motion.[18] It is useful in disposing of a badly chosen main motion that cannot be either adopted or expressly rejected without possibly undesirable consequences. Only a main motion may be postponed indefinitely. If the motion to postpone indefinitely passes, it kills the main motion for the meeting. The only way a main motion that was postponed indefinitely can come up again is as a new motion at a later meeting.

Explanation

The motion to postpone indefinitely takes precedence over nothing except the main motion to which it is applied. It is the lowest ranking of the subsidiary motions. It can be applied only to a main motion, and, therefore, can be made only while the main motion is immediately pending. It requires a second, is debatable, is not amendable, and requires a majority vote to pass. During debate of the motion to postpone indefinitely, discussion can proceed fully on the merits of the main motion.

The motion to postpone indefinitely also has another purpose. It is sometimes used by strategists to test their strength on a particular main motion, according to the vote it receives.

Example

During discussion of the pending main motion "that our FFA chapter have a hayride," a member proposes a motion to postpone the main motion indefinitely.

Member: *Mr. Chairperson.*

Chairperson: *Mr. Speaker.*

Member: *I move to postpone indefinitely the main motion that our FFA chapter have a hayride.*

Chairperson: *Is there a second?*

Another Member: *I second the motion.*

Chairperson: *It has been moved and seconded that we postpone indefinitely the main motion; this motion is debatable but unamendable; is there any discussion?*

Note: A majority vote is taken by the chairperson. If the vote passes, the main motion is lost for the remainder of the meeting. If the motion to postpone indefinitely fails, the assembly returns to discussion of the pending main motion "that our FFA chapter have a hayride."

Suspend the Rules

Rationale

The motion to suspend the rules is desirable when the best interests of the organization are served by a temporary suspension of the written rules that govern its operation. A group can adopt a motion to suspend the rules provided that the proposal does not go against the organization's bylaws (or constitution); local, state, or national laws; or the fundamental principles of parliamentary procedure.

Only rules of procedure can be suspended. Rules that cannot be suspended are

- Common parliamentary law
- Rules in the organization's charter or its constitution
- Rules in the organization's bylaws, unless the bylaws have provisions that allow for the suspension of the rules

Explanation

The motion to suspend the rules is an **incidental motion**. It requires a second, is undebatable and unamendable, and requires a two-thirds vote to pass. The motion to suspend the rules can be made whenever no motion is pending. When a motion is being discussed, the motion to suspend the rules takes precedence over any motion if it is for a purpose connected with that motion.

Example

During discussion of the main motion "that our FFA chapter send a $35 floral arrangement to the funeral of the school secretary," the following business transpires.

Member: *Madam Chairperson.*

Chairperson: *Madam Speaker.*

Member: *There is a rule in our local chapter that we may spend no money without the approval of the Budget and Finance Committees; therefore I move to suspend this rule.*

Chairperson: *Is there a second?*

Another Member: *I second the motion.*

Chairperson: *It has been moved and seconded that we suspend the rule that states "that the Budget and Finance Committees be previously consulted before spending money" to send flowers to the funeral of the school secretary.*

Note: Since the motion is undebatable and unamendable, an immediate two-thirds vote is in order.

If it receives the two-thirds vote required to pass, the rule is suspended temporarily and discussion continues on the main motion. If not suspended, it is because the rule receives less than two-thirds vote. If it fails, the chair should rule the main motion out of order, since it is a notation of a chapter rule.

Rise to a Point of Order

Rationale

A member may rise to a point of order to draw attention to a **violation** of the rules and insist on the enforcement of the proper rules (Figure 10–5). Ordinarily, the motion is used by members to call

Figure 10–5 This member did not need recognition to rise to a point of order.

attention to errors made by other members, which are not corrected by the chair in the course of a meeting. After ruling on the point of order, the chairperson gives reasons for his or her decision or confers with the **parliamentarian** for a ruling.

Explanation

If a point of order arises out of business that is currently pending, the point of order must be ruled on before the business can proceed. If the chair rules incorrectly on a point of order, a member should appeal from the decision of the chair (see appeal from decision of chair in the following section). A point of order is used when a rule has been broken either intentionally or unintentionally. Members should also correct willful violations of accepted procedures by the chair by rising to a point of order.

These are the steps in rising to a point of order.

- A member rising to a point of order should rise without recognition and say, "Mr. or Madam Chairperson, I rise to a point of order." Note: If the point of order needs immediate attention, the member may interrupt another speaker.
- Chairperson replies, "State your point."
- Member explains violation.
- The chairperson then rules on the point of order.
- If the ruling is incorrect or a member does not agree, the member can appeal from the decision of the chair.

Example

There is a main motion pending before the assembly that "our FFA chapter participate in June Is Dairy Month." During discussion, a member moves to lay the main motion on the table. The chair receives a second on the motion to lay on the table and calls for discussion. An observant member realizes the motion to lay on the table is undebatable and unamendable; therefore, the following comments are made (without recognition).

Member: *Mr. Chairperson, I rise to a point of order.*

Chairperson: *State your point.*

Member: *Since the motion to lay on the table is undebatable and unamendable, the chair should not have called for discussion.*

Chairperson: *Thank you, Madam Speaker. Your point is well taken. The chair should not have called for discussion, and we will proceed to vote on the motion to lay on the table.*

Appeal from the Decision of the Chair

Rationale

Any member of the assembly who does not agree with a ruling of the chair may appeal from the decision of the chair. The appeal must take place immediately following the ruling. The purpose of an appeal is to prevent the chair from improperly controlling the meeting.

Explanation

The most common occurrence of an appeal from the decision of the chair is after a point of order. However, an appeal can be made after any ruling the chair makes. Members should not criticize a ruling of the chairperson unless they appeal from his or her decision.

The appeal from the decision of the chair requires a second, is debatable, but is not amendable. A **tie vote** sustains the chair. This is an application of the rule that a motion is automatically lost in the case of a tie vote. Here, the motion is to overrule the chair. In the case of a tie vote, the motion to overrule is lost; thus, the decision of the chair is sustained. Why? Common sense reasons that the chair would vote for him- or herself anyway and time is saved.

Example

A member makes a main motion "that our class have a cookout." Before receiving a second, the chair calls for discussion on the main motion. An alert member notices the breech and rises to a point of order stating, "The main motion did not receive a second." The chair hastily replies, "I am sorry, Mr. Speaker, but your point is not sustained since a main motion does not require a second." At this point, it is obvious that the chair is trying to push this motion through the assembly, since he rules incorrectly on the point of order. The procedure the assembly should use to keep the chair from improperly controlling the meeting is an appeal from the decision of the chair. Therefore, a member rises after the chair rules on the point of order and the following business occurs.

Member: *Madam Chairperson, I appeal from the decision of the chair.*

Chairperson: *Is there a second? (Note: It may be necessary for the chairperson to ask what the appeal is.)*

Another Member: *I second the appeal.*

Chairperson: *Those who agree with the decision of the chair please raise your right hand. Thank you, hands down. Those who do not agree with the decision of the chair please raise your right hand. Thank you, hands down.*

If the appeal from the decision of the chair sustains the chair, the assembly returns to the discussion of the main motion "that our class have a cookout." If the appeal from the decision of the chair does not sustain the chair, the chair must go back and ask for a second on the main motion. A second must be received before the main motion can continue. Remember, a tie vote sustains the chair.

Division of the House (Assembly)

Rationale

Whenever a member doubts the result of a vote that was taken by voice, he or she may request a revote by calling for a division of the house. A member can demand a division from the moment the negative votes have been cast until the question is stated on another motion. A request for a division is ordinarily granted by the chair as long as there is a reasonable doubt about the outcome of the previous vote. When a division of the house is called, the revote should be taken by a more accurate means, such as standing (or rising), ballot, or roll call.

Explanation

A single member cannot order a counted vote by calling for a division of the house. Only the chair or assembly can order a counted vote.

If after a **vote by voice (viva voce)** the chair is unsure which side has won, the chairperson may initiate a division by explaining the circumstance and calling for a revote by standing (Figure 10–6).

Since a single member can request a division, a division of the house does not require a second

Figure 10–6 The chairperson is counting the standing votes. Other methods of voting are voice, raised hand, and roll call. (Courtesy of California's Fullerton FFA.)

and does not receive a vote. There are different ways in which a member may request a division of the house. The most common form is: a member rises without receiving recognition from the chair and states, "Mr. Chairperson, I call for a division." Other phrases that would serve the same purpose include

> Mr. Chairperson, I call for a division of the assembly.
>
> Mr. Chairperson, I call for a division of the house.
>
> Mr. Chairperson, I doubt the vote.

Example

If, after a voice vote is taken, the results are unclear, the following transpires.

Member: *Madam Chairperson, I call for division of the house.*

Chairperson: *A division of the house is called for. We shall vote again with a more accurate means; all those in favor of the motion please stand.*

Reconsider a Motion

Rationale

The purpose of the motion to reconsider is to bring back before the group a motion that has already received a vote, thus making possible a change of vote on the original motion. In a sense, you are voting to see if you want to revote. The motion to reconsider may be made only by a person who voted on the winning or prevailing side. If a person voted against the motion and it passed, she is on the prevailing side. A member may vote on the prevailing (victorious) side for the specific purpose of being in a position to move to reconsider later in the meeting.

The motion to reconsider a motion that previously lost must be made during the meeting at which the main motion was introduced, except when passed and entered in the minutes. If so, it can be considered only in the next meeting. Otherwise, use the motion to rescind, which is discussed in the next chapter.

Explanation

The motion to reconsider may be applied to all motions *except* a motion to adjourn, recess, suspend the rules, affirmative vote to lay the motion on the table, the affirmative vote to take business from the table, and a motion to reconsider. The motion to reconsider requires a second, is debatable (if the original motion was debatable), but is not amendable. It requires a majority vote. When the chairperson receives the motion to reconsider, the first thing to do is ask the speaker if he or she voted on the prevailing side. If so, the motion is in order.

Suppose a motion to sell magazines with the band was moved and properly seconded. During the discussion a majority of the people were in favor of the motion; the motion passed and was recorded in the minutes. Before the next meeting, a person who voted on the winning (prevailing) side learned that the band did not want to sell magazines with the FFA. At the next meeting, this person would gain recognition from the chairperson and say, "I move that we reconsider the vote selling magazines with the band since they do not want to work with us." The chairperson asks, "Did you vote on the prevailing side?" The member answers yes. The chairperson directs the discussion of the motion to reconsider, takes a vote, and announces the results. Once the motion passes, it places the original motion before the assembly at the same place it occupied before a vote was taken.

Example

Chairperson: *Is there any business to be presented at this time?*

Speaker: *Madam Chairperson.*

Chairperson: *Mr. Speaker.*

Speaker: *I move to reconsider the motion to have an FFA fruit sale since we are not selling magazines.*

Chairperson: *Madam Speaker, did you vote on the prevailing side?*

Speaker: *Yes, I did.*

Chairperson: *Then your motion is in order. Is there a second?*

Another Member: *I second the motion.*

Chairperson: *It has been properly moved and seconded to reconsider the motion to have an FFA fruit sale. Is there any discussion? If not, all in favor of reconsidering the motion say "aye."* (Response). *All those opposed say "nay."* (Response). *The motion is carried. Discussion is now open on the motion to sell fruit. Hearing none, we shall proceed to vote. All in favor of the motion please stand. Thank you. Please be seated. All opposed, please stand. Thank you. Please be seated. Mr. Secretary, the motion passed; 20 for the motion and 6 opposed. The motion carries.* (Tap gavel.)

MOTION TO RECESS

Rationale

A recess is a short intermission in the meeting, but it does not close the meeting. After the recess, business starts at exactly the point at which it was interrupted.

Explanation

There are two forms of the motion to recess: *qualified* and *unqualified.*

A qualified motion to recess is a privileged motion. To be in the privileged form, a motion to recess must be made while another question is pending. In its privileged form, the motion takes precedence over all subsidiary and incidental motions and over all privileged motions except the privileged form of the motion to "Adjourn" and to "Fix the time to which to adjourn." It requires a second and is undebatable; amendments must relate to the time limits of the recess. If it receives the majority vote required to pass, the recess must begin immediately.

An unqualified motion to recess is made while no other question is pending. It is treated as a main motion. It requires a second and is debatable and amendable. When made as a main motion, the motion to recess may begin immediately or at a future time.

A motion to recess is sometimes used by members as a strategic action to hold an informal group discussion or seek information on a question during a meeting.

Example

While a main motion is pending that the FFA chapter "hold its annual FFA Banquet on May 13," it is brought out during discussion that permission is required from the school administrator. In order that the motion may be decided on at that meeting and to further enable the members to hold an informal discussion on the topic, a member makes a qualified motion to recess.

Member: *Mr. Chairperson.*

Chairperson: *Madam Speaker.*

Member: *I move that we take a recess for 15 minutes.*

Chairperson: *Is there a second?*

Another Member: *I second the motion.*

Chairperson: *It has been moved and properly seconded to take a recess for 15 minutes. Amendments are related only to time limits.*

If the motion to recess passes, the assembly takes a 15-minute recess. When the recess is concluded, the business is taken up at the point at which it was left. If the motion to recess fails, the assembly continues with discussion of the main motion without taking a recess.

No other business is on the floor, and the following unqualified motion to recess is made.

Member: *Madam Chairperson.*

Chairperson: *Mr. Speaker.*

Member: *I move that we recess.*

Chairperson: *Is there a second?*

Another Member: *I second the motion.*

Chairperson: *It has been moved and seconded that we recess. This motion is open for debate and amendments if needed. Is there any discussion?*

Motion to Adjourn

Rationale

The motion to adjourn is used to legally end a meeting. When a group meets again following an adjournment, they do not take up at the point at which they left off on their agenda (as in the case of a recess). Instead, they begin a new meeting with a new agenda. Business that was not completed or left unfinished by an adjournment comes up on the agenda of the next meeting under "unfinished business."

Explanation

There are two types of motions to adjourn: unqualified (privileged motion) and qualified (main motion).

Unqualified (Privileged) Motion The **privileged motion** to adjourn is a motion to close the meeting immediately. It takes precedence over all subsidiary and privileged motions except to "fix the time to adjourn." The privileged motion to adjourn is made. If another meeting time exists or if no time has been set for the existing meeting to adjourn, the privileged motion to adjourn is used. It requires a second. It is undebatable and unamendable and requires a majority vote to pass.

Qualified (Main) Motion When the motion to adjourn is made in its qualified (main motion) form, it is debatable and amendable just as a main motion. It is used in the following situations as a main motion.

- When the motion is qualified in any way, such as a motion to adjourn at, or to, a future time
- When a time for adjourning is already established
- When the effect of the motion to adjourn would be to dissolve the group (an example would be a last meeting of a convention)

Example

During a meeting, members are discussing an unimportant main motion. Realizing the late hour on a week night, a member moves to adjourn the meeting.

Member: *Madam Chairperson.*

Chairperson: *Mr. Speaker.*

Member: *I move to adjourn.*

Chairperson: *Is there a second?*

Another Member: *I second the motion.*

Chairperson: *The motion has been moved and seconded that we adjourn. This is a privileged (unqualified) motion, which makes it undebatable and unamendable. We will proceed to vote.*

(If the motion passes)

Chairperson: *The meeting is adjourned.*

Note: If the motion to adjourn fails, the assembly returns to discussion on the main motion.

The following is an example of a qualified motion to adjourn.

Member: *Mr. Chairperson.*
Chairperson: *Madam Speaker.*
Member: *I move that we adjourn.*
Chairperson: *Is there a second?*
Another Member: *I second the motion.*
Chairperson: *Is there any discussion?*

Note: A qualified motion to adjourn is treated just like a simple main motion.

Discussion of advanced parliamentary procedure continues in the next chapter. Rules for the National FFA Parliamentary Procedure Contest are discussed in the next chapter and in Appendix F.

ORDER OF BUSINESS (AGENDA)

Order of business and *agenda* are used as synonyms in parliamentary procedure. The agenda is an established order of business or sequence of activities. The following is an example of an agenda for a group that holds regular business sessions that do not have special requirements.

- Reading and approval of minutes
- Reports of officers, boards, and standing committees
- Reports of special committees
- Special orders of unfinished business and general orders
- New business

Some organizations perform ceremonies or have other special requirements that cause them to customize their agenda. The FFA is such an organization. The following is a sample FFA order of business.

- Opening ceremony
- Minutes of previous meeting
- Officer reports
- Special features
- Unfinished business
- Committee reports
- New business
- Ceremonies
- Closing ceremony
- Entertainment

The agenda is expected to be followed at all times to maintain group harmony and organization. If for some reason, the assembly fails to conform to its agenda, the motion to call for the orders of the day may be used to bring the assembly back to its agenda. (To call for the orders of the day is further explained in Chapter 11. Agendas are further discussed in Chapter 12.) Figure 10–7 provides a sample FFA meeting agenda.

COMMON PARLIAMENTARY ERRORS AND MISCONCEPTIONS

Sometimes a single error is made throughout an entire meeting if no one is knowledgeable enough on correct parliamentary procedure to notice the error. Theoretically, there is a right and wrong way to perform every motion. In meetings in which you are a member of the assembly, you should strive not to make any errors. You should also notice errors made by others and point them out to the chair through point of order if necessary. Below, I correct some common errors and misconceptions in parliamentary procedure.

FFA Meeting Agenda

1. Opening ceremony.
2. Minutes of the previous meeting.
3. Officer reports.
4. Report on chapter program of activities (chairpersons of the various sections of the program of activities are called on to report plans and progress).
5. Special features (speakers, special music, and the like).
6. Unfinished business.
7. Committee reports.
   a. Standing
   b. Special
8. New business.
   a. Plateau Experiment Station 50th Anniversary
   b. State FFA Dairy Judging Contest
   c. State Fair
   d. State FFA Livestock Judging Contest
   e. Forestry Conclave
   f. FFA membership dues deadline
   g. Ag Day—Varsity Visit
   h. National FFA Convention
   i. District Parliamentary Procedure Contest
   j. National FFA Week
9. Degree and installation ceremonies (if appropriate).
10. Closing ceremony.
11. Entertainment, recreation, refreshments.

Figure 10–7 A sample FFA meeting agenda. The agenda should be followed at all times to maintain group harmony and organization.

- In presenting a motion before an assembly, a member should use the proper terminology, i.e., "I move . . ." Less skilled members might say, "I make a motion . . ."
- In determining the outcome of a vote, the only votes counted are those cast by members who are legally entitled to vote.
- A member making a motion about something that has just been said by the chair or another member in an informal discussion during a meeting should avoid statements such as "I so move," and should themselves recite the complete motion that they offer.
- If members are required to have recognition from the chair before they speak, the member should wait until the chair has given that person the floor or permission to speak before addressing the assembly.
- Just because a member makes or seconds a motion, does not necessarily mean he or she must vote in favor of it. In a democratic society, a member may vote any way on any motion, or even abstain from voting.

Quick Reference Guides

When this chapter has been successfully completed and further knowledge and parliamentary skills are still desired, go to Chapter 11. Refer to Figure 10–8 for a summary of the motions in this chapter. The chairperson may want to place this in front of him or her as a quick reference while conducting business or briefly review it before serving as chairperson either in a practice or real situation. Figure 10–9 provides a good visual of the order of precedence of the motions. Figure 10–10 shows a sample of how parliamentary procedure can be used in a meeting. As beginning practice, your teacher may wish to assign you a part and go through the script so you can become comfortable with parliamentary procedure.

Basic Parliamentary Procedure Motions Overview

Motion	Recognition from the Chair Required	Second Required	Debatable	Amendable	Vote Required	Class of Motion[1]
Main motion	Yes	Yes	Yes	Yes	Majority	M
Amend	Yes	Yes	Yes	Yes	Majority	S
Previous question	Yes	Yes	No	No	Two-thirds	S
Refer to committee	Yes	Yes	Yes	Yes	Majority	S
Lay on the table	Yes	Yes	No	No	Majority	S
Take from the table	Yes	Yes	No	No	Majority	U
Postpone definitely	Yes	Yes	Yes	Yes	Majority	S
Postpone indefinitely	Yes	Yes	Yes	No	Majority	S
Appeal from the decision of the chair	No	Yes	Yes	No	Majority[2]	I
Division of assembly	No	No	No	No	Revote	I
Reconsider a motion	Yes	Yes	No[3]	No	Majority	U
Recess	Yes	Yes	Yes/No[4]	Yes	Majority	P
Adjourn	Yes	Yes	Yes/No[5]	Yes/No[5]	Majority	P
Suspend the rules	Yes	Yes	No	No	Two-thirds	I
Point of order	No	No	No	No	None	I

1. M = motion; S = subsidiary; I = incidental; P = privileged; U = unclassified.
2. Majority vote. However, a tie vote sustains the chairperson.
3. Yes, if original motion is debatable.
4. Recess as a main motion is debatable and amendable. Recess in privileged form is undebatable but amendable to time only.
5. Adjourn as main motion is debatable and amendable. Adjourn as a privileged motion is undebatable and unamendable.

Figure 10–8 An overview of the basic parliamentary procedure motions covered in this chapter. (Courtesy of the National FFA Organization.)

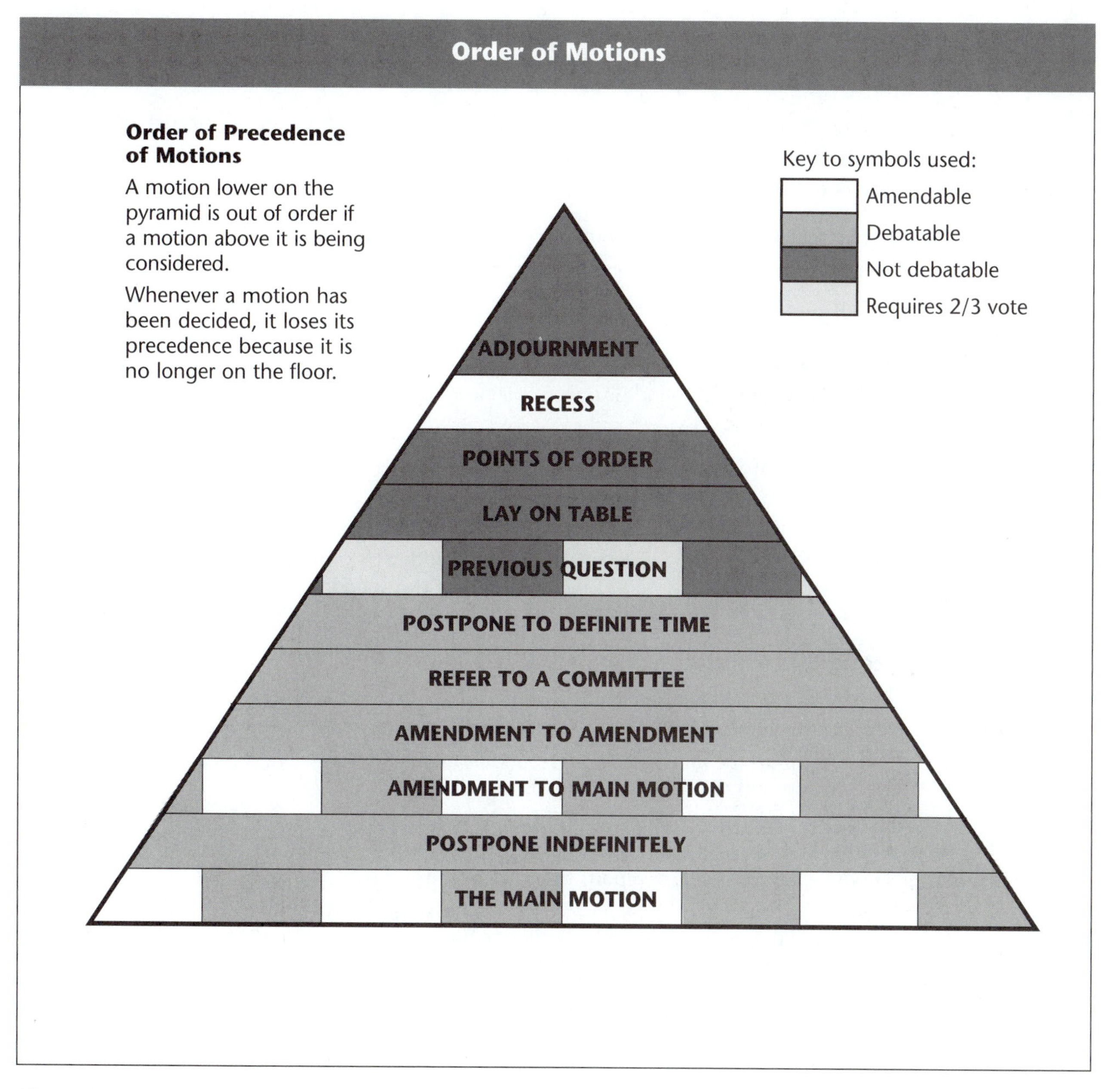

Figure 10–9 The visual order of precedence of motions covered in this chapter. (Courtesy of the National FFA Organization.)

Sample of Parliamentary Procedure Used in a Meeting

Note: We assume that the opening ceremony has been completed and each officer knew their part and did a good job. It is time for the president to ask the secretary for the first order of business. Your meeting could and should proceed along this general pattern:

President: "Mr. Secretary, what is our first order of business?"

Secretary: "Minutes of the previous meeting, Madam President."

After the minutes are read,

President: "You have heard the minutes of the previous meeting. Are there any corrections? If not, the minutes stand approved as read."

(Do not say "of our LAST meeting." It sounds like there will never be another.)

President: "Thank you, Mr. Secretary.
What is our next order of business?"

Secretary: "Officer reports, Madam President."

President: "Are there any officer reports to be presented?"

Treasurer: "Madam President, I should like to present the monthly treasurer's report. We began the month with a balance of $871.35. Our receipts for the period were $786.42 and our disbursements totaled $411.19, leaving a balance of $1,246.58. As the members entered the meeting room, a copy of the month's report was distributed. Madam President, I move the acceptance of this report."

President: "Is there a second to the motion?"

Member: "I second the motion."

President: "You have heard the motion and the second. Is there any discussion? If not, all in favor say 'aye.' All opposed say 'no.' The motion is carried."

(In a chapter meeting, the secretary can easily identify the person who seconds a motion. By the way, one does not say "I second THAT." "THAT" was not the motion under consideration).

President: "Are there any other officer reports to be presented?

Reporter: "Madam President, during the past month we have published a total of 137 column inches of news and put on one five-minute radio broadcast about our summer camp. Madam President, I move acceptance of this report."

President: "Is there a second to the motion?"

Member: "I second the motion."

President: "You have heard the motion and second. Is there any discussion?"

Member: "Madam President."

President: "Julie."

Julie: "I wish to know why the article in *The FFA New Horizons* magazine was not included?"

Figure 10–10 Scripts such as this one provide the opportunity to build your confidence as you learn the correct parliamentary procedure techniques. Your teacher can assign you a part in the script. (Courtesy of the National FFA Organization.)

Sample of Parliamentary Procedure Used in a Meeting (continued)

President: "Will the reporter please respond to the question?"

Reporter: "The article was not included since it did not appear during the month covered in the report. It will be included in the next month's report."

President: "Is there any further discussion on this motion? Hearing none, all in favor of the motion say 'aye.' All opposed say 'no.' The motion is carried and the report is accepted."

"Are there any other officer reports to be presented? Hearing none, Mr. Secretary, what is our next order of business?"

Secretary: "Reports on Chapter Program of Activities committees."

Committee Chairperson: "Madam President, this is the report from the leadership committee. After reviewing the applications, our committee recommends that Todd Jackson, Jason Erickson, Martha Mork, and Donna Anderson be this year's finalists for the Star Greenhand award. I move the acceptance of our committee report."

President: "Is there a second to the motion?"

Member: "I think Tommy Kawalski should be included."

Member: "I rise to a point of order."

President: "State your point."

Member: "Since the motion had not received a second, discussion was not in order."

President: "Your point is well taken. We call again for a second to the motion."

Member: "I second the motion."

President: "The motion has received a second. Is there any discussion?"

Member: "At our last meeting this committee was directed to select four finalists. It was agreed that their decision would be final."

President: "Is there any further discussion? Hearing none, are you ready to vote?"

Members: "Question—Question."

President: "The question has been called. All in favor, signify by saying 'aye.' Those opposed say 'no.' The motion is carried."

President: "Are there any other program activities reports to be heard? If not, Mr. Secretary, what is our next order of business?"

Secretary: "Special features, Madam President."

President: "Our agenda indicated that there are no special features scheduled so we shall proceed to the next order of business."

Secretary: "Madam President, unfinished business is the next order of business."

President: "Is there any unfinished business to come before the chapter?"

Member: "Madam President."

President: "Helen."

Figure 10–10 *(continued)*

(con'd)

Sample of Parliamentary Procedure Used in a Meeting (continued)

Helen:	"At the previous meeting a special committee was appointed to plan our Food for America program. What did they decide to do?"
President:	"This should come up during special committee reports. Mr. Secretary, is that not our next order of business?"
Secretary:	"Yes, Madam President. Special committee reports are next on the agenda."
President:	"Are there any special committee reports to be presented?"
Chairperson:	"Madam President, our committee recommends that our Food for America program be presented to elementary students on the first Monday of April with Tammy Koble as the coordinator. I move the acceptance of this report."
President:	"You have heard the motion. Is there a second?"
Member:	"I second the motion."
President:	"Is there any discussion? Hearing none, all in favor signify by saying 'aye.' Those opposed say 'no.' The motion is carried. Are there any other special committee reports to be presented? If not, what is the next order of business?"
Secretary:	"New business, Madam President."
President:	"Is there any new business to be presented?"
Member:	"Madam President."
President:	"Joan."
Joan:	"I move that our chapter purchase student handbooks for every first-year member in our chapter."
President:	"Is there a second to this motion?"
Member:	"I second the motion."
President:	"Is there any discussion?"
Member:	"Madam President."
President:	"Wilfred."
Wilfred:	"I move to amend the motion by substituting the words 'pay $2.00 of the cost' in place of 'purchase.' "
President:	"Is there a second?"
Member:	"I second the motion."
President:	"Is there any discussion? Hearing none, all in favor of the amendment to the motion, signify by saying 'aye.' Those opposed say 'no.' The amendment is defeated. Is there any further discussion on the original motion? Hearing none, all those in favor say 'aye.' Opposed say 'no.' The motion carried. Is there any more new business to come before the meeting? If not, what is the next order of business?"
Secretary:	"Since we have no degree or installation ceremonies, I have none, Madam President."
President:	(Begins closing ceremony.)

Figure 10–10 (*concluded*)

SUMMARY

Parliamentary procedure has always been a basic curriculum component of agricultural education and the FFA. Since parliamentary procedure uses the democratic process, leadership skills are developed through its use. Parliamentary procedure rules guarantee that a motion proceeds in a purely democratic fashion. The rules of parliamentary procedure are based on consideration for the rights of the majority, the minority, individual members, and absentees.

The presiding officer of a meeting is ordinarily called the president or chairperson. The presiding officer should possess certain qualities and have complete control of the assembly at all times. The presiding officer should be fair, possess good judgment, manifest poise, have a working knowledge of parliamentary procedure, be deliberate, and be tactful.

One of the basic elements of conducting parliamentary procedure effectively is receiving and disposing of motions correctly. These are the eight steps for handling a motion: (1) Member rises and addresses chair, (2) chairperson recognizes the member, (3) member makes a motion, (4) another member seconds the motion, (5) chairperson restates the motion, (6) members discuss and debate the motion, (7) chairperson calls for the vote, and (8) chairperson announces the result of the vote.

There are five major classifications of motions: main, subsidiary, unclassified, incidental, and privileged. Once a motion is made, it receives a second and a vote. Most of the motions require a majority vote, but four motions require a two-thirds majority vote: suspend the rules, previous question, rescind, and the motion to limit debate. Another type of vote is general consent. There are five major methods of voting: voice, hand, standing, secret ballot, and roll call.

After the main motion, the subsidiary motions include amend, refer to committee, lay on the table, postpone definitely, postpone indefinitely, and previous question. Incidental motions include suspend the rules, point of order, appeal from decision of the chair, and division of the house (assembly). Privileged motions include recess and adjourn. Two motions are unclassified: take from the table and reconsider a motion.

The agenda is an established order of business or sequence of activities for a meeting. The agenda is expected to be followed at all times to maintain group harmony and organization.

END-OF-CHAPTER EXERCISES

Review Questions

1. Define the Terms to Know.
2. Discuss five characteristics of an ideal presiding officer.
3. List the eight steps in properly handling a motion.
4. What are eight questions the chairperson should ask when a main motion is received?
5. What is the purpose of seconding a motion?
6. What are five methods of voting?
7. What is the purpose of a main motion?
8. List the order of business for an FFA meeting.
9. Explain five common parliamentary errors and misconceptions.
10. What policy on which our country is built is an application of parliamentary procedure?

11. Whose rights are protected by parliamentary procedure?
12. Name four motions that do not require a second.

Fill-in-the-Blank

1. The ________________ is the person who is actually presiding over a meeting.
2. A ________________ is necessary so the assembly does not waste time on a motion that only one person wants to discuss.
3. If there is no opposition on a question of little importance, it may be adopted by ________________ .
4. The FFA organization has a motion to ________________ built into its closing ceremony.
5. An ________________ is the order in which business is presented in an assembly.
6. If a motion is ________________ , it may have discussion on its merits.
7. Ninety members are voting, which means ________________ are needed for a two-thirds majority.
8. A motion that has been tabled must be taken from the table at the ________________ meeting or the ________________ meeting or it will cease to exist.
9. Motions to postpone definitely must include a ________________ when the original motion will be back on the floor.
10. Every ________________ has a right to express an opinion when parliamentary procedure is being used.

Matching

_____ 1. Recess
_____ 2. Amend
_____ 3. Vote required on suspend rules
_____ 4. Vote required on postpone indefinitely
_____ 5. Postpone indefinitely
_____ 6. Division of assembly
_____ 7. "I make a motion"
_____ 8. Bring back a motion that has already received a vote
_____ 9. A member makes a parliamentary error
_____ 10. You disagree with the ruling of the chairperson
_____ 11. You want to end the debate now
_____ 12. A subgroup can better deal with this matter
_____ 13. Undebatable motion to put off original motion until later
_____ 14. Short or extended break for informal discussion
_____ 15. Add to original motion

A. modify; change
B. motion to amend
C. incorrect terminology
D. majority vote
E. recess
F. short break
G. kills a motion
H. appeal from the decision of the chair
I. rise to point of order
J. reconsider
K. refer to a committee
L. lay on the table
M. previous question
N. revote
O. two-thirds vote

Activities

1. Participate in the following "Parliamentary Procedure" game in class: Stand with your classmates in line in the classroom as you would for a spelling bee. Your teacher names a motion. The first person must state whether or not it requires a second; the next person, whether or not it is debatable; the next person, what vote it requires; and the next person, what other motion is directly above it in the order of precedence. (For incidental motions, the correct answer is "No order of precedence.") A student missing any of these responses must be seated. The teacher continues asking questions about motions until only the winner remains standing.
2. Select one of the motions, and write a script similar to the one in the "reconsider a motion" example. Use these scripts to practice parliamentary procedure. For some motions, write a script for a positive vote and a negative vote. If the teacher assigns or lets students volunteer for different motions, perhaps you could have a practice script for each motion.
3. List the motions that require a two-thirds vote. Write a 100-word essay explaining why these motions require a two-thirds vote but other motions require a simple majority.
4. Write a 100-word essay on why you think some motions are debatable and others are not.
5. Write a paragraph on the circumstances under which it is advisable to table a motion.
6. Write a paragraph on the circumstances under which it is advisable to refer a question to a committee.
7. **Take it to the Net.**

   Explore parliamentary procedure on the Internet. The Web sites listed below contain "lots of" information on parliamentary procedure. Browse the sites and pick one that you find most useful in providing information on parliamentary procedure. Print the first page of the Web site. Write a summary of the site or of an article in the Web site. The summary should include why you feel the Web site is useful. If you are having difficulty with the sites provided, try typing some of the search terms listed in the search field.

   Web sites

   <http://www.csufresno.edu/comm/cagle-p3.htm>
   <http://www.ca.lp.org/lpcn/9602-parliamentary.html>
   <httjp://www.constitution.org/rror/rror-00.htm>
   <http://www.parli.com/>

   Search Terms

   parliamentary procedure
   FFA
   parliamentarians

8. Conduct a mock meeting, with each class member taking a turn serving as chairperson. After the meeting, discuss the problems each of you had while acting as chairperson. The class should be divided into three groups. The chairperson and three to six class members will study the problems in the following "Skills to be Demonstrated" sections. One chairperson can do all three problems with each demonstration or different students can serve as chairperson of each problem.

Skills to be Demonstrated

- Main motion
- Amendment to the amendment
- Previous question
- Motion to reconsider

Problem 1

A. Member moves that each member is to have a minimum of four exhibits at the state fair.
B. Motion receives a second.
C. Motion is discussed by at least two members.
D. Member moves to amend the motion by substituting the number *five* for *four*.
E. Amendment is discussed by at least two members.
F. Member moves to amend the amendment by adding the words *at least one of which will be a crop or horticulture sample.*
G. Amendment to amendment, amendment, and main motion all are passed.

Problem 2

A. Member moves to reconsider action taken at the previous meeting when this motion was defeated: "Jackie moved that FFA dues must be subtracted from state fair premium money before the member receives the balance of the award." Second was by Ruth. After much discussion, motion was defeated.
B. Motion to reconsider receives a second.
C. Motion is discussed by at least two members.
D. Motion is passed.
E. Chair announces that the original motion is now on the floor in its debatable form.
F. Motion carries.

Abilities to Be Demonstrated	Kind of Motion	Second Required	Debatable	Amendable	Vote Required
To receive and dispose of a motion and an amendment to the amendment					
a. Main motion	Main	Yes	Yes	Yes	Majority
b. Amendment	Subsidiary	Yes	Yes	Yes	Majority
c. Amendment to the amendment	Subsidiary	Yes	Yes*	No	Majority
Previous question	Subsidiary	Yes	No	No	Two-thirds
Motion to reconsider	Other	Yes	Yes	No**	Majority

*If the main motion is debatable.
**This motion is not amendable. If the motion is carried, the item to be reconsidered is sent back on the floor in its original form. If the original motion was debatable and amendable, the same situation now exists. Since it is classified as "unfinished business," this should be handled before the other motions in the demonstration.

NOTE: Properly, secretaries introduce *orders* of business. *Items* of business come from the floor.

Problem 3

A. Member moves that we sponsor an all-school harvest dance.
B. Motion receives a second.
C. Motion is discussed by at least three members.
D. Member moves the previous question.
E. Previous question motion is defeated.
F. At least one member discusses the original motion.
G. Motion is defeated.

Skills to Be Demonstrated

- Main motion
- Division of the house
- Lay on the table
- Amendment
- Previous question

Problem 1

A. Member moves that each member donate $1.00 to the FFA Foundation, Inc.
B. Motion receives a second.
C. Motion is discussed by at least three members.
D. Member moves to amend the motion by substituting the amount *$.50* for *$1.00*.
E. Amendment is seconded and discussed by two members.
F. Motion to table the main motion and amendment is presented, seconded, and passed.

Problem 2

A. Member moves that the chapter pay for all meals and lodging expenses for members of the dairy judging team and livestock judging team and the prepared public speaker participating in the state contest.
B. Motion receives a second.

Abilities to be Demonstrated	Kind of Motion	Second Required	Debatable	Amendable	Vote Required
Main motion*	Main	Yes	Yes	Yes	Majority
Amendment	Subsidiary	Yes	Yes	Yes	Majority
Division of the house	Incidental	No	No	No	None
Lay on the table	Subsidiary	Yes	No	No	Majority
Previous question	Subsidiary	Yes	No	No	Two-thirds

*In this problem, when the main motion is voted on as a result of a previous question passage, the main motion requires a two-thirds majority vote. Therefore, a rising vote is necessary.

NOTE: Properly, secretaries introduce *orders* of business. *Items* of business come from the floor.

C. Motion is discussed by one member.
D. Member moves to amend the main motion by striking out the word *meals* from the original motion.
E. Amendment receives a second and is discussed by three members.
F. Amendment is voted on and sounds as if it passes. A division of the house is called for, and the amendment fails.
G. Discussion returns to main motion by two members.
H. Motion for the previous question is made and seconded.
I. Previous question is voted on and passed.
J. Main motion is voted on and passed.

Skills to Be Demonstrated

- Main motion
- Amendment
- Postpone definitely
- Take from the table

Problem 1

A. Member moves that our chapter pay the expenses of one member to the Washington Leadership Conference.
B. Motion receives a second.
C. Motion is discussed by at least two participants.
D. Motion to amend the motion to substitute *two* for *one* member.
E. Discussion by at least two participants.
F. Motion and amendment are disposed with the amendment failing and the main motion passing.

Problem 2

A. Member moves that every member have at least one exhibit at the state fair.
B. Motion receives a second.
C. Discussion by at least two members.
D. Motion to postpone definitely.
E. Motion receives a second.
F. Motion passes.

Problem 3

At the previous meeting, a motion was tabled that would have a committee appointed to select an activity that would earn enough money to pay the dues for all their members.

A. Motion to take the proposal from the table.
B. Motion receives a second.
C. Motion carries.
D. Discussion on the motion (now back on the floor in its debatable form) by at least two participants.
E. Motion carries and president appoints a three-member committee, identifying the committee chairperson.

Note: The three problems appear courtesy of the National FFA Organization.

Notes

1. D. B. Jameson, *Leadership Handbook* (New Castle, PA: LEAD, 1978), p. 31.
2. S. C. Robert, *Robert's Rules of Order, Newly Revised* (Glenview, IL: Scott, Foresman, 1990), p. XLIV.
3. Ibid., p. 439.
4. Jameson, *Leadership Handbook*, p. 40.
5. Robert, *Robert's Rules of Order*, p. 33.
6. R. E. Bender et al., *Mastering Parliamentary Procedure* (Columbus, OH: Ohio State University, Department of Agricultural Education, 1983), p. 5.
7. Robert, *Robert's Rules of Order*, p. 31.
8. Ibid., p. 80.
9. Ibid., p. 53.
10. "One Vote (Your Vote) Could Make a Difference," *The Volunteer Voice* Newsletter of the Tennessee Association of Parliamentarians, February, 1994, p. 12.
11. Robert, *Robert's Rules of Order*, p. 128.
12. K. L. Russell, *The "How" in Parliamentary Procedure*, Fifth Edition (Danville, IL: The Interstate Printers and Publishers, 1990), p. 27.
13. Bender, *Mastering Parliamentary Procedure*, p. 18.
14. Robert, *Robert's Rules of Order*, p. 479.
15. Ibid., p. 212.
16. Ibid., p. 296.
17. Bender, *Mastering Parliamentary Procedure*, p. 18.
18. Robert, *Robert's Rules of Order*, p. 123.

Advanced Parliamentary Procedure

Objectives

After completing this chapter, the student should be able to:

- Describe the duties of the presiding officer
- Explain decorum (propriety) in debate
- Discuss the classification of motions
- Demonstrate each of the following motions:
  - Fix the time to adjourn
  - Question of privilege
  - Limit or extend limits of debate
  - Parliamentary inquiry
  - Division of the question (motion)
  - Motion to withdraw
  - Rescind
  - Call for the orders of the day
  - Object to consideration of a question
- Describe common parliamentary errors and misconceptions
- Take official minutes
- Discuss serving as a qualified parliamentarian
- Explain how parliamentary procedure contests are conducted and scored

Terms to Know

floor
previous question
parliamentary inquiry
withdraw
question of privilege
unstated subsidiary motion
minutes

Parliamentary procedure is based on fairness, rights, and common sense. It is not a difficult subject to learn, but as with any extensive subject, it takes time and practice to master. Once a person has a mastery of basic parliamentary procedure, he or she can join a professional group, based solely on the study of parliamentary procedure. In such a group, a member can achieve certain levels of mastery and offer services as a certified parliamentarian. One such group is the National Association of Parliamentarians.

This chapter is a continuation of Chapter 10. The material covered in these two chapters is by no means exhaustive. Further knowledge of parliamentary procedure can be obtained through the study of parliamentary authorities, such as *Robert's Rules of Order*.

The information contained in this chapter gives basic guidelines on useful motions and topics that are common to most meetings, such as decorum in debate, minutes, and duties of the presiding officer. Several motions are summarized in this chapter, as well as some useful tools and guidelines. Examples include serving as a qualified parliamentarian, parliamentary procedure in governmental bodies, and parliamentary procedure contests.

It is recommended that after completing these two chapters on parliamentary procedure, students continue their learning on the subject. Parliamentary procedure is a useful, vital skill that is mastered by few.

Duties of Presiding Officer

Routine Duties

The presiding officer of an assembly has routine and specific duties to perform that relate to the proceedings of a business meeting. To ensure a smooth, efficient business meeting, the chairperson must fulfill these duties to the best of his or her ability.

It is the duty of the chairperson to open the meeting by calling it to order and to declare it adjourned. It is also the duty of the chairperson to handle all business according to the rules of parliamentary procedure and to expedite the business in every way compatible with the rights of members. Other routine duties are discussed in Chapter 10.

Specific Duties

Receiving Motions The chairperson, either on his or her own initiative or at the secretary's request, can require any main motions, amendments, or instructions to a committee to be in writing before the chairperson states the question. Also, if a motion is offered in wording that requires clarification before it can be recorded in the minutes, it is the duty of the chairperson to see that the motion is put into suitable form before the question is stated.

Discussion on the Floor While a motion is open to debate, there are three important instances in which the **floor** should be assigned to a certain person.

1. If the member who made the motion claims the floor and has not already spoken on the question (motion), he or she is entitled to be recognized in preference to other members.
2. No one is entitled to the floor a second time in debate on the same motion on the same day as long as any other member who has not spoken on this motion desires the floor.
3. In cases in which the chairperson knows that people seeking the floor have opposite opinions on the question (and the member to be recognized is not determined by the previous two cases), the chairperson should let the floor (speaker) alternate, as much as possible, between those favoring and those opposing the measure.[1]

Ending Discussion The chairperson cannot end discussion as long as any member who has not expressed the right to debate desires the floor, except by order of the group, which requires a two-thirds vote. When the discussion appears to have ended, the chair may ask, "Are you ready for the question?" If no one then rises to speak, the chairperson can proceed to a vote. The chairperson must always call for a negative note, no matter how nearly unanimous the affirmative vote may appear. The chairperson announces the result of the vote immediately, or as soon as he or she has paused to permit response to his or her call for the negative vote.

Figure 11–1 After receiving proper recognition from the chairperson, a member can debate a motion.

DECORUM (PROPRIETY) IN DEBATE

The following practices and customs observed by speakers and other members in a meeting assist discussion and debate in a smooth and orderly manner. Unless otherwise defined in the bylaws or any other rule of the group, the following rules of debate apply.

Member Rights in Debate

Each member has the right to speak twice on the same motion on the same day but cannot discuss the motion a second time as long as any member who has not spoken on that motion desires to speak. If the member who made the motion wants to discuss it and has not done so already, he or she is entitled to be recognized in preference to other members. Discussion must be relevant to the pending motion.[2] The member in Figure 11–1 is openly discussing a motion.

Member Responsibility

Members or speakers must address their comments to the chairperson, maintain a courteous tone, and avoid injecting a personal note into the debate. When discussing a motion, makers of a motion, although they can vote against it, are not allowed to speak against their own motion. While discussing a motion, members' remarks must be relevant to the pending motion. More specifically, the comments must have a bearing on whether the pending motion should be adopted.

Presiding Officer Rights

If the presiding officer is a member of the group, he or she has the same rights to discuss a motion as any other member. However, the impartiality required of the chairperson in a group precludes exercising these rights while he or she is presiding.[3] To participate in discussion, he or she must relinquish the chair for as long as that item of business is pending.

Yielding to Presiding Officers

If at any time the presiding officer rises to make a ruling, give information, or make comments within privilege, any member who is speaking should be seated (or should step back slightly if standing at a microphone some distance from a seat) until the presiding officer has finished talking.

Meeting Disruptions

During discussion or comments by the presiding officer, and during voting, no member should be permitted to disturb the meeting by whispering, walking across the floor, or in any other way. The key words here are *disturb the meeting*. This rule does not mean, therefore, that members can never whisper, or walk from one place to another in the meeting area while the meeting is in progress. However, the presiding officer should make sure such activity does not disturb or hamper the transactions of business.

CLASSIFICATION OF MOTIONS

All motions are classified under five classes of motions: main motions, privileged motions, subsidiary motions, incidental motions, and motions that bring a motion (question) again before the assembly.

Main Motion

A main motion is the basis of all parliamentary procedure. It is the method of bringing business before the assembly for discussion and action. It can only be introduced if no other business is pending. Main motions must ordinarily be made first, before any of the other classes of motions would make any sense. They require a second before they may be debated and require a simple majority to pass.

Privileged Motions

Privileged motions are of such urgency or importance that they are entitled to immediate discussion even though they may not relate to the original motion. They take precedence over the pending motion and all other items of business.[4] Privileged motions, in order of precedence, are as follows:

- Fix the time to which to adjourn
- Adjourn
- Recess
- Raise a question of privilege
- Call for the orders of the day

Subsidiary Motions

Subsidiary motions may be applied to another motion for the purpose of modifying it, delaying action on it, or disposing of it. Subsidiary motions are always proposed after the main motion to which they apply. They must be debated and voted on before the group returns to debate on the main motion.[5] The subsidiary motions, in order of precedence, are as follows:

- Lay on the table
- **Previous question**
- Limit or extend limits of debate
- Postpone definitely
- Refer to a committee
- Amend
- Postpone indefinitely

Incidental Motions

Incidental motions either arise out of a pending motion (question), arise out of a motion (question) that has just been passed, or relate to the business of the meeting. Incidental motions usually relate to the way business is transacted rather than to the business itself. They have no rank among themselves because they are incidental to the business of the assembly. Following are the commonly used incidental motions.

- Appeal from the decision of the chair
- Division of the assembly
- Object to consideration of a question
- **Parliamentary inquiry**
- Point of order
- Suspend the rules
- **Withdraw**

Motions to Bring Back

Motions to bring back return a motion to the meeting for consideration. These are sometimes referred to as unclassified motions. Three unclassified motions are commonly used:

- Take from the table
- Reconsider
- Rescind

Ranking Motions

The 13 ranking motions in order of precedence, are as follows:

Privileged motions:
Fix the time to which to adjourn
Adjourn
Recess
Question of privilege
Orders of the day

Subsidiary motions:
Lay on the table
Previous question
Limit or extend limits of debate
Postpone definitely
Refer to a committee
Amend
Postpone indefinitely

Main motion:
Main motion

A main motion may be made only when there is no business on the floor. It is the lowest-ranking motion, yet main motions are the foundation of parliamentary business. A motion to fix the time to which to adjourn can be made at any time a member can legally obtain the floor. It is the highest-ranking motion.

The ranking motions are placed in a logical sequence. Privileged motions are higher ranking than subsidiary and main motions; therefore, privileged motions have privilege over all other motions. Subsidiary motions cannot be made while any privileged motion is pending, because privileged motions are higher ranking. However, subsidiary motions can be made while a main motion is pending. For example, if every ranking motion could be pending at one time, Figure 11–2 shows the order in which they could be made and ultimately voted on.

Ranking of Motions in Order of Precedence

Main motion
Postpone indefinitely
Amend
Refer
Postpone definitely
Limit debate
Previous question
Lay on table
Orders of the day
Question of privilege
Recess
Adjourn
Fix time to adjourn

Figure 11–2 Motions have to be taken in a logical, orderly manner. If every ranking motion could be pending at one time, this is the order in which they would be addresssed.

Once a main motion is made, each motion can be applied in succession, as long as it is higher ranking than the motion in which it is being applied. The only exception is the motion to amend. Amend can be applied to higher-ranking motions, if they are amendable. This is described in the characteristics of each individual motion.

Once a series of motions has been made, they are voted on in the opposite order in which they were made. Thus, the last motion made is the first one voted on and the first one made is the last one that receives a vote. For example, there is a main motion made and a postpone indefinitely made and a motion to refer:

Main motion
Postpone indefinitely
Refer

When the motions receive a vote, the motion to refer is voted on first, then the motion to postpone indefinitely, and finally the main motion.

Fix the Time to Which to Adjourn

Rationale

The purpose of the motion to fix the time to which to adjourn is to set the time for another meeting to continue business of the existing session with no effect on when the existing meeting will adjourn.

Explanation

This motion is privileged only when it is made while a motion is pending in a meeting and there is no existing provision for another meeting on the same or the next day. If a motion to fix the time to which to adjourn is made while no motion is pending, it is handled as a main motion. Therefore, it is debatable and subject to all the other rules just as if it were a main motion.

Whether the motion to fix the time to which to adjourn is a privileged or a main motion, the result of this motion is to establish an adjourned meeting. An adjourned meeting is a legal continuation of an existing meeting to another meeting. This newly scheduled adjourned meeting (whether a privileged or main motion) must be set for a date before the next regular meeting.

If a motion to fix the time to which to adjourn is introduced while no motion is pending, it is treated as a main motion. In this instance, it requires a second, is debatable and amendable, and requires a majority vote to pass.[6]

If a motion to fix the time to which to adjourn is in its privileged form, it takes precedence over all other motions. It requires a second, is not debatable, and is amendable only as to time. The amendment is also undebatable. The motion requires a majority vote to pass.

Example

In an assembly in which there is a motion pending, the following exchange takes place.

Member: *Madam Chairperson.*

Chairperson: *Madam Speaker.*

Member: *I move that when this meeting adjourns, it will adjourn to meet at 10:00 A.M. tomorrow.*

Chairperson: *Is there a second?*

Another Member: *I second the motion.*

Chairperson: *It has been moved and seconded that we fix the time to which to adjourn at 10:00 A.M. tomorrow. This motion is undebatable and only amendable as to time. Is there any discussion?*

If the motion to fix the time to which to adjourn passes, the assembly has set up an adjourned meeting. When the assembly adjourns, they will adjourn to meet at 10:00 AM tomorrow. If the motion to fix the time to which to adjourn fails, the assembly carries on with its regular business. When it adjourns the present meeting, it will adjourn to meet at its next regularly scheduled meeting.

Question of Privilege

Rationale

A question of privilege is a request made by a member and granted or denied by the chairperson. A question of privilege may be made to secure immediate action on some urgent matter that relates to the comfort, convenience, rights, or privileges of the group or assembly, or one of its members. Any member may rise to a question of privilege even though other business is before the chapter, since this is a privileged motion.[7]

There are two types of questions of privilege: privileges relating to the group as a whole and those relating to questions of personal privilege.

Explanation

A question of privilege takes precedence over all other motions except the three higher-ranking privileged motions: fix the time to which to adjourn, adjourn, and recess. It does not require a does not receive a vote. The chair rules on the question. However, if the chair's ruling is challenged through the motion to appeal from the decision of the chair, a vote is taken on the appeal.

Example

In an FFA meeting, a member rises and without awaiting recognition, addresses the chair.

Member: *Madam Chairperson, I rise to a question of privilege.*

Chairperson: *State your question.*

Member: *The temperature should be increased; it's cold in here.*

Chairperson: *Your request is granted. Will the sentinel adjust the thermostat to make the room warmer?*

Note: The chairperson then continues with business.

LIMIT OR EXTEND LIMITS OF DEBATE

Rationale

The motion to limit or extend limits of debate either restricts the time to be devoted to debating a pending motion or removes any limitation placed upon it. Besides the previous question covered in Chapter 10, the motion to limit or extend limits of debate is the only motion a group can make to have special control over debate on a pending motion or series of pending motions.

Note: Neither a motion to limit or extend limits of debate nor a previous question is allowed in committee.

Explanation

The motion to limit or extend limits of debate can be applied to any immediately pending debatable motion, to an entire series of pending debatable motions, or to any consecutive part of such a series, beginning with the immediately pending question. An order limiting or extending limits of debate is exhausted when one of the following situations occur:

- When all of the questions on which it was imposed have been voted on
- When questions affected by the order and not yet voted on have been either referred to a committee or postponed indefinitely
- At the conclusion of the session in which the order has been adopted[8]

The motion to limit or extend limits of debate takes precedence over all debatable motions. It yields to previous question, lay on the table, all privileged motions, and applicable incidental motions. It requires a second, is undebatable, and is amendable. Amendments applied to it are undebatable, and it requires a two-thirds vote to pass.

Example

A main motion is pending "that our FFA chapter sponsor a $1,000 scholarship for a high school senior." Since this important motion deals with such a large amount of money, the members have much to discuss. The following transactions occur:

Member: *I move to extend debate to three times per person on this motion.*

Chairperson: *Is there a second?*

Another Member: *I second the motion.*

Chairperson: *The motion has been made and seconded to extend debate to three times per person on this motion. This motion is undebatable, but amendable, and requires a two-thirds majority vote. Are there any amendments to the motion?*

If the motion to extend debate passes, the times a person may speak on this particular main motion is three. If the motion to extend debate fails, the members may offer the regular amount of discussion, which is only two times per speaker per motion. The members in Figure 11–3 are

Figure 11–3 Parliamentary procedure is used at business sessions at the National FFA Convention. On important issues, debate could go on indefinitely without the motion to limit debate, as evidenced by the intensity of these members.

anxious to discuss a motion. You can see why extending or limiting debate can be so important.

PARLIAMENTARY INQUIRY

Rationale

A parliamentary inquiry is a question that any member may ask the chair regarding the correct usage of parliamentary procedure or the rules of the organization. A parliamentary inquiry usually pertains to the business that is currently pending; however, the question does not have to relate to the pending motion.

The chair may answer the question or direct the parliamentarian to do so. In either case, the question is answered, so no vote is necessary. This is a useful tool when a member is in doubt as to what skills to use or what is the proper parliamentary procedure to accomplish a certain purpose.

Explanation

When making a parliamentary inquiry, a member does not need to obtain recognition from the chair. A parliamentary inquiry takes precedence over any motion as long as the inquiry relates to that motion. A parliamentary inquiry can be made at any time even if no motion is pending. It is in order even if another motion is on the floor if the parliamentary inquiry requires immediate attention. Since a parliamentary inquiry is a question usually answered by the chairperson, it does not require a second, is not debatable or amendable, and is not voted on.

The chairperson's reply to a parliamentary inquiry is not subject to an appeal since it is an opinion, not a ruling. However, a member has the right to act contrary to this opinion.

Example

During lengthy discussion of a main motion, a member rises to a parliamentary inquiry.

Member: *Mr. Chairperson, I rise to a parliamentary inquiry.*

Chairperson: *Madam Speaker, please state your inquiry.*

Member: *Would it be in order at this time to move the previous question to end debate?*

Chairperson: *Yes, a motion to move the previous question would be in order at this time.*

The assembly then returns to discussion of the pending main motion. A member may move the previous question but is not required to do so.

DIVISION OF THE QUESTION (MOTION)

Rationale

If a pending main motion contains two or more parts capable of standing as separate questions, a division of the question (motion) may be used to separate a main motion into separate motions for the purpose of separate debate and voting.

Explanation

To decide if the question (motion) should be divided, each part should be capable of standing alone as a complete motion to receive a vote. When dividing the motion, the manner in which the question is to be divided must be clearly stated.

A division of the question takes precedence over the main motion and over the subsidiary motion to postpone indefinitely. It requires a second, is amendable, is not debatable, and requires a majority vote to pass.

Example

A main motion is pending before the assembly "that our FFA chapter host a Muscular Dystrophy telethon and we donate $200 to the Muscular Dystrophy Association." After some discussion, the majority of the FFA chapter believes it is an excellent idea to host the MDA telethon, but the chapter does not have any money in the budget to donate to the MDA. So, instead of voting down the entire motion, a member moves to divide the question into two parts.

Member: *Madam Chairperson.*

Chairperson: *Madam Speaker.*

Member: *I move to divide the question to consider separately the question of hosting a Muscular Dystrophy telethon and the question of donating $200 to the MDA.*

Chairperson: *Is there a second?*

Another Member: *I second the motion.*

Chairperson: *It has been made and seconded that we have a division of the question. This motion is only amendable. Is there any discussion?*

Note: The chairperson proceeds to take a majority vote.

If the motion to divide the question (motion) passes, the assembly will consider two main motions. The first motion the chapter will consider is "that our FFA chapter host an MDA telethon." The assembly will discuss and vote on this main motion without affecting the second main motion. As soon as the first main motion is disposed of, the second main motion is considered. The second main motion is also treated as a separate main motion. If the motion to divide the question fails, the main motion is still considered as one main motion.

MOTION TO WITHDRAW

Rationale

A member makes a motion that seems reasonable at the time. However, after brief discussion, the member who made the motion realizes the motion was inappropriate. To withdraw a motion enables a member who has made a motion to remove it from consideration before a vote is taken.[9]

Explanation

Before a motion has been stated by the chairperson, it is the property of its maker, who can withdraw it or modify it without asking the consent of anyone. Thus, in the brief interval between the making of a motion and the time when the motion is placed before the assembly by stating it, the maker of the motion can withdraw it.[10]

After a motion has been stated by the chairperson, it belongs to the meeting as a whole and the person who made the motion must request the group's permission to withdraw or modify the motion. A request for permission to withdraw a motion, or motions to grant such permission, can be made at any time before voting on the question has begun, even if the motion has been amended, and even if subsidiary or incidental motions may be pending.[11]

Any member can suggest that the maker of a motion ask permission to withdraw it, which the maker can do or decline to do, as he or she chooses. The motion to withdraw takes precedence over any motion with which its purpose is connected. It requires a second if it has been given to the assembly and the maker of the motion wishes to withdraw it. It is not debatable or amendable, and it requires a majority vote to pass if a vote is taken.

Example

During an FFA meeting, a member makes a main motion "that our group take a trip to Wyoming to go skiing." After lengthy negative discussion, the member realizes the feeling of the group and that no one really wants to go to Wyoming.

Member: *Madam Chairperson.*

Chairperson: *Madam Speaker.*

Member: *I move to withdraw my main motion that our group take a trip to Wyoming to go skiing.*

Chairperson: *Is there any objection to the withdrawal of the main motion? (There is no objection.) The motion is withdrawn.*

If just one member objects, the chair should call for a second and then proceed to vote. If the majority is in favor of withdrawing the motion, the motion is withdrawn. If there is not a majority in favor of withdrawing the motion, the main motion remains on the floor.

RESCIND

Rationale

The motion to rescind repeals or cancels a main motion that has passed at the present or a previous meeting. As situations change, an organization often needs to change. Therefore, this motion is valuable because plans and activities need to be updated. Motions passed previously could hold a chapter back from doing new and creative things to make the organization stronger.

Explanation

There is no time limit for rescinding a motion, but the motion to rescind does not release the organization from any previous commitments. For example, if a chapter had a four-year contract for soft drinks for the concessions, which after three years proved unsatisfactory, rescinding the motion would not cancel the contract.

Rescinding has the effect of voiding a motion from the date it passed up to the time it is rescinded, but it is not retroactive. For example, a motion passed a year earlier to provide free FFA manuals to new members, if rescinded, would apply only from the time the action to rescind took place. Those who had received free FFA manuals would not be asked to pay for them.[12]

The motion to rescind requires a second, it is debatable, and it opens the main motion to debate. Any amendment must apply only to the main motion and requires a majority vote. However, it requires a two-thirds vote unless previous notice was given to all members. Also, if the motion to rescind was scheduled in the agenda before the meeting began, an amendment is out of order. The motion to rescind requires a majority vote.[13]

Example

Speaker: *Mr. Chairperson.*

Chairperson: *Mr. Speaker.*

Speaker: *I move to rescind the motion passed at our April meeting to hold a picnic meeting in June, since so many of our members will be away on vacation.*

Chairperson: *Is there a second to the motion?*

Another Speaker: *I second the motion.*

Chairperson: *It has been properly moved and seconded to rescind the motion to hold a picnic meeting in June. Is there any discussion? Hearing none, we shall proceed to vote. Since no prior notice was given on this motion, it requires a two-thirds vote. All in favor of rescinding the motion stand. Thank you, please be seated. All those opposing the motion, please stand. Thank you, please be seated. The vote is unanimous and the motion is rescinded.*[14] (Rap gavel once.)

CALL FOR THE ORDERS OF THE DAY

Rationale

A call for the orders of the day is a privileged motion by which a member can require the group meeting to follow the agenda, program, or order of business or to follow a general or special order that is to be handled at a specific time. Otherwise, two-thirds of those voting may decide the agenda need not be followed.

Explanation

The call for the orders of the day in a regular meeting is to demand that the group conform to its program and cease talking about other things. If an inappropriate motion, which does not follow the written agenda, were made, a member would rise and say, "I call for the orders of the day." The chairperson would state that the orders of the day (agenda) were being followed or put it to a vote by saying, "Will the meeting follow the orders of the day?" A two-thirds vote against following the orders of the day is required to consider this motion. "All in favor of following the orders of the day raise your right hand; those opposed, same sign. Since there is less than two-thirds against following the orders of the day, this main motion is dismissed," or, "There being two-thirds against following the orders of the day, the main motion is in order."[15]

As a motion, a call for the orders of the day takes precedence over all motions, except other privileged motions and a motion to suspend the rules. It does not require a second, it is not debatable or amendable, and the orders of the day can be suspended only by a two-thirds vote.[16]

Example

In an FFA meeting, a member moves a main motion that is not on the agenda at the time the chapter is in the middle of its meeting. Another member does not believe the matter should be discussed at this time and without awaiting recognition, addresses the chair.

Member: *Madam Chairperson, I call for the orders of the day.*

Chairperson: *The orders of the day are called for. . . . Will the meeting follow the orders of the day? A two-thirds vote in opposition to following orders of the day is required to consider the main motion.*

Note: The chairperson proceeds with the vote. If there is less than two-thirds against following orders of the day, the main motion is dismissed. If there is more than two-thirds in favor of the motion, it becomes part of the orders of the day.

Figure 11–4 includes an agenda of an agricultural education state staff meeting. If items were

AGENDA
Joint State Staff Meetings

Review previous minutes
Evaluate conference
Preliminary plans for conference
Evaluation of summer workshops
Preliminary plans for workshops
Camp Clements report
Discussion of BEP/extended employment
Review Agricultural Mechanics Contest
Review other contests and activities
Report on National Convention
State Alumni Convention
Update on teacher certification
Dates for next meeting
Forestry Camp report
Review east, middle, west calendars
Agriculture in the classroom update

Figure 11–4 Meetings can be conducted with order and harmony if agendas are prepared and followed. If someone brings up an item of business that is not on the agenda, a member can call for the orders of the day.

discussed that were not on this agenda, a member could call for the orders of the day.

OBJECT TO CONSIDERATION OF A QUESTION

Rationale

Objecting to the consideration of a question is used to prevent debate of matters that are not worthy of discussion, such as humorous or ridiculous motions and those in poor taste.

If an original main motion has been made and a member believes it would do harm for the motion even to be discussed in the meeting, he or she can raise an objection to the consideration of the question, provided it is done before discussion has begun or any subsidiary motion has been made. The group members vote on whether the main motion shall be considered. If there is a two-thirds vote against consideration, the motion is dropped.

Explanation

An objection to the consideration of the question must be made before discussion begins on the main motion or before any subsidiary motions have been made. Only original main motions may be the subject of objecting to consideration of a question.[17]

An objection to the consideration of the question takes precedence over original main motions and over an **unstated subsidiary motion**, except lay on the table. It does not require a second, it is not debatable or amendable, and a two-thirds vote against consideration sustains the objection.

Example

A ridiculous main motion is made in the assembly "that each member must give $200 a month to be in good standing with the group." As soon as this motion is made, and before discussion takes place, another member rises without recognition.

Member: *I object to the consideration of this motion.*

Chairperson: *Thank you, a motion has been made to object to the consideration of the question. This motion is undebatable, unamendable, and requires a two-thirds majority vote to sustain the objection.*

If two-thirds of the members are opposed to consideration of the main motion, the main motion is dismissed. If fewer than two-thirds of the members are opposed to considering the main motion, the main motion will be considered.

COMMON PARLIAMENTARY ERRORS AND MISCONCEPTIONS

Many errors are made in business meetings when members attempt to use a motion incorrectly. For this reason, each organization should have a qualified presiding officer to handle every situation correctly and efficiently. Whenever the chairperson or any other member notices a violation of a rule in an assembly, it should be made known to the group. Following are some corrected common errors that occur in business meetings.

- When asking for corrections while approving the minutes, the chair should say, "Are there any corrections to the minutes?" Members of some societies say, "Are there any additions, deletions, or corrections to the minutes?" This is not grammatically correct, since additions and deletions obviously fall under the category of corrections.
- The motion to lay on the table is used to temporarily lay aside a matter so more urgent

business may be handled. Members who often use the motion to lay on the table as a tool to "kill a motion" are incorrect and should be ruled out of order.

- The motion to amend should state only how the motion is going to be changed (e.g., strike out, add) not how it would read after being changed.

TAKING OFFICIAL MINUTES

Content of Minutes

The official record of the proceedings of a meeting is usually called the minutes. In an ordinary meeting, the **minutes** should contain mainly a record of what was done at the meeting, not what was said by the members.[18] In other words, the minutes need only reflect motions introduced by "I move that" indicating an official action. Discussion, opinion, and debate need not be recorded unless the comments lead to an official question (motion) either approved or failed by the group. Official rulings by the chairperson should also be reflected in the minutes, such as general consent.

The minutes should never reflect the secretary's opinion, favorable or otherwise, on anything said or done. The first paragraph of the minutes should contain the following.

- Kind of meeting: regular, special, adjourned regular, or adjourned special
- Name of the society or assembly
- Date, time, and place of the meeting
- The fact that the regular chairperson and secretary were present, or in their absence, the names of the persons who substituted for them
- Whether the minutes of the previous meeting were read and approved as read, or as corrected; the date of that meeting being given if it was other than a regular business meeting[19]
- Whether a quorum is present

Reading and Approval of the Minutes

The minutes of each meeting are normally read and approved at the beginning of the next regular meeting. Corrections, if any, and approval of the minutes are normally done by general consent. If one member objects, formal approval by voting should be done.

The reading of the minutes can be dispensed with (not carried out) at the regular time by a majority vote without debate. If this happens, it can be ordered at any time later in the meeting while no business is pending by majority vote without debate. If the minutes are not read before adjournment, the minutes must be read at the following meeting before the reading of the later minutes. If the group wants to approve the minutes without having them read, it will be necessary to suspend the rules.[20] If an error is found in the minutes, even years later, the minutes can be corrected by means of the motion "to amend something previously adopted," which requires a two-thirds vote. The member in Figure 11–5 is busy taking minutes.

Figure 11–5 Accurate minutes are a crucial element of parliamentary procedure and the democratic process. To prevent authoritarian decisions, the minutes of the previous meeting are always approved by members.

SERVING AS A QUALIFIED PARLIAMENTARIAN

Need for Parliamentarians

A parliamentarian can be a certified professional or an appointed member of the group. For small local meetings, such as the FFA or other youth clubs, a parliamentarian is rarely needed. However, if one is needed, a parliamentarian could be a member of the group appointed by the president. For larger groups, professional organizations, or corporations, the parliamentarian may be a consultant, sometimes a professional, who advises the president, other officers, and committees on matters of parliamentary procedure. According to *Robert's Rules of Order, Newly Revised*, the parliamentarian's role during a meeting is purely an advisory and consultative one, since parliamentary law gives to the chair alone the power to rule on questions of order or to answer parliamentary inquiries.

Some state or national organizations may need parliamentarians throughout the year. They may be needed to assist in interpreting bylaws and rules or to work in connection with the board, officers, or committees. Parliamentarians with these large organizations may have duties including assisting in the planning and steering of business to be introduced.

Duties of Parliamentarians

The president should confer with the parliamentarian before meetings and during recesses about problems that might arise. This helps to avoid frequent consultation during meetings. During the meeting, the following duties or actions of parliamentarians may be called for.

- Give advice to the chairperson and, when requested, to any other member.
- As inconspicuously as possible, call the chair's attention to any error in the proceedings that may affect the rights of a member or do harm.
- Confer with the chair and make suggestions even if it means momentarily causing the chairperson not to give full attention to others.
- The parliamentarian should not wait until asked for advice—it may be too late.
- See a problem developing and be able to head it off with a few words to the chairperson.
- The parliamentarian should avoid speaking to the assembly if at all possible.[21]

The parliamentarian should sit next to the chairperson so as to consult in a low voice when needed. The chairperson should try to avoid consulting too much with the parliamentarian. Parliamentarians are resources; they don't give the chairperson an excuse not to know parliamentary procedure. Whatever advice is given by the parliamentarian, the chairperson has the duty of making the final ruling. The chairperson has the right to follow or disregard the advice of the parliamentarian.

If a member is appointed as a parliamentarian, three things should be remembered: (1) they do not vote on any motion, except in the case of a ballot vote; (2) they do not cast a deciding vote, even if the vote would affect the result since this would interfere with the chairperson's duty; (3) if a member doesn't want to forego the right to vote, he or she should not serve as parliamentarian.

It has often been said the best parliamentarian in a meeting is one who is never seen or heard by the assembly, only by the chairperson.

PARLIAMENTARY PROCEDURE CONTESTS

Parliamentary procedure contests are a useful teaching tool for youth organizations and school clubs. Through participation in these contests, students are introduced and trained in basic and advanced parliamentary laws. In the process, the principles of democratic leadership are instilled in the participating students. The students in Figure 11–6 are practicing for the State FFA Parliamentary Procedure Contest.

Parliamentary procedure team competitions allow students of parliamentary procedure to

Figure 11–6 There are no shortcuts around practice to master parliamentary procedure. These students are preparing for the state FFA Parliamentary Procedure Contest.

become more accurate and efficient while participating in business meetings. These contests vary in format, length, and skill level from group to group and region to region.

The script in Figure 11–7 is an example of a contest script used in some states. It can be used in class as you learn the proper procedure for stating, receiving, and handling a motion. Most of the motions discussed in this book are included in the script. You or your teacher may want to write your own script to make motions relevant to your school or organization.

Contest Format

Many states use a contest format similar to the National FFA Parliamentary Procedure Contest. Students must take a written test, answer oral questions, and perform a parliamentary procedure demonstration. The secretary prepares a set of minutes. For the presentation, the contest official assigns a main motion. Before the contest begins, five members (excluding the chairperson) are given a contest card with five required motions (Figure 11–8 on page 274). There is no limit to the number of subsidiary, incidental, privileged, and unclassified motions to be demonstrated except that the team must demonstrate two subsidiary, two incidental, and one privileged or unclassified motion designated by the officials in charge. Each team consists of six members: a chairperson and five members offering motions.[22] Figure 11–9 on page 275 provides a list of motions a team needs to prepare for the Parliamentary Procedure Contest.

Parliamentary Procedure Practice Script

Chairperson 1:	Receive and dispose of a main motion	Recess
	Lay a motion on the table	Question of group privilege
	Rescind a motion	To postpone definitely

1. I move that our chapter buy new jackets for the officer team.
2. I second the motion.
3. Because we do not know how much money is in the chapter funds, I move we table the motion.
4. I second the motion. (Fail the motion to table, fail motion.)
5. In our monthly meeting, we passed a motion stating that all members must wear official FFA overalls while in the shop. I move we rescind the motion.
6. I second the motion. (Pass the motion)

Figure 11–7 Until you attain self-confidence on parliamentary procedure, your teacher may want to assign you a part in a script so you will know what to say.

Parliamentary Procedure Practice Script (continued)

7. I move that we take a 20-minute recess.
8. I second the motion. (Fail the motion)
9. I move that we have a Sadie Hawkins dance added to our yearly activities.
10. I second the motion.
11. I rise to a question of group privilege. (Wait for response.) There is a draft in here. Could you please close the windows? (Comment from chair; return to item of business.)
12. Since the Recreation Committee chairperson is not present, I move that we postpone this motion until our next chapter meeting.
13. I second the motion. (Pass the motion to postpone.)

Chairperson 2:	Reconsider a motion	Adjourn
	Amend a motion	Rise to a point of order
	Withdraw a motion	Appeal from the decision of the chair

14. Earlier in the meeting, we referred to the Recreation Committee a motion to have a Halloween party. I move that we reconsider the motion to refer to a committee. (Give the chairperson time to ask if he voted on the prevailing side.)
    Yes, Mr. Chairperson.
15. I second the motion. (Pass motion to reconsider.)
16. I move to amend the motion by substituting the words *our chapter officers* for *Recreation Committee.*
17. I second the motion. (Pass the amendment. Pass the motion to refer.)
18. I move we have a "Boots and Boxers" party on Halloween.
19. I second the motion.
20. Halloween is a time for ghosts and goblins, not boots and boxers; besides, it will be cold!
21. I move to withdraw the motion of having a "Boots and Boxers" party on Halloween. (Wait for comment.)
22. I move that we adjourn.
23. I second that motion.
24. I move to amend the motion by adding the words *until our next biweekly meeting on Thursday at 4:30.*
25. I second the motion to amend. (Pass the amendment; fail motion.)
26. I rise to a point of order. You did not get a second on the last motion. (Wait for comment.)
27. I appeal from the decision of the chair. (Wait for comment from chair.) I believe you did not get a second on the last motion.
28. I second the motion to appeal. (Fail the motion to appeal.)

Chairperson 3:	Take from the table	Postpone indefinitely
	Refer to a committee	Object to the consideration of the question
	Suspend the rules	

29. At our last meeting, we tabled the motion to buy a quarterhorse sire for breeding purposes. Since we have found one with good conformation, I move that we take this motion from the table.
30. I second the motion. (Pass motion to take from table. Give chairperson time to state original motion.)

Figure 11–7 *(Continued)*

(con'd)

Parliamentary Procedure Practice Script (continued)

31. I think we need to give this more thought, so I move this be referred to a committee.
32. I second the motion. (Pass a motion to refer.)
33. I move that we buy refreshments for the meeting.
34. I second the motion.
35. The chapter rules state that we cannot spend any money not in the budget. (Wait for comment.)
36. I move that we temporarily suspend the rule.
37. I second the motion. (Pass with two-thirds vote.)
38.* I move that we serve refreshments during the meeting.
39.* I second the motion. (Fail motion.)
40. I call for division of the house. (Fail motion.)
41. I move that we sell beef jerky for our fund-raiser.
42. I move to postpone this motion indefinitely.
43. I second the motion. (Pass motion to postpone.)
44. Since our goal every year is to be the best chapter, I move that we purchase a banner for inspirational purposes that reads "Smash the competition."
45. I object to the consideration of the question. (Two-thirds negative vote prevents discussion.)
46. I move we adjourn.
47. I second that motion. (Pass motion.)

*Numbers 38 and 39 can be eliminated. Continue with main motion to buy refreshments.

Figure 11–7 (*Concluded*)

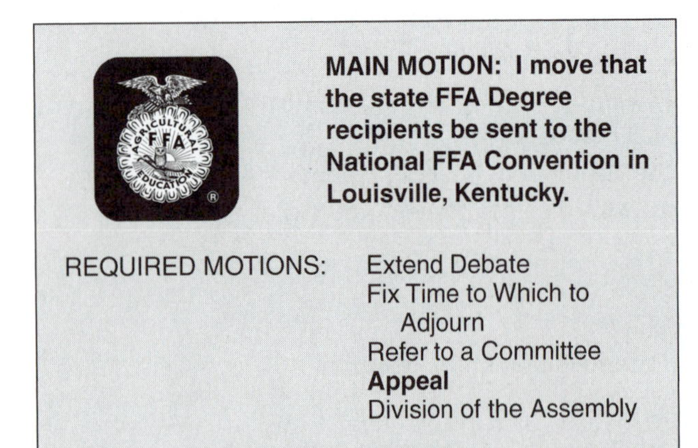

Figure 11–8 A sample parliamentary procedure contest card. The motion to appeal is highlighted, which means that it is the required motion for the holder of this card. (Courtesy of the National FFA Organization.)

Rules of the Contest

The rules of the National FFA Parliamentary Procedure Contest are included in Appendix F. Besides the rules, the format of the written test, oral questions, presentation, presentation minutes, and instruction on minutes are discussed in detail.

Scoring the Official Minutes

The explanation of scoring and instructions on minutes is discussed in the "National FFA Parliamentary Procedure Contest Rules." Form 1, in which the final copy of the minutes is written, appears in Figure 11–10A. The Tabulation Sheet for Scoring Minutes is provided in Figure 11–10B.

Chart of Permissible Motions for the National FFA Parliamentary Procedure Contest

Motion	Debate	Amend	Vote Required	Second	Reconsider	Recognition Required	May Interrupt a Speaker
Privileged							
Fix time to which to adjourn	No	Yes	Majority	Yes	Yes	No	No
Adjourn	No	No	Majority	Yes	No	No	No
Recess	No	Yes	Majority	Yes	No	No	No
Question of privilege	No	No	None	None	No	No	Yes
Call for orders of the day	No	No	2/3 Not to follow	None	No	No	Yes
Incidental							
Appeal	Yes/No	No	Majority	Yes	Yes	No	Yes
Point of order	No	No	None	No	No	No	Yes
Parliamentary inquiry	No	No	None	No	No	No	Yes
Suspend the rules	No	No	2/3, Majority*	Yes	No	Yes	No
Withdraw a motion	No	No	Usually none	No	No	Yes	No
Object to consideration of a question	No	No	2/3	No	Yes, negative vote only	No	Yes
Division of the question	No	Yes	Majority	Yes	No	Yes	No
Division of the assembly	No	No	No	No	No	No	No
Subsidiary							
Lay on table	No	No	Majority	Yes	No	Yes	No
Previous question	No	No	2/3	Yes	Yes before vote	No	No
Extend or limit debate	No	Yes	2/3	Yes	Yes	No	No
Postpone definitely	Yes	Yes	Majority, 2/3**	Yes	Yes	Yes	No
Refer to committee	Yes	Yes	Majority	Yes	Yes	Yes	No
Amend	Yes	Yes	Majority	Yes	Yes	Yes	No
Postpone indefinitely	Yes	No	Majority	Yes	Yes vote only	Yes	No
Main motion	Yes	Yes	Majority	Yes	Yes	Yes	No
Unclassified							
Take from table	No	No	Majority	Yes	No	Yes	No
Reconsider	No/Yes	No	Majority	Yes	No	No	Yes
Rescind	Yes	Yes	Majority, 2/3 if not in writing	Yes	Negative vote only	Yes	No

*Requires only a majority vote if suspending a standing rule of the chapter. Example: "I move to suspend the rule of our chapter that all officers wear official dress at chapter officer meetings."

**Requires a two-thirds vote if it is the special orders of the next meeting. Example: "I move to postpone this item of business to our next regular scheduled meeting and make it the special orders of the day."

Figure 11–9 Summary of permissible motions for the National FFA Parliamentary Procedure Contest. (Courtesy of the National FFA Organization.)

Form 1

NATIONAL PARLIAMENTARY PROCEDURE CONTEST
OFFICIAL MINUTES

____________	____________	____________
Date	FFA Chapter Name	Place
____________	____________	____________
Time Entering Minutes Room	State Name	Time Leaving Minutes Room

____________	____________
Chairperson	Secretary

Figure 11–10A The final copy of the minutes for the National FFA Parliamentary Procedure Contest is written on a form such as this one. (Courtesy of the National FFA Organization.)

Ability	Possible Points	Actual Points
Completeness and Accuracy	25	______
Minutes accurately reflect all business transacted during presentation.		
Format of minutes include: 2 pts. each	10	
Date		______
Time of Presentation		______
Place		______
Presiding Officer		______
Secretary/signature		______
Grammar/Style/Legibility	15	
Complete sentences (0–5 points)		______
Correct spelling (deduct 1 point/mistake)		______
Correct punctuation (deduct 1 point/mistake)		______
Legibility (clarity) (0–10 points)		______
Total Points Earned	**50**	______

Figure 11–10B Tabulation sheet for scoring minutes. (Courtesy of the National FFA Organization.)

Scoring the Contest

A total of 1,000 points is used for the contest. The written test is 100 points, oral questions are 100 points, and the minutes are 50 points. The presentation is worth 750 points.

Discussion Discussion accounts for 500 of the 750 presentation points, which is three-fourths of the contest points. Each of the five members receives 100 points. The required motion presentation receives 30 points, discussion on the topic receives 60 points, and any additional motions made receive 10 points. Thus, each member receives a maximum of 100 points for the discussion part of the contest. Since there are five members, these 500 points represent one-half the total contest points.

Chair Performance accounts for 100 points. The ability to preside counts 80 points. This includes handling of motions, keeping members informed, use of the gavel, and distribution of discussion. Leadership accounts for 20 points. Here, leadership is defined as stage presence, poise, self-confidence, politeness, and voice.

General Effect of Team General effect counts for 150 points. Conclusion of the team, discussion, and the team's voice, poise, and expression are each worth 50 points. Refer to the chapters on speaking (Chapters 7, 8, and 9) for further explanation of voice, poise, and power of expression. Stage presence is also discussed in these three chapters.

The National Parliamentary Procedure Team Score Sheet is provided in the *National FFA Contest*

Use of Parliamentary Procedure in a Meeting

Parliamentary Procedure Motion	Practical Use during Meeting
Main motion	Present business for action
Postpone indefinitely	Kill the main motion
Lay on the table	Set matter aside for a later time
Amend	Change or modify
Commit or refer to a committee	Let a subgroup attend to a matter
Previous question	Stop debate, order immediate vote
Suspend the rules	Action contrary to the rules
Withdraw a motion	Prevent a vote on a motion
Object to consideration	Oppose a foolish motion
Question of privilege	Attend to well-being of individual or the chapter members as a whole
Adjourn	Dismiss meeting
Fix the time to which to adjourn	Set a time for next meeting
Reconsider	Bring a motion back to discuss it and vote again
Parliamentary inquiry	Ask question relating to parliamentary procedure
Point of order	Call attention to parliamentary error
Appeal	Allow member to question the decision of the chair
Take from the table	Bring back an item that was laid on the table
Extend or limit debate	Establish length of time for discussion on a motion
Postpone definitely	Defer a motion to a specified time
Rescind	Cancel action taken by chapter
Division of the assembly	Question the chair's decision on a vote
Division of the question	Divide motion into two separate motions
Call for orders of the day	Want to follow agenda
Recess	Take a break in the meeting

Figure 11–11 A review of the 24 motions and why they are used during meetings.

Bulletin. All of the contest areas totaling 1,000 points are provided. Points for deductions, tie-breakers, and a list of the motions are also provided.

Figure 11–11 lists the 24 motions discussed. Practical use of the motions are also provided.

Summary

Parliamentary procedure is based on fairness, rights, and common sense. It is a useful, vital skill that is necessary to put the democratic process in action.

The presiding officer of an assembly has routine, specific duties to perform that relate to the proceedings of a meeting. Ensuring a smooth, efficient meeting is the responsibility of the chairperson.

Decorum or propriety in debate are observed by members of a group to keep discussion and debate moving in a smooth and orderly manner. Decorum includes member rights in debate, member responsibility, presiding officer rights, yielding to presiding officers, and meeting disruptions.

There are five classes of motions: main motions, privileged motions, subsidiary motions, incidental motions, and motions that bring a motion (question) before the assembly again.

A main motion is the basis of all parliamentary procedure. Privileged motions are of such urgency or importance that they are entitled to immediate discussion even though they may not relate to the original motion. Privileged motions discussed in this chapter are fix the time to which to adjourn, question of privilege, and call for the orders of the day.

Subsidiary motions are those that may be applied to another motion for the purpose of modifying it, delaying action on it, or disposing of it. The subsidiary motion discussed in this chapter was limit or extend limits of debate.

Incidental motions either arise out of a pending motion (question), arise out of a motion (question) that has just been passed, or relate to the business of the meeting. Incidental motions discussed in this chapter are parliamentary inquiry, division of the question, motion to withdraw, and object to consideration of a question (motion).

Motions that bring back a motion before the meeting are sometimes referred to as unclassified motions. The unclassified motion discussed in this chapter is the motion to rescind.

Motions are ranked in order of precedence. Privileged motions are higher ranking than subsidiary and main motions; therefore, privileged motions have privilege over all other motions. Subsidiary motions cannot be made while any privileged motions are higher ranking. However, subsidiary motions can be made while a main motion is pending.

Parliamentary procedure contests are a useful teaching tool for youth organizations and school clubs. Teams are judged on a written test, oral questions, and a parliamentary procedure presentation, which includes participation of members, performance of the chairperson, and general effect of the team. By participating in contests, students learn the principles of democratic leadership.

END-OF-CHAPTER EXERCISES

Review Questions

1. Define the Terms to Know.
2. What are three primary duties of the chairperson or presiding officer?
3. Briefly explain three important situations in which the floor should be assigned to a certain person.
4. Explain two ways the chairperson can end discussion of a motion.
5. List the five classes of motions.
6. What are three common errors in parliamentary procedure?
7. Name six items that should be in the first paragraph of the minutes.
8. When are minutes normally read and approved?
9. What are six duties or actions of a parliamentarian during a meeting?
10. Name four parts of the National FFA Parliamentary Procedure Contest that are given points.

Fill-in-the-Blank

1. Unless otherwise specified, a member may speak ____________ times on a motion each day.
2. Makers of a motion are not allowed to ____________ ____________ their own motion.
3. To participate in discussion, the chairperson must ____________ the chair for as long as that item of business is pending.

4. Whenever the presiding officer is making a ruling or comments, any member who is speaking should be ________________ until the presiding officer is finished.
5. The motion ________________ may be made to secure action relating to comfort of the meeting room or individual member.
6. The motion ________________ is to set the time for another meeting to continue business of the existing session.
7. The motion ________________ may be used to separate a main motion into separate motions for the purpose of separate debate and voting.
8. The motion ________________ is a question that any member may ask the chair regarding the correct use of parliamentary procedure.
9. The motion ________________ enables a member who has made a motion to remove it from consideration before a vote is taken.
10. The motion ________________ restricts the time devoted to debate or removes any time restrictions.
11. The motion ________________ is used to prevent debate of matters that are not worthy of discussion.
12. The motion ________________ can require a group meeting to follow the agenda.

Matching

Terms may be used more than once.

_____ 1. Fix the time to which to adjourn
_____ 2. Limit or extend limits of debate
_____ 3. Introduced if no other business is pending
_____ 4. Reconsider
_____ 5. Object to consideration of a question
_____ 6. Official record of proceedings
_____ 7. Demonstration part of National Parliamentary Procedure Contest
_____ 8. Written test of National Parliamentary Procedure Contest
_____ 9. Minutes of the National Parliamentary Procedure Contest
_____ 10. Oral questions of the National Parliamentary Procedure Contest
_____ 11. A member wishes to remove a motion passed earlier

A. motion to bring back
B. incidental motion
C. subsidiary motion
D. privileged motion
E. rescind
F. 100 points
G. 750 points
H. 50 points
I. minutes
J. main motion

Activities

1. Select one of the motions within this chapter and become an authority on it. Present a report to the class and demonstrate it to the class. When questions arise during class practice, you will be called on to answer any questions about that motion. Know such things as whether the motion is privileged, incidental, subsidiary, debatable, or amendable; the type of vote required; whether a second is needed; or anything special about the motion.

2. Use class time to discuss and practice parliamentary procedure. Take turns serving as a chairperson. Divide the class and compete according to the rules of the parliamentary contest of your state.
3. Give a demonstration of parliamentary procedure at a high school assembly or at local meetings.
4. Demonstrate the following skills with a team of five members and a chairperson:

Skills to Be Demonstrated

- Main motion
- Reconsider
- Division of the house
- Refer to a committee
- Amendment
- Postpone definitely

Problem 1

A. A member moves to build 10 picnic tables for the new commemorative park as part of the chapter's BOAC program. Motion receives a second.
B. The motion is discussed by three members, one of whom moves to amend the main motion by changing the number from 10 tables to 5 tables. Amendment receives a second.
C. The amendment is voted on and passes.
D. A motion is made to refer to a committee the main motion as amended. The motion receives a second, is voted on, and passes.

Judging Summation

Abilities to Be Demonstrated	Kind of Motion	Second Required	Debatable	Amendable	Vote Required
Main motion	Main	Yes	Yes	Yes	Majority
Amendment	Subsidiary	Yes	Yes	Yes	Majority
Refer to a committee*	Subsidiary	Yes	Yes	Yes	Majority
Reconsider	Other	Yes	Yes	No	Majority
Division of the house	Incidental	No	No	No	None
Postpone definitely**	Subsidiary	Yes	Yes	No	Majority

* In referring an item of business to a committee, the member so moving must stipulate the number of people on the committee, who appoints the committee, and when the committee is to report back.

**In postponing definitely, a member must state when the item of business will again come before the chapter, such as the next regular meeting, or a special meeting.

Source: The National FFA Organization.

	Kind	Second Required	Debatable	Amendable	Vote Required	Can Be Reconsidered	Recognition Required	May Interrupt a Speaker
Privileged	Fix the time to which to adjourn							
	Adjourn							
	Recess							
	Question of privilege							
	Call for the orders of the day							
Subsidiary	Lay on the table							
	Previous question							
	Extend or limit debate							
	Postpone definitely							
	Refer to a committee							
	Amend							
	Postpone indefinitely							
	Main motion							
Incidental	Appeal							
	Withdraw a question							
	Division of the assembly							
	Division of a question							
	Objection to the consideration of question							
	Parliamentary inquiry							
	Point of order							
	Suspend the rules							
Unclassified	Reconsider							
	Take from the table							
	Rescind							

Problem 2

A. A member moves to reconsider the item of business passed at the last meeting regarding the increase in the chapter dues from $4.00 to $6.50. Motion is seconded, voted on, and passed.

B. The chairperson announces that the motion is now on the floor in its original debatable form.

C. The motion is discussed by three members, one of whom moves to amend the original motion by substituting *$5.00* for *$6.50*. Amendment receives a second.

D. The amendment is voted on, with a close vote resulting. Division of the house is called, and the amendment passes.

E. Discussion returns to the main motion as amended, with one member moving to postpone definitely this item of business. Motion is seconded, voted on, and passes.

5. Complete the chart on the previous page by stating (1) whether a second is required, (2) whether the motion is debatable or amendable, (3) whether the motion requires a two-thirds or simple majority vote or none, and (5) whether the motion can be reconsidered. (6) Write yes, no, or the vote required in the appropriate blanks. Then, state whether the recognition is required and (7) whether a speaker may be interrupted.
6. To perform well in a parliamentary procedure contest following the national format, a team must be able to discuss and debate motions well. Each member can earn a maximum of 60 points for debating. No more than 30 points can be earned with one recognition of the chair. Examples of quality debate can be found in the national contest rules in Appendix F. Use this exercise to improve debate of a parliamentary procedure team. Give the team a main motion to discuss. Allow them one minute to think of the debate, and then begin. Call the names of the team members and have them stand and debate either for or against the main motion. Students can practice being for and against the main motion. The debate should always be directed to their chair.
7. **Take it to the Net.**

   Browse the sites listed in question 7 in the Activities section of Chapter 10. Find a site you think is useful and print the first page of that site. Write a summary of the site and include why you think it is useful. *(Do not use the same site you used in Chapter 7.)*

Notes

1. S. C. Robert, *Robert's Rules of Order, Newly Revised,* Ninth Edition (Glenview, IL: Scott, Foresman, 1990), p. 30.
2. Ibid., p. 42.
3. Ibid., p. 389.
4. K. L. Russell, *The How in Parliamentary Procedure,* Fifth Edition (Danville, IL: Interstate Printers and Publishers, 1990), p. 65.
5. J. R. O'Connor, *Speech: Exploring Communication* (Upper Saddle River, NJ: Prentice-Hall, 1998), p. 311.
6. Robert, *Robert's Rules of Order, Newly Revised,* p. 242.
7. R. E. Bender et al., *Mastering Parliamentary Procedure* (Columbus, OH: Ohio State University, Department of Agricultural Education, 1983), p. 12.
8. Robert, *Robert's Rules of Order, Newly Revised,* p. 192.
9. O'Connor, *Speech: Exploring Communication,* p. 192.
10. Robert, *Robert's Rules of Order, Newly Revised,* p. 287.
11. Ibid., p. 288.
12. Bender et al., *Mastering Parliamentary Procedure,* p. 31.
13. Ibid.
14. D. B. Jameson, *Leadership Handbook* (New Castle, PA: LEAD, 1978), p. 299.
15. Bender et al., *Mastering Parliamentary Procedure,* p. 13.
16. Ibid., pp. 13–14.
17. Ibid., p. 22.
18. Robert, *Robert's Rules of Order, Newly Revised,* p. 458.
19. Ibid.
20. Ibid., p. 464.
21. Ibid., pp. 456–457.
22. "National FFA Contest," *National FFA Contest* (Alexandria, VA: The National FFA Organization, 2000–2001).

CHAPTER

12

Conducting Successful Meetings

Objectives

After completing this chapter, the student should be able to:

- Identify leadership skills developed by conducting successful meetings
- List skills developed by being an officer
- Discuss basic meeting communication functions
- Describe the effect of meetings on developing leader sensitivity
- Explain the characteristics of a good meeting
- Plan and prepare for a meeting
- Secure attendance at meetings
- Describe the arrangements, paraphernalia, equipment, and supplies needed in the meeting room
- Properly use committees in conducting business
- Select topics for meetings that inform and motivate
- Conduct effective meetings
- Describe ways to get group members involved
- Describe characteristics of mature groups
- Discuss the responsibilities of officers and members
- Describe the FFA Program of Activities
- Explain the need for meeting evaluations

Terms to Know

sensitive
apathy
policy
Executive Committee
Program of Activities
bylaws
agenda
paraphernalia
committees
energy cycles
buzz groups
role-playing
problem solving
group think
sensitivity

Scene 1:

"Are you going to the FFA meeting today?"

"Do I have a choice?"

"Not much, it's either that or homeroom."

"Every second and fourth Wednesday. Just like clockwork."

"One of these days, you'll figure out why."

Scene 2:

"Time for the FFA meeting."

"OK! I'm anxious to find out what is happening next."

"The agenda says Mr. Smith, our advisor, is talking about the spring fling."

"I've got some good information to share that I believe the group will like."

"Good, let's go."

Which meeting scene is most familiar to you? Do you approach an FFA meeting expecting just another annoying waste of time? Do you walk away from it wondering why you wasted your time? If so, you know first-hand how frustrating inefficient meetings are. When unproductive meetings occur regularly, students begin to avoid attending.

The primary reason meetings don't accomplish their objectives is lack of advance planning and organization. Even though executives spend a significant portion of their time in meetings, studies show that 78 percent have never received training on how to plan, organize, and conduct a meeting.[1]

When meetings are well managed, they are an effective communication tool. Important decisions are made, ideas are generated, and information is shared. Meetings are a critical part of building group enthusiasm, and as enthusiasm and excitement grow, the chapter benefits. The chapter also benefits as the group's ability to work together and make decisions grows.

Leadership Skills Developed

Fortunately, the process of conducting and participating in meetings offers many opportunities for developing leadership skills. These skills will be valuable when members graduate and participate in professional, business, and civic organizations. Members can practice public speaking, parliamentary procedure, communications with groups, and various other leadership and personal development skills.

Interesting meetings build members' interest and participation. Arranging for special programs and encouraging involvement in committees are valuable learning experiences. People have always assembled to plan, negotiate, and delegate responsibility. Regular meetings of the FFA chapter are necessary so that activities can be accomplished and members can gain experience in group planning and functions.

Conducting meetings also gives members opportunities to learn how to lead individuals and groups, establish good public relations, and develop cooperation. Members can learn how to plan meetings, write up minutes, establish rapport, lead discussions, and follow democratic procedures. The students in Figure 12–1 are actively involved in a group project at an FFA meeting developing their cooperation skills.

All members of an organization are responsible for electing capable officers. FFA members learn what is expected from officers, what the

Figure 12–1 These FFA members are working on a group project enhancing their cooperation skills.

necessary qualifications are, and how to nominate and elect a person to office. By learning the duties of officers and how to select them, FFA members can better contribute to clubs and civic and private organizations, and as advisors to youth clubs and activities.

Skills Developed by Being an Officer

As an officer, a member can develop better communication skills and further develop his or her public speaking ability. Members can learn what is expected of good officers, how to install officers, and the proper functions of a nominating committee. By understanding each officer's duties and responsibilities, as mentioned in the FFA Opening Ceremony, an officer can learn the following values of the FFA: leadership, willingness to change, cooperation, self-confidence, knowledge, attitude, record keeping, public relations, neatness, honesty, budgeting, alertness, participation, and respect.[2] Many FFA officers get to attend the National Convention in Kansas City, Missouri, each November. Members and officers are permitted to attend sessions at the convention in which official business is conducted.

Basic Meeting Communication Functions

There are five functions that meetings perform better than any other communication technique.

1. *Share knowledge.* Meetings provide a forum in which individual information and experience can be pooled. The group revises, updates, and adds to what it knows.
2. *Establish common goals.* Meetings help every member of the team understand the goals and objectives of the group and how individual efforts affect those objectives.
3. *Gain commitment.* Meetings foster commitment. A sense of responsibility for implementing and supporting the group's decisions is created.
4. *Provide group identity.* Meetings define the team. Those present belong to the team; those absent do not. Attenders develop a sense of collective identity.
5. *Team interaction.* Meetings are often the only time the group works as a team.[3]

Effect of Meetings on Developing Leader Sensitivity

Leaders, supervisors, and administrators trained in effective human relations are sensitive to the needs of others. Sensitivity is a trait of the democratic leader.

Untrained leaders may be so intent on applying rules and maintaining their own positions that they are unaware of individual needs. They may not be **sensitive** to individual needs and frustration within the membership results. When leaders ignore the interests of members, they can block participation and create **apathy** in the group.

Leaders aware of individual needs and differences in group situations recognize the importance of these differences. They can help each individual make a unique contribution to the decision-making process. A leader sensitive to the needs and feelings of individuals accomplishes more than a leader concerned only with the task to be done.

Characteristics of Good Meetings

Meetings held on a regular basis are the lifeblood of a good chapter. Chapter meetings are only as interesting and effective as you make them. The following are characteristics of good meetings.

- They are carefully planned and prepared.
- They are publicized well in advance of the meeting time.
- The agenda is posted.
- Each meeting has a specific purpose.
- A high percentage of members attend.
- Correct parliamentary procedure is used.
- The agenda is followed.
- They proceed in an orderly, businesslike fashion.
- They are conducted by officers who are prepared and know their duties and responsibilities.
- Participating members are well prepared.
- The program offers a variety of business, education, and recreation.
- The members are interested.
- They are held in a pleasant, suitable place.
- They are conducted by members with minimum participation by the chapter advisor.
- Minutes are taken, posted, and distributed.[4]

Meeting Policies

Once you enter the business world, you may see policies on meetings. Why? Efficient use of time saves a company money. Policies help personnel lead and participate in effective meetings. Good meetings improve morale and team spirit because objectives are met. Following is the meeting **policy** used at the Western Center of General Dynamics.[5]

- All meetings start and end on time, with a maximum length of 90 minutes.
- Each participant has a commitment to making the meeting successful.
- A clear objective is established for the meeting and the agenda is followed.
- Common courtesy and caring are emphasized. Issues, not people, are criticized.
- A positive orientation of "We are here to help" is established.

Barriers to Successful Meetings

Unfortunately, some meetings are boring and do not meet your expectations. The following are potential problems or barriers to a successful meeting.[6]

- Poor organization
- Lack of focus on important issues
- Diversion from the agenda by members with selfish motives
- Domination by formal leaders or a few influential people
- Boredom
- Excessive length
- Disruptive behavior by members
- Occur frequently or not frequently enough
- Inconclusiveness

Appropriate Time of Meetings

Most chapters hold regularly scheduled meetings, with the schedule often specified in the constitution or bylaws. One FFA meeting per month throughout the school year and summer is considered the minimum. More active chapters may need more than one meeting per month to conduct needed business effectively. Meetings should be scheduled well in advance. Regular chapter meeting dates and a list of programs for meetings should be planned for the full year as part of the chapter's Program of Activities. This schedule of programs should be posted in view of all members. Major topics should be scheduled on the basis of interest and benefit to a majority of the members.

The major challenge for most chapters is selecting an appropriate meeting time. Many schools allow less than 30 minutes for club meetings. This is not enough time for an effective meeting with a full agenda. An alternative to meeting during school hours is night meetings. Although attendance is somewhat lower, the time attracts a more serious and focused group. However, if meetings are appropriately planned to include business,

Figure 12–2 Chapter meetings will be as interesting and effective as you make them. These FFA members are planning and preparing for the chapter meeting.

educational, and an "extra dose" of recreational activities, members will attend.

A more functional approach may be a combination of day and night meetings. The day meetings could be used for announcements, discussion of upcoming activities, and short motivational talks. The night meetings could follow the full agenda, including the minutes, treasurer's report, committee reports, old and new business, and the program.

There are a hundred reasons why scheduling meetings can be a challenge. Concentrate on the one time that works for your chapter. Remember to have the school administration approve your meeting times. The FFA members in Figure 12–2 are planning an FFA meeting.

PLANNING AND PREPARING FOR MEETINGS

Planning

Planning is the key to good chapter meetings. An order of business should be established and followed at each FFA meeting. The responsibility of developing the order of business for the chapter meeting is given to the **Executive Committee**. This committee is generally composed of chapter officers and those who chair the major committees, often referred to as standing committees. The president presides over the Executive Committee. The Executive Committee needs to work closely with the Conduct of Meetings and Recreation Committees.

Well-planned, regular chapter meetings are essential to maintain member interest, secure attendance, ensure efficiency, and promote the general welfare of the group. Each meeting should be a unit in the series for the year instead of being detached from other chapter interests and activities. Remember, the primary reason for holding meetings is to conduct the business of the chapter; a majority of the business of the chapter should be your effort to carry out a challenging **Program of Activities**. Successful meetings generally consist of the business to be transacted, the program, refreshments, and recreation.[7]

Planning your chapter meetings on a yearly schedule adds unity and purpose. It also ensures that everything is included. Although some revisions may need to be made as the year progresses, a schedule contributes to an effective chapter operation. Having a yearly schedule enables you to schedule important speakers in advance, secure needed school facilities and equipment, and alert members to their responsibilities.

Purpose

During planning, the Executive Committee should define the purpose of each meeting. A clear idea of what to do is the foundation on which everything rests. The planners should have a good idea of what they want to accomplish. Unless the meeting leader is completely clear about the purpose of the meeting, there is no way it will be effective. Meeting leaders should remember to ask for suggestions from the group members. This input helps to ensure that the meetings focus on the members' needs and promotes interest and ownership of the chapter.

Questions for Responsive and Productive Meetings

Meetings can be either great time wasters or very productive. It all depends on how well the officers plan and how responsive they are. Responsive officers know that good meetings hold the key to a good chapter. More than anything else, a productive meeting depends on planning and execution. Following are some questions that can help officers plan productive and responsive meetings.

- Is this meeting really needed?
- What is the reason for holding this meeting?
- What is the goal of this meeting?
- Which officer will preside at the meeting?
- Should all members attend this meeting?
- Will parliamentary procedure be used to guide the meeting?
- Where should the meeting be held?
- How will members be informed before the meeting?
- How will information be presented during the meeting?
- When will the meeting begin?
- When will the meeting end?
- What should be done to follow up on the meeting?

Preparing Members

Before meetings, members should

- Have a thorough knowledge of the meeting subject matter and be ready and able to make a valuable contribution
- Have enough information to make a decision
- Be responsible for implementing decisions or bringing a project to the next stage
- Represent a committee that will be affected by decisions made at the meeting[8]

Preparing a Meeting Agenda

Once the plans have been made, an agenda should be prepared to reflect the planning and preparation for the meeting. Some groups have a clause in the constitution and **bylaws** that no business can be discussed unless it is on the agenda. After the Executive Committee has completed the planning, the secretary should prepare a written agenda for the meeting. The agenda should be posted in a convenient place where all members can see it. In some cases, the agenda can be duplicated and handed out to members.

The agenda is the single most important component of meeting planning. A well-thought-out agenda distributed prior to the meeting provides participants with purpose and direction. It prepares participants and helps to create a solid structure for the meeting. The presiding officer can ask participants to submit agenda items if he or she wants to promote group involvement and avoid surprises at the meeting. A written agenda is useful because it

- Enables participants to come supplied with the right materials
- Provides a framework that supports the flow of the meeting and keeps a time frame for the discussion of each topic
- Keeps the group on track with written reminders and focuses the group's attention
- Lets all participants know who has been invited[9]

Typical FFA Agenda

Make sure the **agenda** is clear and concise, but avoid making it too brief or vague. In general, keep the agenda and meeting short. The following order of business can be used in planning and conducting chapter meetings. Your chapter should feel free to modify this agenda to fit your local needs. However, you should have an order of business that is followed at all regular meetings.

1. Opening ceremony (official ceremonies should always be used)
2. Minutes of the previous meeting (read by the secretary and approved by the membership)
3. Officer reports (treasurer's report at every meeting; others as needed)

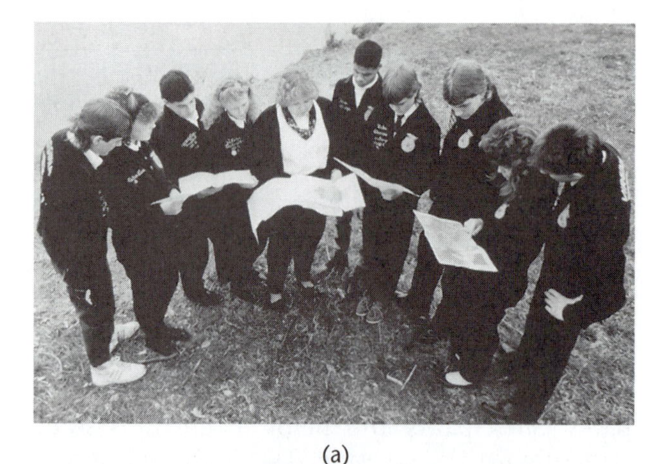

(a)

(b)

(c)

Figure 12–3 Three components are (a) the program, (b) recreation, and (c) refreshments. (Figure 12–3a courtesy of the National FFA Organization.)

4. Report on chapter Program of Activities (chairs of the various sections of the program of activities are called on to report plans and progress)
5. Program (it is often more appropriate to move this to the last item before the closing ceremony)
6. Unfinished business
7. Committee reports
   a. Standing
   b. Special (All committee reports should be in writing. After a report has been accepted, the committee chairperson should note any changes and file his or her report with the secretary for inclusion in the minutes.)
8. New business
9. Degree and installation ceremonies (used only when new members are initiated, when Greenhands are raised to Chapter Degree, and when officers are installed)
10. Closing ceremony
11. Entertainment, recreation, refreshments (Always adjourn the meeting before participating in recreation and before refreshments are served. This helps to divide the formal and informal portion of a meeting.)[10]

Item 11 in the agenda is often ignored, but it can add life and interest to every meeting. Every chapter needs a program director who arranges entertainment, recreation, and refreshments for each meeting in addition to the business or program. Figure 12–3 illustrates the major components of an FFA meeting.

Securing Attendance at Chapter Meetings

If you feed them, they will come. This is referring to feeding the members psychologically, not physically. Although refreshments, especially pizza, have a special appeal to most members, securing a high percentage of attendance at meetings is always a challenge. People tend to make time for what they really want to do, therefore the advisor and Executive Committee should ask, "What is there about our meetings that would make members want to attend?" Members usually want to attend when they have some responsibility for the success of the meeting or when the meetings are interesting and beneficial to them.[11]

Students go where the action is. If motivating business, program, and recreational activities (discussed later in the chapter) are included as a part of each meeting, students will attend. However, there are some logistical hints to improve attendance.

- Schedule meetings at a convenient time, as discussed earlier. Do not schedule to conflict with school athletic events and other school programs.
- Develop car pools.
- The advisor could give a grade for attendance or bonus points since meetings are educational and develop leadership skills.
- Have members address card reminders to themselves to be mailed by the secretary.
- Provide purposeful, beneficial, motivating meetings.

Meeting Room Arrangements, Paraphernalia, Equipment, and Supplies

Arrangements

A well-arranged meeting room that includes all the necessary items of equipment lends dignity to FFA gatherings and creates a desirable spirit of enthusiasm and pride. The sentinel should make sure the meeting room is properly arranged, according to the official FFA manual. Officer stations should be located according to the manual and the paraphernalia prominently displayed. Figure 12–4 shows the room arrangement for an official FFA meeting.

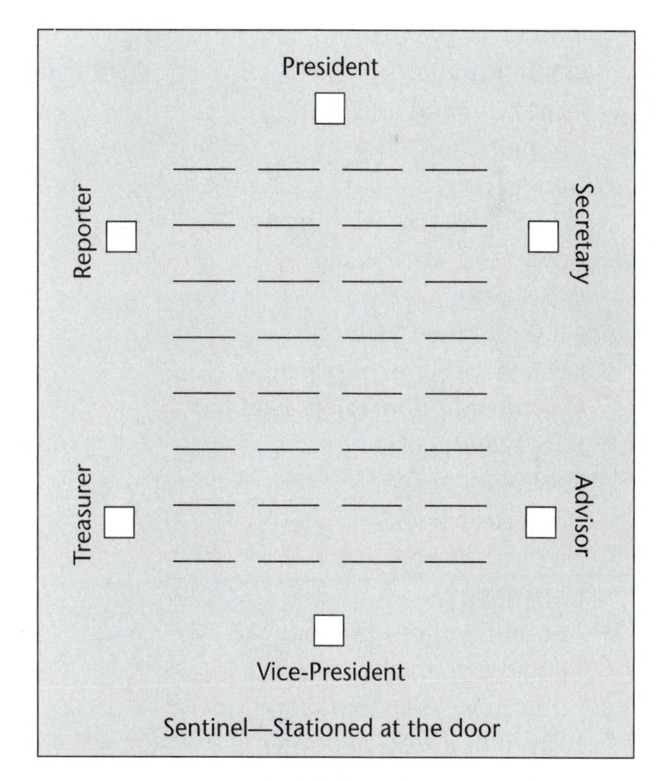

Figure 12–4 Seating arrangement for an official FFA meeting.

Sentinel Duties

Attention should be given to the meeting place by checking the ventilation, heat, and lighting. The sentinel should continue to check these items during the meeting to ensure member alertness and comfort. The sentinel should make arrangements for equipment and supplies needed during the meeting, such as video, TV, slide projector, screen, and speaker's stand. After the meeting, the sentinel should see that all paraphernalia, equipment, and supplies are returned and the meeting room is left in good order.

Official Chapter Paraphernalia, Equipment, and Supplies

The following is a list of chapter **paraphernalia**, equipment, and supplies. They are available from the FFA Supply Service.

1 American flag
1 FFA felt banner (3′ × 6′)
1 plow
1 ear of corn
1 bust of George Washington
1 owl
1 rising sun
1 flag and base (miniature)
1 shield for sentinel station
1 picture of Thomas Jefferson
1 gavel and block
1 secretary's book
1 reporter's book
1 treasurer's book
1 scrapbook
7 or more official FFA manuals
1 charter (framed)
1 official FFA flag (optional)
1 Creed (framed—optional)
1 purposes (framed—optional)
1 profanity order (framed—optional)[12]

Proper Use of the Gavel

The presiding officer should exhibit authority and leadership, become familiar with use of the gavel in chapter meetings, and use the gavel as a symbol of authority to instruct the members in proper procedures.[13]

2 taps of the gavel—Call meeting to order
3 taps of the gavel—Members stand
1 tap of the gavel—Members sit
1 tap of the gavel—After announcing results of a vote or decision of the chair
A sharp tap or series of sharp taps—To restore order

Committees

Committees are a chosen group of members with specified responsibilities. The use of committees is an important part of the FFA organization. In Chapter 14, the committee structure within the FFA will be discussed. Committee reports are a part of most FFA meetings. Committees have their own meetings, and the chairperson reports to the total group. The chairperson makes sure that the committee members have a clear purpose, know what to do, and why they have been selected or appointed.

Organization

The key to effective committees is for leaders to organize them properly. This includes setting goals, identifying steps needed to complete the goals, setting target dates for completion, and estimating resources needed. Committee responsibilities should be divided among the committee members. The Committee Activity Planning Sheet helps committees organize their tasks (Figure 12–5).

Reasons for Using Committees

Involvement is the key reason for using committees. In democratic organizations, the leader delegates. Members need to feel they are a part of the

Committee Activity Planning Sheet

Activity: FFA Week

Division: Public Relations

Members Responsible: Chuck Barstow, Betty Zetlow, and Mary Carlson

Goals	Steps	Target Date	Estimated Budget	Evaluation Notes
(What do we want to do?)	*(How are we going to meet the goals?)*	*(When?)*	*(How much?)*	*(Information for future planning)*
1. Sponsor faculty breakfast.	1. Set date and place on school calendar. Date set: ________	Sept. 15	None	
	2. Arrange to use home economics facilities.	Sept. 15	None	
	3. Have members sign up for the following jobs: room setup, greeters, cooks, servers and cleanup.	Jan. 15	None	
	4. Prepare invitations and place in teachers' school mailboxes.	Feb. 1	None	
	5. Select menu.	Feb. 15	None	
	6. Buy food.	Feb. 20	$50	
	7. Acquaint workers with jobs.	Feb. 20	None	
2. Present a five-minute radio program.	1. Discuss plans with manager of KRNT.	Oct. 15	None	
	2. Select three members for program.	Jan. 15	None	
	3. Develop script using materials from National FFA Center.	Jan. 25	$5	
	4. Review plans with KRNT program manager and set taping date. Date set: ________	Feb. 2	None	
	5. Revise script.	Feb. 8	None	
	6. Rehearse program.	Feb. 15	None	
	7. Tape program.	Feb. 20	None	

Figure 12–5 Committees work best when their activities are planned and organized. The Committee Activity Planning Sheet, which is part of the FFA Program of Activities, helps a committee organize its task. (Courtesy of the National FFA Organization.)

group. Members need to be involved because they offer expertise that can help meet the needs of the group. For example, Bill should be on the Public Relations Committee because he has excellent mechanical abilities and one of the goals of the committee is to enter a float in the homecoming parade.

Committees help develop leadership by bringing out the best in people. Talents can be discovered. Confidence can be built. Members get excited when they know they are a contributing member of the team. Because of the members' involvement, the effectiveness of the group is increased and the team becomes stronger.

Responsibilities of Committee Members

The first responsibility of committee members is availability. Unless you are committed to serving, the goals will not be met. Serving includes attending meetings, contributing ideas, listening, and being tolerant of other ideas.

The second responsibility is involvement. Do not be a spectator. Make decisions based on facts, be innovative, show initiative, accept responsibility, follow through with assignments, and be cooperative. If you do not like an idea or a person, "learn to agree to disagree."

Responsibilities of Committee Chairs

The responsibility of a committee chairperson is basically the same as that of the presiding officer of a larger meeting, such as your FFA meeting. It includes planning; preparing an agenda; collecting background information on an item of business or committee assignment; and selecting a meeting date, time, and place.

Before the committee meeting, the chairperson should supply members with appropriate materials. During the meeting, the chairperson should direct the members' discussion, keep members on the task, and help them develop a sense of responsibility. Each member should be given a chance to express his or her ideas. When necessary, it may be appropriate to select subcommittees.

After the meeting, the chairperson should compliment members for their work when they have done a good job or tactfully motivate if they have not. The chairperson should make sure that committee reports and minutes are complete. Officers of the chapter should be kept informed by the chairperson. Reports should be given to the chapter during FFA meetings when necessary. Committee reports should be recorded in the minutes.

Committee Seating Arrangement

You are probably asking what seating arrangement has to do with anything, much less committee meetings. Verbally, it has nothing to do with anything. Nonverbally, seating arrangements communicate much. In Chapters 1 and 2, we discussed three types of leadership styles: autocratic, democratic, and laissez-faire (participative). Seating arrangements nonverbally communicate a leadership style to the members. Figure 12–6 shows seating arrangements for autocratic, democratic, and participative leadership styles.

In the autocratic arrangement, the chairperson presides at the end of the table. This popular arrangement focuses the attention on the leader and gives him or her the ability to retain tight control over the agenda and meeting. When the leader wants participation, he or she can use one of the two equalizing patterns: democratic or round table. Here, the leader sits among the participants. This is most effective when the leader knows what he or she thinks about a problem and wants to foster an environment in which others can be creative while the leader maintains a low profile, not wanting to lead others with his or her opinions. The chairperson should choose the seating style that best serves the purpose and type of meeting he or she is planning.

In the democratic arrangement, the chairperson sits on one side of the table and the ends are left open. Although the leader can maintain control, there is an appearance of openness. It is easier for participants to communicate with each other rather than directing all comments to the leader.

In the participative, round table arrangement, everyone is equal and encouraged to participate as

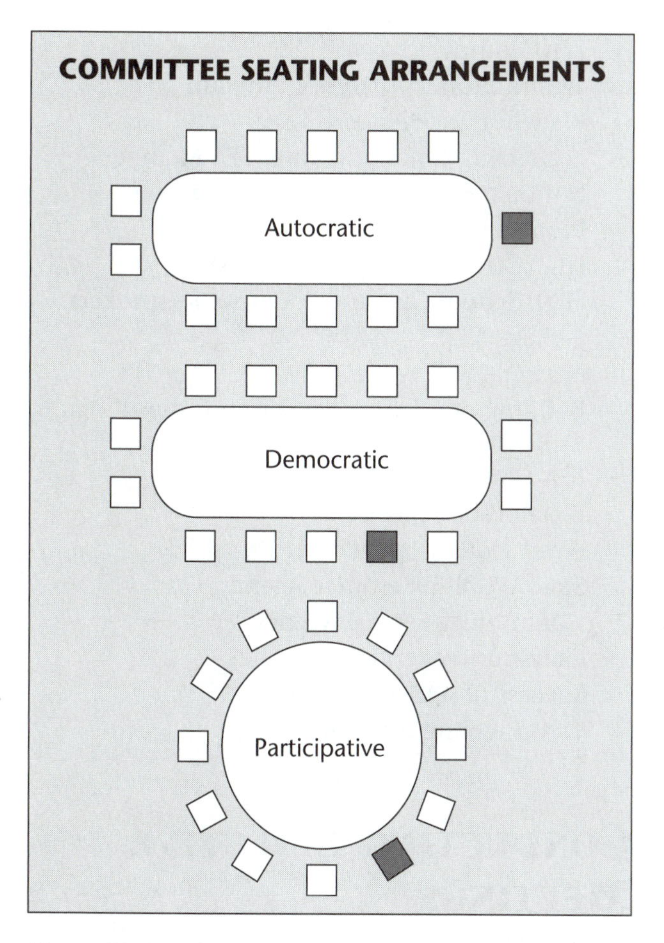

Figure 12–6 Seating arrangements nonverbally communicate your leadership style. Depending on the situation, the committee chairperson may want to vary seating arrangements to meet various objectives.

a full partner, minimizing the effects of status and power.[14]

Meetings That Inform and Motivate

Scene 1:

"What a waste of time!"

"Right! I know Bill was presiding, but he never shut up! The one time I tried to say something, he cut me off."

"And he laughed at Nancy's idea and it sounded great."

"What a jerk!"

Scene 2:

"I can't believe everyone bought our idea!"

"Especially since Bobby wanted to go the other way. He really encourages everyone to say what they think."

"I can't believe the number of ideas that came up."

"Some of them might be implemented in the future."

"Bobby's meetings always seem relevant and exciting."

Exciting

Are your meetings like those discussed in Scene 1 or Scene 2? I hope they are like the exciting Scene 2. Every meeting should be packed with excitement. Remember, we are in an entertainment age, and there are many competing activities every night. With cable television, video rentals, athletic events, and other motivating events available nightly, FFA meetings have to be exciting and entertaining as well as educational.

Meeting Components

Every meeting should contain three key elements: business, program (education), and recreation. The business section should be organized and efficient. It should include a crisp opening ceremony, minutes of the previous meeting read clearly and distinctly, treasurer's report, committee reports by the committee chairs, and proper parliamentary procedure as old and new business is discussed.

Educational (Program) Components

Each meeting should also have a program feature. Variety is the key. The more members are involved, the more motivational the meeting. Programs involving members could be the Chapter Creed Contest, Chapter Public Speaking Contest, Parliamentary Procedure Demonstration, and FFA Talent Contest. Other programs could include

Figure 12–7 FFA camps or leadership conferences provide excellent meeting topics. If conducted well, such meetings should motivate and recruit students for these activities. (Courtesy of *FFA New Horizons Magazine.*)

invited speakers only if they are entertaining and motivational. Special programs could include state officers, FFA alumni, national and state FFA conventions, and FFA Camp (Figure 12–7).

Recreational Components

Recreational activities could include a year-long athletic competition. Teams could be formed around class, year, individual classes, or another format to provide equally talented teams. Athletic competition could include softball, flag football, volleyball, and a variety of minor sports.

Whatever the format, the meetings should inform, motivate, and attract students. Again, students go where the action is. The FFA can be the group in your school "where it's happening." Figure 12–8 contains a suggested list of programs and recreational activities for FFA meetings.

The major topics included in the preceding section are but a few that you might consider in developing your own yearly schedule. Following are some additional topics to consider.

FFA Alumni
Washington Conference Program
Made for Excellence
Leadership Roles of Former FFA Members
National Chapter Award Competition
Proficiency Awards
High School Principal (speaker)
High School Guidance Counselor (speaker)
College Night
National FFA Week
Building our American Communities Program (BOAC)
FFA Contest and Awards
Election of Officers
Open House/Parent's Night
State Wildlife Resource Agency
Community Service Agency
Community Service Activities
Successful Agriculturalist
Thanksgiving Program

CONDUCTING EFFECTIVE MEETINGS

FFA meetings should operate in such a way that every potential member wants to attend. If this is not the case, "go back to the drawing board." However, when a meeting has been carefully planned and officers and members are prepared, a good meeting is almost assured.

As the presiding officer the president plays an important role in having a good meeting. The presiding officer sets the mood for the meeting by being interesting, motivating, and concerned. However, all members are encouraged to take part and share their ideas.

Democratic Decisions

FFA members should have a working knowledge of parliamentary procedure. Chapters 10 and 11 provide the basic skills and understanding needed

Month	Program	Recreational Activity
July	Program of Activities Report	Swimming, boating, games
August	Advisors/Officers Report	Softball (alumni-member game)
September	Initiation of Greenhands	Class flag football competition
October	Parliamentary Procedure Demonstration	Hayride and "weenie" roast
November	Chapter Creed Contest	Class volleyball competition
December	Christmas Program for Underprivileged Children	Christmas party
January	Film of Member's SAEP	Class basketball competition
February	State FFA Officer's/Chapter Public Speaking Contest	Class basketball competition
March	Careers in Agriculture	Skating party
April	State FFA Convention Report	FFA talent contest
May	FFA Camp Program	Class softball competition
June	Film on Year's Activities	Cookout and games

Figure 12–8 Members will attend meetings that are well planned, organized, and full of excitement. Students go where the action is. This sample schedule of FFA meetings balances educational and recreational meetings.

to adequately conduct business meetings. FFA members should be familiar with the chapter's constitution and bylaws. Some presiding officers believe that the speed with which they conduct business is somewhat associated with effectiveness. In reality, the purpose of meetings is to reach sound decisions based on facts. Decisions should reflect everyone's contributions and alert thinking through thorough discussion and exploration of a problem before finalizing the decision with formal group action. Obviously, this type of decision making usually results in a better attitude and group cohesiveness.[15]

Democratic Problem Solving

The presiding officer should focus on problems, not personalities. He or she should look at the quality of the idea, not the person who suggested it. The development of positive attitudes contributes to chapter morale and effective programs.

A good presiding officer is democratic and willing to respond to the group's wishes. He or she believes the democratic process is a way to get other members to think, speak, plan, and work together. The presiding officer is fair and impartial, reserving judgment until asking the opinions of others. A secure presiding officer recognizes individual differences as a potential for the group and not a threat to the leader's position or the program. Following are some steps to help the presiding officer conduct meetings effectively within the framework of parliamentary procedure.

Presiding Officer Abilities

Following is a list of skills the presiding officer should develop.

- Start and end the meeting on time.
- Set a climate for the meeting that encourages order, discussion, and efficiency.
- Prepare for the discussion of business.
- Introduce the topic.
- Keep the discussion on the topic.
- Keep the discussion moving forward.
- Display poise in handling the meeting.
- Speak with a clear voice and loud enough so all can hear.
- Be fair with members and show good judgment.
- Give all a chance to contribute.
- Keep the discussion from becoming too heated.
- Stimulate discussion by asking questions.
- Summarize the discussion at the end of the meeting.
- Formally conclude the business.
- Announce the next topic for discussion.

The presiding officer needs to develop a sense of timing and know when to step in and encourage a response from a nonparticipant. The presiding officer should know when to discourage someone who may be monopolizing the discussion, when to ask for a summary or further explanation if members seem to be lost or confused, and when to move on or table a discussion for a future meeting if an issue cannot be resolved.[16]

Managing the Energy Cycle

Presiding officers who expect to have effective meetings must be aware of **energy cycles**. Meeting energy is affected by attention cycles, interest in topics, complexity of topics, the number of topics to be addressed, the scheduling of those topics, and the level of participation of attenders. Every meeting has an energy cycle that can be managed and enhanced by a presiding officer. Here's a guide for timing and managing the energy cycle.[17]

- The early part of the meeting tends to be more lively and creative than the end of it, so items requiring more imaginative ideas, mental energy, and clear heads should be addressed early in the meeting.
- Any items that absolutely must be addressed should be first on the agenda, avoiding the possibility of getting stuck on low-payoff topics.
- In the absence of critical, high-priority items, one scheduling option is to address first any items that can be brought to closure quickly and easily, leaving the rest of the meeting for lengthier items.
- Consider reserving a controversial, high-interest item until the end of the session. This way, useful work can be accomplished before the topic comes up. The high interest level in that item will keep attention from lagging.
- Items can be grouped in a way that allows people to cycle in and out of the meeting as they are needed. Changing the composition of the group automatically raises the energy level as fresh faces and new voices appear.
- If the meeting will be long, with many agenda items, consider alternating working items with reporting items to avoid boredom.
- Try to find a unifying item to end the meeting. If any of the items on the agenda have been divisive, it is particularly important to bring the group into harmony.

Minutes

If a meeting is important enough to be held, it is important enough to have the decisions and action items recorded. Meeting minutes make sure that important points and decisions are recorded accurately, without misinterpretation. Taking minutes is the responsibility of the secretary. The minutes should be typed and distributed to those who need them. Also, post the minutes in the classroom or appropriate place for all members to see. Minutes should contain the following information.

- Time and date of the meeting, where it was held, and who presided.
- The balance of the chapter's fund, from the treasurer's report.

- All agenda items (and other items) discussed and all decisions reached. If a task was agreed on and assigned, indicate the person responsible for the task.
- The date, time, and place of the next meeting.
- Signature of the secretary who wrote the minutes.

GROUP MEMBER INVOLVEMENT

The presiding officer, leader, and FFA member have many occasions to lead groups in discussions. These may be large groups or small groups. Effective leaders plan the kinds of interaction they want. If member involvement is a serious problem, ask the members for ways to get the whole group involved. Open discussions by letting the members know that the presiding officer is serious about generating discussion and new ideas.

Nothing affects the productivity and outcome of a meeting more than the participation of the attendees. The purpose of a meeting is to benefit from the combined information, wisdom, and experience of the members. If the members are not sharing their ideas and opinions, the purpose of the meeting isn't being met and you run the risk of making inferior decisions.[18] Some techniques for promoting group member involvement follow.

Questions and Questioning Questions stimulate group interaction. They help people generate ideas and promote thinking. Questions can summarize a discussion or move a group to action. Ask specific questions of individuals to encourage them to get involved. Ask for participation and demonstrate rewarding verbal and nonverbal behavior for individual participation. Every member should be treated with courtesy and all ideas and contributors should be acknowledged, even if they run counter to the leader's opinion and beliefs.

Buzz Groups (Small Group Discussion) The group is broken into small groups of six or eight members. A leader and recorder is selected for each group. These **buzz groups** have three advantages.

1. Examine as many ideas as possible and report back.
2. All members get a chance to speak.
3. More members can practice their leadership skills.

Panel Discussions Panel discussions are a good way to present information from experts. Panel discussions can be used in various forms.

- *Question-and-answer sessions.* These give members a chance to ask the panel members questions.
- *Symposium.* Each panel member speaks in an area of expertise.
- *Problem-solving sessions.* Panel members deal with specific problems.

Role-Playing In **role-playing**, two or more people assume parts and act out a situation relating to a problem that the chapter or group is considering. It gives members a picture of how things look from another person's perspective.

Brainstorming Members state every idea that comes into their heads without criticism from other members of the group. After every idea has been given, all can be evaluated.

Parliamentary Procedure Effective parliamentary procedure remains the key to conducting effective meetings. As group members work together, personality types tend to emerge. Certain personality types (the silent member, the talker, the wanderer, the bored one, and the arguer) can cause meetings to bog down. Effective use of parliamentary procedure can help the presiding officer to appropriately deal with each.

Hold Your Opinion Resist the urge to give your opinion up front, which can immediately cause reluctance to participate by some who might disagree. Instead, keep a low profile on issues in which you want the honest opinion of others.[19]

Listen Set an example by listening carefully to each member, taking notes, asking for classification, paraphrasing the remarks, and asking again when necessary to be sure you understand the message.

Group Problem Solving and Thinking

Problem-solving techniques can help members make decisions. Effective problem solving generally follows a pattern. Early in a discussion, the members try to understand the nature of the problem. In this phase, the presiding officer should encourage factual, nonevaluative discussion of the problem. Next, identify all possible solutions without evaluation to be able to select the solution that best fits the needs of everyone involved. After all solutions have been considered, select the best alternative.

Group problem solving is a cooperative process in which members pool their thinking to reach a satisfactory solution. The following abilities are characteristics of good problem solving and thinking.

- Reorganize, state, and comprehend the problem.
- Review the background of the problem.
- Keep the main problem clearly in mind as the solution develops.
- Propose and apply any relevant suggestions.
- Critically examine and evaluate proposed solutions.
- Abandon ideas that prove false.
- Suspend judgment on an idea until all the facts are known.
- Test all conclusions.

Agree on a plan to implement the decision. Group problem solving has several advantages over individual problem solving for the following reasons.

- It uses the contributions of all group members.
- It stimulates, modifies, and refines individual thinking.
- It makes available different points of view and more resources.
- It appeals to collective wisdom and cooperative action.

Figure 12–9 One sign of quality group member involvement is effective problem solving and decision making. The intensity of these members of the Oregon delegation at the National FFA Convention indicates that a good meeting is in progress. (Courtesy of *FFA New Horizons Magazine.*)

The intensity of the group in Figure 12–9 indicates that effective problem solving and decision making is occurring.

GROUP MATURITY

Agree to Disagree

Good presiding officers should maintain a climate of constant inquiry in which all assumptions can be questioned, even theirs. Different points of view, critical thinking, and constructive disagreement should be encouraged. Presiding officers should stimulate creativity and curb members' desires to reach an early decision. Productive meetings often involve conflict, but conflict should be with a clash of ideas, not personalities. In other words, agree to disagree on ideas. Don't take it personally if other members do not agree with you.

Perils of "Group Think"

Tony Alessandra and Phil Hunsaker in their book, *Communicating at Work*, discuss the term **group think**. Once group think begins, the group no longer explores the depths of an issue. Members stay on the surface and go along with easy solutions; they don't challenge ideas and recommendations. Group think is especially prominent in groups that meet over long periods of time. Cooperation is encouraged within the FFA, but cooperation is less important than going along with beliefs, values, and suggestions when we do not agree. Group think is one of the critical pitfalls of FFA meetings, even in meetings that appear to be working well.

Here are some of the causes of group think.

- *Illusion of togetherness.* This is the perception that everyone is in agreement. The group begins to take pride in its lack of disagreement and ability to come to rapid decisions.
- *Conformity pressures.* Dissenters are seen as not being "team players." Dissension is perceived as an unhealthy attack on the group.
- *Self-censorship.* Group members keep quiet about negative feelings about decisions and fail to question the direction of the group.
- *Time pressures.* Time pressures can block deep examination of issues, making people grab at easy solutions and avoid the interpersonal processes that make constructive disagreement possible.

To prevent group think, the presiding officer should promote an atmosphere in which members feel free to disagree. Openness has to be encouraged and minority viewpoints given careful consideration. Members should be encouraged to play devil's advocate, and silence should not be taken as agreement. Ask for different ideas and solicit new views.

Develop Member Sensitivity

Leaders can and should help develop **sensitivity** to their own needs, other members' needs, and the group's needs and goals.[21]

Members sensitive to their own needs should

- Recognize their own needs but try not to meet them at the expense of others
- Share their ideas, feelings, and differences with the group
- Be ready to change their own ideas without emotional upset
- Not dominate or monopolize the meeting
- Recognize the value of other group members, even though they dislike them

Members sensitive to other members' needs should

- Be aware of and be ready to try to understand other's needs
- Encourage fulfillment of others' needs
- Encourage others to participate

Members sensitive to group needs and goals should

- Be ready to assume responsibilities
- Recognize the importance of objectivity
- Be willing to abide by group decisions
- Be flexible and assume the roles recommended by the group
- Put the group's goals first and their own second

Officer Responsibilities

Officer Team Responsibilities

The responsibilities of the officer team in conducting successful meetings are varied. They post the agenda in the agricultural education classroom several days before the meeting. They make sure that other members are prepared for the meeting.[22] The officer team is responsible for guest speakers, room arrangement, refreshments, and entertainment, or at least for seeing that it gets done.

One of the most important officer responsibilities is the FFA opening and closing ceremonies. Impressive ceremonies exhibiting meaning and conviction contribute substantially to a successful meeting.[23] Businesslike ceremonies give the meeting status, purpose, and prestige.

Each officer has some requirements in common with other members of the officer team. All officers must

- Have a genuine interest in being a part of a leadership team
- Be able to lead by example
- Be familiar with the organization's constitution and bylaws
- Be familiar with parliamentary procedure
- Be willing to accept and delegate responsibility
- Be willing to memorize their parts in the various FFA ceremonies[24]
- Be able to work together
- Be willing to share ideas
- Be able to help satisfy members' needs
- Emphasize *we*, not *I*
- Be concerned with what is best for the group rather than themselves

The officers in Figure 12–10 are carrying out their responsibilities and making last-minute preparations before a meeting.

Figure 12–10 A chapter is no better than its officers. These officers are checking last-minute details to make sure everything is in order.

Specific Responsibilities

Specific responsibilities of the individual FFA officers follow.

President

- Presides over all chapter meetings
- Serves as chair of the executive council
- Appoints committees
- Serves as the chapter's official representative
- Coordinates the chapter's activities and keeps in touch with the progress of each division of the Program of Activities

Vice-President

- Assists the president and serves as presiding officer in the president's absence
- Coordinates all committee work
- Assumes responsibility for guest speakers and chapter meeting programs

Secretary

- Prepares and reads minutes of each meeting
- Prepares an agenda for each meeting
- Attends to chapter correspondence
- Prepares, posts, and distributes motions
- Compiles chapter reports
- Keeps member attendance and activity records
- Issues membership cards

Treasurer

- Receives and deposits FFA funds
- Collects dues and assessments
- Helps plan public information programs
- Prepares and submits the membership roster and dues to the national organization in cooperation with the secretary
- Maintains a neat and accurate official FFA Treasurer's Book
- Chairs the Earnings and Savings Committee
- Prepares monthly treasurer's reports for chapter meetings

Reporter

- Prepares a chapter newsletter and a reporter's scrapbook
- Releases news and information to local news media
- Helps plan public information programs
- Sends local stories to area, district, and state reporters
- Sends articles and pictures to the *FFA New Horizons Magazine* and other national and regional publications
- Works with local media on radio and television appearances and FFA news

Sentinel

- Prepares the meeting room and cares for chapter equipment and supplies
- Attends the door and welcomes visitors
- Keeps the meeting room comfortable
- Takes charge of the candidates for degree ceremonies
- Assists with special features and refreshments[25]

Note: The officers mentioned above are required by the National FFA constitution. Optional officers may be added at the local level.

Member Responsibilities

Members should become familiar with the parts of the opening and closing ceremonies and with basic rules of parliamentary procedure. Members should also familiarize themselves with the items of business selected for the meeting and be prepared to discuss them or raise questions to clarify the issues. Members should be on time to meetings and be alert to ways in which they can participate as informed and willing FFA members in all chapter meetings.[26]

The performance of individual FFA members determines the effectiveness of the chapter meeting. When members interact, help set goals, or take part in activities, they are engaged in positive group performance. Members should be involved, and leaders should use the interest and abilities of each member.

The secret of a vital and successful meeting is to involve all members and make the group their own. If students share participation and pride in the meeting, they will want to work hard. The good thing is that they will have fun in the process.

Program of Activities

The FFA Program of Activities is designed to provide FFA chapter officers and agricultural education instructors (FFA advisors) with activities that are appropriate for the chapter. Conduct of Meetings is one area of the chapter Program of Activities. (The chapter Program of Activities is discussed in detail in Chapter 14.) The National FFA Supply Service has excellent materials to help plan the chapter Program of Activities. There are 12 divisions in a chapter's Program of Activities. Refer back to Figure 12–5 to see an example of the public relations Program of Activities area.

Businesses and organizations also develop and follow a plan of action. A properly planned and developed Program of Activities enhances the effectiveness of any organization. Meetings will run more smoothly if they have a functional Program of Activities. Other benefits of a Program of Activities are that it

- Defines what the organization wants to do and how it plans to do it
- Allows the membership to plan
- Helps meet the needs of each member and makes sure all members have a voice in making decisions
- Helps provide consistency among committees and prevents the duplication or omission of duties
- Leads to the development of a workable budget
- Provides the means of evaluating the group's success

- Helps provide continuity from year to year
- Provides direction for meetings[27]

MEETING EVALUATION

End-of-meeting evaluations are an easy way to determine the effectiveness of your meetings. Group members are asked to evaluate selected aspects of meeting leadership, process, and productivity. By doing these evaluations, members can analyze their chapter meetings formally.

Questions for meeting evaluations are often designed to identify strong points as well as weaknesses. This is important for several reasons: A positive approach makes members feel better about an evaluation. It is important to know the strong points so they can be emphasized as future activities are planned. Also, it is possible for much learning to take place while analyzing strong points.[28] Figure 12–11 shows a sample evaluation form.

Every group has problems from time to time. Evaluations should reflect these. One of the common problems, though, is how groups handle these situations.

Meeting Evaluation Form

Rank each of the items. Please circle appropriate number.

	Very Low	Low	Average	High	Very High
1. Physical arrangement and comfort	1	2	3	4	5
2. Orientation	1	2	3	4	5
3. Group atmosphere	1	2	3	4	5
4. Interest and motivation	1	2	3	4	5
5. Participation	1	2	3	4	5
6. Productiveness	1	2	3	4	5
7. Parliamentary procedure used	1	2	3	4	5
8. Agenda followed	1	2	3	4	5
9. Opening and closing ceremony	1	2	3	4	5

Please answer the following questions:

1. How would you rate this meeting No good ____ Average ____ All right ____ Good ____ Excellent ____
2. What were the strong points? ______________________________
3. What were the weak points? ______________________________
4. Suggestions for improvement. ______________________________

Figure 12–11 Democratic leaders listen to members. Evaluations help to make sure members' needs and wishes are met. Using this evaluation form and following the suggestions you receive should improve your meetings.

Being able to conduct a meeting effectively is a critical leadership and communication skill. The benefits of a well planned and conducted meeting are enormous: identification of solutions to problems, shared ideas and information, development of plans that have group commitment, and strong team cohesiveness and morale. Building morale should be a key element in meetings. FFA members need to leave a meeting motivated to be a contributing member of the team.

Summary

Conducting and participating in meetings offer many opportunities for developing leadership skills. These skills will be valuable when members graduate and participate in professional, business, and civic organizations. Even though executives spend a significant portion of their time in meetings, 78 percent never received any training on how to plan, organize, and conduct a meeting.

The basic meeting communication functions are sharing knowledge, establishing common goals, gaining commitment, providing group identity, and team interaction. Conducting meetings in a democratic style and communicating with others helps the leader develop sensitivity.

Good meetings result from careful planning and preparation. They should be well publicized in advance of the meeting time. The agenda should be posted, and the meeting should have a specific purpose. The meetings should be conducted by officers who are prepared and know their duties and responsibilities, use correct parliamentary procedure, follow the agenda, and involve members. The meetings should include business, a program, recreation, and refreshments. Successful meetings are interesting, they are held in a pleasant, suitable place, and minutes are taken.

Committees are an important part of FFA meetings. The key to an effective committee is proper organization by the chairperson. The chair should report to the group as a whole when the committee's task is completed.

Democratic leaders want member involvement. If members are not involved, ask them for ways to get the whole group involved. Some techniques for promoting group member involvement are buzz groups, panel discussions, role-playing, brainstorming, the leader withholding his or her opinion, and listening. Group problem solving and thinking also increase member involvement.

As groups and group members mature, they become more secure. Group leaders should learn to agree to disagree. If disagreement never occurs, a worse thing can develop—group think. The presiding officer should promote an atmosphere in which members feel free to disagree.

Officers of the organization are responsible for conducting successful meetings. Chapter officers should also check to see that other members are prepared for meetings. Evaluations should be made at the end of the meetings to plan and improve future meetings.

End-of-Chapter Exercises

Review Questions

1. List five leadership skills that can be developed by conducting successful meetings.
2. What are 14 values or benefits learned by understanding each officer's duties and responsibilities mentioned in the FFA Opening Ceremony?
3. Name five basic meeting communications functions.
4. What are 15 characteristics of a good meeting?

5. What are nine barriers to successful meetings?
6. List 12 questions to determine responsive and productive meetings.
7. What are eight responsibilities of a committee chairperson?
8. Name 20 educational activities that could be used for FFA programs.
9. Briefly describe seven techniques to manage the energy cycles of a meeting.
10. What are seven techniques to get members involved?
11. Why can group think be a problem within a group?

Fill-in-the-Blank

1. Leaders, supervisors, and administrators trained in effective human relations are more ________________ to the needs of others.
2. Unless the meeting has a ________________, the meeting will not be efficient and effective.
3. ________________ is the key for using committees.
4. Every meeting should contain three elements: ________________, ________________, and ________________.
5. In any meeting, the secretary should record ________________.
6. ________________ assists the president and serves as presiding officer in the president's absence.
7. End-of-meeting ________________ are an easy way to determine the effectiveness of your meetings.
8. Members should develop ________________ to their own needs, other members' needs, and the group's needs and goals.

Matching

_____ 1. Call meeting to order
_____ 2. Members stand
_____ 3. Members are seated
_____ 4. Restore order
_____ 5. Chairperson presiding at end of table
_____ 6. Chairperson sits on one side of table
_____ 7. Round table arrangement
_____ 8. Receives and deposits money
_____ 9. Releases news and information to media
_____ 10. Takes care of the meeting room

A. sentinel
B. one tap of gavel
C. three taps of gavel
D. participative
E. two taps of gavel
F. autocratic
G. democratic
H. reporter
I. series of sharp gavel taps
J. treasurer

Activities

1. Write a policy on the conduct of members at meetings.
2. Write a policy on meeting guidelines for your FFA chapter meetings. If you need help, refer to the meeting guidelines of the Western Center of General Dynamics earlier in this chapter.

3. Prepare an agenda for a meeting. Be creative and professional as you suggest new ideas or items.
4. Develop a plan justifying three recommendations for improving attendance at chapter meetings.
5. You are appointed to a committee to plan an FFA summer recreation event. Use the Committee Activity Planning Sheet in Figure 12–5 as a guide to plan the event. Provide information for each of the categories.
6. Your teacher will present three topics for discussion. You are to lead the group to a decision on all three topics. Use parliamentary procedure if necessary, which you learned in the preceding chapters.
7. Compare the steps in problem solving and the scientific method.
8. Write a one-page essay on the statement "let us agree to disagree," including thoughts from a leader, members, and personal perspective.
9. Videotape a meeting. Watch and evaluate it along with other chapter members. Write at least five suggestions for improvement according to the recommendations in this chapter.
10. **Take it to the Net.**

    Explore the Internet for information on how to conduct a successful meeting. Listed below are several Web sites dealing with conducting meetings. Below each is a description of the site and some directions. You can look at as many sites as you want, but choose one to do as an assignment. If you are having problems with the Web sites, use the search terms listed below to find an article about conducting meetings and summarize the article.

    Web sites

    <http://www.meetingsnet.com>

    This site contains six magazines that contain information on planning meetings. Browse the magazine articles and pick one you find useful. Print the article and summarize it. In your summary include why you feel it is useful.

    <http://www.roseville.ca.us/housing/neighborhood/meetings.htm>

    This site contains an article about conducting productive meetings. Read the article and print it. Write a summary of the article.

    <http://www.4anything.com>

    This site has a lot of information about virtually everything. When you get to this site, a search box appears. In the box type "conducting meetings." Browse the results and find a useful article. Print the article and summarize it. In your summary include why you think the article is useful.

    Search Terms

    meeting
    effective meetings
    how to conduct a meeting
    conducting effective meetings

Notes

1. T. Alessandra and P. Hunsaker, *Communicating at Work* (New York: Simon & Schuster, 1993), p. 188.
2. S. C. Ricketts, *Leadership and Personal Development Abilities Possessed by High School Seniors Who Are FFA Members in Superior FFA Chapters, Non-superior Chapters and Seniors Who Were Never Enrolled in Vocational Agriculture* (Columbus, OH: Dissertation, Ohio State University, 1982). Note: Can be found in ERIC.

3. Alessandra and Hunsaker, *Communicating at Work*, pp. 188–189.
4. R. E. Bender et al., *The FFA and You* (Danville, IL: Interstate Printers and Publishers, 1979), p. 300.
5. Alessandra and Hunsaker, *Communicating at Work*, p. 193.
6. *Personal Development in Agriculture* (College Station, TX: Instructional Materials Service, Texas A & M University, 1989), 8741-I, p. 1.
7. *Agricultural Teacher's Manual* (Indianapolis, IN: The National FFA Organization, 2001).
8. Alessandra and Hunsaker, *Communicating at Work*, p. 197.
9. Ibid., p. 198.
10. *Agricultural Teacher's Manual.*
11. *The FFA and You*, p. 308.
12. *Agricultural Teacher's Manual.*
13. *Personal Development in Agriculture*, 8741-I, p. 3.
14. Alessandra and Hunsaker, *Communicating at Work*, p. 199–202.
15. Bender, *The FFA and You*, p. 311.
16. Alessandra and Hunsaker, *Communicating at Work*, p. 198.
17. Ibid., p. 199.
18. Ibid., p. 209.
19. Ibid., p. 212.
20. Ibid., p. 216.
21. *Leadership for Youth Groups* (Kansas City, MO: Farmland Industries, No Date).
22. Bender, *The FFA and You*, p. 303.
23. Ibid., p. 304.
24. *Official FFA Manual* (Alexandria, VA: The National FFA Organization, 2001).
25. Ibid.
26. Bender, *The FFA and You*, p. 305.
27. *Personal Development in Agriculture*, 8742-B, p.1.
28. G. M. Beal et al., *Leadership and Dynamic Group Action* (Ames, IA: Iowa State University Press, 1962), pp. 291–292.

SECTION

Managerial Leadership Skills

CHAPTER 13

Problem Solving and Decision Making

Objectives

After completing this chapter, the student should be able to:

- Explain the importance of problem solving and decision making
- Differentiate between the terms *problem, problem solving,* and *decision making*
- List mistakes in problem solving and decision making
- List skills involved in problem solving and decision making
- Explain three problem-solving and decision-making styles and identify your own style
- Identify two approaches to problem solving and decision making
- List and describe seven steps to problem solving and decision making and use them to solve problems
- Describe and solve three types of problems
- List advantages and disadvantages of group problem solving and decision making
- List methods used in groups to solve problems and make decisions
- Identify leadership styles used in group problem solving and decision making

Terms to Know

problem
decision making
alternatives
reflexive style
reflective style
consistent style
minimizing approach
optimizing approach
exact reasoning problems or decisions
creative problems or decisions
judgment problems or decisions
conventional method
brainstorming method
devil's advocate method
delphi method
consensus method
nominal group method
synetics
left-brain people
right-brain people
holistic
autocratic leadership style
consultative leadership style
participative leadership style
laissez-faire leadership style

This chapter will help you identify the skills needed to be an effective problem solver and decision maker. It will show you how to develop these skills by making you aware of your own problem-solving and decision-making style, by teaching you approaches to solving problems and making decisions and by illustrating the types of problems and decisions you may encounter. This chapter also focuses on the group problem-solving and decision-making process (Figure 13–1) and the styles leaders use in directing groups.

Importance of Problem Solving and Decision Making

Daily and Life Decisions

In your personal, family, and working life, you solve problems and make decisions every day. You are probably not aware of the many decisions you make or the problems you solve. You decide when to get up, what to wear, where to go, how to get where you are going, and when to come home. These types of decisions are relatively easy to make. Others, however, are of such importance that they require a great deal of thought. Will you attend college, and, if so, where? Whom will you marry? What occupation will you choose? Will you buy a house, and how will you pay for it? The decisions you make in these areas will affect you for the rest of your life. You will have to live with the good and bad decisions you make.[1]

Figure 13–1 These students are solving problems and making decisions. These abilities are signs of effective leadership.

Problem Solving and Decision Making on the Job

The ability to make good decisions and to solve problems is a sign of an effective person and an effective leader or supervisor.[2] Consider the following information from Robert Lussier.[3]

- Solving problems and making decisions are key reasons managers are hired.
- The second most time-consuming task of the supervisor solving problems and making decisions—uses perhaps as much as 13 percent of his or her time.
- Supervisors may make as many as 10 decisions per hour; managers may make hundreds in a given day.
- Problem solving and decision making is one of the top six critical skills for success at the supervisory level.

Obviously, effective problem solving and decision making is important to us if we wish to hold management positions in the workplace; it is also important to our personal financial, educational, and emotional well-being. Fortunately, the ability to solve problems and make decisions is a skill that can be learned, practiced, and improved.[4] Let us begin by examining what we mean when we use the terms *problem, problem solving,* and *decision making.*

DEFINING PROBLEM SOLVING AND DECISION MAKING

Problem and Problem Solving

Most of us are familiar with the term *problem*, but it may be difficult for us to put into words. A **problem** arises when there is a difference between what is actually happening and what an individual or group wants to have happen.[5] For example, if a car dealership has the goal of selling 20 cars per week but is only selling 10 cars per week, a problem exists. The dealership must do something to solve the problem. Problem solving, then, is "the process of taking corrective action in order to meet goals."[6]

Figure 13–2 Mr. Brown is in the process of making the decision to take corrective action to achieve his goals.

Decision Making

In the process of problem solving, a new or different course of action must be taken to correct a problem. The process by which the course of action is selected is called **decision making**. At times, after careful consideration, you may find that your best decision is to do nothing because the problem simply cannot be solved. At other times, the decision may be to change your goals to eliminate the problem.[7] In our example, the dealership could simply change its goal from selling 20 cars to selling 10 cars per week, and the problem no longer exists. However, if the financial success of the dealership depends on selling 20 cars per week, reducing the goal is only going to cause more problems. Therefore, some other decision must be made in order to resolve the situation (Figure 13–2). We want to avoid making similar mistakes by developing our skills in problem solving and decision making.

MISTAKES IN PROBLEM SOLVING AND DECISION MAKING

Phipps and Osborne identify the following mistakes made by individuals and groups when attempting to solve problems or make decisions.[8]

- The problem is not well defined or there is denial that a problem exists.
- An attack is made on the situation resulting from the problem, not on the problem itself.
- Goals and objectives are not clearly defined.
- Possible **alternatives** (solutions) are not carefully considered and evaluated.
- Opinions, emotions, feelings, and self-interest interfere with objective thinking.
- Individuals or groups jump to unwarranted conclusions.
- Individuals or groups are afraid to make mistakes in problem solving or decision making.

SKILLS NEEDED IN PROBLEM SOLVING AND DECISION MAKING

To avoid the mistakes listed in the previous section, Phipps and Osborne suggest that individuals and groups develop certain problem-solving and decision-making skills.[9] Among these are the ability to

- recognize problem situations
- clearly distinguish the problem from the problem situation
- clearly define goals and objectives
- develop creative, imaginative solutions to problems
- gather information related to possible solutions
- be open-minded toward possible solutions offered by others
- carefully evaluate information in accepting or rejecting solutions
- work with others to solve problems
- avoid jumping to unwarranted conclusions (be flexible)
- accept the fact that you may make mistakes
- put aside opinions, feelings, emotions, and self-interest that may interfere with objective thinking
- understand different types of problems and techniques for solving them
- understand and use a systematic approach to problem solving and decision making[9]

How can these skills be developed? The following sections help us answer this question.

DECISION-MAKING STYLES

Determining Your Decision-Making Style

One step toward developing the skills necessary to be a good problem solver and decision maker is to understand your decision-making style. If your style is not conducive to making good decisions, you may want to consider ways to change it. Each person tends to have one of three problem-solving and decision-making styles: reflexive, consistent, or reflective.[10]

1. *Reflexive style.* If you have a **reflexive style** of problem solving and decision making, you probably tend to make quick decisions. As a result, you may not take the time to consider and evaluate all possible solutions to your situation before acting. You tend to be decisive, and you are not likely to put off taking action in a problem situation. However, your quick decisions can result in hastily made decisions you may later regret.

   To improve your problem-solving and decision-making style, take more time to identify possible solutions. Try to gather information concerning each possible solution, and analyze and evaluate them thoughtfully. Follow the steps in the problem-solving and decision-making process described later in the chapter.
2. *Reflective style.* If you are a **reflective style** problem solver and decision maker, you take the time to identify, analyze, and evaluate as many alternatives to solving a problem situation as possible. The advantage to this is that you carefully consider your decisions and do not make them haphazardly. However, you may, at times, take so much time to make a decision that you appear indecisive or fail to act in time.[12]

   To improve your style, continue to be careful in your problem solving and decision making, but attempt to make your decisions more quickly. Andrew Jackson said, "Take time to be deliberate; but when the time for action arrives, stop thinking and go on."[13]
3. *Consistent style.* If you have a **consistent style** as a problem solver and decision maker, you know the appropriate amount of information to consider and evaluate before making a decision and you act in a reasonable

amount of time. You do not make decisions too quickly, as do those of the reflexive style, and you do not act too slowly, as do those of the reflective style. Your decisions are timely, reliable, and consistently sound.[14]

Approaches to Problem Solving and Decision Making

Another way to develop your skills in problem solving and decision making is to understand the approaches you can take to problems and decisions and when to use them. There are two general approaches.

1. *Minimizing approach.* In the **minimizing approach**, the person simply chooses the first solution available. The solution chosen may not necessarily be the best. If this first solution does not work, then a new one is chosen. This process of trial and error continues until an acceptable solution is found. The minimizing approach is useful if the situation is an emergency or if waiting to make a well-thought-out decision could be costly.[15]

   For example, you may choose the minimizing approach when it is raining and water is pouring in from a leak in your roof. Your objective is to stop the leak immediately so that your property is not ruined. There is no time to create an elaborate plan for fixing the roof permanently.[16]
2. *Optimizing approach.* In the **optimizing approach**, you take the time to review many different solutions before making a decision. You choose the most helpful solution. This approach requires a more detailed planning and thinking process than the minimizing approach, but it does tend to be more reliable. The optimizing approach is effectively used when you have ample time to make a decision, when you cannot easily change your decision once it is made, or when the emergency situation that may have required the minimizing approach has passed and you have time to find a more permanent solution.

Using the previous example, once the rain has stopped and water from the leaky roof does not threaten your property, you may wish to use the optimizing approach to deciding which contractor will be the most cost-effective, competent, and reliable to fix your roof.

Steps in Problem Solving

Another way to develop your skills as a problem solver and decision maker is to use a systematic approach to the situations you face. The steps commonly used in problem solving and decision making have their roots in the work of scientists such as Bacon, Newton, Galileo, and their successors, who sought to develop a systematic approach for acquiring knowledge. The work of these individuals and others led to the formulation of the scientific method.[17]

The scientific method consists of (1) a problem situation, (2) defining the problem from the situation, (3) creating hypotheses (solutions) for the problem, (4) gathering data to test the hypotheses, (5) making necessary revisions to the hypotheses or creating new hypotheses to test, and (6) drawing conclusions.

The process of problem solving and decision making is very much like the scientific method. If we attempt to use a logical, systematic approach to solving problems and making decisions, as scientists do, we are more likely to make better choices. Problem solving and decision making are skills that can be learned, practiced, and improved.

Seven Steps of the Problem-Solving and Scientific Method

The steps in problem solving and decision making are presented in Figure 13–3. We examine each of these steps and use our example of the car dealership to clarify the process.

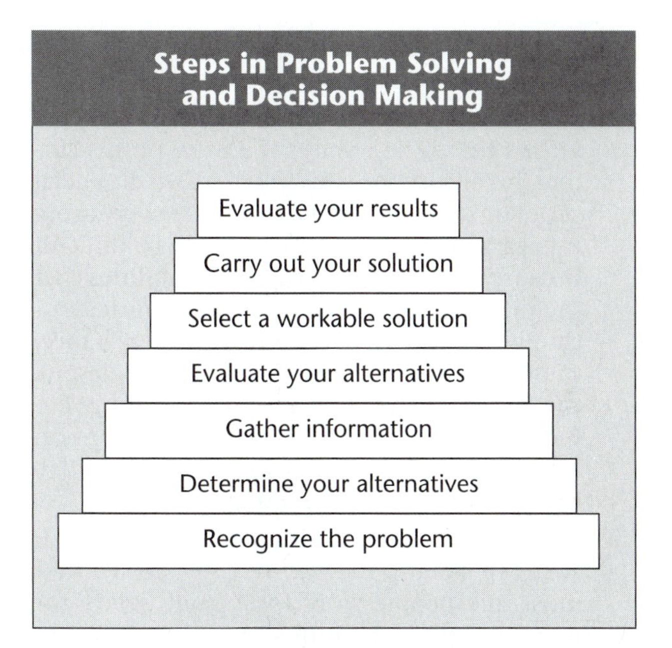

Figure 13–3 The steps in solving problems and making decisions are similar to the scientific method. By following these steps from the bottom up, you are likely to make good choices.

Mr. Brown owns and is the sole employee of a small car dealership. He has determined that he must sell 20 cars per week to make his dealership profitable. Currently, however, he is only selling 10 cars per week. Mr. Brown is determined to be successful in his business and wants to develop a plan for doing so. He decides to use the steps for problem solving and decision making to help him.

Step 1

Recognize the problem.

Remember, problems are inevitable and they must be solved, not ignored, if you are going to be successful. The first step in solving a problem or making a decision is realizing that you have a problem. You must be sure you understand the exact nature of the problem. Ask yourself: What problem am I attempting to solve? What goal am I trying to meet?[18] Do I fully understand the problem? Do I need to take action at all? It may be helpful to write the problem down so that your focus is clear. Remember, not all problems are negative; you may be making a choice between two or more equally good things.[19] Regardless, recognize that you must not ignore problems.

In our example, Mr. Brown realized that if he needs to sell 20 cars to make a profit and cause his business to succeed; he cannot continue to only sell 10 cars per week. He must find a way to sell more cars.

Step 2

Determine your alternatives.

Once you have identified your problem, you need to determine your alternatives. Alternatives are the different courses of action you can take to solve your situation. There are likely to be many alternatives to any problem you have. You need to consider each one carefully before making a decision. It may be helpful to consult other people familiar with your situation to determine alternatives. Be aware that any alternative may directly affect you or a number of people. It may be helpful to list your alternatives on paper to keep a clear focus.

In our example, Mr. Brown lists several alternatives to his problem. We will consider three: (1) He may sell a different type (line) of car, (2) he may lower the price of his current inventory, or (3) he may hire a salesman to help him.

Step 3

Gather information.

Once you have listed the alternatives, you need to gather information relative to each one. Look at the facts. Relying only on opinions, emotions, and feelings may lead to hastily made decisions that you will later regret. In gathering information, ask yourself the following questions: What do I need to know about each alternative?

What materials or information may be needed to implement each alternative? What costs will be involved? Is the alternative feasible?

Don't reinvent the wheel: The library has an abundance of data. There is a good chance that someone has already had the same problem as you and has identified several alternatives.

In our example, Mr. Brown gathers the following information for each alternative, some of it from trade publications on selling cars in the library.

1. If he sells a different type of car, he will need to secure a loan of several thousand dollars, find a manufacturer willing to sell him a new line of cars, and sell or somehow dispose of his current inventory.
2. If he lowers the price of his current inventory, he may sell more cars because he will be selling them for less than other dealerships in the area.
3. If he hires a salesperson, he will need to place an ad in the paper, conduct interviews, pay a salary, and provide a benefit package. He must also be able to work well with the person he chooses.

Step 4

Evaluate the alternatives.

Once you have gathered information for each alternative, you will need to evaluate (and perhaps list) the advantages and disadvantages of each in relation to solving the problem and in relation to one another. Consider also what new problems may be created by adopting each alternative.

Mr. Brown evaluates his alternatives as follows.

1. *Sell a new line of cars.* Mr. Brown has used his full credit line at the bank, so borrowing more money from the bank is not an option at this time. Furthermore, establishing a relationship with a new car manufacturer is a lengthy process that will only hinder his ability to sell more cars now. Mr. Brown has found that he cannot afford to sell his current inventory at little or no profit.
2. *Lower the price on current inventory.* Mr. Brown knows that if he reduces the price on his current inventory, he only furthers his financial problems. He would have to sell even more cars to be profitable, which would be difficult because he has nonselling responsibilities that include paperwork and custodial duties. Although selling cars is of primary importance, these other duties must be attended to despite the fact they take time away from selling. Mr. Brown also is wary of lowering the price on his cars since he is currently very competitive in price with other area dealerships.
3. *Hire a salesperson.* Hiring a salesperson will cost Mr. Brown money, but he knows that most salespeople work for a small salary (or no salary) plus a commission (a percentage of the sale of each car). Thus, the salesperson's pay would depend on the ability to sell cars. Having another salesperson also means more people can sell cars. This is important because Mr. Brown has often been dealing with bookkeeping while customers were on the car lot. When he is unable to help as many people as possible, he may be losing sales. Mr. Brown also thinks that having another salesperson will allow him time to handle his administrative duties.

Step 5

Select a workable solution.

After you have evaluated each alternative and its possible results, choose the one that is the most practical, reasonable, and effective to solve your problem.

After considering all of his alternatives, Mr. Brown decides that hiring a salesperson is his best solution. He can sell more cars, attend to his administrative duties more easily, and afford to pay the salesperson if he pays on a commission basis.

Step 6

Carry out your solution.
Once you have determined your course of action, follow through. If you fail to carry out your solution, you have wasted the time and effort you have used to this point. A solution must be put into place in order for it to work.

Mr. Brown puts an ad in the paper for a salesperson. He interviews a number of people, explaining his goals and his payment plan. He hires a salesperson he feels will be competent as well as compatible with his own personality (Figure 13–4).

Step 7

Evaluate your results.
The problem-solving and decision-making process does not end when the proposed solution is carried out. It ends when you decide if your problem has been solved, if the same problem persists, or if new problems have been created.[20] Evaluation may lead you to accept your solution as a good decision, make further adjustments to improve your solution, or discard your decision and start the process over again.

In our example, several weeks after hiring the salesperson, Mr. Brown evaluates what his decision has meant to him and his business. He finds he is not only selling more cars than he needs to be profitable, but he has fewer worries about his other duties. He also finds that he and the salesperson get along well. Mr. Brown is satisfied with the decision he has made.

Overall, learning and practicing the steps of problem solving and decision making will help you become better at these things. Thus, you will be a better leader. Our example has a happy ending, but not all your decisions may turn out so ideally. As long as you have carefully followed each step, you can feel confident that you made the best decision you could at the time. Do not be afraid to start the process over again if your first decision does not work. You can learn from any mistake you make. Thomas Edison failed numerous times in trying to develop the incandescent light. Rather than being discouraged, he focused on the wealth of knowledge he had gained from his failures. He knew well what *did not* work, and he continued trying until he found what *did* work.

Figure 13–4 Mr. Brown decides to hire a salesperson after considering other solutions to his problem. He believes this decision will help him reach his goal.

Types of Problems and Decisions

As you progress through problem-solving and decision-making steps, you can increase your effectiveness by recognizing the types of problems you are encountering and having some techniques in mind for solving them. Essentially, you can divide the types of problems or decisions you will encounter into three groups: **exact reasoning problems or decisions**, **creative problems or decisions**, and **judgment problems or decisions**.[21]

Exact Reasoning Problems or Decisions

When you encounter this type of problem, you will find there is usually one exact answer. Using mathematical concepts often resolves the problem.

For example, Tom and Kim, a newlywed couple, are considering the purchase of their first home. They must decide if they will finance the house through a local bank or with a mortgage company. They find that the only difference in the terms and services offered by the two institutions is that the local bank charges 8 percent interest over the life of the loan while the mortgage company charges 9 percent. Tom and Kim decide to finance the home through the local bank because they can save 1 percent on the interest charges.

Creative Problems or Decisions

When you encounter this type of problem or decision, you usually need to have a plan or create some sort of design to help you come to a solution.

For example, Phil and Joe are college roommates. They each bring several things from home to put in their 15′ × 20′ dorm room. They must decide how to arrange their room so that it is neat and efficient and contains all their belongings. They must arrange two beds, two desks, a computer, a television and stand, and a stereo. They may want to sketch a room design that is drawn to scale so they can see and decide where everything should go. Such advance planning will save them moving the heavy furniture from one place to another.

Judgment Problems or Decisions

These types of problems or decisions require you to consider many factors, list alternatives, and evaluate the alternatives before reaching a decision. Judgment problems or decisions may fall into four categories: possibilities and factors, improving a situation, steps and key points, and advantages disadvantages. A discussion of each follows.

1. *Possibilities and factors*. In some cases you may find you have to consider many different possibilities and factors to reach a good solution or decision. For example, John has just started a new job and wants to buy a new car. He knows his price range, he knows he wants a red car, and he knows he wants the car equipped with an air bag. He goes to look at three different styles of cars. John outlines the possibilities and factors (Figure 13–5). Based on the information in Figure 13–5, John decides to buy Car style 2 because it satisfies all the factors he considers important.
2. *Improving a situation*. This type of judgment problem involves taking a current situation and using the problem-solving techniques to improve on it.[22] For example, Mrs. Gomez is not happy with the appearance of her house and yard. She is trying to decide ways she can make it look better. She may consider and carry out the plan shown in Figure 13–6. Mrs. Gomez now has a plan and can begin to carry it out.
3. *Steps and key points*. This type of judgment problem requires you to progress through a

Possibilities	Factors		
	In Price Range	Red	Air Bag
Car style 1	No	Yes	Yes
Car style 2	Yes	Yes	Yes
Car style 3	Yes	No	Yes

Figure 13–5 A chart of possibilities and factors can help you decide which of three cars to buy.

series of steps while considering the key points associated with each. For example, Fred and Ginger want to purchase a new home. They have two children. Some steps and key points they may want to consider in their decision are shown in Figure 13–7. With this information, Fred and Ginger will consider only the houses that meet the criteria.

4. *Yes or no; advantages or disadvantages*. In this type of judgment problem, you consider the factors related to the problem or decision and weigh them as positive or negative in terms of their value in solving the problem or making the decision.[23] For example, Mary is considering whether or not to attend State University. She may want to consider the factors in her decision shown in Figure 13–8.

Current Condition	Needed Improvements
Exterior of house is rundown, looks outdated	• Replace aluminum siding with wooden exterior • Paint gutters and downspouts
Yard looks plain, has bare spots	• Plant shrubs along front of house • Plant grass seed on bare spots

Figure 13–6 Sometimes it only takes good planning to improve a situation.

Steps	Key Points
1. Location	Near good schools, stores, and work
2. Cost	Monthly payment must fit in family budget
3. Size	Need a bedroom for each child; two baths

Figure 13–7 Sometimes a decision, such as choosing a house to buy, requires following a series of steps while considering the key points associated with each. Decisions are made only after criteria have been met.

Factors to Consider	Yes	No
1. Close to home	X	
2. Good academic program	X	
3. Offers intended major		X
4. Reasonable cost	X	
5. Scholarships available	X	

Figure 13–8 Factors to consider can be presented as a list of yes-or-no answers or as advantages and disadvantages. One negative can override all the positives, as shown in this chart presenting factors in deciding whether to attend a certain university.

At times, no one factor is more important than another; you can simply pick the option with the most advantages. At other times, one negative factor can override many positive ones in making decisions. The reverse is also true. In this example, there are more positive factors than negative for Mary to attend State University, but Mary realizes that if the university does not offer her intended major, she will not benefit by attending school there.

GROUP PROBLEM SOLVING AND DECISION MAKING

Up to this point, we have discussed problem solving and decision making as it applies generally to both individuals and groups. In the following sections we focus on groups and the various ways they can function in the problem-solving and decision-making process. First, let us consider some of the advantages and disadvantages to involving more people in this process.

Advantages to using groups in the problem-solving or decision-making process are these.

- There is a sense of ownership, shared responsibility, and commitment by group members in the decision made.
- There may be an increase in cooperation, unity, and morale between group members.
- There may be an increase in the number of ideas and possible solutions proposed. This may result in better decisions.
- There may be more people who can implement decisions.

Disadvantages to group problem solving or decision making include these.

- It requires more time.
- Conflicts may arise between group members.
- Leaders may be reluctant to share their power with the group.
- Keeping people working cooperatively on a given task requires great skill.
- People may feel less responsibility for the solution or decision since so many people are involved in the process.
- One person or a small portion of the group may dominate the process, thus causing the solution and decision to be less of a group idea.
- The desire to be accepted by the group may cause individual members to conform to group decisions rather than thinking critically or questioning objectionable solutions or decisions. When conformity to the group negatively affects the problem-solving or decision-making process, the result is group think[24] (Figure 13–9).

Methods

As a group attempts to solve problems and make decisions, it may follow the seven steps in problem solving and decision making outlined earlier in the chapter. As the group progresses through the steps, it can choose from several methods to foster varying levels of group participation or to generate alternatives for consideration. Key to

Figure 13–9 A disadvantage of group decision making is group think. The student on the left in this picture seems more interested in conforming to the group than thinking critically or questioning decisions.

the effectiveness of nearly all these methods is the idea that all group members participate in the problem-solving or decision-making process and that the group be open and nonthreatening in its approach to individual members.

The Conventional Method In the **conventional method**, there is group discussion. It is typically dominated by one or very few individuals. Possible solutions come from the discussion. There is then a vote on a single solution (Figure 13–10). If a majority of the group votes for the proposed solution, it is put into place. A potential problem with this method is that if the vote is very close (51 to 49 percent, for example) a great many of the group members are against the solution. This may affect the cohesiveness of the group or cast doubt on the validity of the solution itself.[25]

The Brainstorming Method Brainstorming "is the process of suggesting many alternatives, without evaluation, to solve a problem."[26] The **brainstorming method** is most effective in determining numerous possible solutions to problems. All members of the group have an equal voice and may offer any alternative, regardless of how unrealistic or unreasonable, to solve a given problem. The group members are encouraged to build on one another's ideas. The alternatives are not judged until all suggestions have been made.

Figure 13–10 Voting is the conventional method of making a group decision.

Devil's Advocate Method In the **devil's advocate method**, a person may propose a solution he or she does not really support just to make others think and react. This method requires the individual to explain and defend his or her position before the group. The group proposes all the reasons that the individual's idea will not work. The group must be careful to judge the ideas presented, not the person presenting them. Likewise, the person presenting ideas must be mature enough to recognize that it is the idea being judged. The purpose of this method is to refine possible solutions to problems so that they will be effective when put into place.

The Delphi Method The **delphi method** polls a group through a series of anonymous questionnaires. After a first round of opinion questionnaires is completed, the opinions are analyzed and the best ones resubmitted to the group for a second round. Several rounds may be necessary before the group reaches a position that is acceptable to all or to nearly all members. This technique is effective for trying to predict the progress that will be made in technology (computers, for example), and how people and industries may be affected by such progress.[27]

Consensus Method The **consensus method** is not to be confused with general consent. In the consensus method, the group is in substantial agreement on a solution. The consensus method is especially effective when time is not a consideration in the decision, when the decision is very important, or when there is an overriding concern for the unity of the group.

In the consensus method, members have the freedom to bring their own ideas to the group. The group reviews all ideas but focuses discussion on the ones it feels are the most important and least important. The discussion determines how important each idea is—not a formal vote or ranking of the ideas. Based on the discussion and the reasons given for each idea, a decision is made that is acceptable to a substantial portion of the group. It is important to realize that reaching consensus does not mean reaching a unanimous (everyone in favor) decision, but rather a decision that is acceptable to a large portion of the group.[28]

Nominal Group Method The **nominal group method** is the process of generating and evaluating alternatives through a structured voting method. This is a group interaction in name only. The group does meet, but individuals formulate their own ideas (solutions) in writing without discussing them with other members. All solutions are presented to each member of the group for review. The proposed solutions are discussed only if there is a need for clarification. The group then votes to determine the top five ideas. After the top five ideas are decided on, they are discussed, and the reasoning behind each is presented. Another vote is held, with the solution receiving the most votes being the one accepted for implementation (Figure 13–11).[29]

Figure 13–11 These students are using the nominal group method. They are discussing and clarifying the top five ideas before the final vote.

Synetics **Synetics** (also spelled synectics) is the process of generating novel alternatives through role-playing and fantasizing. Synetics uses analogies to provide mental images to the brain. A common use of synetics is to project oneself into the essence of the problem. For example, if an agricultural engineer is trying to develop a more efficient combine, the individual might imagine himself or herself as cut wheat going through the internal components of the combine. By actually seeing yourself as wheat, you can visualize what it would take to strip the stem from the chaff and other parts, leaving only the grain.

Prefabricated potato chips were developed by a synetic group. The group wanted to compress potato chips without breaking them. They eventually found an analogy with leaves, which can be compressed without damage as long as they are wet. They tried it, and prefabricated potato chips became a commercial product.

When Nolan Bushnell wanted to develop a new concept in family dining, he began by discussing general leisure activities. Bushnell then moved toward leisure activities having to do with eating out. The idea of a restaurant-electronic game complex where families could play games and purchase pizza and hamburgers evolved and then became a viable business.[30]

Such creativity has become so important to organizations that many of them are beginning to look into left-brain and right-brain thinking.

Left-Brain, Right-Brain Thinking

In recent years, attention has been focused on creativity and brain function. Most people are either left-brain dominant or right-brain dominant. This dominance also dictates the way people do things.

Left-brain people tend to be very logical, rational, detailed, active and objective-oriented. They have a preference for routine tasks or jobs that require precision, detail, or repetition. They like to solve problems by breaking them into parts and approaching the problems sequentially and logically. Most people are left-brain dominant. They tend to be less creative and imaginative than

Comparison of Right-Brain and Left-Brain Dominant Characteristics							
Problem-Solving/ Decision-Making Category	Leadership Style		Personality Type	DISC™ Behavioral Style	Communication Styles		Communication Forcefulness
Right-brain dominant characteristics	Authoritarian Democratic	Situational Leadership	Choleric Sanguine	Dominant Influencing	Directors Socializers	Situational Communication	Aggressive Assertive
Left-brain dominant characteristics	Laissez-faire		Melancholy Phlegmatic	Cautious Steady	Thinkers Relaters		Passive

Figure 13–12 Just as no one is 100 percent of any personality type, there is not a 100 percent correlation between any combination of leadership, personality, and communication style. Sometimes, as in situational leadership, we must vary our problem-solving and decision-making styles as well as our leadership and communication styles to meet the needs and objectives of the moment.

right-brain dominant people. They can be compared to introvert or melancholy personality types discussed in Chapter 2.

Right-brain people are more spontaneous, emotional, **holistic**, nonverbal, and visual in their approach to things. They like jobs that are nonroutine or call for idea generation. They like to solve problems by looking at the entire matter and approaching the solution through hunches and insights.[31] Right-brain people can be compared to extrovert, choleric, or sanguine personality types discussed in Chapter 2.

Figure 13–12 compares left-brain and right-brain leadership styles, personality types, communication styles, and communication forcefulness. With this information, you can almost predict the decision-making style of each leadership style. As we discuss each of these styles, we focus on how decisions are made in each and situations in which each may be used appropriately. We also take one example and show how each style would arrive at a solution or decision.

LEADERSHIP STYLES AND GROUP DECISION MAKING

Leadership Styles

Group leaders, whether consciously or unconsciously, have leadership styles. The leader of a group may be someone who has been appointed as a chairperson of a committee, someone who has been selected by the group to serve as leader, someone who simply takes charge of the group, or a supervisor, foreman, manager, teacher, or administrator. How a leader involves the group in the problem-solving or decision-making process reveals his or her leadership style. These styles were discussed at length in Chapter 1. Good leaders are aware of their style and can vary it based on the situation. You too should be aware of your leadership style as you are put into situations in which you may be the leader of a group or the supervisor of employees. The four leadership styles in group problem solving and decision making are autocratic, consultative, participative, and laissez-faire.

Consider this workplace scenario.

Mark, a supervisor, has been given the opportunity to promote one employee from his group to a foreman's position. Let us see how Mark could use different leadership styles in deciding who should get the promotion.

In the **autocratic leadership style**, the leader makes the decision independent of the group. After the decision is made, the group is informed and may receive an explanation of why the decision was made the way it was. The autocratic style may be effective if (1) there is no time to consult the group, (2) there is enough information available to the leader to make the decision alone, (3) the group is willing to accept and willing to put into action the leader's independent decision, or (4) the group is unwilling to make or is incapable of making the decision.[32]

In our example, using the autocratic style, Mark would not consult anyone; instead he would decide who gets the promotion and inform his superiors. He would then tell his employees who is getting the promotion and why.

In the **consultative leadership style**, the leader goes to individual group members seeking additional information that will help him solve the problem or make the decision. The leader then makes the decision or solves the problem. Before putting the plan into action, the leader explains the solution or decision and its rationale to the group. Questions and discussions may be allowed, but the decision stands.

The consultative style is effective if (1) the leader has only a short period of time to make a decision, (2) he or she needs additional information before making the decision, (3) the group is hesitant about accepting the leader's decision, or (4) the group is willing and moderately capable of providing the leader with useful information.[33]

In our example, Mark has only some of the information that will help him decide whom to promote. Using the consultative approach, he would go to various employees seeking more information. When his information was complete, Mark would choose someone and inform the rest of the employees of his decision and his reasons. He may allow questions and discussion from the other employees.

In the **participative leadership style**, the leader has a tentative decision or solution in mind but goes to the group for its input. The leader is open to change based on the value of the group input. Another option in this style is that the leader may bring the problem or decision to be made to the group and ask for suggestions. The suggestions are the basis for the decision ultimately made by the leader. Again the leader explains the reasoning for the decision to the group. The participative style is effective when (1) there is sufficient time to include the group in the decision, (2) the leader has little or no information on which to base a decision, (3) the group is likely to reject a decision if it has had little or no input, or (4) the group is highly capable and willing to participate or directly affected by the decision.[34]

In our example, Mark would go to the employees and explain that he can promote only one person. He would ask the group to provide him with suggestions and rationale for who should be promoted. When his information is complete, Mark would name the employee to be promoted and explain the reasons why.

In the **laissez-faire leadership style**, the leader presents the problem or decision to be made to the group; the group, not the leader, solves the problem or makes the decision. The leader may even become a group member. The laissez-faire style is effective when (1) there is plenty of time for the decision to be made or the problem to be solved, (2) the leader has little or no information on which to base a solution or decision, (3) the group must have input in the decision or solution, (4) the ability level of the group is outstanding. Notice that of the four styles, the laissez-faire is the only one that permits the group to make the final decision.[35]

In our example, using the laissez-faire style, Mark would tell the group that it needs to decide

Figure 13–13 This group is using the laissez-faire decision-making method, in which the group decides who will lead the group activity.

who is to be promoted. Mark would then become a group member. The group would hold a discussion and decide who gets the promotion (Figure 13–13).

Conclusion

Problem solving and decision making are some of the most important skills needed by leaders. The way we make decisions depends on the situation, our leadership style, our personality type, and our communication style. However, the important thing is the ability to make decisions.

Summary

A problem arises when there is a difference between what is actually happening and what the individual or group wants to happen. Problem solving is the process of taking corrective action to bring about what the individual or group wants to happen. The process by which the new course of action is selected is called decision making.

Problem solving and decision making are skills that can be learned, practiced, and improved. People and groups tend to have certain problem-solving and decision-making styles. Reflexive style problem solvers and decision makers make decisions quickly, without a great deal of thought. Those with a reflective style take a great deal of time to formulate alternatives, evaluate them, and make a decision. People with a consistent style make careful, reliable decisions in a reasonable amount of time.

You can approach problem solving and decision making in two ways. The minimizing approach finds the first, but not necessarily the best, solution that works to resolve the situation. The optimizing approach reviews many different solutions to a situation before choosing the best one.

When you choose the optimizing approach you may wish to use the seven-step process for problem solving and decision making: (1) recognize the problem, (2) determine alternatives, (3) gather information, (4) evaluate alternatives, (5) select a workable solution, (6) carry out the solution, and (7) evaluate the results. You may use these steps to find solutions to problems that have one exact answer, problems that usually involve creating a plan or design, and for judgment problems that call for techniques such as weighing possibilities and factors, trying to improve the current situation, determining sequences of steps and key points, and determining advantages and disadvantages.

There are both advantages and disadvantages to group problem solving and decision making. A group may choose to solve a problem or make a decision through a number of different methods that involve varying degrees of group participation. Some of these are the conventional method, brainstorming method, delphi method, devil's advocate method, consensus method, nominal group method, and synetics method. Leaders of groups may have distinctive styles of solving problems and making decisions. The style they use depends on how much the group members are involved and may vary depending on the situation. These styles, beginning with the least amount of group participation and progressing to the most, are autocratic, consultative, participative, and laissez-faire.

END-OF-CHAPTER EXERCISES

Review Questions

1. Define the Terms to Know.
2. List four reasons why being an effective problem solver and decision maker is important.
3. List seven mistakes individuals or groups make when attempting to solve problems or make decisions.
4. List 13 skills individuals and groups need to develop to effectively solve problems and make decisions.
5. List and briefly describe three problem-solving and decision-making styles.
6. Distinguish between the minimizing and optimizing approaches to problem solving and decision making. When might you use each?
7. List and briefly describe the seven steps to problem solving and decision making.
8. List and briefly describe three types of problems or decisions.
9. List four advantages and seven disadvantages to group problem solving and decision making.
10. List and briefly describe seven methods a group can use in solving problems and making decisions.
11. List and briefly describe four leadership styles used in group problem solving and decision making.

Fill-in-the-Blank

1. The ____________________ decision maker makes quick decisions without taking time to consider and evaluate all possible alternatives.
2. The ____________________ decision maker evaluates all possible alternatives in a reasonable amount of time.
3. The ____________________ decision maker considers all possible alternatives but often takes an unreasonable amount of time in making a decision.
4. In the ____________________ style of group leadership, the leader may have a solution in mind but can be swayed by group input.
5. The ____________________ style of group leadership allows the group to make decisions.
6. In the ____________________ style of group leadership, the leader independently makes a decision and then informs the group.
7. In the ____________________ style of group leadership, the leader seeks only additional information from group members but does not encourage group discussion.
8. ____________________ types of decisions involve listing possibilities and factors, improving a situation, listing steps and key points, or listing advantages and disadvantages.
9. ____________________ types of problems have an exact mathematical solution.
10. ____________________ types of problems require an artistic design or a plan to solve.

Matching

_____ 1. This arises when there is a difference between what is actually happening and what a person or group wants to happen.
_____ 2. The different courses of action you can take to solve a problem.
_____ 3. The process of taking corrective action to meet goals.
_____ 4. The process by which a new or different course of action is selected to correct a problem.
_____ 5. Choosing the first, but not necessarily the best, solution to solve a problem or make a decision.
_____ 6. Method of suggesting as many alternatives as possible to a problem situation without judging the value of the alternatives.
_____ 7. Process of discussing possible solutions to a problem and then taking a vote to decide the course of action. Majority rules.
_____ 8. Reviewing many solutions to a problem situation before choosing the best one to implement.
_____ 9. Results in a substantial portion of a group agreeing on a solution.
_____ 10. Uses a series of anonymous polls to solve problems or make decisions.

A. alternatives
B. brainstorming method
C. optimizing approach
D. delphi method
E. problem solving
F. consensus method
G. problem
H. minimizing approach
I. conventional method
J. decision making

Activities

1. Give an example of a problem you face now. Alternatively, you may choose to create a problem situation like the one in the chapter.
2. Follow the seven steps to problem solving and decision making to solve the problem you listed in activity 1. Be sure to label each step. If you wish, share your thought process with a classmate.
3. Write an example of each of the following problem or decision types.

   a. Exact reasoning b. Creative c. Judgment

   Exchange your examples with a classmate and have him or her design solutions to each. Each person should be able to show or explain how the solution was reached.
4. Write an example of a situation in which you would use the minimizing approach to problem solving or decision making, and one in which you would use the optimizing approach. Read one of your examples to the class. Have the class decide which approach you are demonstrating.
5. Your class needs to raise $500 more than it already has to purchase a computer. Have the class divide into three groups. The first group should attempt to come to a solution on how to raise the money by using the conventional method. The second group should use the consensus method. The third group should use the nominal group method. Report the findings of each group to the whole class.

6. Consider the following situation: A teacher is determining the rules for his or her class. Explain how the teacher would do this using an autocratic leadership style, a consultative leadership style, a participative leadership style, and a laissez-faire leadership style.
7. Have the class divide into four groups. Take the situation given in activity 6 and have each group role-play one of the group leadership styles.
8. **Take it to the Net.**

   Explore decision making and problem solving on the Internet. The following Web sites contain information on problem solving and decision making. Browse the sites. Try to find something you feel would be worthwhile to share with your teacher and classmates. If you are having problems with the listed Web sites or if you want more information, use the search terms listed below.

   Web sites

   <http://www.cdaconsulting.com/decisionmaking.com>
   <http://www.themestream.com/articles/186396.htm/>
   <http://www.demon.co.uk/mindtool/page2.html>
   <http://www.timedoctor.com/lifeskill.htm>

   Search Terms

   problem solving
   problem solving techniques
   creative problem solving
   decision making
   decision making process

Notes

1. *Decision Making*: Skills for Success #18 (Nashville, TN: Tennessee Department of Education, 1993).
2. B. R. Stewart, *Leadership for Agricultural Industry* (New York: McGraw-Hill, 1978), p. 39.
3. R. N. Lussier, *Human Relations in Organizations: Applications and Skill Building* (Columbus, OH: McGraw-Hill Higher Education, 1998.
4. *Decision Making.*
5. Lussier, *Human Relations in Organizations*, p. 276.
6. Ibid.
7. Ibid.
8. L. J. Phipps and E. Osborne, *Handbook on Agricultural Education in Public Schools* (Danville, IL: Interstate Printers and Publishers, 1988).
9. Ibid., pp. 118–119.
10. Lussier, *Human Relations in Organizations.*
11. Ibid.
12. Ibid.
13. Quoted in Lussier, 1998.
14. Ibid.
15. *Personal Development in Agriculture* (College Station, TX: Instructional Materials Service, Texas A & M University, 1989), 8742-C, p. 1.
16. Ibid., 8742-C, p. 2.
17. D. B. Van Dalen, *Understanding Educational Research: An Introduction* (New York: McGraw-Hill, 1979).

18. Stewart, *Leadership for Agricultural Industry*.
19. Ibid.
20. *Personal Development in Agriculture*, 8742-D, p. 2.
21. Ibid.
22. Ibid., 8742-D, p. 3.
23. Ibid., 8742-D, p. 4.
24. Lussier, *Human Relations in Organizations*.
25. *Personal Development in Agriculture*, 8742-C, pp. 4-5.
26. Lussier, *Human Relations in Organizations*.
27. Ibid., p. 282.
28. *Personal Development in Agriculture*, 8742-C, pp. 4–5.
29. Ibid., 8742-C, p. 5.
30. Lussier, *Human Relations in Organizations*.
31. R. M. Hodgetts, *Modern Human Relations at Work* (Fort Worth, TX: Harcourt Brace College Publishers, 1995).
32. Lussier, *Human Relations in Organizations*.
33. Ibid.
34. Ibid.
35. Ibid.

CHAPTER

Goal Setting

Objectives

After completing this chapter, the student should be able to:

- Discuss the reasons for and importance of having goals
- Describe the benefits of having goals
- Explain why people don't set goals
- Explain how to set goals
- Discuss the principles of setting goals
- Describe the steps in goal setting
- Discuss types and kinds of goals
- Set medium-range, long-range, and immediate goals
- Set goals the FFA way

Terms to Know

goals
resources
values
short-term goals
long-term goals
criteria
SMART
ways and means
talent area
tangible
immediate goal
worst-case scenario
psychic income
inertia
momentum

People do not plan to fail, they simply fail to plan. Prior planning prevents poor performance. The ability to set **goals** and make plans is a skill we all need to develop. To be successful and to achieve what we want in life, setting goals and making plans is a necessity. To realize even a fraction of our potential, we need to learn how to set and achieve goals that will move us beyond where we are now.

You must have a goal because it is just as difficult to reach a destination you do not have as it is to come back from a place you have never been. You must have definite, precise, written, clearly set goals if you are going to realize your full potential in life.

REASONS FOR HAVING GOALS

Provide a Target

Only 3 percent of Americans today have defined goals, and that 3 percent outperform the other 97 percent. The 3 percent of Americans who have defined goals are all directed to the specific, whereas 97 percent are shifting back and forth. It is easier for that small goal-oriented population to reach their goals because they have something to strive for, whereas the 97 percent of the population who do not have defined goals cannot reach their target because they have nothing at which to aim.

Provide a Destination

Having goals provides us with a destination to move toward. This, in turn, helps us decide what direction to move in along the way. A person without a goal is like a ship without a rudder or a sail, drifting aimlessly. A person with goals is like a ship with a rudder, sails, a captain with a map, compass, and a port of destination.

Provide Purpose and Meaning

Greater satisfaction can be achieved if we strive toward the accomplishment of something that is important to us. Our mind tends to work like radar. A built-in guidance system keeps us striving and connecting our aims with the goals. Because of this "guidance system," we can accomplish almost any goal we set for ourselves, as long as the goal is clear and we are persistent. Setting the goal and not tapping the natural resources within is a significant problem for some people.

Help Focus Our Energies on Productivity

Defined goals can help us plan our time and focus our energies purposefully and productively. They allow us to make responsible decisions on a daily basis for our thoughts, actions, and behaviors.

Brian Tracy, in *Maximum Achievement*, said, "If your goal is to get through the day and then get home and watch television, you will achieve it. If your goal is to be fit and healthy and to live a long life, then you will achieve that, too."

The same could be said for financial independence or becoming wealthy. Your only limitation is your desire, or how badly you want it (Figure 14–1).[1]

Turn Activities into Accomplishments

Activity should not be confused with accomplishment. People complain that the lack of direction or goals may cause a person to work very hard without accomplishing much. By having **goals**, we can accomplish more because they give us the ability to clarify, plan, have something to aim for, motivate, organize, set levels of achievement, formalize our intentions, and focus and evaluate our progress.

Figure 14–1 The people using the services of the Chicago Board of Trade have financial goals. If their goals are not clear, realistic, and strategically planned, they could be in for some unpleasant surprises.

Help Individual Success, Self-Concept, and Evaluation

Other reasons for setting goals vary for each individual, but they can include these.

1. *Direction.* All people need direction in their lives. Without direction, we are traveling through life without a roadmap or highway signs. It is easy to go down the wrong road. With goals, it is easier to reach our destination because we have directions.
2. *Success.* What does it mean to be successful in life? It is the attainment of your goals; not other goals that are imposed on you, but your own goals. Most people want to succeed, but only you know what you are capable of accomplishing. Success is a key element in setting goals. In reality, it is the basis for goals. When you have finally attained one of your goals, relax and enjoy it. However, don't stop. You are merely at a higher plateau than you were before. There are other heights to be attained. If you stop, you may become bored and unhappy.
3. *Self-concept.* Now that you are headed in the right direction and have become successful, you will develop a feeling of self-worth. This obviously makes you feel good about yourself, or enhances your self-concept. Studies have shown that people who are happy about living and feel good about themselves live longer than people who are unhappy and feel poorly about themselves. The more successes we have, the more positive our self-concept. Therefore, focus your energy on a positive ending.
4. *Evaluation.* Setting goals gives us a standard in determining our progress in life. When you set goals, your level of accomplishment can be evaluated closely. Research has shown that the harder or more challenging the goal, the better the resulting performance. However, there are dangers in setting goals too high or too low. If you set your goal too low, you may reach it and falsely think it is your best. Setting a goal too high may make you discouraged and you may discontinue working toward it. The best goal should be high enough to challenge you but not out of reach.[2]

Do you ever wonder why in football the metal uprights were called goal posts or why the "0" yardline is called the goal line? In basketball, did you ever wonder why the metal hoop is called the basketball goal, as shown in Figure 14–2? Just as with basketball, football, or hockey, you have to have goals in the game of life. Without goals, you never know the score. You never know whether or not you are on target. In these sports, you would not even attempt to play the game without the goals, yet most people attempt to play the game of life without goals, never knowing what

Figure 14–2 You couldn't play basketball without the goals, but most people play the game of life without goals.

the score is or whether they are winning or not. Life is valuable and cherished as long as life has something valuable as its objective.

BENEFITS OF HAVING GOALS

Responsive people have a way of getting their lives sorted out, deciding what really matters, and directing all their creative energies toward the goals they set for themselves. Setting goals is a proven method for high achievement and produces several definite benefits. John R. Noe, in his book, *People Power*, discusses the following five benefits.

1. *Concentrate your efforts.* Goals enable you to concentrate all your efforts and energies in a specific direction. Setting goals enables you to focus all the power available to you into a single focal point. Compare it to what happens when you hold a magnifying glass between a combustible object and the sun. That glass concentrates all the sun's rays that pass through it into a tiny spot for maximum power. If you hold it there long enough, the object will burst into flame. Concentrated power can enable you to do much more with what you have than if you had no goals.[3]
2. *Make the most of your time.* Setting goals enables you to more efficiently use your time. Each of us has only 1,440 minutes each day. You can waste them, spend them, or invest them. Goals enable you to invest your life in things that really matter to you.[4]
3. *Let other people know how to help.* Setting goals helps other people know how to help you. High achievers soon learn that they can reach their full potential only with the help of many others. Many people could help you, if they knew how. When you set and make known your goals, it's like giving other people handles by which they can help you carry your load (Figure 14–3).[5]
4. *Keep your enthusiasm.* Goals can keep your energy level and enthusiasm high. Those around you are also motivated. Desire is the key to all motivation for achieving challenging goals.
5. *Monitor your progress.* Goals enable you to monitor your progress. If you don't know where you are going, how will you know when you get there? When you shoot at a target and miss it, you can determine what adjustments you need to make before the next shot. If you hit the bull's eye, you can zero in on a more challenging target.[6]

Figure 14–3 Former president Clinton would not have reached his goal of being U.S. president if he hadn't had people helping him, even to the details of handling business and tour schedules. (Courtesy of *FFA New Horizons Magazine*.)

Goals Make Mental Laws Work for You

Brian Tracy lists the following nine mental laws in his book, *Maximum Achievement*.

1. *Law of Control.* You feel positive about yourself if you feel you are in control of your life. Goal setting allows you to *control the direction of change* in your life. No one fears a change that represents an improvement. When goals are clear and well planned, you eliminate most of your fear and insecurity.
2. *Law of Cause and Effect.* For every effect in your life there is a specific cause. Goals are causes. Health, happiness, freedom, and prosperity are effects. You sow goals and you reap results. Goals begin as thoughts, or causes, and manifest themselves as conditions or effects. The primary cause of success in life is the ability to set and achieve goals.[7]

   In the workplace, you either work to achieve your own goals or you work to achieve someone else's goals. The best work of all is when you are achieving your own goals by helping others to achieve theirs.[8]
3. *Law of Belief.* You trigger it by intensely believing that you will achieve your goals, and by taking actions consistent with those beliefs. This is the foundation of faith and self-confidence.
4. *Law of Expectations.* You trigger it by confidently expecting that everything, positive or negative, is moving you toward the realization of your goals.
5. *Law of Attractions.* This is activated by thinking continually about your goals. With your goals as your primary thoughts, you invariably begin to attract into your life people with similar goals. You attract ideas, opportunities, and resources that can help you.
6. *Law of Correspondence.* Your outer world corresponds to your inner world. When your mind is dominated by thoughts, goals and plans to achieve the things that are important to you, you become what you think.
7. *Law of Subconscious.* What thoughts you hold in your conscious mind, your subconscious mind works to bring into your reality. More and more of your subconscious mind is dedicated to making your words and actions fit a pattern consistent with what you really want to achieve.
8. *Law of Concentration.* Whatever you dwell on grows. As a person thinks, so he or she becomes. The more you dwell on, reflect on, and think about the things you want and how you can attain them, the more apt you are to attain your goals.
9. *Law of Substitution.* You can substitute a positive thought for a negative one. Whenever something goes wrong, think about your goals. If you have a bad day, think about your goals. It is impossible to think about your goals

continually without being optimistic and highly motivated.[9]

Tracy says that when you begin using all these mental laws behind a clearly defined purpose to which you are totally committed, you become an unstoppable powerhouse of mental and physical energy that will not be denied. With clear, specific goals you develop and use all your mental powers. You then accomplish more in a few years than most people accomplish in a lifetime.

You Tap Your Resources

Each of us have untapped **resources** within us just as nature does. Once scientists learn to tap or further perfect the available resources afforded by the sun, water, wind, the atom, ocean waves, ions, magnetism, and other undiscovered energy sources, great things will begin to happen (Figure 14–4). Great things happen in our lives once we set goals and tap the resources within us. In his *"I Can"* course, Zig Ziglar says that "if we want something badly enough, we must make it our definite goal. When we go after it as if we can't fail, many things will happen to help make certain we won't."

WHY PEOPLE DO NOT SET GOALS

Some people are talkers instead of doers. They are not willing to do what is necessary to be more successful and improve their lives. You can tell what a person really believes by his or her actions, not words. It is what you do that counts. Your behavior expresses your true **values** and beliefs. One person who takes action is worth a hundred talkers who do nothing.

Until people accept that they are fully responsible for their lives and for almost everything that happens to them, they will not take the first step toward goal setting. Such people use all their creative energies making excuses for their failure to make progress.

People who are mentally and emotionally negative are not the kind of people who confidently and optimistically set **short-term goals** and **long-term goals**. People raised in a negative environment do not have attitudes capable of goal setting unless an attitude adjustment occurs (Figure 14–5).

Figure 14–4 These wind generators will tap wind resources for energy. Great things can happen to you once you tap the resources within you. (Courtesy of USDA.)

Figure 14–5 An apparently unmotivated student. Unless she adjusts her attitude, she may never set any personal goals.

Your environment will affect your actions. If your parents don't have goals, you can be raised without even knowing there are such things as goals. It is a fact that 80 percent of the people around you are going nowhere. It is only natural that you go with them. Brian Tracy said, "If people knew that all their hopes and dreams and plans, all their aspirations and ambitions, are dependent upon their ability and their willingness to set goals—if people realized how important goals are to a happy, successful life—I think far more people would have goals than do today."[10]

Goal setting is more important to your happiness than almost any other single subject you could ever learn, yet, until recently, you probably never received an hour's worth of instruction in goal setting. People can even get a college degree and not have instruction in goal setting. Someone who has truly mastered goal setting is probably very successful.

People can be very negative and cruel. Many do not praise you for your successes, but they certainly criticize you for your failures. Children are smart; they learn how to get along. Eventually, a child who is constantly criticized or discouraged stops coming up with new ideas, new dreams, or new goals.

Keep your goals confidential to avoid criticism. Don't tell anybody except your boss, spouse, or people who will encourage you in the direction you want to go. Effective goal setters finally learn to keep their goals to themselves. People can't laugh at you or criticize you if they don't know what your goals are. However, if you need encouraging, encourage others, and they will encourage you.

The greatest single obstacle to success is the fear of failure. It is learned early in life by parents not letting their children try things. They say things such as "if you can't do it right don't do it at all." Little do they realize that their children would probably eventually get it right if they were permitted to fail until they learned to get it right. The negative attitude of the parents becomes entrenched in the subconscious mind and does more to deter hope and kill ambition than anything imaginable.

The major reason for the fear of failure is that most people don't understand the role of failure in achieving success. Those who do nothing make no mistakes. The rule is simply this: *It is impossible to succeed without failing.* Failure is a prerequisite for success. The greatest successes in human history have also been the greatest failures. Babe Ruth was not only the homerun king; he was also the strike-out king. Success is a numbers game. There is a direct relationship between the number of things you attempt and the probability of eventually succeeding. The important thing is to keep swinging and not worry about striking out occasionally. Thomas Edison is remembered as a success for his light bulb, which took over 11,000 failed experiments before he got it right. Edison viewed the experiments as learning opportunities.

Great successes are almost always preceded by many failures. It is the lessons learned from the failures that make the ultimate successes possible (Figure 14–6). Look on temporary setbacks as signposts that say, "STOP, go this way instead." Great leaders never use the words *failure* or *defeat*. Instead, they use words like "valuable learning experiences" or "temporary glitches." You can learn to overcome the fear of failure by being absolutely

Figure 14–6 The inventor Preston Tucker achieved many goals, but he was probably ahead of his time. The car he designed in 1948 (only 100 were built) had many features, such as shatterproof glass and disc brakes, that were ridiculed at the time but later became standard equipment.

clear about your goals and accepting that temporary setbacks and obstacles are the price you pay to achieve any great success in life.[11]

How to Set Goals

Set goals that are clear, challenging, and obtainable. Make sure the goals are measurable. Set immediate goals, goals for the short term, and goals for the long term. When you reach one goal, take a moment to congratulate yourself and cherish the moment. Move on to the next goal with confidence, energized by what you have already accomplished.

The goals must be realistic as well as attainable. Don't think you should, or can, accomplish everything quickly. Set interim goals as you strive to attain the ultimate goal.[12] For example, a person who has the goal of finishing college might become quickly discouraged without short-term goals. Those short-term goals include making it through the week, the next holiday, or until the end of the semester or quarter.

Make sure your goals are specific. Without specific goals, it's far too easy just to drift, never really taking care of your life. Make sure the goals have some challenging time limits.

Make sure the goals are written. Written goals are part of disciplined motivation. Discipline is the process of bringing ourselves under control. When we lose disciplined motivation, we lose the ability to deal with the future. We begin to function based on feelings, which dulls our purpose and obscures our vision. Goals written down become a written vision that helps us reach fulfillment.

Set goals and then strive to meet them. Sometimes you will succeed on schedule, sometimes things will take longer to achieve than you thought, and sometimes you won't attain what you thought you would. The important thing is to keep planning for and pursuing your ultimate goal.[13]

Prioritize your goals. Write down the top 10 things you want to accomplish in your life before you retire. Take those 10 things and prioritize them.

Principles of Setting Goals

Consider Criteria for Goals

Consider the following **criteria** for the goals you make.

- Is it really my goal?
- Is it morally right and fair?
- Are my short-range goals consistent with my long-range goals?
- Can I commit myself emotionally to complete the project?
- Can I visualize myself completing the goal?
- Is the goal specific?
- Is it measurable?
- Is it realistic?[14]

Consider Benefits of Goals

Consider the following benefits of attaining goals. If there are no benefits, perhaps you should consider other goals.

- Have more purpose and direction in life
- Make better decisions
- Be more organized and effective
- Do more for yourself and others
- Feel more fulfilled, enthusiastic, and motivated
- Accomplish more[15]

Use the SMART Principle in Goal Setting

Goals must be meaningful, you must believe in them, and you must have a positive attitude toward achieving them. There are five ways of telling if your goals are **SMART** goals.[16]

Goals should be

Specific—Keep your goals short and concise.

Measurable—Set standards so you know if your goals have been reached.

Attainable—Make sure your goal is realistic and practical so it can be accomplished.

Relevant—Your goal should contribute to the purpose of the group or what you are trying to accomplish.

Trackable—Keep records so you know if you are on target in reaching your goals.

STEPS IN GOAL SETTING

Goals are dreams that you are willing to pursue. To pursue these goals and dreams, you must follow certain steps. Accomplishing goals is much like climbing the steps of a ladder. To get to the top, you have to follow the steps. Before you can accomplish your goals, you also have to follow certain steps (Figure 14–7).

Your thoughts can become reality. You become and you accomplish what you think. The more intense your thoughts, the quicker your goals will become reality. There is a direct relationship between how clearly you can see your goal as accomplished on the inside and how rapidly it appears on the outside.[17] The following steps will help provide the **ways and means** to get from where you are to wherever you want to go.

Step 1

Set goals that agree with your values.

For you to reach your full potential, your goals and values should be in harmony with each other. Your values represent your deepest convictions about what is right and wrong, what is good and bad, and what is important and meaningful to you. High achievement and high self-esteem only happen when your goals and your values are compatible.

Step 2

Develop desire.

Almost every decision you make is made on the basis of emotion, either fear or desire. A stronger emotion overcomes a weaker emotion. If you

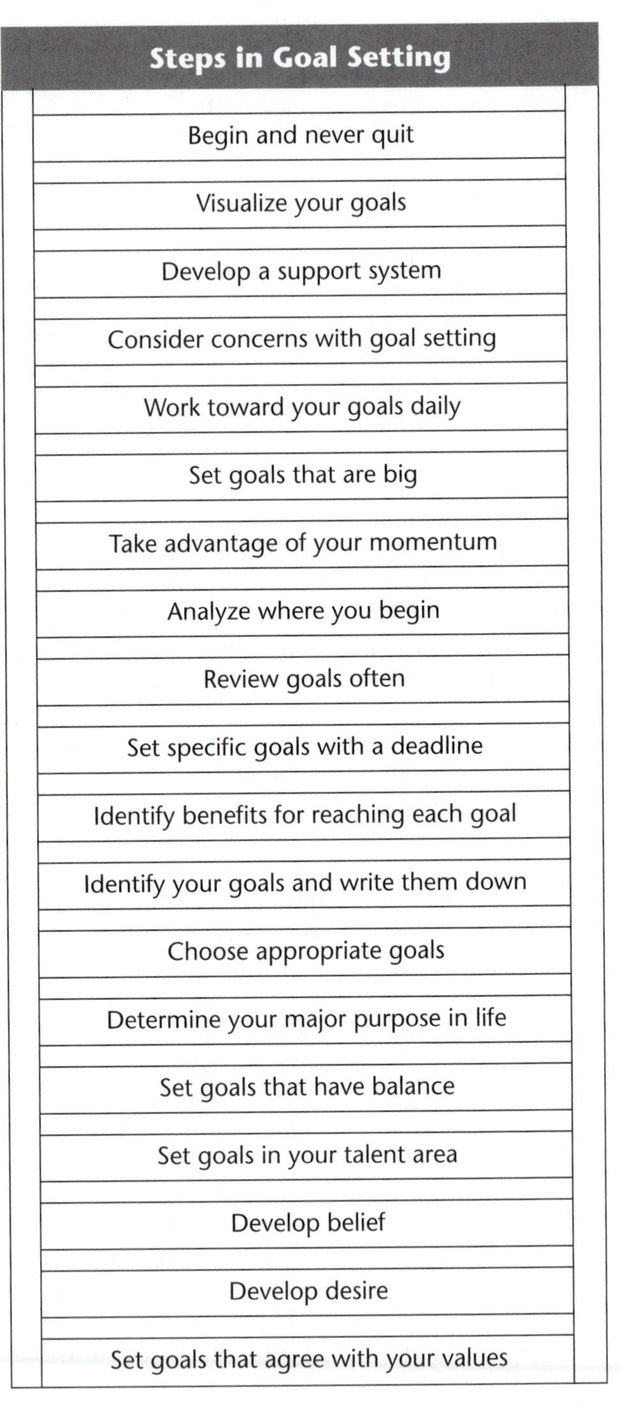

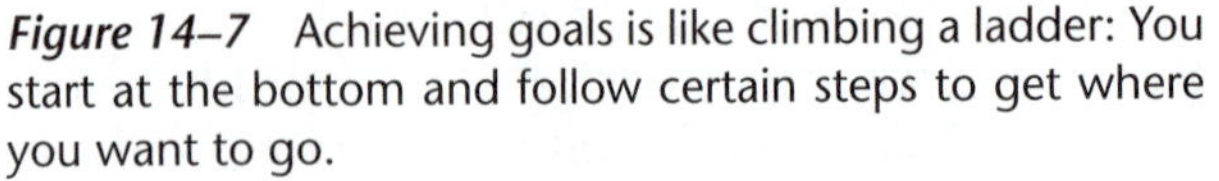
Figure 14–7 Achieving goals is like climbing a ladder: You start at the bottom and follow certain steps to get where you want to go.

focus on your desires, write them down, and make plans to achieve them, your desires will become so intense that your fears will be overcome.

Step 3

Develop belief.
You must absolutely believe that it is possible for you to achieve your goal. You must believe that you deserve the goal. You must have faith and belief that deepens until you know that the goal is attainable. If the goal is worth achieving, it is worth working for patiently and persistently.[18]

Step 4

Set goals in your talent area.
Set realistic goals. Goal setting is necessary for success. However, set goals that you feel you can achieve. If you are 5′1″ tall and weigh 105 pounds, it is inconceivable to set your goal as an offensive or defensive lineman in the National Football League. You could, however, set the goal of riding a Triple Crown Winner.

Each of you has the capacity to be excellent at something. You can achieve your full potential by finding your **talent area** and then putting your whole heart into developing your talents in that area (Figure 14–8).

Don't overlook the talents within you. When considering a major in college, for example, do not overlook the skills and abilities you may have already developed in a particular field. You may be more successful if you pursue educational goals related to a field in which you already have attained knowledge. College may not be able to prepare you in a new field of study to the level that you have already reached in another field.

Start on your goals early. A student may choose to go to college with the goal of being a veterinarian. It is an admirable goal, but the student may not have attained the necessary cognitive skills. College entrance scores must reflect success in high school and suggest that the student has the ability to pass the rigorous college courses needed to be a veterinarian. If this goal is set as a freshman in high school, courses can be taken to gain the proper knowledge and skills.

Figure 14–8 Make sure the goals you set are in your talent area. This FFA member is pursuing a goal in her talent area, singing. She is performing at the National FFA Convention. (Courtesy of *FFA New Horizons Magazine*.)

Make a plan to secure needed skills and abilities. You must learn what you need to know so you can accomplish what you want to accomplish. Make a list of all the information, skills, abilities, and experiences that you need and make a plan to get them. Do not overlook the talents that can be developed to help you reach your fullest potential and happiness. Theodore Roosevelt said, "Do what you can, with what you have, right where you are."[19]

Step 5

Set goals that have balance.
Brian Tracy suggests that you need a variety of goals in seven critical areas of life in order to perform at your best. He suggests the following areas to have goals:

- Family and personal
- Physical and health
- Mental and intellectual

- Study and personal development
- Career
- Financial and material
- Spiritual

By having this balance, you can be working toward goals all the time and be attaining self-fulfillment.

Step 6

Determine your major purpose in life.
This may be most difficult for many. You have to ask yourself, "If all else fails, which goals would make me truly successful and happy?" For many, it is a business goal; for others it is a financial goal or a health or spiritual goal. The ability of a person to choose a dominant goal in life is the primary way to stay focused and make progress.[20]

Step 7

Choose appropriate goals.
Responsive people visualize goals that stretch their resources, but not to the breaking point. Here are some of the guidelines they use.

- Choose goals that can be reached only if you give them your best shot.
- Choose goals that will capture the enthusiasm of the people who must fulfill them.
- Choose relatively higher intermediate goals that will lead to your long-range goals and are within the bounds of your purpose.[21]

Remember: If you set your goals too high, you will become easily discouraged; if you set them too low, you will not be inspired to give them your best shot.

Step 8

Identify your goals and write them down.
When you write down a goal, you make it more **tangible**. One of the most powerful of all methods for instilling a goal into your subconscious mind is to write it out clearly, concisely, in detail, exactly as you would like to see it in reality. Decide what is right before you decide what is possible. Make the description of your goal perfect and ideal in every respect. "When you discipline yourself to write down your goals, the very act overrides your failure mechanism and turns your success mechanism to full power."[22]

As you write it, include *what* you want, *when* you want it, *why* you want it, and *where* you are starting. Make a list of the obstacles you must overcome, the information you will require, and the people whose help you will need.[23] There are many ways to do this. Figure 14–9 shows the way Ben Franklin planned for personal achievement.

Six Steps in Ben Franklin's Formula for Personal Achievement

1. Select 13 principles, virtues, or characteristics that you would like to acquire or develop.
2. Using 3" x 5" cards, prepare a "pocket reminder" for each characteristic. Write a brief summary of the characteristics indicating what it means and what you want to develop.
3. Concentrate on one characteristic every day for an entire week. Carry the reminder card with you and refer to it frequently and try to achieve that characteristic.
4. Start the second week on your second characteristic. Let the characteristic that you have worked on for a week be taken over by your subconscious mind.
5. Continue working on one characteristic each week until the series is completed. Then start again. This allows you to complete the series four times per year.
6. After a year, or when you feel you have acquired the desired characteristics, substitute a new characteristic you wish to acquire in its place.

Figure 14–9 To achieve your goals you must have a step-by-step prioritized plan. Here is one of many plans you could choose. (Used with permission: R. I. Carter, *Leadership and Personal Development*, St. Paul, MN: Hobar Publications, 1990.)

Step 9

Identify the benefits from reaching each goal.
To ensure that your goal is worth the effort, make a list of all the ways that you will benefit from achieving it. When you have significant reasons for achieving your goals, you increase your intensity. If your reasons are important enough, your belief solid enough, and your desire intense enough, you will make your goal more attainable.[24]

Step 10

Analyze where you begin.
The clearer you are about where you begin and where you are headed, the more likely it is that you will end up where you want to be. This gives you a baseline to measure your progress. For example, if you decide to lose weight and you set a goal of 20 pounds, you must first weigh yourself.

Step 11

Set specific goals with a deadline.
An important part of setting goals is setting deadlines for reaching them. Timetables put teeth into your goals. If you have the attitude of accomplishing a goal when you get "around to it," you probably will not. If you state specific goals with definite deadlines, you will be amazed at how often they are met.[25]

Step 12

Review goals often.
In addition to writing down your goals, it is a good idea to have your goals attractively displayed so that you can see them often. The goals you see are the goals you achieve.[26] Accept that your plan will not be perfect. Successful people have detailed plans, but they are not afraid to alter them as the need arises.

Step 13

Take advantage of your momentum.
When you have reached one goal, take advantage of the enthusiasm you've generated and start attacking another goal. The law of inertia also affects people. Resting too long between reaching an intermediate goal and attacking another goal can cause enthusiasm to dampen and actually cause a person to lose momentum.[27]

Step 14

Set goals that are big.
Big goals are set by "big people." It takes big goals to create excitement necessary for accomplishment. There is no excitement in being average or copying others. The excitement comes when you do your best, which you can only do with the proper goals. Consider the following analogy:

> ***Take a bar of iron and use it for a door stop, and it's worth a dollar. Manufacture horseshoes from that iron, and they're worth about fifty dollars. Take the same bar of iron, remove the impurities, refine it into fine steel and manufacture it into mainsprings for precision watches, and it's worth a quarter of a million dollars.***[28]

The way you see life largely determines what you get out of it. The way you see the bar of iron makes the difference, and the way you see yourself and your future makes the difference.

Step 15

Work toward your goals daily.
If you expect to accomplish your objectives, you must work toward your objectives every day. Good students don't wait until the night before to study for an exam. They study daily. Our daily objective should be to improve on yesterday.

Step 16

Consider concerns with goal setting.
Goals can be negative if certain precautions are not taken. First, realize that you are the determining factor in achieving your goals and that luck is not involved. Second, your goal can be negative if it is too big. For example, to accumulate a million dollars in one year is an unrealistic goal. However, to accumulate a million dollars by retirement is very realistic. Third, your goal can be a concern if it is outside your talent area and is set to please someone else.[29]

Prepare for any anticipated roadblocks. Be aware of the obstacles but don't dwell on them. Simply take care of the problem. A roadblock for a weight loss program might be the love of sweets; for climbing a mountain, it could be poor physical condition; for getting a college degree, it could be a lack of self-discipline; for being an elected official, it could be heavy social demands. Whenever great success is possible, great obstacles exist. In fact, obstacles are the flip side of success and achievement. If there are no obstacles between you and your goal, it probably is not a goal at all, merely an activity.[30]

Step 17

Develop a support system.
Find people who will support you and your goals. This could be your family, a friend or co-worker, or certain organizations. To accomplish many of your goals, you will need the help and cooperation of many people. Make a list of all the people whose help and cooperation you will require.

Step 18

Visualize your goals.
In Chapter 4, we discussed the importance of vision and leadership. The same is true for goal setting. Picture your goal as if it were already achieved. Keep this picture in your mind. By doing this, you increase your desire and belief that your goal is achievable. Why? Your subconscious mind is activated by pictures. There are stories of golfers and basketball players who improved their scores or shot percentages merely by visualizing their practice.

Step 19

Begin and never quit.
Once your plan is completed, start immediately. Take one small step at a time. Zig Ziglar makes two excellent comments on reaching goals: "What you get by reaching your goals is not nearly as important as what you become by reaching them." You become a winner. "If we expect to change and improve our circumstances, then we must change and improve ourselves—because

Don't Quit

When things go wrong, as they sometimes will,
When the road you're trudging seems all uphill,
When the funds are low and the debts are high,
And you want to smile but you have to sigh,
When care is pressing you down a bit
Rest if you must, but don't you quit.
For life is queer with its twists and turns,
As every one of us sometimes learns,
And many a failure turns about,
When he might have won if he'd stuck it out.
Success is just failure turned inside out,
The silver tint of the clouds of doubt.
And you never can tell how close you are,
It may be near when it seems so far.
So stick to the fight when you're hardest hit,
It's when things seem worst that *you must not quit!*

—*Anonymous*

Figure 14–10 Many people believe that persistence is more important than intelligence in achieving goals. One thing is certain: You will not achieve your goals if you quit.

we must be something before we can do anything." Reaching goals makes us something. Among other things, we prove that we are not quitters. Strive toward your goals with persistence and determination. Simple logic says that you will eventually be successful if you never quit (Figure 14–10).

Types of Goals

How Do You Begin?

The journey of a thousand miles begins with a single step, but sometimes that first step is the most difficult. Reaching the end of the journey (your goal) is great, but it is the journey itself, the planning and the struggling, that gives you a feeling of confidence and accomplishment.

Visualize yourself living in a valley surrounded by mountains. You can climb any one you choose, but you are not even sure you want to climb a mountain at all. When you finally do select a mountain, make sure that you are prepared with training and the right equipment. Ask yourself, "Is this the right mountain for me?" You may not know for sure. So, although you continue to climb and move ahead and put the best you have into the effort, you keep a part of you flexible. Even if you think you are on the wrong mountain, you have learned things that will help you evaluate, plan, and prepare for making some changes. You might make a total change to another mountain that fits your needs better, or you may look at where you are and realize that the mountain you are on is not all wrong, you just need to find a different path to reach where you want to go.[31] The important thing, whether you begin with the right mountain or the right path the first time or not, is to begin.

Identify Your Goals

We have talked about all the reasons for setting goals. Now, let us select our goals. You may not need any help. You may already have more goals than you think you can accomplish. For those of you who need help coming up with goals, consider the following seven questions.

1. *What are the five most important values to me?* This question helps you clarify what is really important to you and what is less important. Once you have identified the five most important things, organize them in priority from one to five.
2. *What are your three most important goals right now?* Write the answer to this question within 30 seconds. When you have only 30 seconds to write your three most important goals, your subconscious mind sends you many goals quickly.
3. *What would you do, how would you spend your time, if you learned today that you had only six months to live?* This helps you really decide what is important to you.
4. *What would you do if you won a million dollars in cash, tax free, in the lottery tomorrow?* Think of the choices you would make if you had all the time and money you needed. Consider the things you would do differently if you had the ability to choose.
5. *What have you always wanted to do but have been afraid to attempt?* This helps you determine what you really want to do without the fear of rejection.
6. *What do you most enjoy doing? What gives you your greatest feeling of self-esteem and personal satisfaction?* You will always be the most happy doing what you most love to do and that which makes you feel the most alive and fulfilled. Successful people are doing things they like to do.
7. *What one great thing would you dare to dream and do if you knew you could not fail?* Imagine that you could be granted one wish. You are guaranteed that you will be successful in one thing you attempt, big or small, short- or long-term. What one exciting goal would you set for yourself?[32]

Examples of Goals

Goals can include better test taking, graduating, getting a job, personal health, channeling stress, saying no to drugs, tobacco, and alcohol, improving grades, and improved relationships at home.

A group of high school students made a list of all the things they would like to do if they could do anything. A few of their choices are listed below. Are there any that would be on your list?

- Graduate
- Make a million dollars
- Get married
- Travel around the world
- Be a teacher
- Be a marine biologist
- Climb a mountain
- Own a classic car
- Skydive
- Be a doctor
- Drive an 18-wheeler
- Own my own store
- Have my own band
- Build houses
- Live in another country
- Publish a story
- Act
- Learn to ride a horse
- Go somewhere on a plane
- Learn to swim

Brainstorm. Do not stop to think about it. Write down on a sheet of paper all the things you would like to do if nothing were standing in your way.

Categories of Goals

You can have goals in eight areas: physical, personal, mental, school, financial, social, family, and career. Listed below are examples of goal statements for the eight categories. From the goal statements, develop a chart with seven steps that include identifying the goal, listing the benefits, listing the obstacles to overcome, identifying support groups, listing the skills and knowledge required, developing a plan of action, and setting a deadline for achievement.

Family

Spend five hours per week helping my parents.

Spend Friday nights together.

Physical

Walk 30 minutes every day.

Use our health club membership regularly.

Mental

Read two inspiring books per month.

Discover inspired ideas as I concentrate and meditate daily.

Social

Visit the senior citizens home one weekend per month.

Open our home for neighborhood gatherings one evening each month.

Personal Values

Spend 30 minutes each morning reading an inspirational book.

Attend church on a regular basis.

Financial

Save $50 per month.

Earn $400 per month with our part-time family business.

School

Accept more responsibility at school.

Strive for excellence in my classes.

Career

Take courses to prepare me for my career goal.

Travel to get experience with my career goal.[33]

Set goals for myself for each of the eight areas. Refer to activity 6 at the end of the chapter.

MEDIUM-RANGE, LONG-RANGE, AND IMMEDIATE GOALS

An **immediate goal** may be as small as washing dishes immediately after each meal. Another one may be obtaining a *B* on a chemistry assignment. A short-range goal can be part of a long-range goal or it can be independent. Washing the dishes immediately after each meal may be a totally independent goal. Getting a *B* on a chemistry assignment may be part of receiving a *B* for the chemistry class. Short-range goals involve actions to be realized in the next few days or weeks.

Medium-Range Goals

A medium-range goal is similar to the short-range goal only longer, usually 6 to 12 months. Medium-range goals can either be independent, made up of short-range goals, or part of a long-range goal.

Long-Range Goals

A long-range goal is planning where you would like to be months, years, or even a decade from now. Short- and medium-range goals usually make up long-range goals. A high school student may have a long-range goal of being a doctor. His medium-range goals may include finishing high school, college, and medical school. His short-range goals may include getting a *B* in chemistry class or finishing his English paper by next weekend.[34]

Prioritize Your Goals

The key to prioritizing is planning. Evaluate your goals and rank them in each category as well as by timing (immediate, medium-range and long-range). Start work on accomplishing the prioritized goals by generating a favorable climate. For example, you may attend every possible state FFA function just to get exposure and experience if being a state FFA officer is one of your goals. Develop subobjectives for each prioritized goal. Decide on alternatives such as plan B, plan C, or a **worst-case scenario** for each goal. Follow by making a commitment, acting, and reviewing the outcome.[35]

Meeting and Even Exceeding Your Goals

A goal generates commitment. Our identities come out of what we commit to. You can get and maintain the drive and desire to succeed if you will

- Be in a position in which you can use your talents and abilities.
- Make sure you get a "payoff" or **psychic income**. (Some refer to this as planning small wins.)
- Focus on the future.
- Become known as a "doer" by developing a bias for action and focusing on results.[36]

Reach your goals by staying focused. Don't let others rearrange your priorities. Remain committed to the plans and directions you have identified as necessary for success in achieving your goals, and use these to make decisions on a daily basis.

Herman Krannert said, "I sometimes like to measure a man by the things he decides to leave undone." Dwight D. Eisenhower said, "I have found the urgent things are seldom important and, the important things are seldom urgent."[37]

Perseverance Is Important

Goal setting is a continuing process, not a one-time practice. Goals must be pursued on a daily basis.

Keep your momentum going. Your psychological self is similar to Newton's physical principles of **inertia** and **momentum**, which state that a body in motion tends to remain in motion unless acted on by an outside force. It takes a large amount of energy to get a body from a resting position to a state of forward motion; it takes a smaller amount of energy to keep it in motion.

Nothing succeeds like success. Do something every day to keep you striving for your goals. Maintain your momentum, keep up the pressure, and develop a mind-set to become a high-inertia, goal-setting, and goal-achieving person.

Goal Setting the FFA Way

FFA Program of Activities

The FFA organization has had a procedure during most of its existence that provides a tremendous mechanism for goal setting. It is called the FFA Program of Activities. Programs are developed at the local, state, and national levels within the organization. *Chapter Planning and Recognition: A Student Handbook* is available from the National FFA Organization, and most local chapters have a copy. The format presented for developing goals for your chapter can also be used for developing your personal goals.

The FFA helps develop leadership. Learning how to prepare and implement an FFA Program of Activities teaches us how to be goal setters. When we learn how to set and achieve goals, we can become leaders if that is one of our goals. When we achieve our goals, we become successful. Only you, not others, can determine your success.

The Program of Activities is a written plan, developed and published annually, of all activities that the chapter wishes to accomplish during the school year. It serves the chapter much as a road map serves a traveler. It is a guide that identifies the activities necessary to make the FFA chapter meet the personal and occupational needs of its membership.[38] It is divided into 13 program areas, which also constitute the standing committees of the chapter. These areas include supervised agricultural experience, conduct of meetings, cooperation, scholarship, community service, leadership, public relations, recreation, earnings, savings and investments, participation in state and national activities, alumni relations, and membership development.

For any organization to be effective, it must plan its activities. A plan gives a sense of direction—it helps an organization reach its goals. Without a plan, the group's activities will soon be done on a haphazard basis and achievement of the goals will be very limited.

Developing a Program of Activities is part of leadership training essential to the educational process of preparing for a selected occupation. A Program of Activities fosters cooperative spirit and develops individual leadership talent among FFA members by assigning them duties and responsibilities for developing and conducting planned activities. Perhaps the most important benefit of a planned Program of Activities is that it provides a means of evaluating and improving chapter activities each year.[39]

There are several formats a chapter can use in developing its Program of Activities. Although it doesn't make much difference which format is used, it is important that a standard format be used.

A planned Program of Activities needs to be written so that the following questions have answers that are definite, measurable, and understandable to all FFA members.

1. What is going to be done?
2. Who is going to be involved?
3. When are you going to do it?
4. How many are going to do it?
5. How are you going to do it?
6. How much is it going to cost?[40]

The sections of each division of the Program of Activities are shown in Figure 14–11.

The Program of Life Goals

The Program of Life Goals are simply the personal goals that you set in life. They can follow the same basic format or process as the Program of Activities. Review the "Steps in Goal Setting" section of this chapter and you can see the similarities with the FFA Program of Activities. Figure 14–12 adapts the Program of Activities format into

Program of Activities	
Division	The Program of Activities is divided into 11 program areas, which also constitute the standing committees of the chapter.
Activity	An activity is any function that the chapter conducts during the year to meet the stated objectives in each division of the Program of Activities. Be specific. Tell exactly what is to be done. *Example:* "Improve the scholarship of all FFA members."
Goals	Every activity should have one or more goals. Each goal should be written in specific terms so that the committee can determine if the goal was accomplished. Be specific. Tell exactly what is to be done. Avoid generalizations. Use positive words such as *prepare, build, sponsor, use, submit.* Avoid nonspecific verbs such as *encourage, urge,* or *assist.* *Example:* "Have 50 percent of the chapter members raise their overall grade point average by two-tenths of a point."
Ways and Means	These are the methods or the steps that the chapter plans to use in accomplishing its goals. The actual number of steps will depend on the complexity of the goals. 1. Devote one meeting a year to scholarship with a speaker. 2. Present scholarship certificates to the top three students in each class at the chapter award banquet. 3. In cooperation with the school principal, select and provide a scholarship trophy to the member who improves his or her grade point average the most during the year.
Amount Budgeted	Show the amount of money the chapter plans to spend on the ways and means of conducting each activity. In cases in which no cost is involved, this column should be left blank.
Members Responsible	The members responsible are the individuals who will coordinate each goal. Do not include all students who will conduct or participate in the activities.
Date to Be Completed	Projected completion dates are an important part in planning the Program of Activities. A deadline for completing each of the ways and means should be established and adhered to as nearly as possible.
Accomplishments	As each activity progresses and is completed, make comments that will be useful in making that activity more effective should it be conducted another year.

Figure 14–11 The FFA Program of Activities format offers an excellent example of how to achieve organizational, professional, and personal goals. (Courtesy of the National FFA Organization.)

Program of Life Goals	
Areas	Physical School Mental Social Financial Personal Values Family Career
Activity (Objective)	There may be several activities within each area. For career, for example, summer employment, travel, specialized courses, or training may be activities within each area.
Specific Goal	A specific goal can then be set for each activity. Write it down. Be specific. Be realistic.
Ways and Means	Develop a step-by-step prioritized plan. Begin and never quit.
Obstacles, or Price You Will Have to Pay	Identify skills or knowledge required. Identify obstacles. Identify sacrifices.
Support System	Identify people who will support you. Visualize your goals. Identify benefits of these goals.
Deadline to Complete Goal	Set a specific deadline.
Accomplishments/Evaluation	Compare to where you began. Monitor and adjust.

Figure 14–12 The Program of Life Goals format is adapted from the FFA Program of Activities format. See Figure 14–14 for an example of a completed plan of action.

the Program of Life Goals. Instead of the 12 divisions, the eight goal-setting areas of your life are included: physical, personal values, mental, school, financial, social, family, and career. Instead of "Amount budgeted" from the Program of Activities, "Obstacles, or price (personal sacrifice) you will have to pay" is included in the Program of Life Goals. "Support system" is used instead of "Members responsible." As you can see, all other sections are basically the same.

Plan of Action

Figure 14–13 shows two goals from one of the eleven divisions, leadership, of the FFA Program of Activities. All sections are included, with actual data for each.

Career is an example with two goals from one of the eight areas of life (Figure 14–14). Each section provides an example of how you can use this format to plan your personal goals. A complete plan would include approximately eight pages—a page for each area of your life. By using the eight-page, eight-area format, you could have three to five goals for each area, or a total of 20 to 40 goals. Numbers are not important. The important thing is that this gives you a methodical process to set and achieve your goals. Activity 8 at the end of the chapter gives you the opportunity to do this.

Conclusion

By using the Program of Activities format to set personal goals, you use a proven model. The importance of setting goals can be summed up by saying that if you have goals, you know what you want and are likely to achieve success. If you do not have goals, you are less likely to achieve success.

Just as your organization, chapter, or club needs a plan of action, you will find that people need a plan of action in their lives. With a plan of action, they can succeed. You can succeed.

Program of Activity Plan of Action				
DIVISION IV: Leadership				
OBJECTIVE: Each FFA member will take part in leadership development activities sponsored by the FFA chapter and/or by the State and National FFA organizations.				
Goals	**Ways and Means**	**Amount Budgeted**	**Date to Be Completed**	**Accomplishments and/or Evaluation**
Provide every member with the opportunity to improve his or her speaking ability.	1a. All FFA members write and give a five-minute speech as an Ag-Ed class assignment.		January 5	
	1b. Videotape each member's speech and replay for constructive criticism.		January 20	
	1c. Select one member from each class to present his or her speech to local service clubs, radio, etc.		February 15	
	1d. Invite three community leaders to assist in selecting a chapter public speaking winner to participate above the chapter level.	$10.00	March 15	
	1e. Present local chapter winner with FFA Foundation Congratulations booklet and award medal at chapter banquet.		May 12	
	1f. Publicize in school paper and local newspaper.		May 15	
Provide committee work experience for all members.	2a. Through member choice and/or appointment, have each member serve on one of the eleven Program of Activities standing committees.		September 15	
	2b. Provide each member with a copy of the chapter Program of Activities.		September 15	
	2c. All committees and/or subcommittees meet at least once a month.		Second Wednesday	
	2d. Have a different committee member bring and present the committee report to the Executive Committee and chapter meeting each month.		Each month	

Figure 14–13 The FFA Program of Activities has 12 divisions. The division shown here is leadership. By using this format, you can identify chapter goals, the ways and means needed to achieve them, and the budget necessary to accomplish the goal. Note the spaces left for projected completion date and evaluation comments. (Courtesy of the National FFA Organization.)

Program of Life Goals Plan of Action

AREA: Career

ACTIVITY: Obtain work experience to help enhance employment opportunity in an agribusiness after college.

Specific Goal	Ways and Means (Prioritize Steps)	Obstacles	Support System	Deadline	Accomplishment/ Evaluation
Secure a part-time job with Tractor Supply Company (TSC) while in high school	Prepare myself in school Dress appropriately Secure an application Request an interview Type and return application Get the job Work hard and practice good employability skills Learn all I can about the business	Scheduling hours to continue school extracurricular activities	Parents Agriculture Teacher Friends	January of my Junior year in high school	
Secure full-time summer employment while in college and continue part-time during school year	Be an excellent part-time worker Assume more responsibilities as experience is gained	Maintain good grades while working during school year	TSC manager College advisor Friends Parents	Fall of my freshman year in college	

Figure 14–14 This is an excellent format adapted from the FFA Program of Activities to develop a plan of action for your program of life goals. Career is just one of eight areas for planning life goals.

Summary

The ability to set goals and make plans is a skill that we should all develop. If we are to reach our full potential, we must learn to set goals.

There are several reasons for having goals. Goals provide a target, a destination, purpose, and meaning. Goals help focus our energies toward productivity and accomplishment. They help individuals succeed, develop self-control, and evaluate their accomplishments.

Goals let you know whether or not you are on target. Goals provide a benefit to you by helping you concentrate your efforts, helping you make the most of your time, letting other people know how to help, keeping your enthusiasm, and monitoring your progress. Goals make mental laws work for you. These are the Laws of Control, Cause and Effect, Belief, Expectations, Attraction, Correspondence, Subconscious, Concentration, and Substitution. Goals also help you tap your resources.

People who set goals are serious about being successful. They have accepted responsibility for their lives. Some people don't realize the importance of goals. They may not know how to set goals, be afraid of rejection, or be afraid of failure.

There are several things to keep in mind as goals are being set. Goals should be clear, challenging, obtainable, and realistic. Write down your goals, be specific, and prioritize them. Then, plan and pursue your goals.

Some specific principles of setting goals include considering the criteria and benefits, setting goals that are in harmony with your values and talent area, setting goals that have balance, determining your major purpose in life, setting specific goals with a deadline, reviewing goals often, taking advantage of your momentum, setting goals that are high, and working toward your goals daily. There are concerns with goal setting, such as realizing that you control your destiny, setting goals that are too high, and setting goals outside your talent area just to please someone else.

There are 14 steps in setting goals: Develop desire, develop belief, identify your goals and write them down, identify benefits, be specific, be realistic, analyze where you begin, set a deadline, consider the roadblocks, identify skills or knowledge required, develop a support system, develop a step-by-step prioritized plan, visualize your goals, and begin and never quit.

There are several types of goals. Questions to help you identify your goals include asking what your values are, what you want right now, what you would do if you had only six months to live, what you would do if you had a million dollars, what you always wanted to do but were afraid to attempt, what gives you the greatest feeling of self-esteem and satisfaction, and what you would really want to do if you knew that you wouldn't fail.

Categories of goals include family, physical, mental, social, personal values, financial, school and career. There can be several kinds of goals within each category. You should have immediate, medium-range, and long-range goals.

The FFA uses the Program of Activities to accomplish the goals of the FFA Chapter. The Program of Activities format is an excellent format to adapt for setting your own goals in life, called the Personal Program of Life. The Program of Life Plan of Action includes area, activity, specific goal, ways and means (prioritized steps), obstacles, support system, deadline, and accomplishments or evaluation. By using the Program of Activities format to set personal goals, you use a proven model.

End-of-Chapter Exercises

Review Questions

1. What are six reasons for having goals?
2. Compare the physical goals in basketball, football, and hockey with the mental goals of life.
3. What are five benefits of goals according to John R. Noe?
4. What are five reasons people don't set goals?
5. List five key words to remember when setting goals.
6. Name eight principles to consider when setting goals.
7. Give three cases when goal setting can be negative.
8. List 19 steps to use in setting goals.
9. What are seven questions to help you select your goals?
10. What document does the FFA organization use to accomplish goals of the chapter?

Fill-in-the-Blank

1. People don't plan to fail, they simply fail to ________________.
2. It is just as difficult to reach a destination you don't have as it is to come back from a place you have ________________ ________________ .
3. It is impossible to succeed without ________________ .
4. Goals written down become a written ________________ .
5. SMART is an acronym for S________________ , M________________ , A________________ , R________________ , T________________ .
6. Theodore Roosevelt said, "Do what you can, with what you ________________ , right where you ________________ ."
7. Big goals are set by ________________ people.
8. "Make no ________________ plans for they have not capacity to stir men's souls."
9. If you believe it, you can ________________ it.
10. If your goal is worth achieving, it is worth working for ________________ and ________________ .
11. There are few unrealistic goals, only unrealistic ________________ .
12. What you get by reaching your goals is not nearly as important as what you ________________ by reaching them.

Matching

_____ 1. Feeling positive about yourself to the degree that you determine your own destiny.
_____ 2. For every effect in your life there is a specific cause.
_____ 3. What you dwell on grows.
_____ 4. When you think continually about your goals, people with similar goals are drawn to you.
_____ 5. Let a positive thought take the place of a negative thought.
_____ 6. Anticipating that everything that will happen is moving you toward your goals.
_____ 7. Your mind (inner world) communicates to your outer world (things you actually do).
_____ 8. If you think about something with your conscious mind, you will do things without ever thinking about it.
_____ 9. Self-confidence and faith that you will achieve your goals by taking appropriate actions.

A. Law of Belief
B. Law of Substitution
C. Law of Correspondence
D. Law of Control
E. Law of Subconscious
F. Law of Concentration
G. Law of Expectation
H. Law of Attention
I. Law of Cause and Effect

Activities

1. Several benefits of goals were given in the chapter. Select the one that would most apply to you and write a paragraph or short essay supporting your reasoning.
2. Mental laws were briefly discussed in this chapter as they relate to goals. Select and write a paragraph or short essay on one of the mental laws that causes you to think about yourself and how you react in certain goal-setting situations.
3. Read the section on "why people don't set goals." Select and write a paragraph or short essay about the one of which you are most guilty.
4. Write the names of five people (friends, parents, teacher, boss, etc.) who would wholeheartedly support you if you shared your goals with them.
5. Brian Tracy suggests that you answer seven questions to help you come up with your goals. Answer each of the following questions.
   a. What are your five most important values in life?
   b. What are your three most important goals in life right now?
   c. What would you do, how would you spend your time, if you learned that you had only six months to live?

d. What would you do if you won a million dollars cash, tax free, in the lottery tomorrow?
e. What have you always wanted to do but been afraid to attempt?
f. What do you most enjoy doing? What gives you the greatest feeling of self-esteem and personal satisfaction?
g. What one great thing would you dare to dream and do if you knew you could not fail?

6. Complete the "Goal-Setting Chart" on the next page by writing goals within each of the eight areas. Don't feel you need to write five for each. Add more if you wish.
7. Select one of the goals from the "Goal Setting Chart" in activity 6 from one of the areas and break it down into long-range, short-range, and immediate goals using the format provided.
8. Develop a "Program of Life Goals Plan of Action." Study Figures 14–12 and 14–14. How do they compare to the FFA Program of Activities format in Figures 14–11 and 14–13? Complete the provided "Program of Life Goals Plan of Action" on at least one goal for each of the areas. (Note: Your teacher may want you to do a plan of action for each of your goals.)
9. **Take it to the Net.**

   Explore goal setting on the Internet. Listed below are several Web sites that contain information on goal setting. Browse all the sites and pick the one you feel is most interesting. Print the first page of the site. Summarize the site and tell why you feel it is interesting. If you are having problems with the Web sites, use the search terms listed below.

   Websites

   <http://www.topachievement.com/goalsetting.html>

   <http://www.lifedesigncenter.com/Thoughts/settinggoals.htm>

   <http://www.mindtools.com/page6.html>

   Search Terms

   setting goals

   goal setting

   self-improvement

   personal goals

Goal Setting Chart

On a separate sheet of paper, write three to five goals for each of the categories.

Physical 1. 2. 3. 4. 5.	*10 ____ 9 ____ 8 ____ 7 ____ 6 ____ 5 ____ 4 ____ 3 ____ 2 ____ 1 ____	**Financial** 1. 2. 3. 4. 5.	*10 ____ 9 ____ 8 ____ 7 ____ 6 ____ 5 ____ 4 ____ 3 ____ 2 ____ 1 ____
Personal Values 1. 2. 3. 4. 5.	*10 ____ 9 ____ 8 ____ 7 ____ 6 ____ 5 ____ 4 ____ 3 ____ 2 ____ 1 ____	**Social** 1. 2. 3. 4. 5.	*10 ____ 9 ____ 8 ____ 7 ____ 6 ____ 5 ____ 4 ____ 3 ____ 2 ____ 1 ____
Mental 1. 2. 3. 4. 5.	*10 ____ 9 ____ 8 ____ 7 ____ 6 ____ 5 ____ 4 ____ 3 ____ 2 ____ 1 ____	**Family** 1. 2. 3. 4. 5.	*10 ____ 9 ____ 8 ____ 7 ____ 6 ____ 5 ____ 4 ____ 3 ____ 2 ____ 1 ____
School 1. 2. 3. 4. 5.	*10 ____ 9 ____ 8 ____ 7 ____ 6 ____ 5 ____ 4 ____ 3 ____ 2 ____ 1 ____	**Career** 1. 2. 3. 4. 5.	*10 ____ 9 ____ 8 ____ 7 ____ 6 ____ 5 ____ 4 ____ 3 ____ 2 ____ 1 ____

*The numbers 1 to 10 in each category represent an "area" for achievement in your life. Rate your proficiency by placing an "X" next to the number that best states where you are today (1 is poor and 10 is excellent).

Setting Long-Range, Short-Range and Immediate Goals

1. On a separate piece of paper, write one Long-range goal. It needs to be something that will take you more than a year to do. (For example, become a state FFA Officer in two years.)

2. Name one short-range goal. It needs to be something you can do in a few weeks or a few months (less than a year). (For example, make the parliamentary procedure team.)

3. Write down two immediate goals. These are things you can do in a day. (For example, make a good grade on my agricultural education test tomorrow or attend the state fair and see all the agriculture exhibits.)

Program of Life Goals Plan of Action

Area

Activity

Specific Goal	Ways and Means (Prioritize Steps)	Obstacles	Support System	Deadline	Accomplishments/ Evaluation

Notes

1. B. Tracy, *Maximum Achievement* (New York: Simon & Schuster, 1993), p. 140.
2. *Personal Skill Development in Agriculture* (College Station, TX: Instructional Material Service, Texas A & M University, 1989), 8742-E, p. 1.
3. J. R. Noe, *People Power* (Nashville, TN: Thomas Nelson Books, 1984), p. 145.
4. Ibid., p. 146.
5. Ibid.
6. Ibid., p. 147.
7. Tracy, *Maximum Achievement*, p. 142.
8. Ibid., p. 142.
9. Ibid., p. 143.
10. Ibid., p. 144.
11. Ibid., p. 148.
12. S. R. Levine and M. A. Crom, *The Leader in You* (New York: Simon & Schuster, 1993), p. 158.
13. Ibid.
14. Appreciation is extended to Dr. Joe Townsend, Agricultural Education, Texas A & M University for permission to use notes from his Leadership Class at Texas A & M University. Some of the same material can be found in *Personal Skill Development in Agriculture*. Dr. Richard J. Carter, Iowa State University, also had input on the notes originally.
15. Ibid.
16. *Personal Skill Development in Agriculture*, 8742-E, p. 2.
17. Tracy, *Maximum Achievement*, p. 157.
18. Ibid., p. 159.
19. Quote found in Tracy, *Maximum Achievement*, p. 150.
20. Ibid., p. 151.
21. Noe, *People Power*, p. 151.
22. Tracy, *Maximum Achievement*, p. 161.
23. Ibid., p. 167.
24. Noe, *People Power*, p. 152.
25. Ibid.
26. Ibid., p. 153.
27. Ibid., p. 154.
28. Z. Ziglar, *The "I Can" Course Learner's Manual* (Carrollton, TX: The Zig Ziglar Corporation, 1989), p. 176.
29. Ibid., p. 178.
30. Tracy, *Maximum Achievement*, p. 164.
31. "Goal Setting: Self-Esteem in Action" (Lubbock, TX: Creative Educational Video, 1989).
32. Tracy, *Maximum Achievement*, p. 155.
33. Ziglar, *The "I Can" Course Learner's Manual*, p. 190.
34. *Personal Skill Development in Agriculture*, 8742-E, p. 2.
35. *Goals* (Shawnee Mission, KS: Pryor Resources Inc., 1990), p. 10.
36. Ibid., p. 13.
37. *Goals* (Pryor Resources Inc.), p. 6.
38. *Chapter Planning and Recognition: A Student Handbook* (Indianapolis, IN: The National FFA Organization, 2001).
39. Ibid.
40. Ibid.

CHAPTER

15

Time Management

Objectives

After completing this chapter, you should be able to:

- Discuss the four levels of time management
- Differentiate between urgent and important
- Explain ultimate time management for quality leaders
- Set and prioritize personal and professional goals
- Discuss the importance of balancing the major categories of personal and professional time
- Analyze your use of time with a daily log
- Describe time wasters
- Manage time more effectively and efficiently

Terms to Know

time management
prioritize
urgent
important
matrix
coherence
portability
personal time
professional time
socializing
daily log
Pareto's principle
procrastination
qualifier
prime time
accountability

Did you ever get to the end of a busy day, when you were tired and looking forward to going to bed, and realize you had a five-page assignment due for history first thing in the morning? Where did the time go? Looking back over your day you think, "There just wasn't enough time to get everything done!" You might know someone who seems to be involved in everything—student council, basketball team, band, academic Olympics, a local youth group—and yet is on the honor roll every time report cards come out! How do those people do it?

People who seem to have an endless supply of energy don't have any more time to get things done than you do; they have just learned to become good managers of their time. **Time management** is planning how to control your time in order to do the things you need and want to do. Good managers of time have the same 24 hours per day as everyone else. The only difference is they've taken the time to decide how they would control what they do in that 24-hour period. Time is your most valuable personal resource; use it wisely because it cannot be replaced.

Four Levels of Time Management

People are always looking for ways to do things more efficiently. Time management is no exception. Each generation, or level, of time management builds on the one before it and helps us gain greater control of our lives. Each level is important, but the more effective leaders operate at the fourth level. The four levels or generations are as follows.[1]

Notes and checklists are the first level of time management. Keeping track of what you have to do can help you meet and manage the many demands placed on your time and energy.

Calendar and appointment books, the second level, help you schedule events and activities in the future.

Third-generation time management adds to the preceding levels the following important ideas: **prioritization**, clarifying values, comparing the worth of various activities based on your values, and setting goals (long-term, intermediate, short-term). This level also includes daily planning, or making a plan to accomplish specific goals and objectives on a daily basis.

Managing ourselves is the fourth generation of time management. The challenge is not to manage time but to manage ourselves. Rather than focusing on things and time, fourth-level expectations focus on improving relationships, as shown in Figure 15–1, and

Figure 15–1 Fourth generation time management focuses on preserving and enhancing relationships. As demonstrated above, it includes meeting human needs as well as personal and professional goals and objectives.

getting results. These results include meeting human needs as well as personal and professional goals and objectives.

URGENT VERSUS IMPORTANT

Urgent

Two words that can be used to describe an activity are **urgent** and **important**.[2] *Urgent* means requiring immediate attention, such as answering the telephone. Most people can't stand the thought of allowing a phone to ring. Urgent matters are usually visible: They insist on action, they're usually right in front of us, and often they are pleasant and easy to do. But frequently they are unimportant. For example, three phone lines are ringing, demanding immediate (urgent) attention, but one is for a lunch appointment with a friend, the second is to schedule a golf game, and the third is a call asking the time of the charity softball game next month. Urgent matters don't help us reach our goals and are usually time wasters.

Important

Important matters have to do with results. If something is important, it contributes to reaching your goals. We react to urgent matters, but important matters that are not urgent require more initiative and planning.[3] We must act to seize opportunity and make things happen. If we don't have a clear idea of what is important, of the results we desire in our lives, we are easily diverted into responding to the urgent. An *urgent* and *important* matter would be completing a scholarship application in time to meet a deadline. An example of an important matter is daily exercise, such as walking two miles. It is not *urgent* since no one is in a crisis, there is no pressing problem or deadline to be met. However, it is *important* if our long-range goal of maintaining good health is to be met.

Time-Management Matrix

Stephen Covey in his book, *The 7 Habits of Highly Effective People*, presents the time-management box, or **matrix** (Figure 15–2).

Time-Management Matrix

	Urgent	**Not Urgent**
Important	1. Get it done—no debate • Crises • Pressing problems • Deadline-driven projects	2. Relevant in the long run • Relationship building • Recognizing new opportunities • Planning, recreation
Not Important	3. Nagging time wasters that require attention • Interruptions, some calls, mail, reports, meetings, and e-mail	4. Pleasant busywork to stay organized • Trivia, busywork • Some mail, some e-mail • Some phone calls, some answering machine messages

Figure 15–2 The four areas of the time-management matrix. Area 1 activities are important and urgent, such as meeting deadlines. Area 2 activities are important but not urgent, such as exercising for health reasons.

Area 1: Get It Done, No Debate These matters are both urgent and important, being significant problems that require immediate attention (crises). Leaders who operate in area 1 are crisis managers, problem-minded people, and deadline-driven producers. Leaders who focus on area 1 allow it to get bigger and bigger until it dominates them. Leading out of urgency is not good—but often unavoidable. People who operate in area 1 too much need to plan so that they manage their time instead of their time managing them.

Area 2: Relevant in the Long Run This is the area from which real leaders operate. It is the heart of effective personal management. "It deals with things that are not urgent, but are important. It deals with things like building relationships, long-range planning, exercising, preventive maintenance, preparation, and all those things we know we need to do, but somehow seldom get around to doing, because they are not urgent."[4] Effective leaders who operate in area 2 are not problem minded but opportunity minded. They feed opportunities and starve problems. They think preventively. The result is vision, perspective, balance, discipline, control, and few crises.

Area 3: Nagging Time Wasters That Require Attention Some leaders spend a great deal of time in urgent, but not important area 3 activities thinking they are in area 1. They spend most of their time reacting to things that are urgent, assuming that they are also important. The reality is that the urgency of these matters is often based on the priorities and expectations of others. Some call this "majoring in the minors." Examples of area 3 activities are any interruptions, phone calls, mail, reports, and meetings that do not help accomplish any of your long-term goals.

Area 4: Pleasant Busywork to Stay Organized Area 4 provides a place of escape. Some leaders are literally beaten up by problems all day every day. The only relief for these leaders is to escape to the unimportant, not urgent activities of area 4. These include trivia, busywork, and such pleasant activities as reading through brochures or intriguing junk mail. Although leaders should avoid this area as much as possible, there are times when you need to unwind to release stress, especially between major projects.

Reality versus Ideal

Most leaders should spend most of their time in area 2, on important matters that will lead to professional and personal growth. In fact, most leaders spend most of their time in area 1, dealing with crises and deadlines, considerable time in area 3 with nagging time wasters that require attention, and the rest of their time in area 4, regrouping doing trivial tasks. No matter how hard you try, it is probably impossible to keep from spending some time in each area. However, leaders should strive to increase their time in area 2 to be more productive.

Make a Positive Difference in Your Life

Consider this question in relation to the time-management matrix. What one thing could you do (that you aren't doing now) that if done on a regular basis, would make a tremendous positive difference in your personal or professional life? More than likely your answer would fit into area 2—something that is obviously important but is not urgent. Because it is not urgent, it often does not get done.

Whether you are a high school student, a university student, an assembly-line worker, or a professional, the same principles apply. Those who operate in area 2 increase their effectiveness dramatically. Crises and problems shrink to manageable proportions because you are thinking ahead, planning, and taking preventive action to keep situations from developing into crises in the first place. You take control of your time instead of letting your time take control of you.

Ultimate Effective Time Management for Leaders

Leaders put first things first, then they proceed to organize and execute around those things. The best leader and time manager (operating in area 2) needs to meet six important criteria, according to Covey.

- **Coherence** There is harmony between your roles and categories, goals and objectives, things to do, priorities, and scheduling.
- **Balance** Keep balance in your life so that you do not neglect the important areas of health, family, school or professional preparation, and community.
- **Focus** Stay focused on important issues so that you are preventing rather than prioritizing crises.
- **People centered** A good leader deals with people effectively and humanely.
- **Flexibility** Be flexible with your schedule to allow for the unexpected.
- **Portability** Carry a portable weekly organizer.

Goals and the Four Basic Questions

- Goals develop **knowledge**, which answers the question ***what***?
- Purpose develops **understanding**, which answers the question ***why***?
- Planning and strategy develop **wisdom**, which answers the question ***how***?
- Priorities develop **timing**, which answers the question ***when***?

Figure 15–3 Goals are the beginning of your destination. Along the way, you must gain knowledge, plan, and determine your priorities, which means managing time efficiently.

Setting and Prioritizing Personal and Professional Goals

We discussed goals in the previous chapter, but it is hard not to mention goals again in relation to time management. We cannot manage our time correctly and efficiently if we do not know what our goals are.

Goals Provide Direction for Time Management

What do you want to do with your life in the short range—the next 24 hours, the next week, the next month? How about the long range—a year, five years, or 10 years from now? The words *goal* and *objective* are often used interchangeably, but a goal is an end that you strive to attain or reach in the long term. An objective is a specific end that can be reached in the short term. For instance, you will have to maintain certain grades in particular courses to receive a high school diploma. You must work to achieve grades to reach the objective of obtaining a high school diploma. Goals are necessary to give direction to your life; without them it would be difficult, if not impossible, to manage your time. Helen Keller was asked what could possibly be worse than being blind. She replied, "To have sight with no vision." Consider the statements on goals in Figure 15–3.

Periodically Adjust Goals and Time Commitments

Keep in mind that in order to manage your time you will have to update your goals from time to time. Former long-range goals become short-range goals, and short-range goals are reached and therefore are no longer goals because they have been attained. People also change their minds. In high school a student may wish to go on to college and become a veterinarian. But after attending college for a year, she might get interested in another field, such as biotechnology. It is okay to change your mind, as long as you realize that it means setting new goals and changing how you spend and manage your time. A new goal may mean spending additional time on education or preparation for a career.

Balancing Personal and Professional Time

Many of the categories of **personal time** and **professional time** overlap. They are all parts of the whole and are interdependent. To be the best time manager possible you must take care of your whole self. In other words, time management does not apply only to our professional lives, it applies to our personal lives as well. We must manage our time in all these areas to meet our goals in life. Time spent on various activities can be divided into four major categories: health, family, community (personal time), and school or employment (professional time). Each category is important. If any one of them is neglected, an adverse impact can be observed in the other categories.

Health

Taking care of yourself takes up much time, but it is time well spent. No health, no work, no life. You will get more done if your body is healthy, and a healthy body improves your mental, emotional, and spiritual well-being. The way you take care of your health can add years to your life or take them away.

Sleep Most people need a good eight hours of sleep in every 24-hour period in order to function at their best. This amount of time can vary from person to person and might change as you get older. As you lose sleep, your immune system does not have time to recharge, which makes you more prone to infection.

Diet People who eat a well-balanced diet and maintain their proper weight experience fewer energy "burnouts." When you eat better, you feel better, work better, and are better able to achieve your time-management objectives. "Garbage in = garbage out" is a fit adage. Substances such as caffeine can adversely affect your performance; it is a quick stimulant that can eventually act as a depressant. It is wise not to overindulge in any one type of food but instead to eat a balanced diet (Figure 15–4).[5]

Exercise The American Medical Association has said that moderate exercise is all that is needed to dramatically reduce death due to heart disease, cancer, and other diseases. One way to get it is 30 minutes of brisk walking, or some other aerobic exercise, at least every other day.[6] Adequate exercise, a balanced diet, and enough sleep will give you the energy and stamina needed to finish what you start and accomplish your daily goals.

Recreation Good health does not just pertain to the physical; it also includes your mental, emotional, and spiritual self. You need to set aside time for recreation not only for the body but for the mind. Sports, games, reading a good book, watching a play, listening to music, or engaging in the arts, special interests, and hobbies all play an important part in achieving good health. A relaxed body leads to a relaxed mind, which can make your use of time more productive.

Socializing Getting together with friends and others with common interests is important for developing interpersonal skills. These skills can help you as you develop a career and new interests.

Spiritual Needs Many researchers have found that people who spend a certain amount of time on spiritual needs tend to have long, healthy, and happy lives. Morals, ethics, and positive lifestyles, often influenced by various faiths, are important

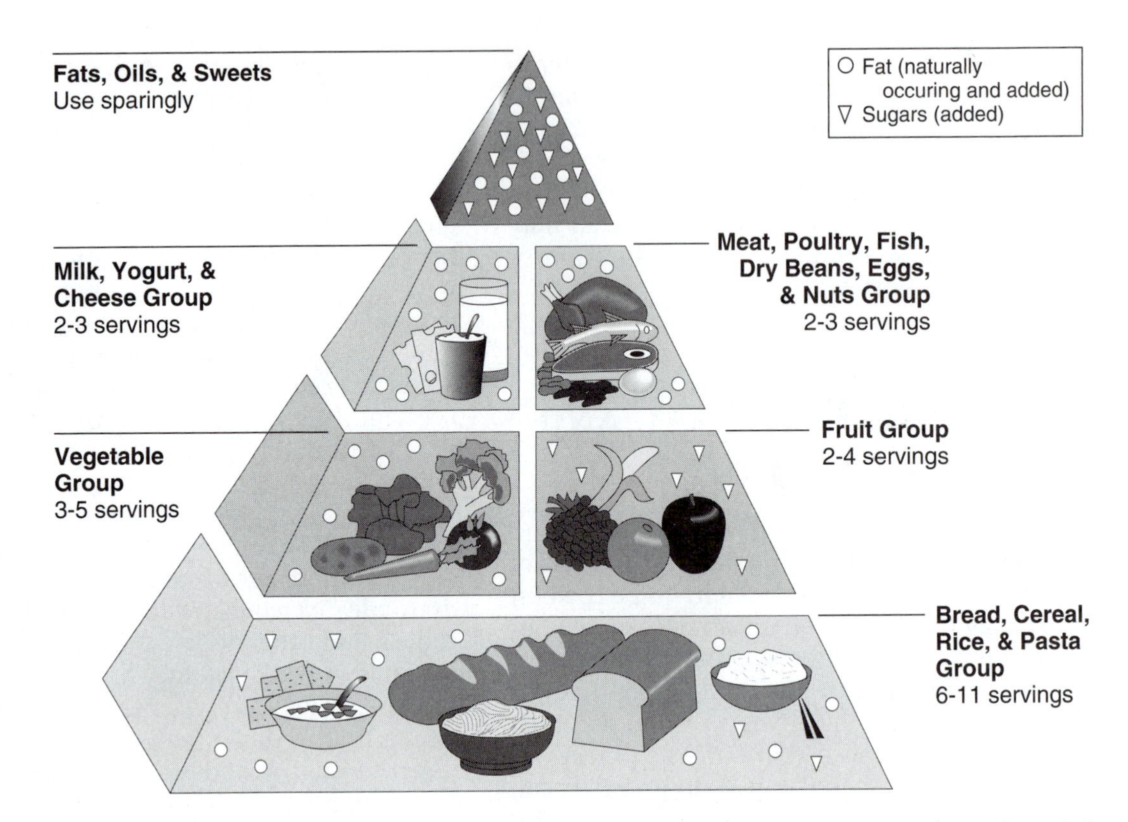

Figure 15–4 Good diet gives you more energy to meet your time-management objectives. This food pyramid can help direct your diet. (Courtesy of USDA.)

for happy, well-rounded perspectives on life. The amount of time spent on spiritual needs depends on the needs of the individual person.

Family

Time spent with your family is also important. A doctor once said, "I have never heard a dying man say that he wished he had spent more time at the office." For most of us, when all is said and done, families are the most important thing. Families instill a sense of belonging and have always been an essential part of human survival. Older people teach younger members of a family how to take care of themselves and how to be part of a strong group.

Responsibility and cooperation are two important skills that can be developed in the family setting (Figure 15–5). Many families divide up the chores of cooking, cleaning, and home maintenance. These are skills that high school youth need for independence when they leave their home to start a life of their own as adults.

Community

We have obligations as community members. It is part of the American dream to make our communities better places to live. Well-rounded individuals participate in groups and activities whose goals include community improvement. This may include being a part of a neighborhood watch, helping to establish and maintain parks and libraries, or helping with youth groups. Charitable work for nonprofit organizations is another means of bettering your community. What if no one budgeted

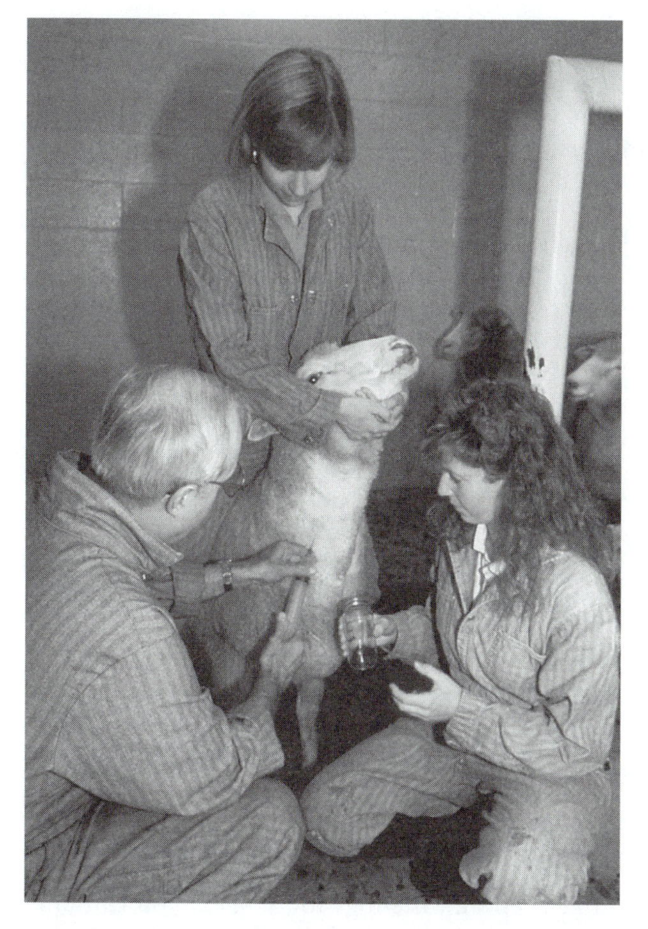

Figure 15–5 This farm family is working together. Families are closer when they have similar goals and objectives. (Courtesy of USDA.)

time in this area? The man in Figure 15–6 is contributing his time to a community project.

Employment

Employment is also an important time category. Aside from filling our need for income, employment can provide for personal growth and development. The luckiest people are those who have a job doing something they enjoy. It is best to choose a career that you enjoy and find interesting. Much of the helpful time-management advice in this chapter centers around the workplace.

Figure 15–6 Being a contributing member to your community should be a goal for all of us. Our communities will be better places to live if we all contribute.

ANALYZING YOUR USE OF TIME WITH A DAILY LOG

Before you decide to improve your use of time, you may want to find out where your time is going. Seeing how you have used your time in the past helps to make planning your future use of time easier.

Keep a Daily Log

If you keep a **daily log**, or diary of everything you do for at least one week, you can then analyze how you have been mismanaging your time and make the necessary changes. Take this log with you everywhere you go for one week. Mark down in 15-minute intervals how you spend your time. Do not wait until the end of the day to record in this log—record activities as you do them. For example, "6 AM, woke up, took shower, and dressed; 6:15, ate breakfast; 6:30, caught school bus; 6:45, at school, studied for test first period." Continue making short, brief notes every 15 minutes until you go to sleep that night. Continue this daily log for at least one week. You may want to use a log sheet similar to the one in Figure 15–7.

Daily Log Sheet

Day ____________	Date ____________
6:00	2:00
6:15	2:15
6:30	2:30
6:45	2:45
7:00	3:00
7:15	3:15
7:30	3:30
7:45	3:45
8:00	4:00
8:15	4:15
8:30	4:30
8:45	4:45
9:00	5:00
9:15	5:15
9:30	5:30
9:45	5:45
10:00	6:00
10:15	6:15
10:30	6:30
10:45	6:45
11:00	7:00
11:15	7:15
11:30	7:30
11:45	7:45
12:00	8:00
12:15	8:15
12:30	8:30
12:45	8:45
1:00	9:00
1:15	9:15
1:30	9:30
1:45	9:45
	10:00

Figure 15–7 Completing a daily log such as this one will help you analyze how you are spending your time.

Pareto's Principle

You may be surprised to find out where all your time went! You might have had the best of intentions to get an outline done for a research paper one night only to find out 10 PM had come and gone and you had been on the phone since 7 PM. Watch out for time wasters—they can keep you from reaching your goals. You will probably find that 80 percent of the things you do are low-priority tasks. The "80-20" rule, or **Pareto's principle**, can be applied to many different situations in your life. The 20 percent of tasks that are high priority can generate 80 percent of the results you need to reach your goals.[7] Pareto's principle is also called the time-value ratio. We need to put first things first. Is what I am doing worth the time it is taking to do it? You increase the return on your investment of time and energy if you work on the important things. Something is important when it leads to the achievement of your goals. We have to be efficient (doing the task the right way) and effective (doing the right task). Figure 15–8 is a chart illustrating Pareto's principle.

Analyzing Your Time Logs

After keeping time logs for a week, you can analyze them by doing the following.

- Review the time logs to determine how much time you are spending on your primary responsibilities. How do you spend most of your time?
- Identify areas in which you are spending too much time.
- Identify areas in which you are not spending enough time.
- Identify major interruptions that keep you from doing what you want to get done. How can you eliminate them?
- Identify tasks you are performing that you do not have to be involved with. If you are a manager, look for nonmanagement tasks. To whom can you delegate these tasks?

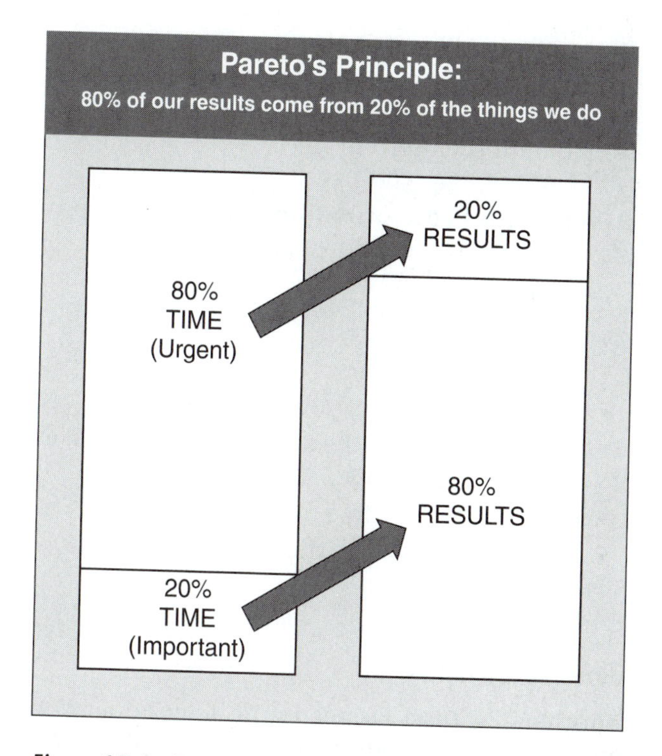

Figure 15–8 Dwight D. Eisenhower said, "Urgent things are seldom important." Urgent things do not lead to achieving your goals. Something is important when it leads to achievement of your goals. Twenty percent of your high-priority tasks can generate 80% of the results you need to reach your goals.

- How much time is controlled by others? How much time do you actually control? How can you gain more control of your own time?
- Look for crisis situations. Were they caused by something you did or did not do? Do you have recurring crises? How can you plan to eliminate recurring crises?
- Look for habits, patterns, and tendencies. Do they help or hinder getting the job done? How can you change them to your advantage?
- List your three to five biggest time wasters. What can you do to eliminate them?
- Determine how you can manage your time more efficiently.[8]

You can then use this daily log to help you better plan your weekly and daily activities (discussed later in the chapter). First, let us examine some of the time wasters that may hinder you.

Time Wasters

Procrastination

It is one thing to identify a time waster and an entirely different matter to get rid of it. One of the worst and most widespread is **procrastination**. Procrastination can mean doing a low-priority and often enjoyable task instead of a high-priority and difficult or boring task. Why do people procrastinate? If you're avoiding doing something you do not need to do, no problem. But if the task you are avoiding is one you need to complete in order to reach an important and necessary goal, procrastination can really hold you back.

Lack of Understanding and Direction When I was a child working on the farm, my father told me, "The hardest part of any job is getting started." Once you can get beyond procrastination, you have learned much. Many people procrastinate because they don't understand what it is they are supposed to do. These seemingly undesirable tasks are often imposed on the procrastinator by an outside party. A common example for students is facing a major research paper for English, history, science, or a foreign language. Often these are long-term projects that require many hours of work over several weeks To make this task less intimidating, the procrastinator needs to get specific directions from the instructor on how to complete the task. A large project such as this can then be broken down into smaller tasks that can each be completed in one sitting or work session. These smaller tasks might include selecting a topic, doing library research, compiling a bibliography, drafting an introduction and a conclusion, outlining the research paper, writing paragraphs to go under the various subheadings in the outline, and writing footnotes or endnotes.

Identify Short-Term Rewards for Accomplishing Tasks Another aid in eliminating procrastination is to sit down and list the rewards for finishing each stage of the research paper on time and the consequences for not finishing or even doing each stage on time. B.F. Skinner, a famous psychologist, found out through original laboratory research that positive reinforcement or rewards for desirable behavior (finishing each stage of the paper on time) worked better than suffering punishment or the consequences for undesirable behavior (not finishing each stage of the research paper on time). Promising yourself that you can watch a half-hour situation comedy program on TV if you first complete all of the library research for your paper is much more likely to change your behavior in a positive manner than not eating dessert that night because you did not do your homework.

Identify Long-Term Rewards for Accomplishing Tasks If you do well on your research paper, you will receive a good grade on the assignment, which will help you to get a good grade in the course. If you learn to make good grades in one class, that could lead to making good grades in other subjects. If you graduate with a high overall grade point average, you are more likely to get scholarships to attend the college of your choice. If you attend college and do well in your studies, this could help you get started in a career you enjoy and make a desirable salary. On the other hand, if you do not learn how to complete papers well and on time, the ultimate consequence may be that you do not get the job you wanted.

Characteristics of a Procrastinator A procrastinator will always have more time tomorrow. When faced with a difficult task, procrastinators wait for a better time. They do not start a project unless they can finish it. When procrastinators are facing an important but difficult project, they spend their time doing other activities rather than getting started. Procrastinators tend to be perfectionists. They also have trouble with priorities and lack self-confidence; they don't plan; and they overcommit themselves.

Overcoming Procrastination To overcome procrastination, do the following.[9]

- Do it now.
- Take small steps.
- Do the easy parts first.
- Tell someone else of your commitment.
- If possible, reduce the amount of time it takes to accomplish a task.
- Plan work at the time of least fatigue.
- Decide when you are the most productive.
- Reward yourself.

The Telephone

Another big time waster is the telephone. You might say, "But I didn't call him, he called me!" Make it known to family and friends that you must spend so many hours a week doing homework and taking care of other responsibilities. You might want to set aside a specific time each day for study and ask people ahead of time not to interrupt you. With tact, this can be handled appropriately, and your request will be respected.

Terminating Conversations When you have more important things you need to be doing, there are several things you can say in a polite manner to terminate phone conversations. Refer to Figure 15–9 for help on terminating conversations.

Another way to minimize your time on the phone is to stand while you are talking; this helps you be aware of how long you have been talking so you can keep the call brief. You can use a clock, the timer on the microwave oven, or any other timer, set for no more than 15 minutes. When you hear the timer go off, it is time to say goodbye. If you cannot afford the time even to tell someone you will call back later, use an answering machine. Listen in on the call in case it is an

Polite Conversation Terminators

"I can't talk right now . . ."

"Can I get back to you?"

"I have company right now, so I need to call you back later."

"I have to go now . . ."

"I can't talk much longer . . ."

"I was just on my way out the door . . ."

"I know you're busy, so I'll let you get back to what you were doing."

"Well, I'm going to have to let you go . . ."

"Right now is my study time, but I can call you back later."

Figure 15–9 Sometimes we just do not have time to talk on the telephone, but we do not want to be rude. The tips above should help you end telephone conversations in a tactful way.

emergency; if not, do not pick up. You can return the call later.

Positive Tool The telephone can also be used as a positive tool. You might not be able to remember how to complete a certain math problem, and you need to call a friend for help. In this way the telephone can be a time saver. Instead of going to your friend's house, a quick phone call saves you time. It has been suggested that some business people not have a ringer on their phone. The phone is for their convenience for outgoing calls—it is not an instrument by which to be disturbed. The telephone is a tool for your use. Run your telephone, do not let the telephone run you. Consider the following.

A Manager's Guide for Handling Incoming Calls

Incoming calls

- Log your calls according to their priority.
- Redirect calls to others better suited to handle them. "Coach" callers to do it themselves.
- Automate a callback system.
- "Mr./Mrs. Doe is not available right now, may I help you?"
- Gather name, number, best and worst times to return the call, and the nature of the call.

Outgoing calls

- Have the number at hand, or consider a quick-card phone number retrieval device. Post a list next to the telephone.
- For group or conference calls, good times are usually after lunch, midmorning, or midafternoon. This obviously depends on your work (e.g., sales).
- Have materials at hand, with the agenda for conversation outlined.
- Ask "Are you free to talk?" Beware of the presumption of legitimacy; remember the power of public relations, and diplomacy.

Alternative communication systems

- Answering machine
- Phone message center
- Cell phone, beeper device, or answering service[10]

Television

Television is another thing that can help or harm you. If you have attained all your higher-priority goals for the day, you might watch TV as a form of recreation or socialization. Television can be entertaining and educational if used wisely. It can also turn you into a "couch potato" who is risking poor health by not exercising and a numb mind by not expanding your horizon beyond what the advertisers and TV moguls want you to know. Some studies report teenagers watching TV 42 hours per week. If you are spending several hours a day watching TV, you may need to reassess how your time is spent and get back to a schedule that is balanced between the five major categories of personal and professional time.

Inability to Say No

If you think you can't get everything done that you are supposed to because you have too much to do, that is probably true. Good time managers keep a pocket calendar planner with them and write down all their obligations. If you have got too many things to do, then you need to back up and regroup. Reprioritize your goals. Do what is most important to you; do what you need to achieve to reach *your* goals, not someone else's. You also have to realize that there are only so many hours in a day; there will always be times when someone wants you to do something and you will have to tell them no. When you say no, say it early, be willing to take a risk, and understand the benefits of saying no.

Say No in a Positive Way Saying no can be done in a positive and nonthreatening way that prevents hurt feelings and angry misunderstandings. You need to be honest when people ask you to help them and you just do not have the time to do it. For instance, you have been asked to work on a committee to decorate for an upcoming school party, but you have already obligated yourself to fund-raising for the school party and practicing for an upcoming debate contest. You can say no politely with, "I'm flattered you want me to help you, but I'm already committed to doing two other things." Most people will understand and respect your honesty and integrity.

Use Qualifiers You could also tell them no with a **qualifier** such as, "No, I can't help you this time with the fall dance, but I can help you later with the spring dance." You don't want to offend anyone and burn your bridges behind you. Learning how to deal with situations like this in high school can help you deal with similar situations when you start working. There may come a time when you need to call on those same people to help you with some project. People are more inclined to say yes to helping you when you have told them yes in the past. But do not say yes when you know you will not follow through. If you do, you will lose your credibility and dependability. If a task or job is worth doing, it is worth doing to the best of your ability.[11] Do not let people pressure you into overextending yourself. The stress related to trying to do too many things at once is not worth it.

Other Time Wasters

There are many other time wasters. Some of these are

- Making work expand to fit the time available
- Inertia
- Fear of failure
- Unnecessary correspondence
- Haste (take time to get it right)
- Socializing
- Junk mail
- Complaining about having too much to do instead of doing something about it
- Lack of priorities
- Poor communication
- Overcommitment
- Daydreaming
- Failure to listen
- Lack of organization
- Committees
- E-mail
- Unclear goals and objectives
- Chat rooms
- Crises
- Lack of delegation
- Instant messaging
- Perfectionism

Managing Time More Effectively and Efficiently

There is never enough time to do all the things we want and need to do. We must choose between competing demands on our time. The choices we make mean that we are ultimately responsible for how we spend our time. Some small tasks do not need to be completed to perfection, and some do not need to be completed at all. Learn to recognize tasks you are over-completing (i.e., spending too much time on). You may spend too much time for too little result.[12] Manage both effectively (doing the right task) and efficiently (doing the task the right way). We have already discussed the importance of plans, goals, and objectives. We must also keep schedules, construct to-do lists, set priorities, and delegate.

Scheduling

Look at the week ahead with your goals in mind, and schedule time to achieve them. Having identified roles and categories and set goals and objectives, you can translate each goal and objective to a specific day of the week, either as a "thing to do" or, even better, as a specific time or appointment. Refer to Figure 15-10 for an example of a scheduling calendar with categories, things to do, and priorities. Places are also available for people/relationship goals, creativity/ideas, and greatest accomplishment of the day.

Scheduling Weekly If your goal is to stay in shape through exercise, you may want to set aside 30 minutes every day (or 3.5 hours per week) to accomplish that goal. There are some goals that you may be able to accomplish only during working hours and some that you can do only on weekends. Be sure to schedule other things, such as reading, family time, entertainment, or whatever is important in your life. As you see, there are some advantages to organizing the week instead of the day.

Scheduling Daily Successful leaders also have daily schedules. At the end of each day, you should complete your schedule for the next day. Fill in time slots that are not already scheduled. This should take 15 minutes or less. Schedule your goals and objectives of the week and your "things-to-do" list each day. Allow enough time to do each task. Many leaders find that estimating the time it will take to perform a nonroutine task and then doubling it works well.[13]

Schedule High-Priority Items During Your Prime Time **Prime time** is the period of time when you perform at your best. For most people it is early in the morning. However, some people are slow starters and perform better at later hours. Determine your prime time and schedule the tasks that need your full attention. Do routine things outside of your prime-time hours.[14]

Schedule a Time for Unexpected Events Have people call you—and call them—during a set time. Regardless of your planning, unplanned things will come up. Therefore, expect the unexpected. Do not do an unscheduled task before a scheduled task without prioritizing it first. If you are working on a high-priority item, and a medium-priority item is brought to you, let it wait. As discussed earlier, often the so-called urgent things can wait. Of course, if your superior says "Do it," I advise you to do it regardless of any time-management plan.

Daily Adapting

Since the best leader organizes weekly, daily planning becomes more of a function of daily adapting: prioritizing activities and responding to unanticipated events, relationships, and experiences in a meaningful way.[15]

Using "Things-to-Do" Lists to Reach Daily Goals

In essence, a "things-to-do" list is a list of objectives. If we reach our objectives, we reach our

WEEKLY PLANNING, ORGANIZING, SCHEDULING CALENDAR

Roles or Categories	Weekly Goals/ Objectives
Professional or School	
Family	
Community	
Health	
My Time	
Meaningful People/ Relationship Goals	
Creativity/ Ideas	

Week of:	Sunday	Monday	Tuesday	Wednesday	Thursday	Friday	Saturday
Weekly Priorities	Today's Things to Do and Priorities						
	Schedule of Appointments/Commitments						
	8	8	8	8	8	8	8
	9	9	9	9	9	9	9
	10	10	10	10	10	10	10
	11	11	11	11	11	11	11
	12	12	12	12	12	12	12
	1	1	1	1	1	1	1
	2	2	2	2	2	2	2
	3	3	3	3	3	3	3
	4	4	4	4	4	4	4
	5	5	5	5	5	5	5
	6	6	6	6	6	6	6
	Evening	Evening	Evening	Evening	Evening	Evening	Evening
	7	7	7	7	7	7	7
	8	8	8	8	8	8	8
	9	9	9	9	9	9	9
	10	10	10	10	10	10	10
	My Greatest Accomplishment Today						

Figure 15–10 Many types of time-management calendars are available. This example is somewhat different because it provides space for goals and objectives for professional or school, family, community, health, your personal time, meaningful people or relationship goals, and creativity or ideas.

goals. If we reach our goals, we are successful. Being successful makes us happy, and if we are happy, we feel good about ourselves. Preparing and completing "things-to-do" lists give you a sense of accomplishment that affects your attitude. They help you keep focused and finish the things that you need to do daily.

In using the "things-to-do" list, write down each activity that needs to be done and assign it a priority (Figure 15–11). Remember that priorities can change during the day due to unexpected events that must be added to your to-do list. Start with the high-priority (H) activities by performing the most important one. When it is completed, cross it off and select the next, until all high-priority activities are completed. Then do the same with the medium (M) priorities, then the low (L) priorities. Be sure to update the priorities.

Things to Do by Priorities					
DATE:					
ACTIVITY	Delegate	High (H)	Medium (M)	Low (L)	Deadline

Figure 15–11 In reality, a "things-to-do" list is a list of objectives. By delegating or assigning priorities to each, it helps us to reach those objectives more efficiently.

As deadlines come nearer, priorities will change. With time, low priorities often become high priorities.[16]

Each day you write up a new to-do list. Yesterday's low priorities, if not completed, become the next day's high priorities. The important thing to remember is to leave room for flexibility. Failing to account for unforeseen circumstances can lead to Murphy's Law: (1) Nothing is as simple as it seems; (2) Everything takes longer than it should; (3) If anything can go wrong, it will. Good planners are rarely tripped up by Murphy's Law.[17]

Setting Priorities

At any given time we are faced with many different tasks. One thing that separates successful from unsuccessful people is their ability to do the important things (priorities) first and the less important things later.

Priorities should be determined by a person's major responsibilities. Do what is really important. This means assigning a priority to each task. To determine priorities, a good leader should answer the following questions.

1. Do I need to be personally involved because of my unique knowledge or skills? (yes or no) There are times when you are the only one who can do the task, and you must be involved.
2. Is the task within my major area of responsibility?
3. When is the deadline? Is quick action needed? (yes or no) Should I work on this activity right now, or can it wait? Time is a relative term. In one situation, months or even a year may be considered quick action, but in another situation a matter of minutes may be considered quick action. For example, the decision to earn a college degree may have to be made close to four years in advance. It often takes several months before applicants are told if they are accepted. To the admissions personnel, this may be quick action.[18]
4. Can I delegate this?

Delegation

We accomplish all that we do through delegation—either to time or to other people. If we delegate to time, we think efficiency. If we delegate to other people, we think effectiveness.[19] Many people refuse to delegate to other people because they feel it takes too much time and effort, and they could do the job better themselves. Also, other people do not delegate because they have no experience in delegating work, they have no confidence in subordinates, and they are perfectionists. Some perceived leaders are afraid that someone else may do it better, and they would feel threatened.

Why Delegate? Besides helping others to develop, delegation is the key to the leader's sanity. You cannot do it all yourself. The more you try, the greater the pressure and tension. Extensive delegation also frees valuable time for you to concentrate on your goals. Get into asking yourself if what you're doing can be handled by someone else.

The late J. C. Penney was quoted as saying that the wisest decision he ever made was to "let go" after realizing that he could not do it all by himself any longer. That decision, made long ago, enabled the development and growth of hundreds of stores and thousands of people.[20]

Delegate with Deadlines Work expands to fill the time available for its completion. In other words, people tend to stretch out jobs as long as they can (that is, if you have three hours to complete a task, you will take three hours, while in reality it could have been done in two hours or less). To avoid procrastination and delay, directives to your group members should *never* be open-ended. Reasonable completion dates should always be specified.

Besides helping the leader, delegating with deadlines benefits those who are given assignments. Leaders can then concentrate on tasks without interference or undue pressure. Be sure to keep deadlines realistic. If your personnel keep missing them for reasons beyond their control, they will soon start ignoring them completely.[21]

Two Types of Delegation Stephen Covey in his book, *The 7 Habits of Highly Effective People*, states that there are basically two kinds of delegation: "gofer" delegation and stewardship delegation.

Gofer delegation means "go for this, go for that, do this, do that, and tell me when it's done." This is not a good practice because the leader does not allow creativity, imagination, and self-direction or growth.

Stewardship delegation is more appropriate for leaders in most cases. It is based on appreciation of the self-awareness, imagination, conscience, and free will of other people. It is focused on results instead of methods. It gives people a choice of method and makes them responsible for results. Stewardship delegation involves clear, up-front mutual understanding and commitment regarding expectations in five areas:

1. *Guidelines.* Identify the parameters within which the individual should operate. These should be as few as possible to avoid methods delegation, but they should include any formidable restrictions.
2. *Desired results.* Create a clear, mutual understanding of what needs to be accomplished, focusing on *what*, not *how*; on *results*, not *methods*.
3. *Resources.* Identify the human, financial, technical, or organizational resources the person can draw on to accomplish the desired results.
4. *Accountability.* Set up the standards of performance that will be used in evaluating the results and the specific times when reporting and evaluation will take place.
5. *Consequences.* Specify what will happen, both good and bad, as a result of the evaluation. This could include such things as financial rewards, psychological rewards, different job assignments, and natural consequences tied into the overall mission of an organization.[22]

Figure 15–12 illustrates a clear and concise way to handle delegation. When leaders delegate, they should clarify what degree of initiative is expected. There should be no excuses for lack of communication if you use this form.

Delegating and Clarifying Degree of Initiative Expected

From: ______________________

To: ______________________

Re: ______________________

❑ 1. Look into this problem. Give me all the facts. I will decide what to do.

❑ 2. Let me know the alternatives available with the pros and cons of each. I will decide which to select.

❑ 3. Recommend a course of action for my approval.

❑ 4. Let me know what you intend to do. Delay action until I approve.

❑ 5. Let me know what you intend to do. Do it unless I say not to.

❑ 6. Take action. Let me know what you did. Let me know how it turns out.

❑ 7. Take action. Communicate with me only if your action is unsuccessful.

❑ 8. Take action. No further communication with me is necessary.

Comments: ______________________

____

____

Signature Date

Figure 15–12 When leaders delegate, they should clarify what degree of initiative is expected. The use of this form in delegating should remove all doubts.

Clear instructions will save time and embarrassment. If you aren't clear and consistent about delegation, those who answer to you may be afraid to take the appropriate initiative for fear of being criticized.

Helpful Hints for Managing School- and College-Related Time

Learn to Use a Computer Not only will it help you to save time instead of writing out assignments in longhand, but it is also an important skill that can help you get a good job. Computers can also help you save time if you use features such as editing, spelling- and grammar-checking, and the thesaurus.

Find Out the Test and Assignment Schedule Find out at the beginning of the school year when exams are. Reserve a week before exams to study. Ask all your teachers for deadlines on major grades, such as projects and major tests. Mark these dates on your pocket planning calendar so you will not create scheduling conflicts.

Get a syllabus from each teacher on what is required to make good grades, a list of materials needed for a particular course, and dates for turning in work for each grading period. Write the important dates and deadlines down in your monthly/yearly planning calendar so that you can schedule your time in order to attain your goals.

CONCLUSION

Leaders who manage their time are more productive leaders. Time management involves setting goals, balancing parts of your life, setting priorities, and exercising self-discipline. By failing to plan, we plan to fail. If we do not know where we are going, we will not know when we get there. If we do not have objectives to accomplish during the day, we can work all day without doing anything productive or, at least, anything that moves us toward reaching our goals.

SUMMARY

Time management is planning who will control your time in order to do the things you need and want to do. Good leaders sit down and decide what they are going to do in a 24-hour period. Time is your most valuable personal resource. Use it wisely because it can't be replaced.

Time management has basically gone through four generations or levels. The first generation includes notes and a checklist. The second generation uses calendars and appointment books. The third generation includes setting priorities, clarifying values, setting goals and objectives. The fourth generation is more concerned with managing ourselves than managing time. It includes enhancing relationships and meeting human needs as well as personal and professional goals and objectives.

Two ways to describe an activity are *urgent* and *important*. Urgent matters require immediate action, but they do not move you toward accomplishing any of your goals. Important matters have to do with results; they contribute to your goals and objectives. We react to urgent matters, but important matters that are urgent require more initiative and planning.

Ultimate time management for good leaders includes putting first things first. The ultimate leader and time manager must meet six important criteria: coherence, balance, focus, people-centeredness, flexibility, and portability. We cannot manage our time correctly, effectively, and efficiently if we do not know what our goals are. Goals provide direction for time management. Periodically, your goals and time commitments have to be adjusted.

You must balance the major categories of personal and professional time. Time spent on various activities can be divided into four major categories: health, family, community (personal), and school or employment (professional time). Each category has its own importance. Health includes diet, exercise, recreation, socializing, and spiritual pursuits. Family should always be a high priority. We want to make our communities a better place to live. We are more concerned with productivity in the employment category.

Analyze your use of time with a daily log. Before you decide to improve your use of time, you may want to find out where your time is going. Pareto's principle, or the 80-20 rule, says that 20 percent of tasks that are high priority can generate 80 percent of the results you need to reach your goals.

There are many time wasters: procrastination, the telephone, television, the inability to say no, unexpected visitors, lack of planning, and emotional conflicts. Some other time wasters include making work expand to fit the time available, disorganization, overcommitment, worrying, fatigue, lack of organization, unclear goals and objectives, and perfectionism.

There is never enough time to do all the things we want and need to do. Therefore, we must manage our time effectively and efficiently. This includes using weekly planning, organizing, and scheduling calendars; "things-to-do" lists; setting priorities; and delegating. There are two types of delegation: gofer and stewardship. Gofer delegation means "go for this, go for that." It does not enhance self-worth. Stewardship delegation is based on appreciation of self-awareness, the imagination, the conscience, and the free will of others. It involves guidelines, desired results, resources, accountability and consequences. There are many helpful hints for school and college students to save time. Some of these include learning to use a computer and finding out when exams and assignments are scheduled.

End-of-Chapter Exercises

Review Questions

1. Define the Terms to Know.
2. Name at least two types of things included in each of the four levels or generations of time management.
3. Within the time-management matrix (Figure 15–2), which area is best? Explain why.
4. Briefly explain what the other three areas in the matrix represent.
5. What six important criteria must be met for ultimate time management by good leaders?
6. What is the relationship between goals and time management?
7. What are the four major categories in which you spend your time?
8. What does diet have to do with time management?
9. List eight ways to overcome procrastination.
10. List four major time wasters.
11. What are 10 ways to deal with phone calls once you are in a leadership or management position?
12. What five areas does stewardship delegation involve?
13. List two helpful time-management tips for school and college students.
14. In determining priorities, what four questions should be answered?
15. What are the seven major categories to schedule with a scheduling calendar?

Fill-in-the-Blank

1. ________________ is one of a leader's most valuable resources.
2. Effective leaders are not problem-minded, they are ________________ minded.
3. Helen Keller said that worse than being blind is to have sight with no ________________ .
4. Health, family, and community are types of ________________ time, whereas school and employment are types of ________________ time.
5. The "80-20" rule or Pareto's principle can be stated thus: The ________________ percent of tasks that are high priority generate ________________ percent of the results you need to reach your goals.
6. B. F. Skinner found out that ________________ ________________ , or rewards for desirable behavior, worked better than negative consequences as a tool for changing behavior.
7. Work expands to fill the ________________ available for its completion.
8. ________________ delegation means "go for this, go for that, and tell me when it's done."
9. ________________ delegation is based on appreciation of the self-awareness, imagination, and free will of other people.
10. Schedule goals and objectives for the ________________ and "things-to-do" list each ________________ .

Matching

_____ 1. Schedule your high-priority items
_____ 2. Could cause you to be a "couch potato"
_____ 3. Can cost time because you don't take time
_____ 4. More time tomorrow
_____ 5. Time-value ratio
_____ 6. "I was just on my way out the door"
_____ 7. Putting first things first
_____ 8. "I can do it better"
_____ 9. Daily objectives
_____ 10. "I'm flattered you want me to help you, but I'm already booked"

A. procrastination
B. lack of planning
C. prime time
D. telephone etiquette
E. saying no
F. things-to-do list
G. television
H. lack of delegation
I. setting priorities
J. Pareto's principle

Activities

1. What one thing could you do (that you are not doing now) that if you did on a regular basis, would make a tremendous positive difference in your personal or professional life? Explain why you are not doing it.
2. Before you start a daily log for one week, take a few minutes to write down everything you think you do in a typical day or week and estimate how much time you spend on each activity. After you keep a daily log for one week, record how much time you actually spent on each activity. Then answer the following questions:

- On what activities did you spend more time than what you had estimated?
- On what activities did you spend less time than what you had estimated?
- How could you use this information to manage your time?

3. Your teacher can provide you a daily log similar to the one in Figure 15-7. Complete the daily log. Mark down in 15-minute intervals how you spent your time. Record activities as you do them. A more accurate picture of your time would result if you did this every day for a week.
4. Analyze the daily log in activity 2 (above) and complete it as if you did all the things you should do whether you want to or not. Write down your ideal day from the minute you wake up to the minute you fall asleep. Do not forget to include activities from each of the four categories of personal and professional time.
5. List your own personal five biggest time wasters and what steps you will take to eliminate them.
6. Make a "to-do" list of goals you want to work on and achieve tomorrow. Be sure to prioritize. You may want to use a form similar to Figure 15-11.
7. Your teacher can provide copies of a weekly planning, organizing, scheduling calendar similar to the one in Figure 15-10. Complete one of these for the week. Write a report and share with the class what you learned by completing and following the weekly or daily calendar.
8. Select five things you learned that will best help you manage your time in the future. Write a paragraph justifying each.
9. **Take it to the Net.**

   Explore time management on the Internet. Listed below are Web sites that contain information on time management. Some of the sites contain only one article and some contain various articles. Browse the articles and find one you feel is useful. Print the article and summarize it. In your summary describe why you feel the information in the article is useful. If you are having problems with the sites listed, use the search terms to find an article.

   Web sites

   <http://www.mindtools.com/page5.html>

   <http://www.pvc.maricopa.edu/lac/quest/time.html>

   <http://careerplanning.about.com/careers/careerplanning/library/weekly/aa111697.htm>

   <http://www.ncf.carleton.ca/~an588/time_man.html>

   Search Terms

   time management
   effectively managing time
   time management skills

Notes

1. S. R. Covey, *The 7 Habits of Highly Effective People* (New York: Simon & Schuster, 1997).
2. Ibid., pp. 150–152.
3. Ibid.
4. Ibid., pp. 153–154.
5. M. E. Douglas and D. N. Douglas, *Manage Your Time, Manage Your Work, Manage Yourself* (New York: AMACOM, A Division of American Management Associations, 1993).

6. E. C. Bliss, *Getting Things Done: The ABC's of Time Management* (New York: Macmillan, 1991), p. 30.
7. "How to Get Things Done" (Shawnee Mission, KS: Fred Pryor Seminars, 1989), p. 5.
8. R. N. Lussier, *Human Relations in Organizations: Applications and Skill Building* (Columbus, OH: McGraw-Hill Higher Education, 1998).
9. "How to Handle Multiple Priorities" (Shawnee Mission, KS: Fred Pryor Seminars, 1989), p. 7.
10. Ibid., p. 45.
11. J. L. Barkas, *Creative Time Management: Be More Productive and Still Have Time For Fun* (New York: Prentice-Hall, 1984), p. 24.
12. "How to Handle Multiple Priorities", p. 40.
13. Lussier, *Human Relations in Organizations.*
14. Ibid.
15. Covey, *The 7 Habits of Highly Effective People*, p. 165.
16. Lussier, *Human Relations in Organizations.*
17. Douglas and Douglas, *Manage Your Time, Manage Your Work, Manage Yourself*, p. 94.
18. Lussier, *Human Relations in Organizations.*
19. Covey, *The 7 Habits of Highly Effective People*, p. 171.
20. Ibid.
21. F. A. Manske, Jr. *Secrets of Effective Leadership* (Columbia, TN: Leadership Education and Development, Inc., 1990), p. 81.
22. Covey, *The 7 Habits of Highly Effective People*, p. 174.

CHAPTER

16

Positive Reinforcement and Motivation

Objectives

After completing this unit, you will be able to:

- Explain motivation
- Describe the characteristics of motives
- Discuss various theories of motivation
- Explain four types of reinforcement
- Explain reinforcement scheduling
- Discuss positive reinforcement
- Explain the various forms of positive reinforcement

Terms to Know

motivation
need
behavior
internal motivation
external motivation
content theories
process theories
hierarchy
hygienes
learned needs theory
equity
inequity
operant conditioning
reinforcement
positive reinforcement
avoidance learning
extinction
punishment
reinforcement scheduling
fixed interval schedule
variable interval schedule
fixed ratio schedule
variable ratio schedule
physical strokes
verbal strokes
stroke deficit
confirmation behaviors

Whether it is on the basketball court or on the stage of an auditorium, Rick Pitino, former head coach of the University of Kentucky Wildcats and Boston Celtics, has all the qualities of an effective coach and an effective motivational speaker. Not only does he motivate others, he also reinforces the actions he motivates others to take through positive reinforcement. For example, when a member of the Kentucky Wildcats team did well, Rick Pitino was often the first to congratulate and praise that player for his good work. By praising, Pitino reinforced that behavior or performance.

MOTIVATION

Have you ever wondered what makes a person do what he or she does? Why does a student strive for good grades? Why does an athlete give her all for the team? Why do some workers excel and others do not? The best answer to these questions is simply motivation. Leaders must know what makes people tick.

Motivation in Everyday Life

Motivation is a key component of the classroom and the workplace. As a matter of fact, it is a key to life. We do homework because of motivation. We go to jobs every day because of motivation. We go to college because of motivation. We take part in religious activities due to motivation. *Motivation* comes from the word *motive,* meaning the reason for doing something or behaving in a certain way. Much of what motivates us depends on our goals, discussed in Chapter 14. Our goals depend on our needs.

As a leader, you must be aware of what makes people tick. Needs, motives, and motivation are a big part of why people act the way they do. Why do you like some of your teachers better than others? The answer probably lies in their ability to motivate you. Good teachers understand your needs and motives. Therefore, they tend to be alive and exciting in the classroom, inspiring students to reach their full potential.

Leaders in the workforce also must motivate people. The major goal of most companies and organizations is productivity. When a leader is able to motivate his or her followers, that leader has achieved a major objective of leadership.

Leadership is a process in which an individual influences members to work toward the group's goals. Good leaders build positive relationships among members by using motivation. They understand that the group's success depends on a leader's ability to use influence effectively.

Good leaders rely more on influence than on their authority from power. Maybe you think of actual people, such as Washington, Lincoln, Churchill, or Napoleon. Influence combines enthusiasm, excitement, charisma, and wisdom. When leaders use influence they appeal to a person's values, skills, and knowledge. Keeping followers motivated and enthusiastic is basic to effective leadership. The way a leader uses power determines whether his or her influence will motivate followers.

The FFA awards program is motivation driven. Our motive is the attainment of awards: the Greenhand degree, chapter FFA degree, state FFA degree, and American FFA Degree. There are many proficiency awards, several speaking events and many contests (discussed in previous chapters). One of the reasons that the FFA is so successful is that it meets students' needs. These needs motivate students.

Motivation Defined

Motivation reflects willingness to expend effort; it makes us do what we do. For example, Sarah is hungry (a **need**) and has a drive or a motive to get food. When she finally eats (**behavior**), her need for food is satisfied. The hunger she felt was her motivation to get something to eat.

Types of Motivation

Internal motivation is an inner force or power that helps a person achieve a goal.[1] Internal motivation deals strictly with the individual and the internal rewards that he or she receives. Feelings of responsibility, personal growth, or the satisfaction of achievement are some internal rewards. A person with internal motivation is sometimes referred to as self-motivated.

External motivation is the second type of motivation. External motivation is the outer force or power that causes a person to want to achieve a goal.[2] External motivation can come from rewards, such as FFA awards, given by others that encourage a person to behave a specific way (Figure 16–1). Whereas internal motivation has to do with the individual and inner rewards, external motivation involves rewards offered by outside sources for achieving a certain goal.

If you are the individual or leader who is responsible for motivating those around you, you must recognize people, include them, encourage them, and train them in the desired behavior. You must make them feel valued. You must show them they are trusted, respected, and cared for. Above all, you must motivate them to excel for themselves and for the organization.

Figure 16–1 The awards program of the FFA is one of the many reasons for its success. Awards meet the need for achievement and recognition. (Courtesy of Fullerton High School, California.)

Characteristics of Motives

The motivators that cause us to act in certain ways are often referred to as motives. To know how to motivate someone, we must understand five important characteristics of motives.

Motives Are Individualistic

Each person has a set of needs (or motives), which may be completely different from that of another person. One student may be motivated to "ace" a test by the need to earn a scholarship for college. Another student taking the same final exam may be motivated by the chance to make honor roll. Each student's motives are different, although their actions are the same. Sometimes, it is difficult to understand another individual's actions; looking for a motive may help in understanding.

Motives Can Change

During a person's lifetime, motives can change. What motivated us as children may no longer motivate us as adults. For instance, early in one boy's academic career, he may study for a test just so he can get out of the class. Later, he may study so that he can get a good scholarship for college. In college, he begins to study so that he can graduate with honors and receive good job offers. One student's motives for studying continually change throughout his academic career.

Motives Can Be Unconscious

Often we are not aware of or do not fully understand the inner forces and needs that shape our behaviors. Those inner forces or needs may be feelings of inadequacy, loneliness, or lack of acceptance by peers. Feelings like these may motivate people to strive for top awards or higher positions in an organization.

Motives Can Be Inferred

In most cases, we notice people's behavior, but we can only guess what motivates them to act the way they do. The best way to learn what motivates another person is through discussion. There is no more practical method to obtain information about motives than simply asking the question "What motivated you to act in that way?"

Motives Are Hierarchical

Motives come in various strengths. The stronger the motive, the more likely a person is to achieve a goal. A student who works part-time to earn money for a trip during spring break knows she must make car payments, so she is motivated to work by the fact that if she does not keep up her car payments she will lose the car. For her, being able to drive to Florida for spring break overshadows the prospect of spending her free time with friends. Behavior, therefore, is ruled by those motives or needs we deem most important.[3]

Here is where the issue conflict of values arises. You have a new job, your brother is having a bachelor party, you have an FFA parliamentary procedure contest, your coach has scheduled an extra, previously unannounced practice—and they are all on the same night. What do you do? My choice would be different than yours. Your values determine your motives and thus your choices. A person cannot do everything in life; life involves trade-offs and choices. By following the problem-solving and decision-making process outlined in Chapter 13, you can come to a satisfactory decision. Regardless, your decision will depend on your motives.

THEORIES OF MOTIVATION

Several theories about motivation exist, but they all fall into two types: content and process theories. **Content theories** focus on factors within the individual that cause the individual to behave in a particular manner. Basically, a content theory looks at individual needs that motivate behavior. **Process theories**, on the other hand, examine how the behavior is motivated. For a process theory to be effective, it must identify the processes surrounding the behavior. The first content theory to be discussed is the hierarchy of needs theory.

Maslow's Hierarchy of Needs

Clinical psychologist Abraham Maslow theorized that all behavior is based on the needs of the individual.[4] These needs are arranged in a **hierarchy** that denotes the relative importance of each. An individual begins with the lowest level of need and cannot move to another level until the previous level's need is satisfied. Human needs, as determined by Maslow, are (1) physiological needs, (2) needs for safety and security (3) social or belonging needs, and the need for (4) self-esteem and (5) self-actualization (Figure 16–2).

Physiological Needs We all have basic survival needs, such as air, food, clothing, shelter, and sleep. A person must have all these needs satisfied before they worry about meeting the next level of needs. Most people in the United States are able to meet their physiological needs.

Safety and Security Needs Once physiological needs have been met, we become aware of the need for freedom from the threat of harm. People yearn for a feeling of order, safety, and predictability. Students in a school building need to feel secure and safe from personal harm. Families need to feel protected from the violence that plagues our streets. Workers need to feel assured that their working conditions are safe. Only when people feel secure can they focus on the third level of needs. The police officer in Figure 16–3 provides many people with a sense of safety and security.

Social and Belonging Needs People have social needs, too. We need to have the feeling of belonging to a group, and we need friendship, interaction with others, and love. These needs affect our mental health and well-being, whereas the previous two levels of needs dealt with physical health and well-being. When an individual is trying to fulfill one of these needs, he or she may join a team or an organization in order to feel like he or

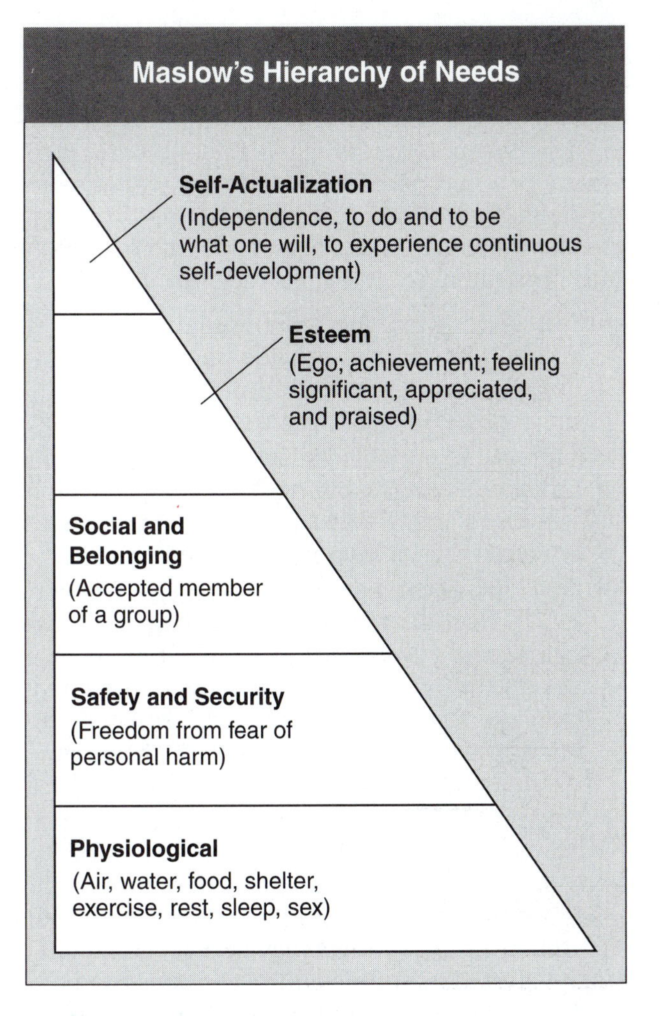

Figure 16–2 Basic physiological needs have to be met before we are particularly concerned about psychological needs. We do not worry about self-concept if we have no food or water. (Adapted from Abraham H. Maslow, *Motivation and Personality,* New York: Harper & Row, 1954.)

she belongs, to find friends, and to interact with others.

In high school, the FFA may meet some social and belonging needs. In college, fraternities, sororities, and friends can meet this need. Earlier in life, the 4-H Club, Boy Scouts, a youth group at church, family, and a variety of other groups can meet people's social needs.

Figure 16–3 Police officers help people feel secure and safe from personal harm.

Self-Esteem Individuals need to feel good about themselves. How a person feels about himself is referred to as self-esteem. Individuals also need to be recognized by others for their achievements. Esteem needs refer to a person's need for self-respect as well as respect and recognition from others. Status symbols, personal titles, and awards that denote achievement can help meet esteem needs. Self-esteem can be affected by words of praise and appreciation from peers and others. Self-concept and self-esteem are discussed in Chapter 17.

Self-Actualization The need for self-actualization is often the most difficult need to satisfy and to understand. Self-actualization represents the maximum use of an individual's talents, abilities, and skills. A self-actualized person has realized his or her fullest potential as a human. Each person's search for self-actualization is different because of the uniqueness of each person. Many people find it difficult to satisfy their self-actualization needs because they are limited by circumstances. If you fulfill this need, you no longer have any challenges to be met. Perhaps this is the reason so many people never completely satisfy this need;

they do not want to reach the end of a challenging lifelong journey.

ERG Theory

A second major theory based on needs was proposed by C. P. Alderfer. ERG stands for three types of needs: existence, relatedness, and growth. Alderfer defined them as follows.[5]

Existence Needs These are material and are satisfied by environmental factors, such as food, water, pay, benefits, and working conditions.

Relatedness Needs These involve relationships with "significant others," such as co-workers, superiors, subordinates, family, and friends.

Growth Needs These involve the desire for unique personal development. They are met by developing whatever abilities and capabilities are important to the individual.

Alderfer's theory differs from Maslow's in three important dimensions.

1. Alderfer proposed three need categories in contrast to Maslow's five.
2. Alderfer arranged his needs along a continuum, as opposed to a hierarchy. Existence needs are the most concrete and growth needs the most abstract.
3. Alderfer allowed for movement back and forth on the continuum, in contrast to Maslow's moving up the need hierarchy. A person who becomes frustrated in satisfying higher needs would regress toward fulfilling lower needs.

Alderfer's ERG theory has received more support than Maslow's hierarchy of needs theory. However, many of the same problems plague both: Exactly what constitutes a need is unclear. In addition, people engage in behavior that is seemingly unrelated to need fulfillment. Alderfer's theory, although less general than Maslow's, is quite removed from reality. Needs theories may be of little value in day-to-day leadership.

Expectancy Theory

V. H. Vroom (1964) pushed expectancy theory into the arena of motivation research. According to this cognitive theory, each person is assumed to be a rational decision maker who will expend effort on activities that lead to desired rewards. The theory has five major parts: job outcomes, valence, instrumentality, expectancy, and force.[6]

- **Job outcomes.** Job outcomes are the things an organization can provide, such as pay, promotions, and vacation time. Theoretically, there is no limit to the number of outcomes. They are usually thought of as rewards.
- **Valence.** Valences are one's feelings about job outcomes and are usually defined in terms of attractiveness or anticipated satisfaction.
- **Instrumentality.** Instrumentality is the perceived degree of relationship between performance and outcome attainment. It exists in a person's mind. If a person thought that a pay increase depended solely on performance, the instrumentality associated with that outcome would be very high.
- **Expectancy.** Expectancy is the perceived relationship between effort and performance. In some situations there may not be any relationship between how hard you try and how well you do. In others there may be a very clear relationship: the harder you try, the better you do.
- **Force.** The amount of effort or pressure within someone is referred to here as force. The larger the force, the greater the motivation.

Expectancy theory explains why some jobs seem to create high or low motivation. On assembly lines, group performance level is determined by the speed of the line. No matter how hard one works, one cannot produce any more until the next object moves down the line. Thus there is no relationship between effort and performance. However, salespeople who are paid on commission realize that the harder they try (more sales calls), the better their performance (sales). Expectancy theory is very good at analyzing the components of motivation.

Herzberg's Two-Factor Theory

Frederick Herzberg and his associates interviewed accountants and engineers to describe situations in which they were satisfied or motivated and dissatisfied or unmotivated. Herzberg concluded from these data that the opposite of satisfaction is not dissatisfaction, as had been traditionally believed, and argued that removing dissatisfying characteristics from a job does not necessarily make the job satisfying.[7]

Herzberg classified needs into two categories. He called the lower-level needs **hygienes**. They include salary, job security, working conditions, relationships, status, company procedures, and quality of supervision. When these conditions are not present, employees become dissatisfied; however, if they are present, they do not guarantee motivation. The second category, higher-level needs, were called motivators. They include achievement, recognition, responsibility, advancement, the work itself, and the possibility of growth. When these conditions are present, strong levels of motivation can result in good job performance.

Once hygiene factors are adequate, employees can be motivated through their jobs. The best way to motivate employees is to build challenge and opportunity for achievement into the job.[8]

Learned Needs Theory

Another content theory of motivation is called the **learned needs theory**, which was developed by David C. McClelland.[9] Learned needs theory states that there are needs that are acquired from society itself, and when the need is strong enough in a person, the person is motivated to a behavior that will satisfy the need. According to McClelland, an individual learns what his needs are by interacting with his environment. The needs essential to this theory are the needs for achievement, power, and affiliation. Once an individual understands his or her needs, behavior, personality, and performance will be affected.

Achievement The need for achievement is the need to excel and to strive to succeed. People with a high need for achievement are usually goal oriented. They seek challenges and take moderate risks. They also desire personal responsibility for solving various problems. People with a low need for achievement do not perform as well as people with a high need in challenging, competitive, or nonroutine situations. People with a strong achievement need tend to enjoy entrepreneurial positions.

Power The need for power is often associated with a need to make others behave in ways that they would not behave otherwise. People with a great need for power want to control situations and other people. They like competition but only if they can win—they hate losing. They also do not mind confrontation with others. People with a strong need for power seek positions of authority and personal status.

Affiliation The need for affiliation is the desire for interpersonal relationships and friendships. People with a high need for affiliation want to be liked by others and enjoy social activities. They also tend to join organizations and groups to have a feeling of belonging. These individuals seek positions or situations in which they can help and teach others. The FFA members in Figure 16–4 are meeting their need for affiliation by

Figure 16–4 These FFA members are having their need for affiliation met by being members of the FFA and participating in the homecoming parade. (Courtesy of the Eagleville High School FFA.)

participation in the group. They prefer to be part of a group rather than leaders.

Each individual has varying degrees of each need. The degree is based on what a person has learned by dealing with his or her environment. One of the three needs is usually dominant in each of us and, thus, motivates our behavior in a dominant way.

You may see a correlation between the personality types discussed in Chapter 2 and the learned needs theory. Sanguine personality types tend to have a high achievement need. Choleric personality types have the need for power and control. Melancholy personality types like affiliation but do not particularly have a great need for power or achievement.

Equity Theory

Process theories, as stated earlier, describe how behavior is motivated. The first process theory of motivation to be discussed is the equity theory.

The equity theory was developed by J. Stacy Adams.[10] It is based on the concept that all individuals want to be treated fairly. **Equity** is an individual's belief that she is being treated fairly in relation to her relevant others. (Relevant others may be peers, co-workers, or a group of people with whom an individual can compare herself.) **Inequity** is an individual's belief that she is being treated unfairly in relation to her relevant others. J. Stacy Adams contended that people are motivated to seek social equity in the rewards they receive for the performance they give.

Individuals compare their inputs (what they contribute to the situation) and their outcomes (what they receive in return for their input) to that of others. Inequity usually occurs when the perceived input-outcome ratio is out of balance. For example, a student does an extra-credit project and receives five extra points on his final grade. If that student put hours of work into his project but finds out that another student received five extra-credit points for 30 minutes of work, the hard-working student will feel a sense of inequity. Therefore, the next time extra credit is offered, the student's behavior will be motivated by the inequity he initially felt. A perceived inequity causes tension that motivates a person to make the input-outcome ratio balance or to make a situation equitable.

There are several ways to reduce a perceived inequity so as to balance the input-outcome ratio. The most common methods follow.[11]

Change inputs. You can reduce your level of effort.

Change outcomes. You can request more rewards, such as pay, vacation time, or extra credit points.

Misrepresent self-perception. You can change the way you think of your own inputs. You may rethink exactly how much you put into a situation.

Misrepresent perception of others. You can change your view of others' inputs, outcomes, or overall situations. One student may begin to think that the other student's project was indeed better than his own.

Change comparisons. You can change your comparison group. A student may focus her comparison on another student who worked as hard but received only three extra credit points.

Abandon the situation. You can choose simply to leave the situation or to not participate next time. You may not do extra credit the next semester so that you don't have to experience the feelings of inequity.

Reinforcement Theory

Another process motivation theory is reinforcement theory. With its basic assumptions developed by psychologist B. F. Skinner, reinforcement theory states that behavior is motivated and controlled through the use of rewards. Reinforcement theory uses Skinner's **operant conditioning** or behaviorism as a way to modify an individual's behavior through the appropriate use of immediate rewards or punishments.[12] This theory is also concerned with maintaining a behavior over time;

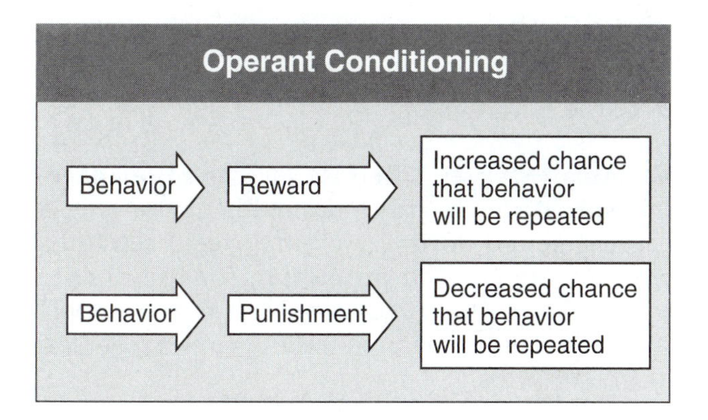

Figure 16–5 Psychologist B. F. Skinner theorized that a person's behavior is motivated and controlled through the use of rewards or punishments.

it contends that rewarded behavior will be repeated, whereas unrewarded behavior will not be repeated (Figure 16–5). Two important aspects of controlling behavior are the type of reinforcement and the schedule of the reinforcement.

TYPES OF REINFORCEMENT

There are four types of **reinforcement**: positive reinforcement, avoidance learning, extinction, and punishment.[13]

Positive Reinforcement

Positive reinforcement is a method of using pleasant consequences (rewards) for a desired behavior. For example, after a student studies several hours for a test and receives an *A*, it is more likely that those study habits will occur again because of the reward of the good grade. Positive reinforcement is the best motivator of people. Positive reinforcement also works for animals. The dog in Figure 16–6 will keep retrieving the ball as long as positive reinforcement continues.

Avoidance Learning

Avoidance learning is often called negative reinforcement. By using avoidance learning, an individual can avoid a negative consequence. It entails the threat of punishment with no actual punishment. A student hurries to class so that he is on time to avoid a reprimand by the teacher or a visit to the principal's office. The student has learned what to do to avoid an unpleasant situation. This is the basic premise of avoidance learning.

Extinction

Extinction is the removal of a positive reward. Extinction occurs when one tries to eliminate an undesirable behavior by withholding rewards when the behavior occurs. A student who is late for class misses the teacher's reward for promptness that day. By withholding the reward, the teacher can eventually eliminate the chronic tardiness (the tardiness becomes extinct). However, extinction can also work on good, desirable behaviors when they are not rewarded. A student who works hard during class but whose hard work goes unnoticed may soon stop working as diligently. Can you imagine receiving the chapter, state, or American FFA degree with no awards, no banquet, no State Convention or National Convention to attend? After a while, these awards would become extinct without any type of recognition.

Figure 16–6 If the dog is given praise (positive reinforcement), he will continue to retrieve the ball. If he is scolded, this behavior will eventually cease.

Punishment

Punishment provides a negative consequence for an undesirable behavior. It is used after the behavior has occurred. Punishment may reduce the undesirable behavior, but it may increase other unwanted behaviors. A student who is publicly reprimanded for harassing his teacher may stop harassing the teacher but begin disrupting the class in other ways. Punishment is the least effective way of modifying one's behavior but it may be the only way for irrational people.

REINFORCEMENT SCHEDULING

The second consideration in the use of reinforcement is when to reward behavior. The timing of rewards or punishments is known as **reinforcement scheduling**. Reinforcement scheduling has two classifications: continuous and intermittent.

Continuous Reinforcement

With continuous reinforcement, every time the desired behavior occurs the individual is rewarded. Continuous reinforcement is better at sustaining the desired behavior, but it is not always as practical as using intermittent reinforcement.

Intermittent Reinforcement

With intermittent reinforcement, the positive consequence occurs after a passage of time or after an output. An interval schedule is based on the amount of time between rewards. A ratio schedule is based on the number of outputs between rewards. Intermittent reinforcement has four different scheduling alternatives from which to choose.[14]

Fixed Interval Schedule A **fixed interval schedule** rewards behavior at regular, fixed intervals. An employee is rewarded with a weekly paycheck for performing a desired behavior. This alternative can lead to irregular behavior and often rapid extinction of the desired behavior once the rewards stop.

Variable Interval Schedule A **variable interval schedule** rewards a desired behavior after a varying period of time. An employee is rewarded with an unexpected promotion for the desired performance. This alternative leads to moderately high and stable performance and slow extinction of the behavior.

Fixed Ratio Schedule A **fixed ratio schedule** rewards the behavior after a predetermined number of actual outputs. A student receives five extra-credit points after bringing her completed homework to class three times. This alternative quickly leads to very high and stable performance and slow extinction of the desired behavior.

Variable Ratio Schedule A **variable ratio schedule** rewards behavior after a varying number of outputs. A student receives an unexpected award for turning in his completed homework. This alternative leads to very high performance and slow extinction of the desired behavior. This is often the most powerful method because individuals maintain the desired behavior due to possibly long times between reinforcements.

POSITIVE REINFORCEMENT

Positive reinforcement is the best motivator of people. It is also a good method of improving interpersonal relationships. People who understand and use positive reinforcement are more likely to be successful in their careers and in their relationships. By understanding that all humans need some kind of positive feedback, we can better understand ourselves and our behavior.

Positive Feedback

Individuals yearn for some type of appreciation or positive feedback (positive reinforcement). Included in Maslow's hierarchy of needs are the needs for security, belonging, and self-esteem. It would seem impossible to fulfill these needs without some type of positive feedback.

Security. An individual might never feel secure without someone recognizing his accomplishments and truly making him feel secure in what he is doing.

Need to belong. With the need to belong, an individual would never feel like he totally belonged until others showed him somehow that he was, in fact, a member of the group.

Self-esteem. The need for self-esteem definitely requires positive reinforcement before it can be satisfied. The student in Figure 16–7 apparently is not lacking in self-esteem. It would be difficult to have a very high level of self-esteem without some recognition for one's accomplishments.

Figure 16–7 Positive feedback meets people's needs for security, sense of belonging, and self-esteem. These needs appear to have been met for this student.

Best Motivator of a Person's Behavior

B. F. Skinner's work examined the importance of reinforcement and reinforcement scheduling. He, too, theorized that positive reinforcement was the best motivator. Thus, the best way to make a person perform well is to recognize and acknowledge his or her accomplishments. By having his or her work praised, the person's need for reinforcement is satisfied, and he or she will likely repeat the praised behavior.

This is basic knowledge for every leader. Democratic leaders lead by positive reinforcement. They also use the theory Y assumption that people do not have to be told everything to do and how to do it. People respond best to praise, not ridicule.

Strokes

Eric Berne's work also identifies the need for positive reinforcement in what he calls *strokes*.[15] *Stroking* is a term used to describe various ways in which a person may recognize another person's behavior. **Physical strokes** are any recognition that occurs physically, such as an acknowledging smile or a pat on the back. **Verbal strokes** are any recognition that occurs verbally, such as words of thanks and appreciation.

Berne explains that stroking is important for physical and mental well-being. A person who is lacking in strokes from others has a **stroke deficit**. A person with stroke deficit may actually ask for strokes; this is typically called "fishing for a compliment." It is important for people to get positive feedback in the form of strokes so they can feel worthwhile and needed. People who suffer from stroke deficit may damage their relationships by criticizing or undermining the work of others; they seem to have to put others down in order to lift themselves up. Obviously, this is insecurity.

In extreme cases, people with stroke deficit may criticize themselves in an exaggerated manner by finding fault with their work or their own personality. With the use of positive reinforcement, individuals who have a stroke deficit may overcome their constant hunger for recognition of their behavior. Leaders should be aware of this problem and compliment workers or followers. Just a little attention can bring a big reward in terms of increased productivity.

FORMS OF POSITIVE REINFORCEMENT

Positive reinforcement comes in various forms to be used at the appropriate times. No matter which form is used, it must be noted that positive reinforcement must be used continually for it to be effective. Effective use of positive reinforcement effectively means varying the form of the reinforcement. If you always praise someone with the same words, your praise may appear rehearsed and therefore not genuine.

Confirmation Behaviors

Confirmation behaviors, as Evelyn Sieberg calls them, are positive behaviors that affect the recipient's feelings of self-worth. The six confirmation behaviors include orientation, praise, courtesy, active listening, positive written communication, and performance reviews.[16]

Orientation A proper orientation consists of an adequate introduction to an organization's procedures, policies, and environment. An individual who is well acquainted with his or her surroundings is more likely to perform well and to feel like an important part of the organization.

Praise The least difficult and most powerful form of recognition to give an individual is praise. Praise means to express approval of a behavior and to commend its worth. The use of praise as a reinforcer ensures the repetition of the desired behavior. It also reassures the individual that his or her behavior is being respected.

In *The One-Minute Manager*, Kenneth Blanchard and Spencer Johnson stated that praising works well when you

- Tell people up front that you are going to let them know how they are doing.
- Praise people immediately.
- Tell them what they did right; be specific.
- Tell them how good you feel about what they did right and how it helps the organization and others who work there.
- Stop for a moment of silence to let them feel how good you feel.
- Encourage them to do more of the same.
- Shake hands or touch people in a way that makes it clear that you support their success in the organization.[17]

Courtesy Another confirmation behavior that recognizes people's need for acknowledgment is courtesy. Courtesy is simply respecting others and being polite, helpful, and considerate. It also means showing each individual that he or she is important.

Active Listening Active listening consists of showing the speaker that you were indeed listening by interpreting what the speaker has just said. This shows the speaker that you do care about what he or she has to say and makes the speaker feel important.

Positive Written Communication Notes and letters that express positive recognition can reinforce behavior. Positive written communication refers to any written letters that express appreciation for a well-performed behavior. Positive written communication lets the recipient know that the writer of the letter cares enough to take time to express his or her consideration. This form of confirmation behavior can lift an individual's spirits, raise his or her self-esteem, and increase the likelihood that the desired behavior will occur again. Refer to Figure 16–8.

Dear Carlos,

What a great job you did organizing the food drive for those laid off from the factory. Many people have told me how impressed they were with your compassion and dedication. I totally agree. Your leadership ability continues to emerge daily.

Keep up the good work.

Sincerely,

Juan Clemente, Principal

Grove City High

Figure 16–8 Positive letters such as this one should raise self-esteem and cause the behavior to be repeated.

Performance Reviews A timely performance review by a higher-up is an insightful way of reinforcing employee behavior. Instead of criticizing the performance, an effective review focuses on the positive aspects and recognizes them. This meaningful recognition motivates the individual to strive for more positive feedback in future performance reviews; thus, it reinforces the behavior.

Barriers to Positive Reinforcement

Although recognition is easy to give, it is often avoided because of certain barriers. People like to hear how well they are doing, and they appreciate the recognition that they may receive for a job well done. Why then is positive reinforcement so often neglected?[18]

Preoccupation with Self Self-preoccupation interferes with an individual's ability to acknowledge the performance of others. A healthy ability to take care of your own business should not interfere with your relationships with others. You should be able to set goals and follow your dreams without completely ignoring other people. Preoccupation with self can make a person focus on him- or herself to the exclusion of others who deserve attention.

Misconceptions Misconceptions are false ideas or thoughts that arise from lack of information. Some common misconceptions about positive reinforcement are that it takes power away from the giver, that receivers of positive feedback will want more rewards, and that individuals do not deserve any reward beyond what they work for, such as a grade or a paycheck. Others believe that once a person receives positive reinforcement, she will not work as hard. In reality, the opposite is true. These misconceptions can be damaging to relationships and to the motivation to perform the desired behavior. By understanding how positive reinforcement works, individuals can overcome this obstacle.

"Too-Busy" Syndrome Some people feel that they are too busy to recognize the work of others or even to review their performance. To overcome this barrier, individuals must put time aside to plan for reviews and recognition. Just a few minutes is all that is necessary to offer verbal praise or to write a note letting someone know that her behavior is respected and acknowledged. Good leaders will find the time. Even a short note can do wonders (Figure 16–9).

Fear of Sounding Rehearsed Sometimes it is difficult to know what to say or do to recognize behavior. When an individual says or does the same thing over and over again, it seems rehearsed and insincere. It will eventually fail to reinforce the desired behavior. There are several verbal and nonverbal ways to show appreciation and approval. For example, you could say "Great thinking" or "Keep up the good work." You could also make direct eye contact, pat someone on the back, or simply smile to let someone know you

Nadine,

You did a great job on this project. I appreciate your hard work and attention to details. You are really an asset to the company.

Thanks again,

SCR

Figure 16–9 It does not take much time to write a note like this. The results are amazing.

approve of his or her work. Some actions demonstrate approval, such as asking for advice, displaying a person's work, or recognizing a person's work in public. Any of these ideas, along with a variety of others, can help an individual give positive feedback to others.

Lack of Role Model Some people have never encountered an individual who can give effective positive reinforcement. They have no idea where to begin because they lack role models who use positive reinforcement. This can be a barrier, but it can be overcome by individuals using positive reinforcement themselves. Positive attitudes can be contagious. By becoming a role model, an individual ensures that positive reinforcement will continue to be a powerful source of motivation.

Environment The environment must be a positive one for motivation to flourish. People who work together should be respectful of each other and supportive of a positive environment. Positive reinforcement will soon disappear in an atmosphere of negativity and dismal attitudes.

As a leader, make the necessary changes to build a positive work environment. A positive work environment includes, in order of importance, interesting work, full appreciation of efforts, involvement, good pay, job security, promotion and growth, good working conditions, loyalty to employees, help with personal problems, and tactful discipline. Good leaders have control over most of these.

Conclusion

Motivation comes from the word *motive*. It is a key component of the classroom and the workplace. Our needs result in a drive or motive to perform certain behaviors. For example, hunger (need) is the motive for us to eat (behavior).

Positive reinforcement is the best way to motivate people. It is also a good method of improving interpersonal relationships. People who understand and use positive reinforcement are more likely to be successful in their careers and in their relationships. By understanding that all humans need some kind of positive feedback, we can better understand ourselves and our behavior.

Summary

Motivation is a key component of the classroom. As a matter of fact, it is a key to success throughout life. We do homework because of motivation. We work because of motivation.

Motivation is the individual's desire to demonstrate constructive behavior and reflects willingness to expand effort. There are two types of motivation: Internal motivation, which is an inner force or power that helps a person achieve a goal, and external motivation, which is any outer force or power that helps a person achieve a goal.

To understand what motivates a person, we must understand five important characteristics of motives: Motives are individualistic, motives can often change, motives can be unconscious, motives can often be influenced, and motives are hierarchical.

Theories about motivation fall into two types: content and process theories. Content theories focus on factors within the individual that cause the individual to behave in a particular manner. Process theories examine how the behavior is motivated. Theories of motivation include Maslow's hierarchy of needs, ERG (existence, relations, growth) theory, expectancy theory, Herzberg's two-factor theory, learned needs theory, equity theory, and reinforcement theory.

There are four types of reinforcement: positive reinforcement, avoidance learning, extinction, and punishment. Positive reinforcement is the use of pleasant consequences or rewards for a desired behavior. Avoidance learning is often called negative reinforcement. Extinction is the removal of a positive reward. Punishment provides a negative consequence for an undesirable behavior.

The timing of rewards or punishments is known as reinforcement scheduling. Reinforcement scheduling has two classifications: continuous and intermittent. With continuous reinforcement, every time the desired behavior occurs, the individual is rewarded. With intermittent reinforcement, the positive consequence occurs after a passage of time or after an output. Four types of intermittent scheduling from which to choose an effective reinforcement schedule are fixed interval schedule, variable interval schedule, fixed ratio schedule, and variable ratio schedule.

Positive reinforcement is the best motivator. It is also a good method of improving interpersonal relationships. People who understand and use positive reinforcement are more likely to be successful in their careers and relationships. Individuals yearn for some type of appreciation or positive feedback. Closely related to positive feedback are the need to belong and self-esteem.

Eric Berne's work also identifies the need for positive reinforcement, in the form of strokes. *Stroking* is a term used to describe various ways in which a person may recognize another person's behavior. There are physical strokes and verbal strokes; a lack of strokes is called a stroke deficit.

Confirmation behaviors are positive behaviors that affect feelings of self-worth in the receiver of the confirmation. Six confirmation behaviors are orientation, praise, courtesy, active listening, positive written communication, and performance reviews.

There are also six barriers to positive reinforcement. These are preoccupation with self, misconceptions, "too busy" syndrome, fear of sounding rehearsed, lack of role models, and a negative environment.

End-of-Chapter Exercises

Review Questions

1. Define the Terms to Know.
2. Explain the relationship between (a) motive and motivation, and between (b) need and behavior.
3. What is the difference between internal and external motivation?
4. What are five characteristics of motives?
5. What are the five levels of Maslow's hierarchy of needs?
6. How does the ERG motivation theory differ from Maslow's motivation theory?
7. What are the five major parts of the expectancy theory of motivation?
8. Explain the difference between hygienes and motivators in the Herzberg two-factor theory of motivation.

9. What are the three essential needs of the learned needs theory?
10. List the methods used to reduce a perceived inequity.
11. Name the four types of reinforcement.
12. Compare avoidance learning and punishment in reinforcement theory.
13. Explain the four types of intermittent reinforcement scheduling and their effects on behavior.
14. What are six forms of confirmation behaviors?
15. What are five barriers to positive reinforcement? Give one example of each and how they can be overcome.

Fill-in-the-Blank

1. By using a ________________ ________________, you can recognize the particular behavior by smiling or by patting the person on the back.
2. ________________ ________________ ________________ is any letter that expresses appreciation for a job well-done.
3. Any exterior elements that make up your appearance are referred to as ________________.
4. Orientation, praise, and active listening are examples of ________________ ________________.
5. ________________ comes from the word motive.
6. ________________ ________________ represents the maximum use of an individual's talents, abilities, and skills, according to Maslow.
7. ________________ ________________ is the best motivator.
8. ________________ are false ideas or thoughts that people form because of the lack of information.
9. ________________ may be the least difficult and most powerful form of recognition to give to an individual.
10. ________________ is the removal of a positive reward.

Matching

_____	1. To express approval of a behavior	A.	verbal stroke
_____	2. Outer forces that help a person achieve a goal	B.	equity
_____	3. Belief that one is being treated fairly in relation to others	C.	extinction
_____	4. Removal of a positive reward	D.	praise
_____	5. Undesirable consequence for an undesirable behavior	E.	punishment
_____	6. Any recognition that occurs verbally	F.	external motivation

Activities

1. Select two people you know well. On a piece of paper, write three factors that you believe would motivate each of them. Then ask the same two people what would motivate them to do their best.

Compare their answers to yours. Did you accurately judge their motives? How could knowing what motivates others be helpful to you?

2. Analyze Maslow's hierarchy of needs. At which level are you presently? Explain your answer.
3. Demonstrate the use of each of the reinforcement tools listed below for reinforcing a behavior.
   a. Avoidance learning
   b. Extinction
   c. Punishment
   d. Positive reinforcement
4. Refer to the needs associated with the learned needs theory. Write a paragraph for each need describing a person for whom that need is dominant.
5. Practice giving someone "physical strokes."
6. Practice giving someone "verbal strokes."
7. Review the confirmation behaviors. Practice each one on a classmate. Then reverse roles.
8. **Take it to the Net.**

   Explore motivation on the Internet. The Web sites listed below contain information on motivation, including motivational quotes, articles, stories, and moments. Browse the sites and find something you find motivating, whether it be a quote or an inspirational story. Print out or write down what you found and share it with the class. If you are having problems with the Web sites or want more information, use the search terms below.

   Web sites

   <http://www.motivationalquotes.com/>
   <http://www.quotationspage.com/mgotd.php3>
   <http://www.motivateus.com/>
   <http://www.motivational-messages.com>

   Search Terms

   motivation
   motivational quotes
   motivational theories
   inspirational

Notes

1. B. L. Reece and R. Brandt, *Effective Human Relations in Organizations*, Fifth Edition (Boston: Houghton Mifflin, 1993), p. 152.
2. Ibid., p. 153.
3. C. Beryman-Fink, *The Manager's Desk Reference* (New York: AMACOM, 1989), pp. 156–157.
4. A. H. Maslow, Motivation and Personality (New York: Harper & Row, 1954).
5. C. R. Alderfer, *Existence, Relatedness, and Growth: Human Needs in Organizational Settings* (New York: Free Press, 1972).
6. V. H. Vroom, *Work and Motivation* (New York: Wiley, 1964).
7. F. Herzberg et al., *The Motivation to Work* (New York: Wiley, 1959).

8. R. N. Lussier, *Human Relations in Organizations: Applications and Skill Building* (Columbus, OH: McGraw-Hill Higher Education, 1998).
9. D. C. McClelland, "Business Drive and National Achievement," *Harvard Business Review*, July-August 1978, pp. 99–112.
10. J. S. Adams, "Toward an Understanding of Inequity," *Journal of Abnormal and Social Psychology*, November 1963, pp. 422–436.
11. D. F. Jennings, *Effective Supervision: Frontline Management for the '90s* (St. Paul, MN: West Publishing Company, 1993), pp. 154–155.
12. B. F. Skinner, *Science and Human Behavior* (New York: Macmillan, 1953).
13. R. L. Daft and R. M. Steers, *Organizations: A Micro/Macro Approach* (Glenview, IL: Scott, Foresman, 1986).
14. R. D. Pritchard et al., "The Effects of Continuous and Partial Schedules of Reinforcement on Effort, Performance, and Satisfaction," *Organizational Behavior and Human Performance* 25, 1980, pp. 336–353.
15. Berne quoted in Reece and Brandt, *Effective Human Relations in Organizations*, p. 271.
16. E. Sieberg, "Confirming and Disconfirming Organizational Communication," *Communication in Organizations*, ed. James L. Owen et al. (St. Paul, MN: West Publishing, 1976), p. 130.
17. K. Blanchard and S. Johnson, *The One-Minute Manager* (New York: William Morrow and Company, 1982).
18. Reece and Brandt, *Effective Human Relations in Organizations*, p. 282.

Personal Development

CHAPTER

17

Self-Concept

Objectives

After completing this chapter, the student should be able to:

- Discuss the importance of self-concept
- Discuss the ingredients of self-concept
- Discuss the factors that affect the development of self-concept
- Develop a positive self-concept
- Describe the characteristics of people with a high self-concept
- Explain how leaders can raise the self-concept of others

Terms to Know

self-concept
conceit
self-esteem
self-image
self-confidence
self-determination
motivation
self-responsibility
resilient
fear
doubt
anxiety
surface analysis
desire
action
self-fulfilling prophecy
Pygmalion

An individual's self-concept is the core of his personality. It affects every aspect of human behavior: The ability to learn, the capacity to grow and change, the choice of friends, mates and careers. It is no exaggeration to say that a strong, positive self-image is the best possible preparation for success in life.

—Dr. Joyce Brothers

An effective leader possesses a wide variety of skills, as you have learned in previous chapters. The effective leader must use his or her own abilities to master these skills to achieve personal goals. The leader also encourages and motivates others to achieve their goals as well as the goals of a group or organization.

Where do these abilities come from? How can you gain these abilities? In this chapter you will learn that self-concept is a key ingredient in mastering the abilities of an effective leader. You will learn the definition of self-concept, why it is important, what affects a person's self-concept, and what you can do to improve your self-concept.

Figure 17–1 A person with a positive self-concept is pleasant, emotionally secure, and content, like the person shown here.

Becoming an effective leader begins with self, and that self is important and valuable. There is only one you. You are unique! You can make a difference!

To have a positive **self-concept** is to respect yourself, to be aware of your weaknesses as well as your strengths. You must believe in yourself and accept yourself to have a positive self-concept. A person with a strong self-concept is able to accept mistakes or setbacks without shame, guilt, or blame. He or she accepts the weaknesses and mistakes of others. As you develop a positive self-concept, the feeling that you need to prove yourself to others will lessen. You can approach tasks without fear of failure or defeat. Relationships with others will become stronger. More people will want to be around you (Figure 17–1).[1]

IMPORTANCE OF SELF-CONCEPT

When your self-concept improves, your performance improves. You cannot consistently perform in a manner that doesn't reflect the way you see yourself.[2]

Having a positive self-concept is the single most important factor in achieving success. If you do not believe in yourself, how can you expect to gain the confidence of your peers? To operate in this world, you must have this type of belief system. For example, if you apply for a job that you don't think you can do, an employer is not likely to hire you. To gain the respect of others, a leader must be able to prove himself or herself through a positive self-concept.

Self-acceptance is one thing that none of us can live without. If we cannot accept ourselves for what we are, we lessen our chances of being happy, productive, and successful. This does not mean that we as individuals should stop striving for improvement; it simply means that we should be able to realize our faults and weaknesses. For example, if we cannot accept the fact that we are not good spellers, we may never be good spellers because we will not spend the extra time and effort to achieve this goal. Furthermore, self-acceptance involves understanding our strengths and weaknesses and gives us the ability to reach our goals and become successful.

Individuals do not have to play false roles when they develop positive self-concepts. These false roles may include trying to convince others everything is going well or blaming others for your present situation. A person with a positive self-concept will not be bitter, but better.[3] Striving for self-esteem brings out the potential for leadership in an individual. A positive self-concept can lead to many rewards.[4]

- Stronger self; more confidence
- Trust in one's own ideas, skills, and knowledge
- Ability to turn opportunities into realities
- Using mistakes to learn and improve
- Gain in endurance and fortitude
- Fears and obstacles do not stop progress
- More dynamic and interesting personality
- Gain in social approval
- Ability to focus thoughts on bigger aspects of life
- Emotional security
- Inner courage
- Ability to control one's future, to create circumstances instead of following circumstances
- Ability to handle success; keeping your head in the clouds and your feet on the ground at the same time
- Having positive feelings for and from others

Positive Self-Concept vs. Conceit

Since no one likes an arrogant person, it is necessary to explain the difference between self-concept and **conceit**. Conceit is the excessive regard for one's own worth. Although self-esteem also means feeling good about one's worth, the key difference is the amount. Excessive self-regard leads to bragging about how great we are. A conceited person is boastful and arrogant. On the other hand, a person with a positive self-concept does not brag or voice his or her own self-satisfaction. Because these people naturally shine through to others, it is usually others who make public the worth of individuals with strong self-concepts. Although there may seem to be a fine line between conceit and positive self-concept, when attaching qualities to individuals, the differences between the two are apparent.[5]

Related Terms

At first glance, self-concept is like many terms. It may or may not have a specific meaning to you. Figure 17–2 lists similar terms.

Confusing? Which word means what? Obviously not all of these words have the same definition, but they do relate to a common idea or concept. This chapter uses the term self-concept because it will best combine the general idea or definition of all the words in Figure 17–2. You will first learn a general definition aided by the use of

Terms Associated with Self-Concept	
self-actualize	self-image
self-acceptance	self-importance
self-adjustment	self-improvement
self-affected	self-knowledge
self-asserting	self-perception
self-awareness	self-pride
self-belief	self-realization
self-concern	self-recognition
self-confidence	self-reflection
self-content	self-regard
self-determination	self-respect
self-discipline	self-responsibility
self-esteem	self-scrutiny
self-expression	self-trust
self-fulfillment	self-worth

Figure 17–2 Self-concept is like many other terms. Whatever term we use, the more we value ourselves, and the happier we are.

an analogy. You will then learn the individual parts that, when put together, result in self-concept.

Ingredients of Self-Concept

A positive self-concept can be achieved only through the understanding of its ingredients. Those ingredients are self-esteem, self-image, self-confidence, self-determination, and self-responsibility. Each part is important in creating a positive self-concept; just with the different ingredients that are needed to make a cake, each has its own role.

Self-Esteem

Self-esteem is an internal feeling. How do you feel about yourself? **Self-esteem** relates to how much you accept yourself and how you perceive your worth and value as a human being.

Self-esteem is the core or beginning of a positive self-concept. Low self-esteem tends to suppress the ability to explore and take on the challenges of life. Low self-esteem does not render an individual totally disabled. Many people have a low opinion of themselves: They have jobs, get married, have children, buy homes, and more. The difference is within themselves. They are generally unhappy or dissatisfied with life and usually have a measure of hopelessness and despair much like the person in Figure 17–3.

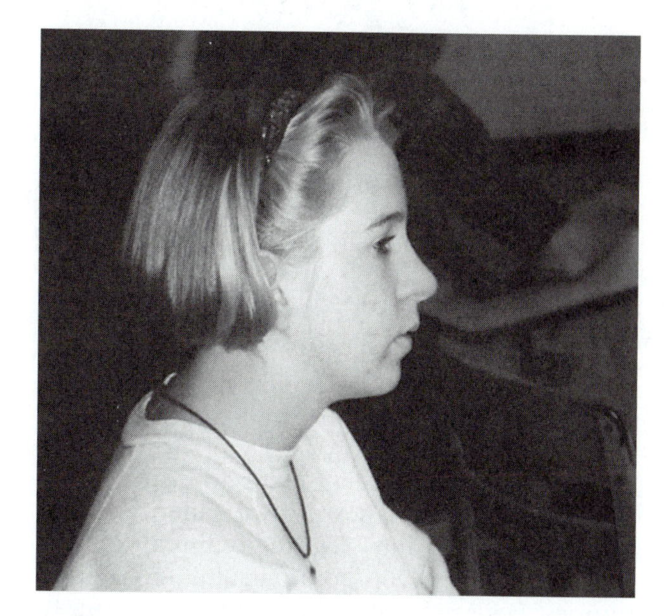

Figure 17–3 A person with low self-esteem is generally unhappy or dissatisfied with life. Friends can help such a person overcome hopelessness and despair and start working toward building a positive self-concept.

To enjoy life is what life is all about, but the enjoyment begins with you. You must feel good about the person within your body. If you have low self-esteem, you will struggle through life. If you have positive self-esteem, you can use it as a driving force and fully realize the pleasure of life.

Nathaniel Branden, in his book, *How to Raise Your Self-Esteem*, made the following comment: "Apart from problems that are biological in origin, I cannot think of a single psychological difficulty—from anxiety and depression, to fear of intimacy or of success, to alcohol or drug abuse, to under-achievement at school or at work, to spouse battering or child molestation, to sexual dysfunctions or emotional immaturity, to suicide or crimes of violence—that is not traceable to poor self-esteem. Of all the judgments we pass, none is as important as the one we pass on ourselves. Positive self-esteem is a cardinal requirement of a fulfilling life."[6]

Self-Image

Self-image relates to accepting yourself and presenting yourself in such a manner that you show confidence. Outward appearance is not a measure of a person's worth, just as money is not the measure of a person's success.

John Foppe was born without arms. As he grew up he had to deal with the pain of looking different. He had to endure the jokes and cruel remarks of others. Although life wasn't easy, John accepted his appearance and developed a positive self-image. Today, John is a very successful lecturer and leads a happy and rewarding life. John has a positive self-concept that has enabled him to overcome obstacles and achieve success. His self-image is a part of that overall self-concept.

Self-Confidence

Self-confidence is being secure in your ability to take on new tasks and develop new skills. Self-confidence is knowing that you can accomplish tasks that are presented you. People with a high level of self-confidence are eager to display abilities they know they possess. Confidence is also the ability to not allow fear to dominate your decisions or prevent you from pursuing new opportunities. Possessing confidence allows a person to grow and expand his or her horizons. It is amazing what can be done with self-confidence and persistence. Consider the accomplishments of the "late bloomers" below, who exhibited an abundance of persistence and self-confidence.[7]

- Beethoven's music teacher said, "As a composer, he is hopeless."
- Isaac Newton's work in elementary school was rather poor.
- Einstein couldn't speak until the age of four, and he couldn't read until age seven.
- Edison's teacher told him he was unable to learn.
- F. W. Woolworth's employer refused to allow him to wait on customers because he "didn't have enough sense."
- Louis Pasteur was given a rating of "mediocre" in chemistry at Royal College.
- Admiral Byrd was deemed "unfit for service" before he flew over both poles.
- Winston Churchill failed sixth grade.
- Walt Disney was fired by a newspaper editor because he had "no good ideas."
- Henry Ford was evaluated as "showing no promise."

Your confidence level can greatly determine how much success you enjoy in life. Have you ever said, "I just don't think I can do that"? That statement speaks to a low confidence level and is either based on fear or lack of motivation. Self-confidence is a state of mind that exists regardless of ability. Many people possess adequate skill but lack the confidence to use that skill. Many people with low self-esteem have been "taught" that they are not good enough and so they never take chances. Other people with low self-confidence have experienced a failure and do not want to experience failure again. In short, they are afraid and are willing to allow that fear to steer their decision-making process.

Skill level is increased when accompanied by a feeling of self-confidence. Most people develop certain skills that are almost second nature to them, allowing them to perform those skills with little thought to what they are doing. Remember when you learned to ride a bicycle? At first you had to concentrate on certain fundamental skills, and it was not easy. As skill level increased, so did your confidence. After a while you rode your bike and never thought about how you were doing it. The display of those skills requires confidence, and most people possess a high skill level in areas in which they feel confident.

Self-confidence helps overcome fear of rejection. The true test of self-confidence takes place when you are confronted with a situation where your skill level is not high. Will you be able to learn new skills and complete unfamiliar tasks? A high level of self-confidence allows you to confront the challenge and overcome the fear of failure and rejection.

Self-confidence is also the ability to understand your limitations. Confidence is a positive trait, but you must not allow yourself to become overconfident. When a person becomes overconfident, he is more vulnerable to failure. Take driving a car, for example. As you first learn to drive, you build confidence in your ability to perform that task. If you reach the point of overconfidence, you are more likely to take risks you would not normally take. You then become an unsafe driver because you create an environment for failure. Be confident, but understand your limitations.

Self-Determination

Self-determination deals with motivation from within. **Motivation** is a driving force that encourages you to seek out and accept new challenges and explore different areas that life will present. You will encounter certain outward motivations. Parents, teachers, friends, and others may provide motivation through words or actions and cause you to achieve and grow. The obvious rewards of a situation may motivate you to action. Money and prestige are examples of outward motivation. Everyone is influenced, to a degree, by outside motivation, but all people possess inward motivation or self-determination. Having the drive and determination from within is what allows you to continually push yourself beyond normal levels of accomplishment. Self-determination helps you overcome doubt and insecurity.

To be self-determined is to be in charge of your own fate. Nobody can be in total control of all situations and events, and many things will happen to you that you have little control over, but a self-determined person has the ability to work through those obstacles. This ability allows you to remain on a positive course. You will experience failure and disappointment, but do not let that deter you from moving forward. Who better to determine your fate in life than yourself?

Self-Responsibility

Self-responsibility is the ability to accept the consequences of any effort, good or bad. This ability is difficult to master because society has placed such a high value on success. Thomas Edison invented the light bulb, but his ultimate success was preceded by tremendous failure. He accepted the consequences of his failures but did not allow those failures to prevent him from reaching his goal. You will be faced with failure, and how you use that failure will in part determine what direction your life will go. Failure is not always a negative: You can use failure to evaluate yourself. Mistakes can be used in a positive way. Thomas Edison evaluated his failures, corrected mistakes, and eventually found success.

You need the ability to be **resilient**, or to bounce back. You need the ability to accept the consequences of any decision or venture regardless of failure or success. As you make choices, you must expect the unexpected and avoid placing blame on others. Blaming others is a way of avoiding responsibility, and it seems to be an increasing trend in our society. People seem less willing to accept responsibility for their own actions and want to blame somebody else. Self-responsibility is also the ability to know when you need help or assistance. Do not hesitate to seek assistance if you need to, but do not develop a habit of always depending on others. Seeking help from others is not a weakness; dependence is.

Factors That Affect the Development of Self-Concept

Factors that affect the development of an individual's self-concept can be classified as chronological, internal, or external. These three areas are closely connected. Many things that affect your self-concept originate from your environment, an external factor. All external factors affect how you feel, think, and react. First, we examine the chronological factors, then the external factors. Last, we examine the internal factors. Figure 17–4 shows the relationship between chronological, external, and internal factors.

Chronological Factors

You are not born knowing who and what you are to become. Carl Rogers, a well-known psychologist, believed that self-concept is the guiding principle that structures our personality.[8] You acquire your image of yourself over time by

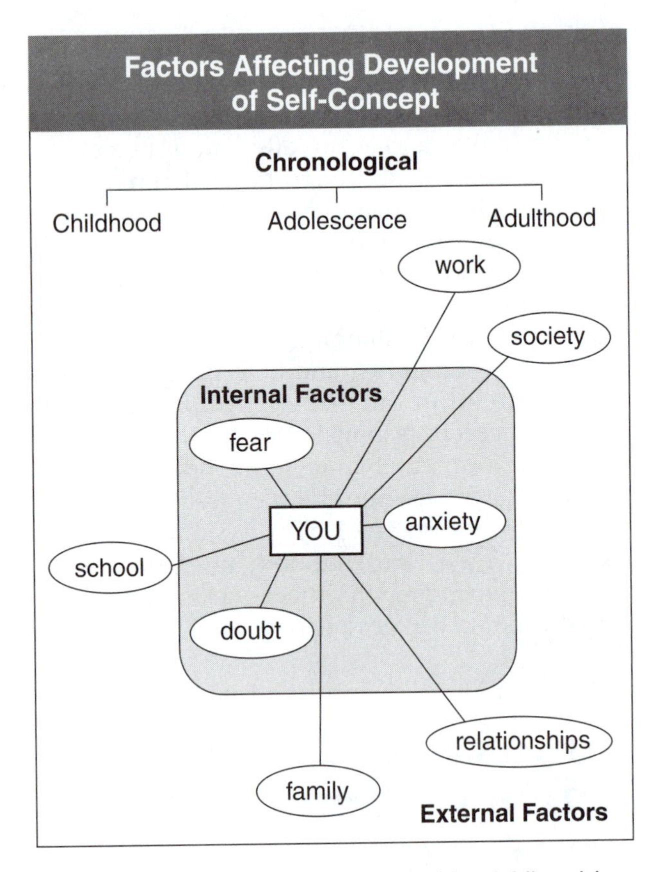

Figure 17–4 The self-concept formed in childhood lays the foundation. As you get older, you start comparing yourself to other people. By the time you reach adulthood, your self-concept has been molded by many internal and external factors.

realizing your natural abilities and by constantly receiving messages about yourself from the people around you. Your self-concept is developed in your childhood mostly by your parents. As children approach adolescence and adulthood, teachers, coaches, friends, and others also have an influence on their self-concept.[9]

Childhood The self-concept formed in childhood lays the foundation for your attitude toward work, your future success, your personal abilities, and the roles you play. Your family is the first source of information about yourself. Every person a child comes in contact with leaves a mark. Parents do not particularly teach their children self-concept, but they shape it with positive and negative messages,[10] as shown in Figure 17–5:

Little did you realize you accepted these messages as truths and recorded them in your memory. Your subconscious mind gradually developed a self-concept, whether accurate or distorted, that you came to believe was real.

Adolescence As you get older, you start comparing yourself to others. Typically, you become less happy with who you are. You may wish you were more like others you perceive as better. Some may begin to use putdowns as an equalizer. Often, teenagers criticize or ridicule others in an attempt to remove insecurities about themselves and their negative self-concept.[11]

The media plays a strong part in how adolescents perceive themselves. Television and movie characters present unrealistic images to adolescents, which they may use to measure their own attributes and lifestyles. Unfortunately, it is easy to feel deficient in the multi-million-dollar fantasy media world.

The ages of 12 to 18 are critical years in developing your self-concept. You are moving away from the close bond between your parents and are attempting to establish ideals of independence and achievement.[12] You must also deal with physical changes; relationships with peer groups; an emerging, often confusing identity; the loss of childhood; and the assumption of some adult responsibilities. In fact, many people never move beyond the self-concept they had of themselves while in high school, and they continue into adulthood trying to prove to their classmates that they can "make it."[13]

Adulthood When you reach adulthood, your self-concept has been formed by those in your environment from all your past experiences. You may compare yourself to others, as you did during adolescence, or you may focus on your own inner sense of self-worth. There are three ways that adults tend to define themselves.[14] Each can influence self-concept. We may define ourselves

Negative	Positive
Bad boy! Bad girl!	You're great!
You're so lazy!	You can do anything!
You'll never learn!	You're a fast learner!
What's wrong with you?	Next time you'll do better.
Why can't you be more like . . . ?	I like you just the way you are.
It's all your fault.	I know you did your best.

Figure 17–5 As children we were told, "Sticks and stones may break my bones, but words will never hurt me." This is simply not true. Words play a big part in psychological development, and negative words can damage self-concept.

- *In terms of the things we possess.* This is the most primitive source of self-concept. We buy material things such as cars, houses, and land to achieve a greater self-concept. People who define themselves in terms of what they have may have difficulty deciding how much is enough and may spend their life in search of more material possessions.[15]
- *In terms of what we do for a living.* Too often our self-concept depends on our job titles. We look to external forces such as a corporation, the school or university, the media, counselors, friends, or our parents in selecting a job or career. People who have been pushed into jobs they don't like may make ample money but hate going to work, even though their job is admired by others.[16]
- *In terms of internal value system and emotional make-up.* Emmett Miller, in his book, *The Healing Power of Happiness*, says this is the healthiest way for people to identify themselves. Miller also states that if you don't give yourself credit for excellence in other areas of life, besides your job and material possessions, you have nothing to keep your identity afloat in emotionally troubled waters.[17]

It is important to learn how to protect your self-concept against those who criticize you and listen to those who encourage and challenge you. This can help you determine the difference between what is helpful and what is destructive, what is correct and what is incorrect. It can also help you expand the range of what you believe you can be and do in the future.[18]

External Factors

Family, relationships, school, work, and society all have an effect on your level of self-concept.[19] The people and situations that confront you daily have either a positive or negative effect on you. Learning more about these external factors will give you a better understanding of what you face. People have a chance to resolve problems when they fully understand the problems. Dealing with the unknown is always difficult.

Family The home environment plays a large role in the development of a person's self-concept. The years from birth to age 18 are considered the formative years. During these years the family environment is highly influential in a person's development. The family environment satisfies the very basic human needs of survival, safety, and security. A child must depend on family for food, shelter, and clothing, but beyond the basic needs, the family should be a place in which a child experiences love and acceptance, praise, and constructive criticism.

Praise children for small displays of competence, beginning at an early age. Arrange "success experiences" for young children so that they come to expect positive results. Avoid negative comparisons among siblings. Let children hear you praise others for behaviors they themselves demonstrate. Pass on to a child compliments you hear from others, and give positive support to children with a smile, a pat on the back, or any other appropriate gesture. The child should establish certain values, such as respect, responsibility, discipline, and cooperation, as well as the ability to distinguish between right and wrong.

Relationships The people that you come in contact with outside of the family constitute your network of past, present, and future relationships. These people exert influence on you and either consciously or unconsciously place on you certain demands and expectations. Some people you may encounter only briefly: a camp counselor, a guest speaker at some public function, or even someone you meet in a department store. Today's world is mobile, and people don't always stay in one spot for long. The length of time is not always a gauge of the impact others can have on you.

Other relationships are ones that last and remain a real part of your everyday life. The friends you choose reflect your level of self-concept. More than likely, your friends are individuals with whom you have many things in common. You and your friends probably have common likes and dislikes, enjoy common activities, and share common interests. Friendships can present a mirror of yourself—not necessarily a physical image but an image of your feelings, thoughts, likes, and dislikes. This is not to imply that friends always share a common interest in everything. You might enjoy viewing a basketball game with certain friends and go to a movie in the company of others. The key in your relationships with friends is that you tend to build friendships with people who share your level of self-concept.

Your relationships will expand beyond that of immediate friends. Eventually you will branch off into the area of love, marriage, and possibly the beginning of your own family. This area of relationships can be extremely rewarding. Mixed with the rewards are new responsibilities and demands that will test and build you and your self-concept.

School As a small child you expand your environment to more than just the home. You will spend a major part of the next 12 to 16 years of life in school. This new environment offers an entirely new set of challenges to the individual. Education today is asking more and more from students. Students must develop and use skills that are different from those of a previous generation.

School can be an exciting place, filled with newfound challenges and experiences. It can awaken your mind in ways you never dreamed possible. The opportunity exists to discover a new and wonderful you that you may never have been aware of. The benefits and new discoveries you make will be your responsibility. Your teachers provide the opportunity, but you must take that opportunity and turn it into reality. The responsibility is on your shoulders and will have profound effects on you later on in life, as well as now. You can decide to accept challenges or allow them to pass you by. The school environment, because of the new challenges, is an excellent place to begin to build your self-concept. Have you ever passed up the chance to go somewhere because at the time it didn't interest you? Have you ever regretted that choice when others returned from that event with great stories about how wonderful it was? Get involved—don't allow this opportunity to pass without taking advantage.

Peer pressure can be intense, and the demand on you to follow the crowd may be unavoidable. To follow is not always bad, just be careful of who and why you follow. If you must follow, the best course to follow is that of being a leader. Leaders can be found in the forefront and are not lost in the crowd. The groups in Figure 17–6 will have an influence on each other's self-concept.

Figure 17–6 Although these two groups of people are from different cultures and live thousands of miles apart, their kind actions toward each other will help build their positive self-concept. (Courtesy of *FFA New Horizons Magazine*.)

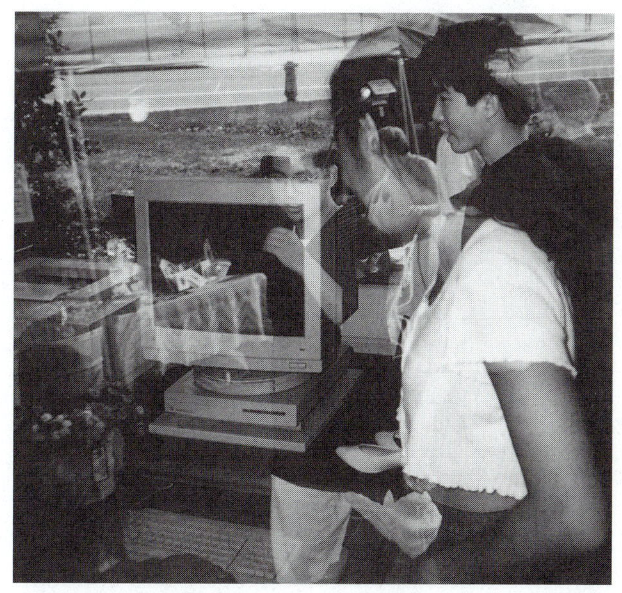

Figure 17–7 Developing your self-concept involves family, relationships, school, work, and society. They are all interrelated and affect you as a person.

Work When the time comes to leave your school environment, you will be thrust into yet another environment—the world of work. Companies compete for people who possess a high self-concept because they have the necessary tools to be effective employees. If you are a good employee, a company profits from your contributions. The stronger your self-concept, the more tools you possess. The more tools you possess, the more in demand you will be, which gives you many choices to select from. Finally, the more choices you have, the more likely your chances of enjoying your job. Figure 17–7 illustrates the complexity of developing self-concept.

Internal Factors

How you think, feel, and act are affected by the world around you. Some internal factors you must deal with are fear, doubt, and anxiety.

Fear is overwhelming anticipation or awareness of danger. Overcoming fear of failure or rejection requires determination. People shelter themselves from fear through avoidance. Employees pass up advancement opportunities because they are fearful they won't be able to succeed. Opportunities for relationships are passed up because of the fear of rejection. Fear is a basic human emotion that serves a purpose—it forces us to be cautious. The avoidance of fearful situations altogether stems from a low self-concept. Our emotions should assist us, not control us.

Doubt is a state of questioning our ability to learn, think creatively, make decisions, accomplish, and succeed. "I can't do that" is a phrase many people use too often. The phrase should be, "I question my ability to do that." When you are faced with a situation you have no control over, you may not be able to succeed, but that's no reason not to give it a try.

Anxiety is having an uncomfortable feeling or uneasiness about a situation or event. The phrase "having butterflies" defines anxiety in a way you can probably relate to. Of the three

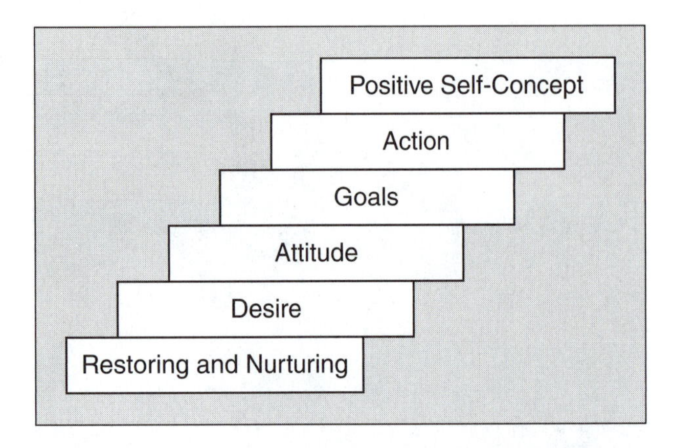

Figure 17–8 There are five steps to achieving a positive self-concept. Once each step is secure, you are ready for the next step.

negative internal factors affecting self-concept, anxiety is the easiest to overcome.

DEVELOPING A POSITIVE SELF-CONCEPT

Five areas contribute to developing a positive self-concept. Think of these five areas as a set of steps. To reach the top you must go up each step, one at a time (Figure 17–8). At the top of the steps is the desired destination, a positive self-concept.

Step 1

Restoring and nurturing a healthy self-concept.

The first step toward developing your self-concept is to accept yourself as you are now. The past cannot be changed, but the future is determined by how you think and act. The practices presented below have been developed to help people restore and nurture a healthy self-concept.

- Identify and accept your limitations.
- Learn to accept others.
- Make a list of your greatest talents.
- Make decisions.
- Stop procrastinating.
- Develop expertise in some area.
- Find a mentor.
- Avoid **surface analysis** of yourself and of others.
- Dress as if you are already successful.
- Use positive self-talk.
- Continue to learn and be observant.
- Tackle the things you fear.
- Choose your friends and associates carefully.
- Learn from failures.
- Do quality work; then compliment yourself.
- Smile and compliment others.
- Speak up and share your views.
- Forgive yourself as you would others.
- Seize the opportunity to learn new skills.
- Stand up for others.
- Go the extra mile to do your best.
- Believe in yourself.
- Do good for others.
- Make a list of your positive qualities and occasionally refer to it.
- Make a victory list of your past successes.
- Join an organization that requires your involvement.
- Look people in the eyes when speaking.
- Refuse to give permission to anyone to make you feel inferior.
- Finish every job you start.
- Change enemies into friends.
- Practice good manners.
- Learn to love to read.

Step 2

Desire.

You must begin your development of a positive self-concept with a genuine **desire** to change. Desire is a state of mind, a longing or hope for something. Have you ever wanted something so much that you couldn't imagine what life would be like without it? The desire to possess is strong in humans. Advertising is such a big industry because companies must strive to provoke a desire for their products. Examine the benefits of a positive self-concept and decide for yourself if you possess a genuine desire to change. When you make that decision, you will normally be faced with some doubts and anxiety. Change breeds uncertainty and is not to be feared.

Step 3

Attitude.
When you have completed the second step you will need to create the proper environment to allow your desire to continue to grow. Attitude is a state of mind; it is feelings, beliefs, and outlook, and it can be either positive or negative. If your attitude is positive, then your level of desire can be maintained. If your attitude is negative, your desire diminishes. Previously, we stated that your attitude determines your altitude. What usually happens to a boy who is made to take piano lessons when he would rather be doing something else? The majority of the time he develops a negative attitude and ends up having little desire to continue to play the piano when he no longer has to.

Be aware of the attitudes of those with whom you associate. A positive attitude may turn negative simply by accepting another person's attitude. Remember, when people try to discourage you it is usually because they are insecure with themselves. Surround yourself with people who display a positive outlook. People with positive attitudes are a source of encouragement.

You determine your attitude. Being positive creates an environment for growth and helps you maintain a clear vision of where you want to go and what you want to be. A negative attitude is like driving a car in dense fog: You are unsure of what lies ahead and judgments will have to be made quickly. A positive attitude is like driving on a clear day when your vision is clear: You can see obstacles coming and are able to react in time to avoid disaster. You also can enjoy the scenery around you as you travel.

Step 4

Goals.
There is a relationship between your goals and your self-concept. Being equipped with desire and proper attitude is good, but now you must develop some direction or purpose. This is accomplished through the establishment of goals. Use grades as an example. If you set a goal of all A's, then the end to that goal is either reached or not reached when you receive your report card. In between is the effort put forth in attaining your goal. If your goal is reached, your self-concept is enhanced.

Goals are usually categorized into short-, medium-, and long-term. A short-term goal may be to make an *A* on a six-week test. A medium-term goal may be to maintain an *A* average for the semester. A long-term goal may be to graduate from school with a 3.0 grade point average. The feeling of success from achieving short-term goals enhances your self-concept, and this spurs you on to set and attain higher medium- and long-term goals.

Setting goals is best done through an organized process. Sit down with pen and paper and write out a step-by-step procedure so that you can have a map to keep you on track. Figure 17–9 is a good guide, or you can refer to Chapter 14 for other models. Organization is a key part of being able to set and reach goals.

Step 5

Action.
You have now reached the last step. **Action** represents a state of motion, either physical or mental. Without action, all you have are good intentions. Have you ever set out to do something

Procedure for Setting Goals

- **Statement of Goal**—A single-sentence statement that fully defines the goal.
- **Establish Time Frame**—Categorize the goal as short-, medium-, or long-term. Set a completion date.
- **Make an Agenda**—List all possible steps involved in reaching the goal and list possible obstacles to overcome. This is the how-to of the process.
- **Be Realistic**—Try to set goals that are attainable. If you have a long-term goal, it might be beneficial to set some short-term goals to be accomplished along the way.
- **Evaluate**—Take time to check progress periodically. A more efficient method is to set evaluation points in the agenda to check the level of progress.
- **Adjust**—Be flexible after evaluation. If the goal-reaching process you've designed isn't working, alter some aspect, such as time frame or steps involved. Add to or take out as necessary, but do not reach a point of stagnation.
- **Reward**—When the goal is reached, celebrate in some way the accomplishment that has been made. Reward yourself for a job well done.

Figure 17–9 Goals give direction and allow you to organize and set priorities in your life. A life void of goals is a life without direction. Without direction, you cannot reach your destination. Reaching goals enhances your self-concept.

and failed to accomplish it? Maybe you wanted to learn to play the guitar. You had the desire to learn to play. You had a positive attitude and were confident you had the ability. You even set a goal of learning to play. What's left? Action. You have to physically sit down and practice. You must put your good intentions into motion. Failure to practice would leave you with only good intentions.

Action is by far the hardest step to achieve because it requires energy, both physical and mental. It requires discipline to keep from getting off track. It requires determination to keep from getting discouraged. It requires persistence to try again when faced with setback and disappointment. It requires patience when things do not happen as fast as you want. When you take action you become vulnerable to the internal and external factors that affect self-concept.

Be prepared at all times to overcome the negative and maintain the positive. Remember that change is not magical. You cannot wish it and make it so. Achieving a positive self-concept is hard, but the benefits far outweigh the work.

You should constantly take on new challenges and be open-minded to new ideas. Everybody has certain talents and natural abilities. Build on those and then begin to expand yourself and see if you do not enjoy the new you. Keep remembering that self-concept is a natural part of everyone's existence. The key difference between individuals is whether their self-concept is positive or negative. A positive self-concept enables you to take full advantage of the exciting possibilities and opportunities life has to offer. Your self-concept development will encounter difficulty, but the reward can be a full, happy, and successful life.

A positive self-concept equips you with qualities and abilities that can be used in your education, career, and society. Becoming a leader requires qualities such as being innovative, creative, determined, responsible, and confident. These traits will enable you to be successful in life—successful by your standards. Success measured *by your standards* is an important distinction, because the world has an abundance of standards and guidelines. You may agree with and accept some of society's standards, but allow the decision on success to be yours and not that of society or other individuals. All aspects of your public life are in one of three areas (education, career, society), and a positive self-concept is a key part of each area.

Self-Concept and the Workplace

The advances in technology and science are creating tremendous progress as well as new problems. Today's world is more complex, more challenging, and more competitive than that of the last generation. The increased demands of our world create increased demands on the individual. Today, organizations need a higher level of knowledge and skill among all those who participate. In addition, there is a need for higher levels of independence, self-reliance, self-trust, and the capacity to exercise initiative—in a word, self-concept.[20] Others take you at the value you put on yourself. If you believe in your own ability, your co-workers and others will also believe in you and treat you with the respect that you have provided for yourself.[21]

Increases Productivity In the workplace, those with a positive self-concept are inclined to form nourishing relationships. These workers tend to do more than what is strictly required on the job. They are receptive to new experiences, new people, taking on responsibility, and making decisions.[22] Such people contribute to the well-being and productivity of the workplace.

Results of Low Self-Concept Employees with low self-concept manifest many of the following characteristics at work.[23]

- Negative thinking
- Fear of taking action
- Blaming others for failures
- Lying and covering up
- High absenteeism
- Rebelliousness
- Clock watching
- Increased illness
- Fear of asking for help

Reason for Hostility People with a low self-concept are likely to feel hostile, show a lack of respect for others, and attempt to retaliate against others to save face in a difficult situation. A hostile response to others is a natural outcome of a low self-concept. Our emotional system is controlled by a balancing mechanism. When our self-concept is threatened, the mechanism is knocked off balance and we start to feel hostile and anxious.[24]

Results of Leaders with a Low Self-Concept A person in a position of leadership who has a low self-concept is not likely to treat group members, co-workers, or peers fairly. Leaders with a low self-concept can affect the efficiency and productivity of a group because they tend to exercise less initiative, hesitate to accept responsibility or make decisions on their own, ask fewer questions, and take longer to learn procedures.[25]

Conditions Needed to Empower Self-Concept in the Workplace A workplace where the leaders and co-workers operate at a high level of self-concept would be a workplace of extraordinarily empowered (self-fulfilled) workers (Figure 17–10). The following conditions allow leaders to help their co-workers reach their full potential.[26]

- People feel challenged when given assignments that excite, inspire, test, and stretch their abilities.
- People feel recognized when acknowledged for individual talents and achievements and rewarded monetarily and nonmonetarily for extraordinary contributions.
- People receive constructive feedback. They hear how to improve performance in nondemeaning ways that stress positives rather than negatives and that build on their strengths.
- People see that innovation is expected of them. Their opinions are solicited, their

Figure 17–10 A workplace where the leaders and co-workers operate at a high level of self-concept is one of extraordinarily empowered (self-fulfilled) workers, like those above. (Courtesy of USDA.)

brainstorming is invited, and the development of new and usable ideas is desired of them and welcomed.

- People are given clear-cut and noncontradictory rules and guidelines. They are provided with a structure appropriate to their job description and they know what is expected of them.
- People see that their rewards for successes are far greater than any penalties for failures. In too many companies, in which the penalties for mistakes are much greater than the rewards for success, people are afraid to take risks or express themselves.
- People are treated fairly and justly. They feel the workplace is a rational universe they can trust.
- People are able to believe in and take pride in the value of what they produce. They perceive the result of their efforts as genuinely useful and their work as worthwhile.

Workers Perform to Expectations Workers tend to behave in a way that supports what they believe about themselves, often referred to as **self-fulfilling prophecy**. Your career successes and failures are directly related to the expectations you hold about your future. However, people can also be greatly influenced by the expectations of others. The **Pygmalion** effect sometimes causes people to become what others expect them to become.

Robert Rosenthal, a Harvard University professor, developed a theory based on a Greek legend about Pygmalion, the King of Cyprus. Pygmalion carved and then fell in love with a statue of the goddess, Aphrodite, who brought the statue to life. Pygmalion saw the statue as real, thus she became real. From this story, Rosenthal and others formed what has become known as the expectation theory ("you get what you expect"). Consider the following example.[27]

> ***An insurance sales manager was given a group of "average producers." They were motivated and instilled with high expectations. The salespeople accepted the challenge and outperformed another group of peers with "above-average" performance. Why? The sales manager insisted they had the ability to outperform others and did not permit them to think of themselves as "average" salespeople.***

When workers are expected to do great things, they do great things. Their self-concept is enhanced; thus, they believe in themselves. When people believe in their ability and have faith, great things can be accomplished.

Self-Concept in School and the Social Arena

You spend most of your time in school. School and society make many demands on each individual. As a member of the social arena, you will be confronted with many situations that will force you to make choices. When confronted with many of society's demands, you will come face to face with doubt, fear, despair, rejection, confusion, and more. People with a low self-concept will try to avoid these feelings and emotions. A positive self-concept will better enable a person to face society's demands by making sound decisions and judgments.

People with a positive self-concept tend to be outgoing. People with a positive self-concept tend to be more receptive to new experiences and meeting new people. They are willing to tolerate differences in others. They tend to go out of their way to greet people and meet their needs. Generally, people with positive self-concepts have more friends than people with negative self-concepts.[28] They are more willing to express emotions and share ideas with others, as the people in Figure 17–11 are doing.

People with a positive self-concept are humble. A positive self-concept does not mean being arrogant and boastful. If you know any people like that, you probably realize that their behavior could be improved by a change in attitude. People don't like others to flaunt their imagined, or real, superiority.[29]

People with negative self-concepts often have trouble relating to others. They tend to be pessimistic, and they don't trust others. Many people with negative self-concepts feel that others are out to get them. They are not able to accept a compliment without looking for an ulterior motive. Their lack of self-confidence often holds them back from being involved with the group. The group, in turn, often misinterprets the reason for lack of involvement as unfriendliness.[30]

Figure 17–11 People with positive self-concepts are outgoing and enthusiastic, like this group. (Courtesy of the National FFA Organization.)

A positive self-concept helps you cope. A positive self-concept does not shield you from difficult situations, rather it provides a method to battle them. A healthy immune system does not guarantee that a person will never become ill. It does make one less vulnerable to disease and better equipped to overcome it. The same is true of self-concept. Life's difficulties will cause you a degree of pain and anxiety, but you will rebound faster if you have a positive self-concept.

If self-concept is so important, why do so many people have a poor one? Several things contribute to this, such as the following.[31]

- We live in a society in which being negative and ridiculing others seems to be the popular thing to do.
- We confuse failure on a project with failure in life.
- We confuse being unintelligent with an untrained memory, when in reality we are not spending enough time studying.
- We tend to compare our experience with another person's experience and confuse experience with ability and intelligence.
- We compare our worst features to someone else's best features rather than concentrating on our best features or abilities to reach our goals in life.
- We set standards of perfection that are unrealistic and unreachable rather than concentrating on what we can do.[31]

When we have a poor self-concept, we may exhibit the characteristics or behaviors in Figure 17–12.

Critical and jealous nature	Too much emphasis on material things
Involvement in gossip	Lack of genuine friends
Improper reaction to criticism	Senseless and erratic actions
Improper reaction to laughter	Excuses to justify failure
An uncomfortable feeling when alone	Spur-of-the-moment, impossible promises
An "I don't care" attitude	Rebellion against authority
Breakdown in decency	Foolish and impulsive actions

Figure 17–12 Many of these behaviors are the result of not being secure within ourselves. As we become more secure and our self-concept improves, many of these behaviors will diminish.

Characteristics of People with a Positive Self-Concept

Once your self-concept is attained, improved, or enhanced, you will exhibit many positive characteristics. Barry Reece and Rhonda Brandt, in their book, *Effective Human Relations in Organizations*, list six characteristics of people with a high self-concept.[32]

1. *Future-oriented.* People with a positive self-concept look forward to the future and are not overly concerned with past mistakes and failures. These people believe every experience has something to teach if a person is willing to learn from them. Falling down does not mean failure. Staying down does.
2. *Able to cope with life's problems.* People with a positive self-concept are able to cope with problems and disappointments. Successful people realize that problems need not depress them or make them anxious. It is their attitude toward problems that makes all the difference.
3. *Able to deal with emotions.* People can't help the way they feel, but they can help the way they act. Those with a positive self-concept are able to feel all dimensions of emotion without letting those emotions affect their behavior in a negative way. By exhibiting self-control, people with a positive self-concept are able to establish and maintain effective human relations with people around them.
4. *Able to help others and accept help.* People with a positive self-concept are not threatened by helping others to succeed, nor are they afraid to admit weaknesses. Secure people in leadership roles surround themselves with good people. As a matter of fact, they surround themselves with people so good that these co-workers are capable of doing the leader's job. (First-rate people hire first-rate people. Second-rate people hire third-rate people.) In helping others, people benefit themselves as well.
5. *Able to accept people as unique, talented individuals.* People with a positive self-concept learn to accept others for who they are and what they can do. Acceptance of others is a good indication that you accept yourself.
6. *Able to exhibit a variety of self-confident behaviors.* They accept compliments or gifts by saying, "Thank you," without making self-critical excuses and without feeling obligated to return a favor (Figure 17–13). They can laugh at themselves without self-ridicule. They let others be right or wrong without attempting to correct or ridicule them. They

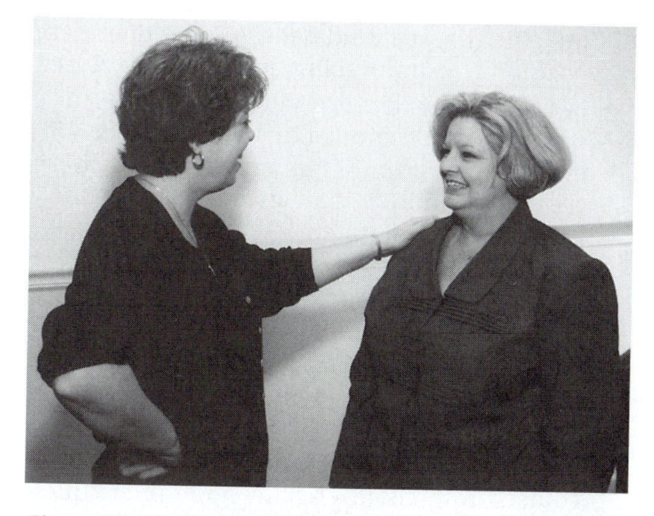

Figure 17–13 People with a positive self-concept accept compliments appropriately.

feel free to express opinions even if they differ from those of their peers or family. They enjoy being by themselves without feeling lonely or isolated. They are able to talk about themselves without excessive bragging.

Other Characteristics of People with a Positive Self-Concept

- Your face, manner, and way of talking and moving naturally project the pleasure you take in being alive.
- You are open to criticism and comfortable about acknowledging mistakes because your self-concept is not tied to an image of perfection.
- Your words and movements have a quality of ease and spontaneity, since you are not in conflict with yourself.
- There is harmony between what you say and do and how you look, sound, and move.
- You have an attitude of openness to and curiosity about new ideas, new experiences, and new possibilities because for you, existence is an adventure.
- Feelings of anxiety or insecurity, if they present themselves, are not likely to intimidate or overwhelm you, since it is possible to manage them and rise above them.
- You enjoy the humorous aspects of life in yourself and in others.
- You are flexible in responding to challenges and are moved by a spirit of inventiveness and even playfulness, since you trust your mind and do not see life as doom or defeat.
- You are comfortable with assertive (not belligerent) behavior—you are quick to defend and advocate yourself.
- You tend to preserve a quality of harmony and dignity under conditions of stress, as feeling centered comes naturally to you.[33]

Ways to Present a Positive Self-Concept

When you have developed a positive self-concept, you must exhibit it in your behavior. Most of the behaviors are natural nonverbal responses; others need some work. You should

- Smile on a day-to-day basis.
- Dress neatly and for the occasion.
- Be polite and considerate of others.
- Be an active listener.
- Take pride in your work.
- Be independent and make wise decisions.
- Make the best possible situation out of everything you do and have a positive mental attitude.

Even on the physical level, there may be noticeable changes as you gain a positive self-concept, such as the following.[34]

- Your eyes may well become more alert, bright, and lively.
- Your face will at some point become more relaxed and (barring illness) tend to exhibit natural color and good skin vibrancy.
- Your chin will probably be held more naturally and more in alignment with your body.

- Your jaw will tend to become more relaxed.
- Your shoulders typically will become more relaxed yet erect.
- Your hands will be relaxed, graceful, and quiet.
- Your arms will hang in a relaxed, natural way.
- Your posture will be relaxed, erect, and well balanced.
- Your walk will be purposeful, without being aggressive and overbearing.
- Your voice will be modulated with an intensity appropriate to the situation, and with clear pronunciation.

How Leaders Can Raise the Self-Concept of Others

For leaders to bring the best out in people, they must relate to them appropriately. In large part, this includes contributing to a positive self-concept in others. In doing this, the leader stimulates active and creative participation that allows for innovation. Nathaniel Branden suggests several things that you as a leader can do to raise the self-concept of group members or co-workers.[35]

- Work on your own self-esteem: Commit yourself to raising the level of consciousness, responsibility, and integrity you bring to your work and your dealings with people.
- Give people opportunities to practice self-responsibility: Give them space to take the initiative, volunteer ideas, attempt new tasks, expand their range.
- Give the reasons for rules and guidelines when they are not self-evident, explain why you cannot accommodate certain requests; do not merely give instructions.
- If you make a mistake in your dealings with someone, such as being unfair or short-tempered, admit it and apologize. Do not imagine that it would demean your dignity or position to admit taking an action you regret.
- Let people see that it is safe to make a mistake or say "I don't know, but I will find out." To evoke fear of error or ignorance is to invite deception, inhibition, and an end to creativity.
- Describe undesirable behavior without blaming: Let someone know if his or her behavior is unacceptable, point out its consequences, communicate what kind of behavior you want instead, and omit character assassination.
- If someone does superior work or makes an excellent decision, invite him or her to explore how and why it happened. Do not limit yourself to praise; by asking appropriate questions, help raise the person's consciousness about what made the achievement possible and thereby increase the likelihood that others like it will occur in the future.
- Praise in public and correct in private. Acknowledge achievements in the hearing of as many people as possible while letting a person absorb corrections in the safety of privacy.
- When the behavior of someone creates a problem, ask him or her to propose a solution. Whenever possible, avoid handing down solutions. Instead, give the problem to the responsible party, thereby encouraging self-responsibility, self-assertiveness, and intensified awareness.
- Give people the resources, information, and authority to do what you have asked them to do. Remember that there can be no responsibility without power, and nothing so undermines morale as assigning the first without giving the second.
- Find out a person's central interests and, whenever possible, match tasks and objectives with individual dispositions. Give people an opportunity to do what they enjoy most and do best; build on people's strengths.
- Ask people what they would need to feel more in control of their work and, if possible, give it to them. If you want to promote autonomy,

excitement, and a strong commitment to goals, empower people.

- "Stretch" people: Assign tasks and projects slightly beyond their known capabilities.
- Educate people to see problems as challenges and opportunities; this is one perspective clearly shared by high achievers and by people of positive self-concept.
- Support the talented nonteam player. In spite of everything we can say about the need for effective teamwork, there is a place for the brilliant hermit who is moving to different music; even team players benefit from seeing this respect for individuality.
- Write letters of commendation and appreciation to high achievers and ask other leaders to do likewise. When people see that the company values their *minds*, they are motivated to keep pushing at the limits of what they feel capable of achieving.

CONCLUSION

The person with a positive self-concept is one whose lower-level needs have been met, or are being met, and who is thus self-motivated—motivated from within by the need for continued self-development and the fulfillment of his or her potential as a human being.[36] Such people accept themselves because they know their strengths and build on them. They know they are human, less than perfect, and therefore take advantage of constructive criticism and suggestions. They value their mistakes and failures for what they can teach. They are realistic. People with a positive self-concept are aware of their own and others' capabilities and can cope well with pressure because they know what can be done.

People with a positive self-concept are tolerant of uncertainty. They are not afraid of new ideas and conditions. They are willing to innovate and function within new parameters. They accept other people and are open with them. They work wholeheartedly with new personnel. People with positive self-concepts are not afraid to make independent decisions. They are committed to their work. They go beyond the call of duty because they believe in what they are doing. They are willing to work overtime if needed. People with positive self-concepts are appreciative and grateful. They have a genuine interest in others and in the infinite richness of life. They are appreciative and grateful to be, to feel, to know, to do, to create, and to become.[37]

What can people with a positive self-concept become? They can become energetic, enthusiastic, understanding, creative, confident, caring, disciplined, positive, happy, organized, nice, consistent, good, likeable, capable, loving, helpful, humble, ethical, loved, obedient, trustworthy, kind, appreciative, persistent, forgiving, knowledgeable, communicative, goal-directed, talented, fair, motivated, determined, loyal, honest, healthy, intelligent, sensitive, truthful, dependable, friendly, sincere, hopeful, independent, patient, polite, cooperative, trusting, considerate, self-controlled, generous, and hard-working.[38]

SUMMARY

Self-concept is more than a single idea or ability. It is the product of several factors that affect how a person feels and acts. It deals with esteem, image, confidence, determination, and responsibility of self. All these factors affect you as an individual and establish the level of worth and acceptance that you possess in the effort called life. Each factor is an important ingredient in self-concept, for without each there is an unfinished product.

Self-concept is an important part of any individual's life. Success and happiness are the ultimate goals of all people, but those goals are attained only through effort. The person who sits on the bench all the time is missing out. You, as a player in the game of life, must get off the bench and participate to actually experience all the good opportunities life has to offer. A positive self-concept does not mean that your life will be totally void of disappointment. You may encounter many situations that are unpleasant and you may experience disappointment, discouragement, fears, doubt, and anxiety. A positive self-concept equips you with the ability to overcome the negative aspects of life. It helps you accept new challenges and find success in family, school, work, society, and relationship.

Several external and internal factors affect your self-concept. External factors include family, relationships, school, work, and society. Internal factors are fear, doubt, and anxiety.

There are five steps in developing a positive self-concept: restoring and nurturing, desire, attitude, goals, and action. Restoring and nurturing your self-concept includes such things as identifying and accepting your limitations, learning to accept others, making decisions, developing expertise, dressing as if you are successful, continuing to learn, being observant, tackling the things you fear, choosing your friends and associates carefully, learning from failures, and maintaining or altering your physical appearance or condition.

There are several causes of a poor self-image: a negative environment, ridicule, confusing failure on a project with failure in life, an untrained memory, comparing your worst features to someone else's best features, and setting standards of perfection that are unrealistic and unreachable.

Once your self-concept is attained, improved, or enhanced, you will exhibit many positive characteristics. These include being future oriented; able to cope with life's problems; able to deal with emotions; able to help others and accept help; able to accept people as unique, talented individuals; and showing a variety of self-confident behaviors.

Leaders can raise the self-concept of others in many ways. Nathaniel Branden offered several ways. Once people have achieved a positive self-concept through excellent leadership, they accept themselves and are realistic, tolerant of uncertainty, and committed to work. They are appreciative and grateful to be, to feel, to know, to do, to create, and to become.

End-of-Chapter Exercises

Review Questions

1. Define the Terms to Know.
2. List 15 rewards of a positive self-concept.
3. Differentiate between a positive self-concept and conceit.
4. List five ingredients of a positive self-concept.
5. Give 10 examples of people who were "late bloomers."
6. List four critical things in developing your self-concept during adolescence.
7. What are three ways that adults tend to define themselves in relation to self-concept?
8. What are five external factors that affect self-concept?
9. What are three internal factors that affect self-concept?
10. What are the five steps to achieving a positive self-concept?
11. List 33 things you can do to restore and nurture a healthy self-concept.
12. What are two main reasons that self-concept is important in the workplace?
13. List nine characteristics of workers with a low self-concept.

14. Explain how a low self-concept can lead to hostility.
15. What are six causes of a poor self-concept in the school or social arena?
16. List 10 characteristics exhibited by a person with a poor self-concept.
17. List 16 characteristics of people with a positive self-concept.
18. What are seven ways to present a positive self-concept?
19. What are 10 physical changes that can occur as you gain a positive self-concept?
20. What are 16 things leaders can do to raise the self-concept of others?

Fill-in-the-Blank

1. Having a positive ________________ is the single most important factor in becoming a success.
2. If you do not ________________ in yourself, how can you expect to gain the confidence of your peers?
3. If we cannot ________________ ourselves for what we are, we lessen our chances of being happy, productive, or successful.
4. A person with a positive self-concept will not be ________________ , but better.
5. An increasing trend in our society is the inability of people to ________________ responsibility for their own actions.
6. In the social arena, people with a positive self-concept are more ________________ , ________________ , and a positive self-concept helps you ________________ .
7. We often confuse experience with ________________ and/or ________________ .
8. We often confuse failure on a project with failure in ________________ .
9. We confuse being unintelligent with an ________________ ________________ .
10. The ages of ________________ to ________________ are among the most critical in developing your self-concept.

Matching

_____ 1. Music teacher said "as a composer, he is hopeless."
_____ 2. Rating of "mediocre" in chemistry at Royal College.
_____ 3. Mysterious power of expectations.
_____ 4. Fired by a newspaper editor because he had "no good ideas."
_____ 5. You perform the way others believe you can.
_____ 6. Greek king in a legend that Robert Rosenthal related to expectancy theory.
_____ 7. Failed the sixth grade.
_____ 8. Critical and jealous nature.
_____ 9. Teacher told him he was unable to learn.
_____ 10. Could not speak until age four and could not read until age seven.

A. Walt Disney
B. Louis Pasteur
C. Einstein
D. Edison
E. Winston Churchill
F. Beethoven
G. expectations
H. self-fulfilling prophecy
I. Pygmalion
J. poor self-concept

Activities

1. Several illustrations of "late bloomers" were given in this chapter. Identify a late bloomer whom you know (friend or celebrity). Explain why you believe that person achieved success.
2. Select the external factor that you believe had the most impact on your self-concept. Write a paragraph explaining why.
3. List five things you will do when you become a parent to enhance the self-concept of your children.
4. In the section, "Restoring and Nurturing a Healthy Self-Concept," several practices were given. Write a statement beginning with "I will" for 10 of the practices, describing how you will personally improve your self-concept.
5. Write three good things or positive characteristics about each member of your class. Read these out loud to the class. Write down the positive things that were said about you.
6. Suppose you were given a supervisory or leadership position in the workplace. Read the sections, "Conditions Needed to Empower Self-Concept in the Workplace," and "How Leaders Can Raise the Self-Concept of Others." List 10 practices you would use to enhance the self-concept of workers. Share these with the class and get their responses.
7. Select a friend, identify five negative self-concept characteristics exhibited by the individual, and write suggestions on how they could be improved. Note: Don't share these unless they are done under the supervision of your teacher or are handled carefully, as suggestions for improvement. Our purpose is to enhance a positive self-concept, not lower a self-concept.
8. Read "How Leaders Can Raise the Self-Concept of Others." Select five things that would be easy for you to do as a leader, and select five that would be a challenge to you. Read these in class and get your classmates' reactions.
9. Think of a past experience when you had to deal with someone with an unhealthy self-concept. Answer the following questions.
   a. How was the poor self-concept manifested?
   b. How did it affect your relationship?
   c. Do you enjoy being around that person?
   d. What steps can (could) you follow to enhance the relationship?
10. Suppose your local newspaper is doing a story on each graduating senior. They plan to print a comprehensive review of you. Write the story as you would like it to appear in the paper. List your positive characteristics and accomplishments. Remember, if you cannot occasionally "toot your own horn," nobody else will toot it for you.
11. **Take it to the Net.**

    Explore self-concept on the Internet. Several search terms are listed below. Go to a search engine, such as Yahoo (www.yahoo.com). Type the search terms in the search field. Browse various Web sites and find something related to this chapter. Print what you find and prepare to share it with the class.

    Search Terms

    self-concept
    self-assessment
    self-esteem
    self-concept assessment
    self-image

Notes

1. *Personal Skill Development in Agriculture* (College Station, TX: Instructional Materials Service, 1989), 8736-A, p.1.
2. Zig Ziglar, *The "I Can" Course* (Carrollton, TX: The Zig Ziglar Corporation, 1989), p. 90.
3. *Personal Skill Development in Agriculture*, 8736-A, p. 1.
4. Ibid.
5. Ibid., p. 2.
6. N. Branden, *How to Raise Your Self-Esteem* (New York: Bantam Books, 1988), p. 5.
7. Ziglar, *The "I Can" Course*, Teacher's Guide, p. 37.
8. S. Rathus, *Psychology*, 5th ed. (Fort Worth, TX: Harcourt, Brace, and Jovanovich, 1993), pp. 587–588.
9. L. H. Lamberton and L. Minor, *Human Relations: Strategies for Success* (Chicago: Irwin, Mirror Press, 1995), p. 27.
10. B. L. Reece and R. Brandt, *Effective Human Relations in Organizations* (Boston: Houghton Mifflin, 1993), p. 95.
11. Ibid., p. 96.
12. M. Henning and A. Jordan, *The Managerial Woman* (New York: Anchor Books, 1977), pp. 106–107.
13. Reece and Brandt, *Effective Human Relations in Organizations*, pp. 96–97.
14. E. E. Miller, *The Healing Power of Happiness* (Emmaus, PA: Rodale Press, 1989, pp. 12–13.
15. Ibid.
16. A. Saltzman, *Down-Shifting* (New York: HarperCollins, 1990), pp. 15–16.
17. Miller, *The Healing Power of Happiness*, pp. 12–13.
18. Reece and Brandt, *Effective Human Relations in Organizations*, p. 98.
19. N. Branden, *The Six Pillars of Self-Esteem* (New York: Bantam Books, 1994).
20. Ibid.
21. "How to Gain Power and Support in the Organization" *Training/HRD*, January 1982, p. 13.
22. Reece and Brandt, *Effective Human Relations in Organizations*, p. 101.
23. J. Canfield, "Self-Esteem in the Workplace," *Self-Esteem Newsletter*, Fall 1990, p. 1.
24. M. Layden, "Whipping Your Worst Enemy on the Job: Hostility," *Nation's Business*, October 1978, pp. 87–90.
25. Reece and Brandt, *Effective Human Relations in Organizations*, p. 102.
26. Branden, *The Six Pillars of Self-Esteem*, pp. 252–253.
27. J. S. Livingston, "Pygmalion in Management," *Harvard Business Review*, July-August 1969, p. 83.
28. R. N. Lussier, *Human Relations in Organizations: Applications and Skill Building* (Columbus, OH: McGraw-Hill, 1998).
29. Ibid.
30. Ibid.
31. Ziglar, *The "I Can" Course*, pp. 101–109.
32. Reece and Brandt, *Effective Human Relations in Organizations*, pp. 99–101.
33. Branden, *How to Raise Your Self-Esteem*, pp. 158–160.
34. Ibid., p. 160.
35. Branden, *The Six Pillars of Self-Esteem*, pp. 254–257.
36. Dr. Joe Townsend, Notes shared with me from his Leadership Class (College Station, TX: Texas A & M University).
37. Ibid.
38. Ziglar, *The "I Can" Course*, p. 120.

CHAPTER

18

Attitudes

Objectives

After completion of this unit, the student will be able to:

- Explain the importance of attitude
- Discuss the types of attitudes
- Describe how attitudes are formed
- Discuss how attitudes affect behavior and human relations
- Describe attitudes in relationships
- Explain how attitudes can help you personally
- Compare attitudes and skills
- Discuss how to change attitudes
- Discuss attitudes valued by employers
- Describe the effect of attitude on career success

Terms to Know

attitude
aptitude
cognitive attitudes
affective attitudes
behavioral attitudes
socialization
peer groups
reference groups
role model
culture
condescends
tunnel vision
psychosomatic
symbiotic
skill

Attitude affects our lives. William James said that we can alter our lives by altering our attitude. We can choose to have a positive attitude, or we can choose to have a negative attitude. Positive attitudes will have positive results because they are contagious. We cannot always have good things to happen to us, but we can control our attitudes when challenges arise. When lemon situations arise, we simply have to make lemonade out of them.

If you think you can, you can. If you think you cannot, you cannot. Generally we really control our own destiny. We can achieve whatever we want to if we are willing to pay the price of persistence, sacrifice, and dedication. Whether we are successful depends in large part on our attitude.

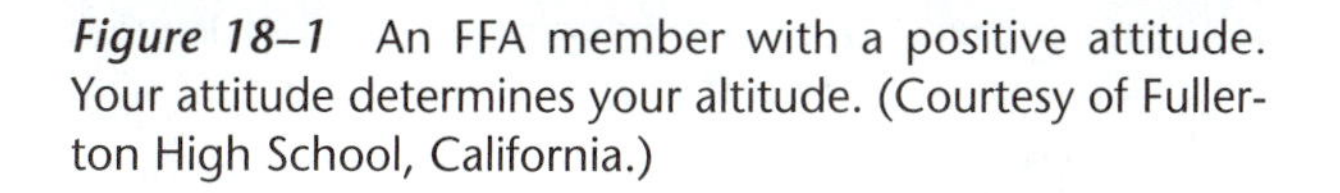

Figure 18–1 An FFA member with a positive attitude. Your attitude determines your altitude. (Courtesy of Fullerton High School, California.)

There are many definitions and approaches to attitudes. The focus of this chapter is discussing attitude as a positive outlook on other people, our job, and our surroundings (Figure 18–1). Ralph Waldo Emerson said, "What lies behind you and what lies before you pale in significance when compared to what lies within you."

IMPORTANCE OF ATTITUDE

What Is an Attitude?

An **attitude** is a strong belief or feeling toward people, things, and situations. All of us have positive or negative attitudes about life, human relations, work, school, and everything else. Attitudes cannot be changed easily. Our friends and those around us usually know how we feel about things. Our attitudes are shown by our behavior. If you make a disrespectful gesture behind someone's back, your friends will assume you have a negative attitude toward him or her.[1]

Attitude and Self-Concept

Our attitude is usually related to our self-concept. Those with a low self-concept often exhibit attitudes that are not based on the way things really are but rather on their own feelings of inadequacy. The way we feel about ourselves affects every other attitude we express, consciously or accidentally. If we don't like ourselves, everything we see will be affected by that feeling.[2]

Attitude versus Aptitude

Attitudes have a powerful influence on our life, future, and careers. In reality, our attitudes hold us back more than our aptitudes do. Jesse Jackson said, "Your attitude, not your **aptitude** will determine your altitude in life." People who posses a positive mental attitude and an optimistic view of life are more apt to achieve personal and economic success. Such success does not happen by chance. People with a positive view of life are more apt to develop a plan for their lives that includes a series of goals that guide them daily.

Doctors, lawyers, teachers, salespeople, parents, and students all share the opinion that your attitude is the dominant factor in your success. A study by Harvard University revealed that 85 percent of the reasons for success, accomplishments, promotions, good grades, happiness and many of the "good things" of life are based on our attitudes, and only 15 percent because of our technical expertise.[3]

Attitude and Control

The more control people exert over their lives, the more likely they are to be optimistic. Individuals who have a negative mental attitude simply hope that somehow things will work out for them. These negative people usually have no long-term goals and no definite future plans. They often feel out of control because they have made little attempt to influence the future.[4]

Attitudes Can Greatly Influence Our Mental and Physical Health

In *Love, Medicine, and Miracles*, Bernie Siegel stated that there is a strong connection between a patient's mental attitude and his or her ability to heal. Siegel says that when a doctor can instill some measure of hope in the patient's mind, the healing process begins. Siegel has observed this as a doctor. As part of his treatment plan, Siegel attempts to give patients a sense of control over their destinies. He wants his patients to believe in the future and know that they can influence their own healing.[5]

Other people also believe there is a relationship between good health and good thoughts. Although certain segments of our society have always believed it, a growing number of health-care specialists are taking the position that a positive mental attitude and outlook about the future can be potent treatments for a variety of mental and physical health problems. These people say that our thoughts are the primary causes of the effects or conditions in our lives. To change our future lives, we must change our present attitudes.[6]

Attitude and Job Satisfaction

Several factors affect our attitude toward job satisfaction. These include attitudes toward the company's benefits, promotions, co-workers, supervision, safety, and the work itself. Job satisfaction is usually a concern to an employer because management knows there is a link between attitudes and productivity. Employees who don't like their jobs are more likely to be late or absent from work, become unproductive, or quit. A positive attitude toward work can reduce tardiness, absenteeism, and employee turnover.[7]

Attitude and Organizations

People shape their attitudes about you by what they see and hear. They interpret your attitudes through your behavior. How you feel about something is usually no secret to friends and schoolmates. If your FFA does not have regular meetings, your committees do not function, or your chapter doesn't participate in contests, it conveys the attitude that your advisor and officers have about the value of the FFA. These attitudes will attract or not attract people to the organization with the same commitment.

Attitudes represent a powerful force in any organization. An attitude of confidence and value can pave the way for improved meetings and greater participation from the officers and members (Figure 18–2). A sincere effort by the chapter officers to improve, filtered through attitudes of confidence, trust, and hope, results in the type of organization that everyone desires.

Attitudes and Research

Attitudes and their effect on relationships, health, and intelligence are backed up by research. Although many studies have been done, only one will be reported here, by psychologist Carl Rogers in a paper titled "The Characteristics of a Helping Relationship."

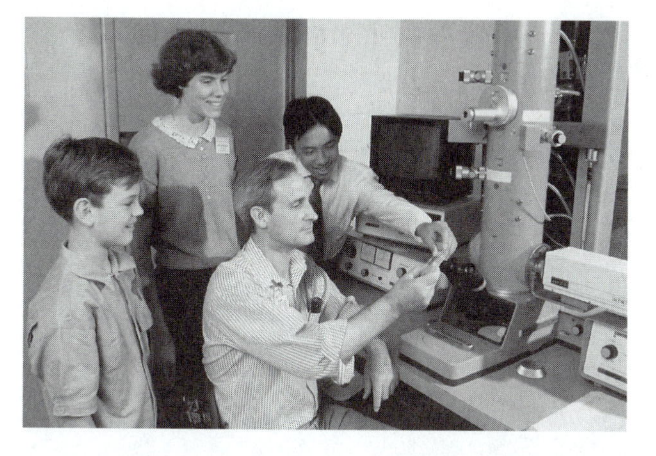

Figure 18–2 These students are exhibiting an attitude of confidence as their teacher explains the use of a powerful microscope. (Courtesy of USDA.)

The children of accepting parents and of rejective parents were compared on various traits. The children of accepting and democratic parents developed better attitudes. Children of these parents with their warm, caring attitudes showed a higher intelligence, more originality, more emotional security and control, and less excitability than children from other types of homes. Though somewhat slow initially in social development, they were, by the time they reached school age, popular, friendly, nonaggressive leaders.

The children from rejective parents showed slightly decelerated intellectual development, relatively poor use of the abilities they did possess, and some lack of originality. They were emotionally unstable, rebellious, aggressive, and quarrelsome. The children of parents with other attitudes tended to fall between the characteristics of the two groups discussed.[8]

Attitudes and Individual Success

In Chapters 1 and 2 we told several success stories of winners who succeeded because of their positive attitudes. These included Lincoln, with his several defeats before winning the presidency, and Roger Bannister, who broke the four-minute-mile barrier. Glenn Cunningham was told he would never walk again after a near-fatal accident he had while starting a fire, but because of his attitude, he became a world class runner. Other ways that attitudes relate to individual success are as follows.

- Academic achievement is directly related to an individual's attitude toward school and authority figures.
- Job performance and promotion are directly related to individual attitudes toward work, supervisors, work ethic, and self.
- Athletic abilities are affected by attitude as well as talent.
- Success in family relationships is affected by attitude.
- Responsible citizenship is directly related to an individual's attitude toward authority figures, neighbors, and self-worth.
- Personality, success in human relations, and sound judgment are based on positive attitudes.

Types of Attitude

We do not see things as they really are; rather, we view things through perceptual filters, or the way we think they are. Perception is also influenced by one's attitudes and understanding of others. Attitudes are a combination of an individual's feelings, thoughts, beliefs, and values.

Attitude has three components or elements. The first is the cognitive component, or the thoughts, ideas, and beliefs of a person. The second is the affective, or feeling, component. The last is the behavioral component or the tendency to act consistently with attitude.[9] To put it simply, an attitude may be considered a way of thinking, feeling, and behaving.

Cognitive Attitudes

Cognitive attitudes are the set of values and beliefs that a person has toward a person, object, or event. For example, a fellow student tells you, "I don't like my teacher. He has it in for me." This cognitive attitude is the belief that the teacher is unfair.

Affective Attitudes

Affective attitudes are the emotions attached to an attitude that are felt about a person, object, or event. Happiness, anger, and disappointment are affective attitudes. When our FFA team loses in district, regional, or state competition and we feel sad, this also is a result of affective attitudes.

Behavioral Attitude

Behavioral attitudes are the tendency to act in a particular way toward a person, object, or event.[10] For example, when a student learns that an undesirable teacher is being transferred, the student smiles. When your parliamentary procedure team or public speaking contestant wins a contest, you are excited. If your agricultural classes are enjoyable, you are likely to participate more often.

How Attitudes Are Formed

Our attitudes are formed primarily through our experiences. As we develop from child to adult we interact with parents, family, teachers, and friends. From all these people we learn what is right and wrong and how to act.

When encountering new people or situations we usually have not had time to form an attitude. This is when we are most open and impressionable. Before entering new situations we often ask others with experience about it. This begins the development of attitudes before the encounter.[11] In *Effective Human Relations in Organizations*, Barry Reece and Rhonda Brandt report that attitudes are learned from the numerous circumstances of each person's life, such as **socialization**, peer and reference groups, rewards and punishment, role model identification, and cultural influences. A brief explanation of each follows.

Figure 18–3 This little girl is forming attitudes from daily interactions with her father. (Courtesy of *FFA New Horizons Magazine*.)

Socialization

Impressionable young children interact with parents, teachers, and friends, thus they often believe what these authoritative figures say (Figure 18–3). Children who observe their parents recycling, using public transportation instead of cars to get to work, and saving electricity may develop a strong concern for protection of the environment.

Peer and Reference Groups

As children begin to break away from their parents, they become more closely associated with children their own age. These **peer groups** can often be stronger influences than parents as children become young adults. With the passing of years, **reference groups** replace peer groups as sources of attitude formation in young adults. The reference groups, such as college fraternities or sororities, may act as a point of comparison and a source of information for individuals. The FFA is a reference group for many. Reference groups can and should have a positive influence on positive attitudes. If not, find a different and better reference group.

Rewards and Punishment

Authoritative figures generally encourage (reward) some attitudes and punish others. It is natural for a child to want to maximize rewards and minimize punishments. A student who is praised for participating and asking questions in class is likely to repeat the behavior. Adults continue to have their attitudes shaped by rewards and punishment at work.

Role Model Identification

Young people often achieve their goals of increasing status or popularity by identification with a **role model**. These role models, whether they be parents or stars, have a tremendous influence on developing attitudes. Television, movies, and radio also influence people's selection of role models. Why? By the time students graduate from high school, they will have spent 50 percent more time in front of a television than in the classroom or with parents, family, or friends. Albert Schweitzer, the French philosopher and Nobel Peace Prize winner, said, "Example is not the main thing in influencing others, it is the only thing."

Cultural Influences

Culture includes the values and a broad range of behaviors that are acceptable within a specific society.[12] People strive to define themselves in every culture. Every country and even sections within a country have different cultures. We become what we grow up with. Most Catholics remain Catholics. Most Protestants remain Protestants. Most members of the other religions of the world remain that religion. Most of those who are raised with no religious training usually do not change their culture either.

As discussed in Chapter 1, the FFA has its own symbols and culture. Generally speaking, agricultural education and the FFA represent a set of beliefs, customs, symbols, values, and norms that binds members of the organization together. FFA culture has a strong influence on member attitude.

How Attitude Affects Behavior

Attitudes affect your everyday behavior, but they are not the only cause of behavior, nor are they always a reliable predictor of behavior. Attitudes are complex. Even if you have a negative attitude toward your job you may still work hard because of other attitudes, such as a positive attitude toward your boss, your peers, or promotion. The attitudes listed in Figure 18–4 affect behavior.

Attitude and Positive Thinking

Every thought you have has some effect. People get sick more easily when they are emotionally down or depressed. Do not program your mind with bad thoughts or you will get bad reasoning and attitudes. Remember, even when you cannot

Moody	Neat	Jealous	Afraid	Honest
Open-minded	Thrifty	Popular	Kind	Happy
Unreasonable	Even-tempered	Shy	Modest	Friendly
Demanding	Dependable	Clumsy	Proud	Sad
Sensitive	Angry	Show-off	Lazy	Serious

Figure 18–4 Our attitudes affect our behavior. Which of these attitudes do you possess?

do anything about a situation, you can do a lot about your attitude toward the situation. Walter Wintle (Figure 18–5) says it well with his poem, "The Man Who Thinks He Can."

You Reap What You Sow

Presenting a positive attitude and conducting yourself in such a manner will cause others to treat you with respect. The psychological mind can be compared to the physiology of plants. If you plant corn, you will get corn, not soybeans. Whatever a person sows, he or she will reap. I once had a student come to my office and start making negative comments about one of his former teachers. After a few minutes, I stopped him and said, "Ronnie, look at these two blocks on the wall. Every time you say something negative about somebody, you get a mark in this block. Every time you say something good about somebody, you get a mark in this other block. Ronnie, you reap what you sow. For every mark in the negative block, somebody says something negative about you. For every mark in the good block, somebody says something good about you."

The Man Who Thinks He Can

By Walter D. Wintle

If you think you are beaten, you are.
If you think you dare not, you don't.
If you like to win, but you think you can't
It's almost a cinch you won't.

If you think you'll lose, you're lost;
For out of the world we find
Success begins with a fellow's will—
It's all in the state of mind.

If you think you are outclassed, you are.
You've got to think high to rise.
You've got to be sure of yourself before
You can ever win a prize

Life's battles don't always go
To the stronger or faster man,
But sooner or later, the man who wins,
Is the man who thinks he can.

Figure 18–5 This poem by Walter Wintle illustrates what many people believe about the effects of a positive attitude.

Attitude and Motivation

Our attitudes can motivate ourselves and others. Attitudes can be shaped to be positive. Over time you can condition your mind so that you will automatically respond positively to the negative situations you encounter in life. Psychologist David McClelland of Harvard University found that you can change your attitude and self-motivation by changing the way you think about yourself and your circumstances. Needs create goals. Pursuing goals results from motives. Our motives motivate us to action. Some say you cannot motivate others. Nonsense! The old saying "You can lead a horse to water, but you can't make him drink" is really not true. Give the horse some salt and watch it drink. The salt, creating a need for water, motivates the horse to drink. Positive attitudes can motivate us to positive thoughts and actions.

Winners Find a Way

Winners find a way. Others find an excuse. Be willing to grow into greatness. A small acorn grows into a great oak tree. Seeds of greatness can grow within our lives. It takes work and discipline. It takes proper nurturing. It takes time and may not happen overnight.

Constantly hold before you the goals toward which you work. Avoid a complaining attitude. A winner never **condescends** but lifts others around him or her to a higher level of encouragement. Help others attain success, and you will help yourself.[13]

Enthusiasm Is an Attitude

Most people let conditions control their attitude instead of using their attitude to control conditions. If things are going well, their attitude is upbeat. If things are going badly, their attitude is bad. It is best to build a solid attitude, so that when things are bad your attitude is still good.

Enthusiasms can make a difference in your attitude. How we respond to failure and mistakes is one of the most important decisions we make. With enthusiasm, failure does not mean that nothing has been accomplished. It simply means we figured out another way something does not work. There is always the opportunity to learn something. With enthusiasm, what is within you will always be bigger than whatever is around you.[14]

Attitudes and Failure

We all experience failure and make mistakes. In fact, successful people have more failure in their lives than average people do. Only those who do not expect anything are never disappointed. Only those who never try avoid failure completely. Anyone who is currently achieving anything in life is simultaneously risking failure. However, it is always better to fail in doing something than to excel in doing nothing. A flawed diamond is more valuable than a perfect brick. People who have no failures also have few victories. Everybody gets knocked down, but those with a good attitude get back up fast.[15]

Attitudes and Relationships

A positive attitude or a change in attitude can dramatically improve relationships with others. We must look for the good in others. People will build relationships if they live by the saying "If you can't say something good about somebody, don't say anything at all." Our attitudes will be enhanced when we show sincere love, care, and concern for others. "No man is an island." The more we understand others, work cooperatively with others, and show love and respect toward others, the better our lives will be.

Find Good in Others

As a child, I used to go to a hill on our farm and start shouting to hear echoes. I would shout, "You are great." The echo would shout back, "You are great." I would shout, "You are greater," and the echo would return with, "You are greater." Life is an echo. What you give comes back. Regardless of who you are or the type of occupation you have, if you are looking to find the good in others, you are looking for the way to reap the most rewards in all areas of life, including enhancing your attitude. Adopt the Golden Rule (do unto others as you would have them do unto you) as a way of life. Find the good in others, and they will find the good in you.

Sculpturing Special Abilities

As a teacher, I visualize every student as a block of wood. Within that block of wood are special traits or talents. If I keep carving, I will eventually discover that special trait; that is, I will not only find the good, but I will also find the special traits or talents that each student has. Then, I will help direct and cultivate those talents. The student's attitude and self-esteem begin to develop.

The psychologist, Arthur Combs, relates the following story.[16]

> ***A young man whom the writer tested at an induction center during the war illustrates the point very well. This young man was a newsboy on the streets of a West Virginia city. Although he had failed repeatedly in grammar school and was generally regarded as "not bright," he appeared on a national radio hook-up as "The Human Adding Machine." He was a wizard at figures. He could multiply correctly such figures as 6,235,941 × 397 almost as fast as the problem could be written down. He astounded our induction center for half a day with his numerical feats. Yet, on the Binet Test given by the writer he achieved an IQ of less than 60! People in his home town, who bought his papers, amused themselves by giving him problems to***

figure constantly. When not so occupied this young man entertained himself by adding up the license numbers of cars that passed his corner. He was a specialist in numbers. Apparently as a result of some early success in this field, he had been led to practice numbers constantly, eventually to the exclusion of all else. This was one area in which a poor underprivileged boy could succeed, and he made the most of it. His number perceptions were certainly rich and varied, but other things were not. Although he was capable of arithmetic feats not achieved by one in millions, he was classified as dull!.

The film, *Rain Man*, starring Dustin Hoffman and Tom Cruise, told a similar story. An autistic man was practically socially dysfunctional, yet his math ability was so great that he was banned from the casinos in Las Vegas. Are all people created equal? I do not know, but as we learn more, we are beginning to find that there is a genius in each of us be carved into a masterpiece sculpture.

When gold is mined, several tons of dirt must be moved to get an ounce of gold, but we do not mine for dirt; we mine for gold. Do not look for the negatives, look for the gold or positive aspects of a person. The harder you look, the more positives you will find.

Praise and Success

Compliments A sincere compliment is one of the most effective attitude adjusters and motivational methods there is. Some say compliments are just hot air, but hot air can make you fly high, or like the air we use in automobile tires, it helps us ease along the highway.

Tunnel Vision Praise and success can break us out of **tunnel vision**.[17] Many students believe that they are *A* students, and this attitude *makes* them an *A* student. The same is true with many *B* and *C* students. They become what they believe they are. I was a victim of tunnel vision. I perceived myself as a *B* student, therefore, I was a *B* student. I did not think I had the capability to become more. Once I began to realize that I was a victim of tunnel vision, I began to expand my self-confidence and attitude toward learning. My grade point average as an undergraduate was 2.78. Once I broke out of tunnel vision and found that I could make *A's* just like anyone else, I finished my doctorate degree with a 3.83.

Success Breeds Success Appendix A explains a process to break students out of tunnel vision based on praise and success. Once students feel success, they like it. Success breeds success. One ounce of praise and successful experiences is worth more than a pound of criticism. Unfortunately, many people do not realize this. As a high school teacher, I occasionally used a technique called the "Shotgun Approach to Teaching." This technique involved introducing the unit, teaching it, reviewing, and giving a test all within one class period. By teaching short segments of information, using several principles of learning, and reviewing extensively, I helped over 80 percent of the class to get *A's*. When the students who were not accustomed to making such grades got *A's*, they experienced a positive attitude shift. The success broke them out of their tunnel vision and they saw themselves as better students. Therefore, they became better students because they enjoyed the taste of success and wanted to experience it again.

Expectations

People Rise to the Expectations Others Have A new teacher at the beginning of the school year looked up IQ scores in the students' files. She had two classes. She was amazed at the difference in the intelligence level of the two classes. For the class with IQ scores in the 120 range and up, she taught challenging classes at a high ability level. The students responded according to her expectations. For the class with IQ scores in the 90 range and lower, she taught at a slower rate, being sure not to cover the material too quickly. She didn't give much homework and practically "spoon fed" the students. This class also responded as she expected.

An interesting thing happened later in the year. She was looking for something in one of the student files and discovered that she had inadvertently gotten locker numbers rather than IQ scores. In reality, there was absolutely no difference in the level of intelligence of the two classes. The difference in performance was the direct result of the difference in the teacher's expectations. In short, the teacher treated the two classes differently because she saw them differently, and different treatment brings different results. From the students' point of view, they also performed the way the teacher expected them to perform. The way you see others is the way you treat them, and the way you treat them is often the way they become.

Expectations and Attitudes Can Affect Health

The famous Harvard psychologist James Allen wrote, "The greatest discovery of my generation is that human beings, by changing the inner attitudes of their minds, can change the outer aspects of their lives." Physician Bernie S. Siegel said, "Years of experience have taught me that cancer and indeed nearly all diseases are **psychosomatic**."[18] The body knows only what the mind tells it. If one has taken part in getting sick, one can also take part in getting well.[19]

Belief in Others

Our attitude of belief in others can push individuals toward greatness. Many of you have had teachers, parents, friends, or somebody else who has given you the spark of belief that got you going. I challenge you to do the same for others.

When Helen Keller was given England's highest award as a foreigner, Queen Victoria of England asked her, "How do you account for your remarkable accomplishments in life? How do you explain the fact that even though you were both blind and deaf, you were able to accomplish so much?" Without a moment's hesitation, Helen Keller said that had it not been for Anne Sullivan, the name of Helen Keller would have remained unknown. Anne Sullivan was Helen Keller's teacher. Sullivan herself had been nearly blind during childhood until two surgeries restored her sight. However, Anne Sullivan taught Helen Keller the alphabet by touch, how to connect words with objects, how to read and write in Braille, and how to type.[20]

Attitude of Cooperation

Canada geese instinctively know the value of cooperation. The geese fly in "V" formation and regularly change the lead goose. Why? The lead goose, in fighting the headwind, helps create a partial vacuum for the geese flying behind it (Figure 18–6). In wind tunnel tests, scientists have discovered that the geese as a group can fly 72 percent farther than an individual goose can fly. Could it be that people could fly higher, farther, and faster by cooperating with, instead of fighting against, those around them? As you practice your attitude of cooperation, remember that people pay more attention to what you do than to what you say.

Figure 18–6 Canada geese demonstrate the value of cooperation. The "V" formation fights the headwind, creating a partial vacuum than allows geese to fly 72 percent farther than they would if each flew alone. (Courtesy of Tennessee Wildlife Resource Agency).

A POSITIVE ATTITUDE CAN HELP YOU

Attitude During Job Interviews

Attitude is critical during the job interview, especially attitude toward work. Employers look for signs of how you will work with others and deal with problems. You communicate attitude as well as words during an interview. (See Chapter 21 for more about job interviews.)

Achieving Competence in Attitude

A person can become competent in attitude just as one can become competent in a technical skill. However, it is far more difficult to measure. Nevertheless, people respond in a positive way when you have a good attitude and in a negative way when you do not. Even if your attitude cannot be measured, it is being observed.

Positive Attitude Leads to a Brighter Future

People who concentrate on good attitude get the best jobs and eventually rise to the top in most organizations. All organizations are built around people, and when you build healthy relationships with your fellow workers and supervisors, you open doors that would otherwise be closed. Whether it is right or wrong, other people control your job future, and the better your attitude to them, the better things will go for you.

Better Attitudes Lead to Becoming a Better Leader

With a good attitude, not only will you become a better supervisor or leader, you will become one sooner. Work on your attitude now will strongly influence your progress later, as well as your willingness to work.

Relationship Between Attitude and Learning

There definitely is a relationship between attitude and learning. The expression "openness to learning" is used to communicate that when a mind is open (free of blocks, fears, prejudices), it will more readily accept and retain new data and ideas. If a student dreads taking a required course (such as algebra, chemistry, or English), the chances of success are less because the fear causes a block to learning. However, through an "attitude adjustment" such as talking with the teacher, the block can be partially eliminated. Learning will be easier because the student's attitude toward learning is more positive.[21] Would you say that the attitude of the teacher in Figure 18–7 could enhance a student's learning?

Figure 18–7 This teacher and student appear to have good rapport. Learning is much easier when the student's attitude toward learning is positive. (Courtesy of USDA.)

Connection Between Attitude and Personality

There is a **symbiotic** relationship between attitude and personality. Personality is generally considered a combination of special physical and mental characteristics that allow you to transmit a unique image to others. When you have a positive attitude, the image communicated is at its best. Of all the physical traits (such as eyes, size, hair color), and mental traits (such as ability to learn, patience, determination), attitude is the only characteristic that surpasses other traits and pulls them together into a more attractive image. A positive attitude can overshadow both physical and mental characteristics (remember Helen Keller). When you are positive, your whole body becomes positive. Even your less favorable traits appear more attractive. Your total personality is appreciated and enjoyed more by others.[22]

ATTITUDE OR SKILL

What are the characteristics of successful people? If you and a friend made a list of qualities of the most successful person you know, your lists would be similar. It doesn't matter whether this person is an employer, employee, neighbor, or friend.

Several desirable qualities or characteristics for success that would probably be on your list are shown in Figure 18–8. Study the 50 qualities in the figure and decide which are attitudes and

Helpful	Prepared	Cooperative
Alert	Common sense	Good moral character
Honest	Knowledge	Dedicated
Enthusiastic	Hard-working	Humble
Understanding	Self-respect	Organized
Confident	Loving	Dependable
Patient	Humor	Decisive
Discipline	Friendly	Good listener
Loyal	Motivated	Communicator
Trust	Compassionate	Learner
Goals	Integrity	Empathy
Assertive	Commitment	Creative
Caring	Consistent	Energetic
Supportive	Responsible	Faith
Personable	Prompt	Optimistic
Sensitive	Teachable	Thoughtful
Positive mental attitude		

Figure 18–8 Although these are desirable qualities or characteristics for success, most are not learned in a formal academic setting. Could it be true that attitude is as important as aptitude?

which are skills. If your answers are consistent with those of most people, you probably identified most of these characteristics as attitudes. Besides knowledge and creativity and two or three that could be a combination of attitude and **skill**, most are attitudes.

Attitudes Are as Important as Aptitudes

Ask any group what it takes to make people successful and 90 to 95 percent of the answers will be in the form of an attitude. However, we constantly hear headlines about our academic weaknesses compared to other countries. Yes, academics are very important, but our attitudes are as important as our aptitudes.

Learning Qualities of Successful People

Can these qualities of successful people be learned? Yes. Unfortunately, they are not typically taught in public schools. Since you are reading this, however, you are being exposed to the proper attitudes for success. Texas Commissioner of Education William Kirby reported that in Japan, from kindergarten through high school they teach a course one hour each day on the importance of honesty, character, integrity, hard work, positive mental attitude, enthusiasm, responsibility, thrift, free enterprise, patriotism, and respect for authority.[23]

We do learn many of these attitudes but only in the military, at home, or in church. At one time, out of the chief executive officers of Fortune 500 companies, 175 were former marines. The motto of the Marine Corps is "Always Faithful," and the Marines are strong on discipline, commitment to excellence, loyalty, and accepting responsibility for whatever task they are assigned. Five of our last 10 presidents served in the U.S. Navy and 26 presidents served in the military. Apparently, discipline, commitment, and loyalty are desirable qualities for achieving success in top positions.[24] Attitude is important in life regardless of what you are doing or plan to do.

CHANGING ATTITUDES

Changing Your Attitude

You can gain control of your attitude and change your outlook on life.[25] Usually, we cannot control our environment, but we can control and change our own attitudes. We can choose to be optimistic or pessimistic. We can look for the positive and be happier and get more out of life. The following suggestions can help you change negative attitudes.

- **Be aware of your attitudes.** Research has found that people with the personality characteristic of optimism have higher levels of job satisfaction.[26] Consciously try to maintain a positive attitude. As mentioned earlier, if a situation gives you lemons, make lemonade.
- **Think for yourself.** Develop your own attitudes based on others' input; do not simply copy others' attitudes. Defend your reasoning.
- **Do not harbor negative attitudes.** Realize that there are few, if any, benefits to harboring negative attitudes. A negative attitude, like holding a grudge, can only hurt your relations with other people, and hurt yourself in the end.[27]
- **Keep an open mind.** Listen to other people's input. Use it to develop your positive attitudes.
- **Alter your thinking.** Letting go of attitudes that are no longer appropriate can be difficult. Decide whether your negative attitudes toward people and situations are still valid or not. I can remember in the early 1970s I would not let girls take vocational agriculture. Shame on me! My daughter is my pride and joy, along with my two sons. Agricultural education and the FFA is her life. She

is now an agricultural education instructor. As you can see, I had an attitude adjustment.

- **Change behavior.** One of the most difficult approaches to dealing with an undesirable attitude is to attempt to change the accompanying behavior. In the words of the famous commercial, "Just do it."
- **Change ideas and beliefs.** In many cases, an undesirable attitude may be the result of insufficient or misleading information. Simply becoming aware of new facts is very likely to help you modify your attitude.[28]
- **Change feelings.** The most promising approach for dealing with feelings or emotions in others involves listening. This can be extremely difficult. It requires listening for understanding and being sympathetic and nonthreatening so that the individual feels free to express his or her feelings, problems, and attitudes.
- **Change the situations.** When possible, change the situation that is the source of unfavorable attitudes, such as working conditions.

Although changing others is especially difficult, we can work on ourselves. We can smile, say something pleasant every hour, change negative statements to positive statements, change a negative problem to a positive situation, and keep a mental picture of the kind of people we want to be. We can change our attitude if we have high expectations of ourselves.

Changing Followers' Attitudes

It is difficult to change your own attitudes, but it is even more difficult to change other people's attitudes. Nonetheless, it can be done. The following hints can help you as a leader change the attitudes of your followers.[29]

- **Give followers feedback.** Followers must be made aware of their negative attitudes if they are to change. The leader must talk to the follower about the negative attitude. The follower must understand that the attitude has negative consequences for the individual and the group. The leader should offer an alternative attitude.
- **Accentuate positive conditions.** Followers tend to have positive attitudes toward the things they do well. Make conditions as pleasant as possible; make sure followers have all the necessary resources and preparation to do a good job.
- **Provide consequences.** Followers tend to repeat activities or events that have positive consequences. On the other hand, they tend to avoid things that have negative consequences. Encourage and reward followers who have positive attitudes. Try to keep negative attitudes from developing and spreading.
- **Be a positive role model.** If the leader has a positive attitude, followers are more apt to have a positive attitude (Figure 18–9).

Figure 18–9 The positive attitude of the supervisor causes a positive attitude in the employee, which will enhance the workplace. (Courtesy of USDA.)

ATTITUDES VALUED BY EMPLOYERS

Several attitudes valued by employers have been discussed indirectly throughout this chapter. The following characteristics relate more specifically to the workplace.[30]

Willingness to Assume Self-Leadership

Self-leadership emphasizes self-sufficiency. People need to be leaders of themselves. They need to set their own goals and monitor their own progress toward those goals.

Willingness to Learn How to Learn

Many employers agree that employees who are willing to learn are worth more because they decrease the time and money spent on resources.[31]

Willingness to Be a Team Player

Team players are in greater demand as employees are organized into teams. Team players work well with others in building products, solving problems, and making decisions.

Concern for Health and Wellness

Many employers provide programs on health and wellness. The employees who attend these meetings usually take fewer sick days and have a higher level of energy.

Enthusiasm for Life and Work

Employees who show enthusiasm for life and work are more likely to be a positive influence on their co-workers. Enthusiasm displayed by employees is contagious. People with enthusiasm about life and work look for the bright spots and can find the good in almost everything. They avoid negative people whenever possible and refuse to be persuaded by negative thinkers, who see only problems, not solutions (Figure 18–10).

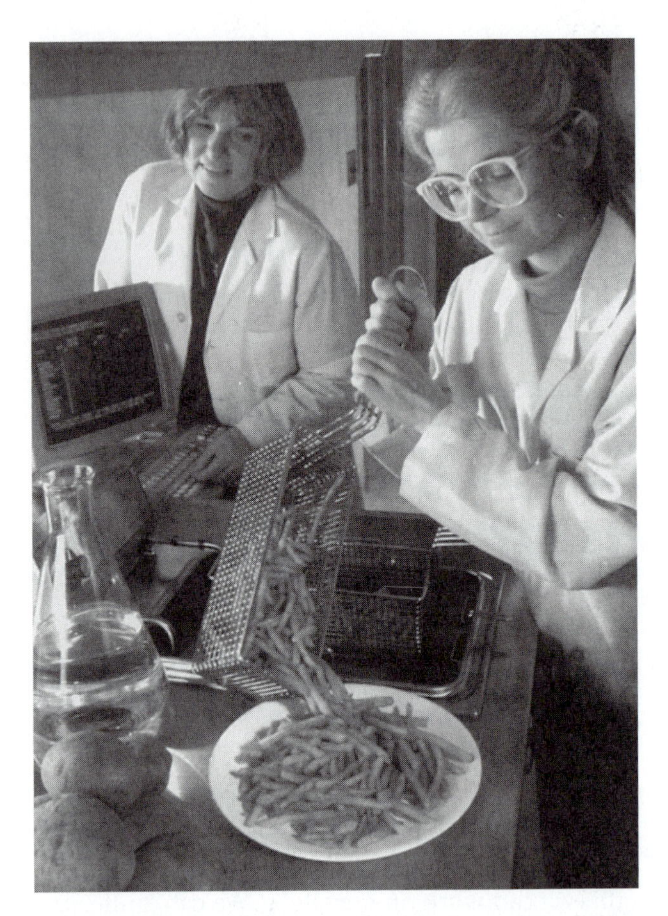

Figure 18–10 Two food technologists above appear to be enthusiastic about their research. Employees who show enthusiasm at work are more likely to be a positive influence on co-workers. (Courtesy of USDA.)

ATTITUDE AND CAREER SUCCESS

Self-Evaluation of Attitudes

Robert Lussier, in his book *Human Relations in Organizations*, compiled the following positive job attitude statements.[32] How do you rate on this list? Each of these attitudes would enhance your career success.

- I smile and am friendly and courteous to everyone at work.
- I make positive, rather than negative, comments at work.
- When my boss asks me to do extra work, I accept it cheerfully.
- I avoid making excuses, passing the buck, or blaming others when things go wrong.
- I am an active self-starter at getting work done.
- I avoid spreading rumors and gossip among employees.
- I am a team player willing to make personal sacrifices for the good of the work group.
- I accept criticism gracefully and make the necessary changes.
- I lift co-workers' spirits and bring them up emotionally.

Positive Attitudes Are Essential for Career Success

The moment you can no longer be positive about your career with your company, your chances for success diminish. A positive attitude is essential to career success for the following reasons.[33]

- **Energy.** When you think positively, you usually display a more energetic attitude. You also are more motivated, productive, and alert. A positive attitude opens a gate and lets your inner enthusiasm spill out.
- **First impressions.** The initial impression you make on people is extremely important to employers, especially if you display a positive attitude. Co-workers will receive a friendly, warm signal, and be attracted to you. Otherwise, they will try to avoid you.
- **Productivity of others increases.** An employee who is positive makes other employees more productive. Attitudes are caught more than they are taught. People with positive attitudes are more successful because they have (1) the ability to shrug off bad news, (2) a willingness to take risks, (3) a desire to assume personal control of events rather than just allowing things to happen, and (4) willingness to set ambitious goals and pursue them.[34]
- **Positive and productive work environment.** When you show a positive attitude, co-workers enjoy your presence more, thus allowing for a more productive and positive work environment. This makes your job more interesting and exciting because you are in the middle of things and not on the outside complaining.
- **Promotion opportunities.** Your future success will depend on the kind of attitude you express toward authority figures. Your mental attitude is constantly being read even though you may think it doesn't show. By projecting a positive attitude, you are more likely to be considered for special assignments and promotion opportunities.

CONCLUSION

A positive attitude will have positive results because attitudes are contagious. Emerson said, "When a happy person comes into the room, it is as if another candle has been lit." If you think you can, you can; if you think you cannot, you cannot. Why? Your attitude determines your altitude. We can alter our lives by altering our attitudes because, among other reasons, your attitude is just as important as your aptitude.

You cannot control every situation you encounter, but you can control your response to the situation. Furthermore, you cannot create the ideal situations for yourself, but you can tailor your attitude to fit them before they arise. Most of the reasons for success, accomplishments, promotions, good grades, happiness and many of the good things of life are based on attitude.

Summary

Attitudes affect our lives. We choose the type of attitudes we have. Your attitude determines your altitude, or how high you go in life. If you think you can, you can; if you think you cannot, you cannot.

An attitude is a predominant belief or feeling about people, things, and situations. Cultivating a positive attitude is important because it gives you more control of your life, influences your mental and physical health, increases job satisfaction, increases the strength of your organization, and helps you accomplish individual success.

There are three types of attitudes: cognitive, affective, and behavioral. Attitudes are formed by socialization, peer and reference groups, rewards and punishment, role model identification, and cultural influence. Positive attitudes cause us to be more motivated, find a way to win, be more enthusiastic, and overcome failure.

A positive attitude can dramatically enhance your relationships by helping you find the good in others, bringing out the best in others, praising others, creating high expectations for and belief in others, and enhancing cooperation.

Positive attitudes can help you personally: during job interviews, by leading to a brighter future, by becoming a better leader, by increasing your learning ability, and by enhancing your personality.

When the qualities of successful people are listed, 90 percent of those qualities are attitudes. Attitudes can be learned, as evidenced by the fact that a large percentage of chief executive officers of Fortune 500 companies are former Marines and five of the last 10 U.S. presidents served in the Navy. The military teaches positive, constructive attitudes.

We can change our attitudes if we become aware of our attitudes, think for ourselves, be careful not to harbor negative attitudes, alter our thinking, change our behavior, change our ideas and beliefs, change our feelings, and change situations. Leaders can help change their followers' attitudes by giving followers feedback, accentuating positive conditions, providing consequences, and being a positive role model.

Attitudes valued by employers include the willingness to assume self-leadership, willingness to learn how to learn, willingness to be a team player, concern for health and wellness, and enthusiasm for life and work. Positive attitudes that are essential for career success include being energetic, creating good first impressions, increasing productivity of others, and enhancing a positive and productive work environment.

End-of-Chapter Exercises

Review Questions

1. Define the Terms to Know.
2. What are four broad areas in which attitude is important?
3. What are five specific ways that a good attitude can contribute to individual success?
4. Give an example of a research finding that positive attitudes really do make a difference.
5. Name and explain the three types of attitudes.

6. Name five ways attitudes are formed.
7. What are some behaviors that a positive attitude can affect?
8. What are five things we can do to improve relationships?
9. What are five ways attitudes can help you personally?
10. What attitudes are taught for at least an hour per day to Japanese schoolchildren?
11. Name seven ways to change your attitudes.
12. Name four ways a leader can change the attitude of followers.
13. What are five attitudes valued by employers?
14. Give five reasons why positive attitudes are essential to career success.

Fill-in-the-Blank

1. We can alter our lives by altering our ________________ .
2. Positive attitudes will have positive results because they are ________________ .
3. When lemon situations arise, you should make ________________ out of them.
4. If you think you can, you can. If you think you cannot, ________________ ________________ .
5. We can become just about whatever we want to in life, if we are willing to pay the price of ________________ , ________________ , and ________________ .
6. What lies behind you and what lies before you pales in significance when compared to what lies ________________ ________________ .
7. Your attitude is more important than your ________________ .
8. Your attitude determines your ________________ .
9. You reap what you ________________ .
10. Attitudes are a posture or position assumed to serve a ________________ .
11. It is better to fail in doing something than to excel in doing ________________ .
12. Everybody gets knocked down, but those with a good ________________ get back up fast.
13. Visualize yourself as a block of wood; within that block of wood is an ________________ .
14. Do unto others as you would have them ________________ unto you.
15. Find the "good" in others, and they will find the ________________ in you.
16. You can have everything in life you want if you will just help enough people get ________________ ________________ ________________ .
17. People can fly higher, farther, and faster by ________________ with, instead of fighting against, those around you.
18. People pay more attention to what you ________________ than to what you say.

Matching

_____ 1. Find the good in others
_____ 2. One of the most effective attitude adjusters and motivational methods in existence
_____ 3. Bad deeds to others produce bad deeds to you
_____ 4. Those who never try
_____ 5. Less important than attitude in keeping most jobs
_____ 6. People with personality characteristics of optimism have higher levels of
_____ 7. Like holding a grudge, these can only hurt your human relations
_____ 8. One within every block of wood
_____ 9. Worth more than a pound of criticism
_____ 10. Students perform better when this is present

A. negative attitudes
B. never fail
C. praise and success
D. skills
E. high expectations
F. ounce of praise
G. sculpture
H. job satisfaction
I. They will find the good in you.
J. You reap what you sow.

Activities

1. The chapter discusses five ways that attitudes are formed. Select the one that you believe has had the most effect on your life and write 100 words (about one page) explaining why.
2. Refer to the list of 50 attitudes and skills (90 percent are attitudes) in Figure 18–8. Select the attitudes that you personally need to work on in order to be successful. Write a brief action plan for each.
3. Pretend that you are an employer and you want to hire the best people possible. Identify the 10 most important attitudes and skills on your priority list. Defend each of these selections.
4. You have been asked to give a speech on attitudes. By using the information in this chapter, write a five- to seven-minute speech and present it to the class. Use your own stories or other information. (You may wish to refer to Chapters 7 and 8 on writing and presenting speeches.)
5. You are the CEO (chief executive officer) of a large corporation. Prepare a worker evaluation form to be used by the supervisor in each department.
6. Think of a role model in your life who has shaped many of your attitudes. Identify and explain attitudes that this person helped to form.
7. Including the stories, sayings, facts, and other information from this chapter, list five things you learned from this chapter. Share these with the class.
8. **Take it to the Net.**
   Explore attitudes on the Internet. Several search terms are listed below. Go to a search engine, such as Yahoo (www.yahoo.com). Type the search terms in the search field. Browse various Web sites and find something related to this chapter. Print what you find and prepare to share it with the class.

   Search Terms

   attitudes
   positive attitudes
   improving attitudes
   attitude types

Notes

1. R. N. Lussier, *Human Relations in Organizations: Applications and Skill Building* (Columbus, OH: McGraw-Hill, 1998).
2. L. H. Lamberton and L. Minor, *Human Relations: Strategies for Success* (Chicago: Irwin, Mirror Press, 1995), pp. 63–64.
3. Zig Ziglar, *The "I Can" Achievers Course Learner's Manual* (Carrollton, TX: The Zig Ziglar Corporation, 1989), p. 52.
4. B. L. Reece and R. Brandt, *Effective Human Relations in Organizations* (Boston: Houghton Mifflin, 1993), p. 69.
5. B. S. Siegel, *Love, Medicine, and Miracles* (New York: Harper & Row, 1995).
6. Ibid.
7. J. Kagan, *Psychology: An Introduction* (New York: Harcourt Brace Jovanovich, 1984), p. 548.
8. C. Rogers, "The Characteristics of a Helping Relationship," in *The Helping Relationship Book*, ed. D. L. Avila et al. (Boston: Allyn & Bacon, 1977), pp. 5–6.
9. C. R. Milton, *Human Behavior in Organizations: Three Levels of Behavior* (Englewood Cliffs, NJ: Prentice-Hall, 1981), p. 28.
10. R. M. Hodgetts, *Modern Human Relations at Work* (Chicago, IL: Harcourt College Publishers, 1995).
11. Lussier, *Human Relations in Organizations*, p. 60.
12. W. M. Pride and O. C. Ferrell, *Marketing*, Sixth Edition (Boston: Houghton Mifflin, 1994), p. 139.
13. Ziglar, *The "I Can" Achievers Course*, pp. 49–83.
14. Ibid.
15. Ibid.
16. A. W. Combs, "Intelligence from a Perceptual Point of View," in *The Helping Relationship Book*, ed. D. L. Avila et al., (Boston: Allyn & Bacon, 1977), p. 215.
17. Ibid., p. 218.
18. Siegel, *Love, Medicine and Miracles*, pp. 111–112.
19. Ibid., p. 112.
20. "Helen Adams Keller," *The World Book Encyclopedia*, (Chicago: World Book, Inc., 1989).
21. E. N. Chapman, *Your Attitude Is Showing* (Upper Saddle River, NJ: Prentice-Hall, 1998).
22. Ibid., p. 15.
23. Z. Ziglar, *Raising Positive Kids in A Negative World* (New York: Ballantine Books, 1996), pp. 42–43.
24. Ibid.
25. J. Allen, *As a Man Thinketh* (New York: Lush, 1999).
26. The research was conducted at the University of California at Berkeley, as reported in *The Wall Street Journal*, April 18, 1988, p. 27.
27. Lussier, *Human Relations in Organizations*, p. 63.
28. Milton, *Human Behavior in Organizations: Three Levels of Behavior*, pp. 33–34.
29. Lussier, *Human Relations in Organizations*, p. 63.
30. Reece and Brandt, *Effective Human Relations in Organizations*, pp. 76–78.
31. A. P. Cornwale et al., *Workplace Basics Training Manual* (San Francisco: Jossey-Bass, 1990), p. 3.
32. Lussier, *Human Relations in Organizations*.
33. Chapman, *Your Attitude is Showing*, pp. 21–24.
34. J. Neimark, "The Power of Positive Thinkers," *Success*, September 1987, pp. 38–41.

CHAPTER

19

Ethics in the Workplace

Objectives

After completing this chapter, you should be able to:

- Explain values in our society
- Describe terminal and instrumental values
- Explain the development of values
- Determine your personal values
- Discuss basic societal values
- Discuss workplace ethics
- Determine correct ethical decisions
- Discuss the importance of workplace etiquette
- Discuss workplace etiquette concerning the office phone, cellular phones, and e-mail
- Discuss meeting etiquette

Terms to Know

values
truth
belief
philosophy
terminal values
instrumental values
value system
mass media
prudence
temperance
justice
vitality
fidelity
integrity
etiquette
ethics
value conflict

Why is it important for leaders to understand values, ethics, and cultural diversity? For one thing, if we do not understand others values and cultures, we probably will not get along very well with other people. We will most likely not be able to hold jobs, build strong relationships with other human beings, or even function in society. It is important for us, as beings who must live together, to understand what is expected of us. We must know how to treat each other, take care of ourselves, and find meaning in our lives.

People are not born with an internal set of values. Values and ethics must be taught, just as culture is taught. Problems occur between people of different cultures primarily because each side tends to assume that its own culture is the right way.[1] They mistakenly believe that specific behavior practiced in their own cultures are universally valued.

In the past, the teaching of values and ethics has been the responsibility of parents and religious institutions. Increased rates of violence, teen pregnancy, teen suicide, and vandalism, however, indicate that this society's traditional values are not being passed on. The acceptance and application of values and ethics is necessary for the harmonious functioning of society. Without values and ethics, individuals' purpose and direction in life can become unfocused, and they along with society begin to degenerate and self-destruct.

The task of teaching values, ethics, and cultural diversity is one that everybody must take on. People of good character are not all going to come down on the same side of every political and social issue, but they will come down on the same side of the truth.

Values in Our Society

Values Defined

Values are the principles and beliefs that you consider important. Your values influence your decisions and actions, and that is why it is important for you to become aware of your values.[2] To say that something is a value is to imply that at least one other person holds the same value and that anyone else must be able to recognize it as a possible principle for regulating behavior. Value in this sense means belief, but it is more than simply what one believes. Values are social elements, principles or standards accepted by a large number of people, usually over a long period of time. Several values are being exhibited in Figure 19–1.

Figure 19–1 These students are showing their values of honoring the flag and their country, respect for others, honoring tradition, and respect for authority. (Courtesy of the National FFA Organization.)

Values and Human Behavior

Values are especially important for understanding human behavior. Conflicts on the job as well as conflicts between leaders and followers are often due to a difference in values. When you do not get along very well with another person, you may want to take a look at how your basic values differ.

International differences in values also need to be understood.[3] In the workplace, you will deal with people from many different belief systems, political systems, cultural backgrounds, and ethnic groups. We tend to think that our American value system is the correct one. People from other countries also have the same tendency to believe that their national and cultural values are the best. Be aware of differences in values and cultural differences, never make assumptions, and always be courteous and willing to adapt.[4]

Differences Between Beliefs, Truths, and Values

Although the terms *belief*, *truth*, and *value* have different meanings, they are related. An individual's beliefs are normally based on that individual's perception of accepted truths. If an accepted truth is disproved, an individual's belief may change. Individual values are usually the product of beliefs and truths. Therefore, individual values remain constant as long as the individual's beliefs and perceptions of truth remain the same. Consider the following definitions.

Belief is the state of believing, or the conviction or acceptance that certain things are true or real.

Truth is the quality or state of being true, agreement with a standard or rule, or an established or verified fact or principle. *Values* can be defined as the social principles, goals, or standards held or accepted by an individual, class, or society.

Influence of Philosophy on Values

Leaders must be aware of different philosophical beliefs and values even if they don't agree with them. Respect for individual differences means more cooperation and understanding in a group.

Different values arise from different philosophies. Basically, everyone believes his or her own philosophy is the truth. **Philosophy** has great value in our complicated world. Many people have no real foundations, or sets of beliefs. Philosophy can provide them with a reasoned framework within which to think. By accepting a particular philosophy, people can begin to seek certain goals and to direct their behavior, which affects their values. For example, those who have a stoic philosophy try to remain master of their emotions. Epicureans seek happiness through pleasure. Rationalists attempt to gain knowledge through reason. Each philosophy leads to a particular way of thinking and behaving.[5]

Values can be examined by evaluating the following six categories: theoretical, economic, aesthetic, social, political, and religious.

- *Theoretical values*. Individuals with theoretical values seek to discover the truth. They observe what happens and try to systematize things.
- *Economic values*. People with economic values are concerned with the production of goods and services and the accumulation of wealth rather than social or artistic values.
- *Aesthetic values*. The aesthetic person sees highest value in form, harmony, and artistic works and opposes people who repress individual thought.
- *Social values*. People with social values value and love other people. Kindness, sympathy, and unselfishness are very important values.
- *Political values*. The political person is interested primarily in power. He or she seeks leadership, influence, competition, and fame.
- *Religious values*. The highest value for the religious person is unity. Such people try to understand the universe as a whole and relate to it with meaning.[6]

People of different professions tend to have different values within the six value categories. For example, scientists tend to be highest in theoretical values, business people have strong economic values, artists place great significance on aesthetic values, social workers have high social values, and politicians have strong political values. In reality, two or three of these types of values may be present and important to each one of us.[7]

Your basic values may not change much after your high school years, but some may become less important As you mature, your priorities change because your needs and goals change. As our needs are met, our desires change. For example, after high school, getting an education may be more important than earning money. However, once you graduate, marry, and get a full-time job, your family, achieving material success, and being accepted by co-workers may dominate your values. Therefore, we have somewhat different values at different stages of our life.

Terminal and Instrumental Values

Terminal Values

Some values, such as security, self-respect, and spiritual growth, may remain the same all through a person's life. Milton Rokeach, in his book, *The Nature of Human Values*, calls these **terminal values**, representing goals you strive to accomplish before you die (where you want to be). Other terminal values include these.

- Comfortable, prosperous, stimulating life
- Sense of lasting contribution and accomplishment
- Equal opportunity for all
- Family security, loved ones taken care of
- Freedom of choice and independence
- Enjoyable, leisurely life
- Social respect and admiration
- An exciting, active life
- A world of peace
- Being happy and content
- Inner harmony
- True friendship
- Wisdom
- Self-respect

Instrumental Values

Instrumental values reflect the way you prefer to behave (ways to get there). Each individual determines which of his or her values are instrumental and which are terminal.[8] Some examples of instrumental values follow.

- Ambitious and hard-working
- Capable, competent, and effective
- Cheerful, creative, courageous
- Independent, self-reliant, self-controlled
- Loving, affectionate
- Respectful, obedient, forgiving
- Responsible and dependable
- Neat and tidy
- Polite and well-mannered
- Open-minded
- Courageous
- Forgiving
- Helpful
- Honest
- Imaginative
- Intellectual
- Logical
- Self-controlled

Development of Values

People are not born with a set of values. We watch how our parents, grandparents, and other relatives act, and our observations help to start our **value system**. Values set early in your life motivate and develop a value system that carry you through life. Early values help develop self-worth and a positive influence that aids in your success. Later on, values are formed further from interacting with and observing people from different backgrounds, racial groups, and cultural groups; and we need to recognize that each group has a part in today's society. We must accept others and work harmoniously together, and we can only do this if we have developed good values early in life.

As you can see, the development of values follows approximately the same course as the development of knowledge or manners. Several factors influence the development of values:

family, historical belief system, peers, school, and the media. The most important factor is the family.

Family Influence

Parents pass on to their offspring the set of values they have accumulated. This process begins at birth and happens indirectly as well as directly. Parents indirectly influence their children when the children observe their behavior in different situations (Figure 19–2). For example, when children hear their parents arguing, they may conclude that that is how to resolve differences of opinion. When children see their parents help each other with household chores, they learn the value of cooperation. There is no escaping the influence of parents on children, no matter how little time children may spend with parents.

The School

At one time values were taught in school to differentiate between right and wrong. As students learned cooperation, teamwork, the importance of meeting deadlines, following directions, etiquette, sharing, and how to use their minds, teachers were a great influence. Later, people felt that schools should be "value neutral." This movement led to a society that did not want anyone's values expressed or taught to their children. However, the public is now asking for some type of guidance for children because children do not seem to learn appropriate values at home. Many church schools took on the teaching of values, and now some public schools have taken on some of the responsibility.[9] Young people spend almost as much time with their teachers, coaches, sponsors, and friends as they do their parents. All of these people influence students in some way. Teachers, coaches, and sponsors can demonstrate how to work together, how to accomplish goals, and how to meet deadlines, for instance. The teacher in Figure 19–3 is indirectly communicating values as he instructs the student.

Figure 19–2 Parents pass on to their offspring the set of values that they have accumulated. These children are getting an early start, as evidenced by their intense listening and observation.

Figure 19–3 Teachers, coaches, and other educators instill values directly and indirectly as they work with students daily.

Figure 19–4 These FFA members are learning the value of participating in their community. (Courtesy of the National FFA Organization.)

Historical Values

Some value systems exhort people not to steal, lie, or kill, to avoid greed and respect authority, to be honest and to honor marital fidelity. Western morality can be said to have at its foundation ten very old, very good rules for living. Three of these rules cannot be mentioned due to governmental standards. Could it be that certain governmental standards contribute to some of our society's problems? Even this answer is determined by your values.

Peer Group Influence

Peer groups also influence values. Suppose you want to be a member of a particular club. In order to belong to that club, you may have to accept the members' values. This would be a direct influence on your values. If a young person spends time with friends who are active in civic organizations, that person may develop the value that helping the community is a good thing and also become active participants. The students in Figure 19–4 are developing values as they participate in a community activity.

Mass Media

The **mass media** may have a great influence on values.[10] Mass media include any form of communication that reaches a very large audience, such as television, newspapers, radio, magazines, and outdoor advertisements. It is not hard to see the influence that the mass media has had on fashion, music, and recreational choices today. Figure 19–5 shows that many factors contribute to the development of a person's values.

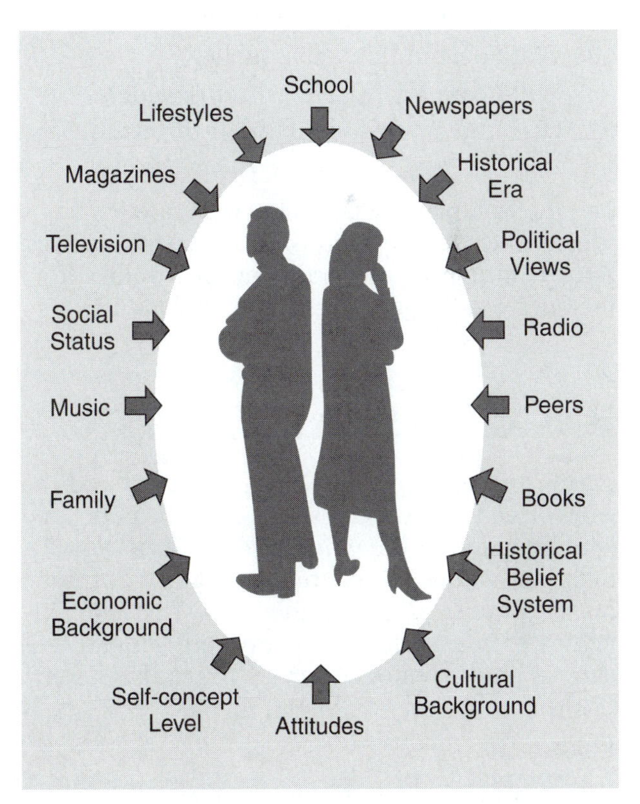

Figure 19–5 Many factors contribute to the development of a person's value system.

Determining Your Personal Values

Sometimes we are not sure how strong our values are. Sometimes, our values are compromised when we are around certain people; if this happens, perhaps they are not really your values.[11] Louis Roth, in his book *Values and Teaching*, has developed a test you can use to find out the strength of your values. To take the Roth test, think of any value you consider important and ask these seven questions about it.[12]

1. Did I choose this value freely, with no outside pressure?
2. Did I choose this value from several alternatives?
3. Did I consider the consequences of my choice?
4. Do I like and respect this value?
5. Will I defend this value publicly?
6. Will I base my behavior on this value?
7. Do I find this value persistent throughout my life?

The workplace values in the next section can be subjected to these questions. Answering these questions will help you separate your real values from the ones you thought were strong in your life.

Basic Societal Values

Certain values are widely held in society and the workplace. These values are usually associated with people who are trying to do what is best. Remember, values, unlike truths, are not absolute; not everyone's values are the same. The following values presented in this section are general tendencies and may not be part of your value system.

Honesty

Not only is "honesty the best policy," it is better than all policy. All a person's values cancel out if that person is not honest. To be honest is to be real, genuine, authentic, and bona fide. Honesty expresses both self-respect and respect for others. Honesty fills lives with openness, reliability, and candor. Mark Twain said, "If you tell the truth, you don't have to remember anything." Honesty comes in many forms:

- Telling the truth
- An honest day's work, or giving the employer fair work for the pay
- Fair actions, such as treating people in an honest and fair manner
- Expression of true feelings, such as making sure your compliments are genuine rather than flattery

Developing an Honest Character To develop an honest character, the individual must first think about how he or she will react to a situation. The more the individual reacts with honesty, the more it becomes a way of life. One develops an honest character by following this progression.

- Thoughts becoming actions.
- Actions becoming deeds.
- Deeds become habits.
- Habits become a way of life.

Dishonesty Dishonesty is a willful perversion of truth in order to deceive, cheat, or defraud. It comes from the intent to mislead. Dishonesty comes in many forms.

- Cheating on a test
- White lies
- Changing the truth just a little
- Exaggerating
- Shoplifting
- Lying
- Making a false call in sports
- Withholding parts of the whole truth in order to fit your needs
- Double dealing
- Misleading
- Mischievous deceit

Prudence

Prudence simply means common sense. It means thinking about things before you act. Being prudent is the opposite of being foolish. A prudent person uses all his or her intellectual ability, but that does not mean you have to have a high IQ score to be prudent. You do, however, have to think about the results of your actions before you act.

Temperance

Temperance means just the right amount, at the right time, of any pleasure. Pleasure means any activity that makes you happy for the moment, including eating, sleeping, exercising, reading, working, and playing. Temperance does not mean abstinence. Temperate people are those who have control over the pleasure in their lives; it does not control them.

Justice

Justice includes everything we usually call fairness. It is honesty, compromise, truthfulness, and keeping promises. It is easy to see why justice is a value. No one respects a person who is dishonest, doesn't share, tells lies and breaks promises.

Courage

Courage includes standing up for what you believe when others are putting pressure on you to do wrong. It is not going along with the crowd. The mark of a real man or woman is not doing what everybody else is doing but doing what he or she believes to be good.

Courage does not mean that you are not afraid. Having some fear is safer than being fearless. Aristotle said that courage is a settled disposition to feel an appropriate degree of fear and confidence in challenging situations. Courage, said Plato, is knowing what to fear. Courage is acting bravely when we don't really feel brave, standing our ground, and advancing or retreating, as wisdom dictates.

Courage is facing a dangerous situation when necessary. It takes courage to speak your mind when you know you will be scorned for doing so. It means having guts—the guts to do what is right, when you know all your friends will oppose you. This is one of the hardest virtues to uphold. Peer groups exert pressure on students to conform, and it requires a great deal of fortitude to take another path from that of your peers.

Perseverance

In the fable of the turtle and the rabbit, slow and steady wins the race. Perseverance is an essential quality of character in high-level leadership. Much good that might have been achieved in the world is lost through hesitation, faltering, wavering, vacillating, or just not sticking with it. With practical intelligence, in the right context, occurring in the right combination with other values, perseverance is an essential ingredient of life (Figures 19–6 and 19–7).[13]

Loyalty

Loyalty is the willingness to support a person or a cause. It affects family, friends, religious, political,

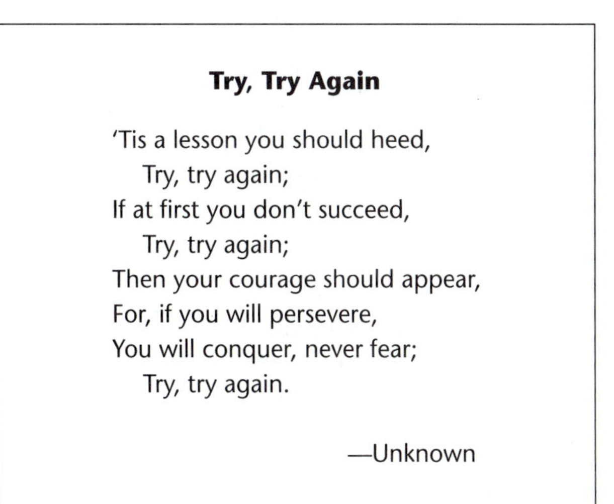
Try, Try Again

'Tis a lesson you should heed,
Try, try again;
If at first you don't succeed,
Try, try again;
Then your courage should appear,
For, if you will persevere,
You will conquer, never fear;
Try, try again.

—Unknown

Figure 19–6 Most goals can be accomplished through perseverance and persistence.

It Can Be Done

The man who misses all the fun
Is he who says, "It can't be done."
In solemn pride he stands aloof
And greets each venture with reproof.
Had he the power he'd efface
The history of the human race;
We'd have no radio or motor cars,
No streets lit by electric stars;
No telegraph nor telephone,
We'd linger in the age of stone.
The world would sleep if things were run
By men who say, "It can't be done."

—Unknown

Figure 19–7 Some believe that vision and persistence are as important as intelligence. Many things have been accomplished by persistent people in spite of other people telling them it can't be done.

and professional lives. Our loyalties reflect the kind of person we have chosen to become. They are a kind of consistency or steadfastness in our attachments. To be a loyal worker, citizen, or friend means to care seriously about the well-being of one's employer, country, or associate. Loyalty is like courage in that it shows itself most clearly when we are operating under stress. Real loyalty endures inconvenience, withstands temptation, and does not cringe under assault.[14]

Commitment

Commitment is a pledge or promise to do something. In today's world, commitment is an important quality that is necessary for success in many aspects of life. In a marriage, husband and wife must be committed to their marriage; a lack of commitment can contribute to problems in the marriage. In the workplace, commitment to quality and to performing a job correctly can result in goods or services that can compete in today's markets. A lack of commitment on the job can contribute to poor-quality products and problems among employees. Commitment means doing what you say you are going to do and following through.

Pride

It feels great to accomplish a job or task with such quality that you are proud for others to see it. Pride may be seen negatively when an individual becomes boastful or arrogant. However, a certain amount of pride is healthy. A lack of pride in one's job can result in goods or services that are below standard. Lack of pride in one's school can result in lackluster academic and athletic performance, vandalism to buildings and property, and behavioral problems. A lack of pride in oneself can cause poor hygiene and grooming, poor job performance, poor school performance, and problems in relationships with people. Pride in work and a job well done are basic values. The students in Figure 19–8 appear to have pride in their work.

Figure 19–8 Pride is a basic societal value. These people above are exhibiting pride in their work.

Conviction

Conviction may be defined as belief or acceptance that certain things are true or real. A person must stand for what he or she believes or risk being swayed by every belief that comes along. Aaron Tippin wrote a song that said, "You have to stand for something or you'll fall for anything." A person who has convictions is dependable because people know where he or she stands. Standing up for your beliefs makes you honest because you have something to believe in. Standing up for what you believe helps to create personal happiness and satisfaction.

Responsibility

To be responsible is to be answerable and accountable. Many believe one of the biggest problems in society today is that people do not take responsibility for their own actions. It is a truism that everything ever done in the history of the world has been done by somebody; some person has exercised some power to do it.[15] Aristotle was among the first to insist that we become what we are as people by the decisions we make. Therefore, "responsible persons are mature people who have taken charge of themselves and their conduct and are accountable for their actions."

Self-Discipline

In self-discipline, you make a disciple of yourself. You are your own teacher, trainer, coach, and disciplinarian. There is much unhappiness and personal distress in the world because of failures to control tempers, appetites, passions, and impulses.[16] "If only I had stopped myself" is heard often. "If it is going to be, it is up to me." Achieving our goals requires much self-discipline.

Compassion

Compassion is a form of love or empathy with other people, their inner lives, their emotions, and their external circumstances. Compassion also involves fellowship, sharing, and supportive companionship to friends when they need you. Although former President Richard Nixon made mistakes, Americans were touched when at his funeral members of both political parties showed compassion and remembered the great things he had done.

Humility

Humility is the opposite of bragging. It is okay to know you have done a good job; but, it is not a very good thing to go around telling everyone about it. "Show-offs" are not humble people. Sometimes, though it is okay to tell about your accomplishments, but not to tell about them over and over again.

Vitality

Vitality is physical or mental strength or force. This does not mean everyone has to be a bodybuilder and astro-physicist. It means that you should use all your physical and mental abilities as hard as you possibly can. Being vital is the opposite of being lazy.

Dedication

Dedication is similar to loyalty except that it is usually thought of as more of an inner motivation. You can be dedicated to anything you do. People are dedicated to their jobs, to taking care of other people, to losing weight, to learning to play an instrument, to an organization, or to doing a good job at everything they attempt.

Assertiveness

Assertive people are confident about their abilities, but not pushy. Assertive does not mean aggressive. Assertiveness means you are willing to step forward and make a statement because you believe it to be true, whether or not anyone else agrees with you. Assertive people find a way to get a job done even in the face of many obstacles.

Obedience

Obedience means a willingness to submit to authority and knowing which people are the authority figures. It means following orders, or doing what you are directed to do.

Fidelity

Fidelity means simply being loyal or dedicated or vowing allegiance. But, in our society today, it has come to signify being faithful in marriage.

Forgiveness

Forgiveness as a societal value means that you not only forgive people for mistakes they have made, but you also forget about it and do not let the incident affect your relationship with that person or your perception of that person. It means wiping the slate clean and starting all over again. Do not forget to forgive yourself also.

Integrity

Integrity refers to striving to live up to a code of rules or moral values. When we say that someone has integrity, we usually mean he or she possesses, in some degree, all the values we have been discussing.

Respect for Authority

Respect for authority involves doing what we are told to do by the appropriate person. It may be your parent, guardian, teacher, or boss. When we do what we are told to do, we gain respect. Relationships are better; we learn more; we stay out of trouble; and we perform with greater effectiveness.

Etiquette

Some other values that smooth relationships between people are sometimes called rules of **etiquette**, to be discussed later in this chapter. Qualities such as having good manners, knowing how to make introductions, write thank-you notes, and make and answer telephone calls constitute valuable social behavior called etiquette.

Work

Work is applied effort; it is whatever we put ourselves into, whatever we expend our energy on for the sake of accomplishing or achieving something. A person who places a high value on work wants to do a job to the best of his or her ability. He or she takes pride in completing a task correctly. The most satisfying work involves directing our efforts toward achieving ends that we ourselves endorse as worthy expressions of our talent and character. Those who missed the joy of a job well done have missed something very important.[17] As you evaluate your own work values, consider these sayings.[18]

"The most practical, beautiful, workable philosophy in the world won't work—if you won't."

"If you don't have time to do it right, when will you have time to do it over?"

"Good enough is usually not good or enough."

"If you do more than you are paid to do, you will eventually be paid more for what you do."

Appreciating Cultural Diversity

In working with people a leader needs to perceive and respect the uniqueness of each individual. Each culture has its own customs that, when understood, can enhance our appreciation for that value system. Treat each person you meet as an individual. When you judge a person by his or her race, class, or ethnic background, you stereotype that person, which can lead to many problems.[19] People tend to assume wrongly that their own culture does things the right way and should be universally valued (Figure 19–9).

Other Human Relationship Values

Values can build strong ties with other people, including cooperation, consideration, dependability, generosity, kindness, patience, respect, tolerance, keeping promises, and defending others from verbal attacks. If a person is found to be lacking in any of these qualities, their relationships with others are likely to suffer.

Figure 19–9 Even though these students are from diverse cultures, their differences don't matter as they work together and strive for common goals.

Workplace Ethics

Ethics, as distinguished from values, encompass the rules of conduct that reflect the character and sentiment of the local community. Kickbacks and payoffs may be acceptable practices in one part of the world but unethical elsewhere.[20] Ethics are very important to businesses (Figure 19–10). Approximately one-third of corporations having more than 100 employees provide some type of ethics training.[21] As a result, many corporations have decided to put their ethics in writing. More than 90 percent of the Fortune 1,000 companies have a written code of ethics.[22]

Figure 19–10 Can you imagine the chaos throughout the financial world if the employees at the Chicago Board of Trade did not adhere to a strict code of ethics?

Complexity of Ethical Behavior

Many ethical issues are complex and cannot be considered in absolute terms, which explains why so many companies offer ethics training. Employees participate in in-depth discussions on topics such as methods of gathering information about competitors, fairness in hiring, and the issues of receiving gifts and entertainment from customers.[23] How do we know what is right? Consider the following:

- A piece of merchandise that was accidentally broken is returned with the claim that it was damaged before it was purchased.
- A woman routinely completes her income taxes by overstating the value of contributions or claiming vacation expenses as deductible business expenses.
- You copy a computer software program so that you can complete assignments on your own computer rather than enduring the inconvenience of using a computer lab at school or at work.[24]
- A businessman who travels extensively feels cheated that telephone calls are not reimbursed travel expenses. He therefore overstates car mileage to cover the telephone calls.
- An investment broker writes an enthusiastic letter to a client stating that she has experienced a 24 percent growth in her investment, without mentioning that the stock market as a whole experienced 36 percent growth in the same period.
- In order to increase profits, a company increases the complexity of a product so that repairs have to be done by an authorized service department.[25]

- A company quotes a price for a job, but the actual time needed to complete the job is significantly less than estimated.

Depending on your value system, some of these examples may be blatantly wrong and some may not be wrong. Your values lay the foundation for your ethical choices. Minor compromises with your values can gradually weaken your foundation.

Making Ethical Decisions

Is there a way to determine a good ethical decision? Is there a way to determine what our values are on a specific issue? Many methods are available to help individuals and businesses analyze ethical issues. One such way of determining whether an issue is ethical is the Pagona model. According to this model, to determine the correct ethical choice on an issue, answer the following six questions.[26]

1. Is the proposed action legal?
2. What are the benefits and costs to the people involved?
3. Would you want this action to be a universal standard, appropriate for everyone?
4. Does the action pass the light-of-day test? That is, if your action appeared on television or the front page of a newspaper, or if others learned about it, would you be proud?
5. Does the action pass the Golden Rule test? That is, would you want the same to happen to you?
6. Does the action pass the ventilation test? Ask the opinion of a wise friend with no investment in the outcome. Does this friend believe that the action is ethical?

Even though a decision may be legal, it could be unethical. For example, many food products were labeled "lite" or "low-fat" when in fact most of the calories in the food were derived from fat. For certain people, this misinformation could cause illness or even death. James E. Purella, executive vice-president of Ingersoll Rand Company, said "Good ethics, simply, is good business." Good ethics will attract investors. Good ethics will attract good employees. You can do what is right, not because of conduct codes and not because of rules or laws, but because you know what is right.[27]

The Importance of Workplace Etiquette

Proper etiquette is critical to success in the workplace. It is often assumed that people learn etiquette at home or that they learn through experience or observation. Unfortunately, this is often not the case.

Workplace interactions contain a large social component. Not conforming to etiquette in purely social situations might cause some embarrassment, a loss of social status, or loss of invitations to future events. Not conforming to etiquette during business interactions can cause you to lose a sale—or it can cost you a job or a promotion. Whether it is right or not, people often look at your behavior and lack of conformity to etiquette and relate this to your ability (or inability) to do a job. Therefore, it is in your best interest to become thoroughly familiar with the behavior that may be expected of you.[28]

Business Phone, Cellular Phone, and E-Mail Etiquette

The Office Phone

That first impression, whether it is of the organization or of you, sends a powerful message and image to the caller. Answering the phone should always be done in a professional manner. You should never answer with just "Hello" or by simply stating your phone or extension number. You should not answer the phone with a nickname because this is unprofessional and lacks authority. Nor should you immediately place a caller on hold. If you must place callers on hold, ask their permission and allow them to answer. They may not be able to wait, or there may be an emergency. Try never to leave someone on hold for more than 30 or 40 seconds, but if this becomes necessary, be sure to apologize when you return to the call. If you think someone will be on hold for more than a minute, offer to call back with the information requested. When you answer a phone and the caller is unknown to you, immediately write down his or her name. This allows you to address the person by name later in the call, which is far more impressive than having to ask the caller's name.

Never eat while on the phone, and try not to talk to other people in the room when you or they are on the phone. If your phone system has call waiting, it is best to finish the first call and then answer the second, rather than interrupting the first to answer the second and then returning to the first. Finish one call, then move on to the next. If you make a call and get a wrong number, always apologize. If you make a conference call, immediately tell the other parties that you are not the only person on the line. Always ask permission before using a speakerphone, and use it only when necessary because they are quite annoying to those not on the speaker.[29] Other guidelines for telephone etiquette include these.[30]

- Answer the phone in three rings or less.
- Answer with your name, your department name, or the company's name.
- Make business calls only during normal business hours whenever possible.
- Return calls in 24 hours or less. Taking 48 or more hours is considered rude.
- Never give out personal information about co-workers over the phone.
- When someone calls you, do not be the first to hang up.
- When you call someone, hang up the phone gently.
- When you call, ask the person if she or he has time to talk.
- If the phone connection is lost, the caller should be the one to call back.

Cellular Phones

When calling someone on a cellular phone, be as brief as possible, even if this means calling back later on a wired phone. If you have a cellular phone provided by your company, it is your ethical responsibility to minimize your time by restricting calls to clients and company business. In any event, you should not talk on a cell phone while walking down the street; while in a theater, restaurant, or classroom; in a meeting; or in other public places.[31] Remove yourself from the presence of others to make a cellular phone call, and for your own safety and that of others, make sure you can drive and talk on the phone before placing a call from your car. In some areas, cell phone use while driving is illegal. Pagers should be of the vibrating type rather than the audible type. Even with a vibrating pager, you should not allow meetings or other activities to be interrupted with false emergencies.[32]

E-mail

Be Brief The increase in usage of the Internet has made e-mail very popular. E-mail is best for shorter messages. If you have a long message, send it as an attachment.

Have Compatible Software Be sure that your recipients have the software for which the attachment was created. For example, don't send a Microsoft Word attachment when everyone else is using WordPerfect, or vice versa.

Be Professional Messages should contain nothing that you wouldn't want everyone to know because it is possible for employers to monitor (read) e-mail messages, and it is sometimes possible to retrieve messages that you believe you have deleted.[33]

Do Not Use All Caps Sending messages and words in capital letters makes you appear to be shouting at the recipient.

Set the Return Receipt Function Do not assume that everyone in the organization uses e-mail. If you need to know that someone received your message, set the Return Receipt function, which sends you a message when the recipient has opened your message.

Use Special Functions Sparingly Always set the Priority function carefully. If all your messages are sent at the highest priority, then High Priority will become meaningless to the recipients of your messages. Similarly, do not abuse the broadcast and "cc:" functions. Do not broadcast your every thought to everyone in the company, nor should you "cc:" (carbon copy) the CEO or others on every message just to show you are working; it appears as if you are trying to make brownie points with them.[34]

MEETING ETIQUETTE

Proper meeting etiquette calls for the chairperson to arrive prepared and on time. If you are late, you should apologize but should not offer an excuse.[35] Excuses delay the meeting even more and are often not believed anyway.

The Chairperson Should Arrive Early

The chairperson should arrive early enough to review the arrangements, ensuring that everything is prepared, that any food or equipment has arrived, and that all the equipment is working properly. The more high-tech equipment being used, the earlier the chairperson should arrive, especially if the chairperson is unfamiliar with the equipment's operation.

Phone Calls and Leaving Early

Meeting should take precedence over phone calls. If some people need to be present for only part of the meeting, it is acceptable to allow them to present their material and leave.[36]

After the Meeting

After the meeting, you are responsible for returning the room to its premeeting condition or ensuring that this is done by others and for ensuring that the minutes are typed and distributed to all those in attendance and all those unable to attend.[37] Refer to Chapter 12, Conduct of Meetings, for more information on proper and etiquette conduct of meetings.

CONCLUSION

You may experience many kinds of **value conflicts** in your personal life and on the job. As a leader, you may be torn between loyalty to workers and higher-ups. As a worker, you may experience conflict between fulfilling family obligations and doing what you need to do to be successful at work. Your values determine the choices you will make. How you resolve these value conflicts, make ethical choices, and conduct yourself will greatly affect your attitude toward yourself and your career.[38]

SUMMARY

Values are principles or standards for behavior accepted by a large number of people, usually over a long period of time. Values are what make it possible for human beings to live together, develop strong relationships, and accomplish goals. Values are especially important to understanding human behavior. Individual values are usually the product of beliefs and truths.

Leaders must be aware of different philosophical beliefs and values. People have different values because their philosophies are different. Philosophy provides us with a reasoned framework within which to think. Philosophical values can be examined by evaluating six categories: theoretical, economic, aesthetic, social, political, and religious. Terminal values represent goals you strive to accomplish before you die. Instrumental values reflect the way you prefer to behave.

Values are developed by a number of influences. The main influence on the development of one's values is the family, but schools, belief systems, peers, and the media all influence people to some extent. Some influences are detrimental while some are quite productive. Sometimes we are not really sure how important our values are. Taking the Roth test can help you find out.

Certain values are appreciated in society and the workplace, including honesty, prudence, temperance, justice, courage, perseverance, loyalty, commitment, pride, conviction, responsibility, self-discipline, compassion, humility, vitality, dedication, assertiveness, obedience, fidelity, forgiveness, integrity, respect for authority, etiquette, work, and appreciating cultural diversity.

Human relationship values include cooperation, consideration, dependability, generosity, kindness, patience, respect, tolerance, keeping promises, and defending others from verbal attacks.

Ethics, as distinguished from values, constitute the rules of conduct that reflect the character and sentiment of the community. Many ethical issues are complex and cannot be considered in absolute right or wrong terms. Even though a decision may be legal, it could be unethical. The Pagano model poses six questions to help you make ethical choices.

Being a person with high values and ethics is important for several reasons. Our behavior influences other people; influence is leadership; and we are judged by our actions every day. What others see in us, they may try to imitate, especially people who are younger and searching for guidance and role models.

Proper etiquette is critical to success in the workplace. The phone should be used in a professional manner. Cellular phones should have limited use in selected areas, and pagers should not interrupt a meeting. E-mail messages should be brief and professional. Proper meeting etiquette calls for the chairperson to arrive prepared and on time and to make sure the meeting room returns to its proper order.

END-OF-CHAPTER EXERCISES

Review Questions

1. Define the Terms to Know.
2. Name and explain six categories in which values can be evaluated.
3. List 10 examples of terminal values.
4. List 10 examples of instrumental values.
5. List five factors that influence the development of values.
6. List seven historical values or belief systems.
7. What are six questions that you can ask yourself to determine the strength of your personal values?
8. List 10 basic societal values you strongly agree with.
9. Name four forms of honesty.
10. Name 10 forms of dishonesty.
11. Name eight human relation virtues that can build strong ties with other people.

12. List three examples of questionable ethical practices.
13. What six questions can you ask yourself to help determine the correct ethical decision?
14. Explain the importance of understanding and appreciating cultural diversity.
15. Why is etiquette important in the workplace?
16. List nine proper uses of the office (business) phone.
17. List five rules about the proper use of cellular phones.
18. List five rules about the proper use of e-mail.
19. List three rules of good meeting etiquette.

Fill-in-the-Blank

1. Several factors influence the development of values, but the most important factor is the ________________ .
2. Values are especially important to understanding ________________ behavior.
3. We have different values in society because our ________________ are different.
4. If it is going to be, it is up to ________________ .
5. ________________ includes standing up for what you believe when others are putting pressure on you to do wrong.
6. A thought becomes an action, an action becomes a deed, a deed becomes a habit, and a habit becomes ________________ .
7. Mark Twain said, "If you tell the truth, you don't have to ________________ anything."
8. ________________ is a form of love or empathy with other people.
9. With practical intelligence, in the right context, and occurring in the right combination with other values, ________________ and persistence are essential ingredients of life.
10. Even though a decision may be legal, it could be ________________ .

Matching

_____ 1. Happy with accomplishing a job or task with quality
_____ 2. Willingness to submit to appropriate authority
_____ 3. Strong belief or acceptance that certain things are true or real
_____ 4. Opposite of bragging
_____ 5. Forgetting about mistakes of others and not letting them affect relationships
_____ 6. Confident of your abilities, but not pushy
_____ 7. Inner motivation
_____ 8. Unblemished or near perfect ethical behavior
_____ 9. Pledge or promise to do something
_____ 10. Willingness to support someone or some cause
_____ 11. Answerable or accountable for our actions
_____ 12. Besides being needed for economic livelihood, it can build self-esteem

A. responsibility
B. humility
C. work ethic
D. dedication
E. integrity
F. commitment
G. obedience
H. forgiveness
I. pride
J. conviction
K. loyalty
L. assertiveness

Activities

1. Review the six categories that influence values: theoretical, economic, aesthetic, social, political, and religious. Write a short paragraph on how you view or value each of the six areas.
2. Rank the five factors that have influenced the development of your values. Write a paragraph explaining each.
3. Write down as many other values or ethical behaviors that you can think of that were not mentioned in this chapter. Be prepared to explain why you selected them.
4. Make a list of five famous people, and give one reason for selecting each as a model of integrity.
5. Write a 300-word essay entitled, "The Perfect Person." Include as many values as you can from this chapter in your essay without being awkward. You may add others.
6. In the section Complexity of Ethical Behavior, seven ethical issues were given. Select three and write a paragraph (or more if needed) on how you view the issue. Be prepared to share your ideas with the class.
7. Write an ethical issue that you have faced or wondered about. Share it with the class and get your classmates' responses. Your teacher may want you to use the six questions of the Pagano model to determine an ethical choice.
8. Review the chapter, and add any more characteristics that you consider important personal values or ethical behavior. Complete the accompanying chart with 25 values entitled, "These Things I Value." Identify which group influenced these values.
9. **Take it to the Net.**

   Explore workplace ethics on the Internet. Several Web sites containing information on workplace ethics are listed below. Browse the sites. Choose one site you feel contains the most useful information on workplace ethics. Write a summary of the site and in your summary include what was useful about the site.

   Web sites

   <http://www.eth-ics.com/>

   <http://www.business-ethics.com/>

   <http://www.ecampus.bently.edu/dept/cbe/>

   <http://www.character-ethics.org/>

   Search Terms

   ethics

   workplace ethics

   ethics in the workplace

   business ethics

   ethical

These Things I Value

Value	Home/ Family	Friends	School	Historic Belief System	Media	Other
1.						
2.						
3.						
4.						
5.						
6.						
7.						
8.						
9.						
10.						
11.						
12.						
13.						
14.						
15.						
16.						
17.						
18.						
19.						
20.						
21.						
22.						
23.						
24.						
25.						

Notes

1. W. C. Himstreet et al., *Business Communications* (Belmont, CA: Wadsworth, 1993), p. 98.
2. J. J. Littrell, *From School to Work* (South Holland, IL: The Goodheart-Wilcox Company, 2000).
3. L. H. Lamberton and L. Minor, *Human Relations: Strategies for Success* (Homewood, IL: Richard D. Irwin, 1995), pp. 71–72.
4. T. Moore, "For the Leninists, It's Mac in the U.S.S.R.," *U.S. News and World Report*, February 1990, pp. 10–11.
5. "Philosophy," *The World Book Encyclopedia*, (Chicago: World Book, 2000).
6. G. W. Allport et al., *Study of Values* (Boston: Houghton-Mifflin, 1960).
7. R. M. Hodgetts, *Modern Human Relations at Work* (Fort Worth, TX: Harcourt Brace College Publishers, 1995).
8. M. Rokeach, *The Nature of Human Values* (New York: Free Press, 1973).
9. S. L. Nazario, "Schoolteachers Say It's Wrongheaded to Try to Teach Students What's Right," *Wall Street Journal*, April 6, 1990, p. B-1.
10. B. L. Reece and R. Brandt, *Effective Human Relations in Organizations* (Boston: Houghton Mifflin, 1993), p. 184.
11. Lamberton and Minor, *Human Relations: Strategies for Success*, p. 78.
12. Louis Roth et al., *Values and Teaching* (Columbus, OH: Charles Merrill Publishers, 1976).
13. W. J. Bennett, ed., *The Book of Virtues* (New York: Simon & Schuster, 1993), pp. 527–528.
14. Ibid., pp. 665–666.
15. Ibid., pp. 185–186.
16. Ibid., pp. 21–22.
17. Ibid., p. 347.
18. Z. Ziglar, *The "I Can" Course* (Carrollton, TX: The Zig Ziglar Corporation, 1989).
19. Himstreet, *Business Communications*, p. 98.
20. Reece and Brandt, *Effective Human Relations in Organizations*, p. 197.
21. B. L. Thompson, "Ethics Training Enters the Real World," *Training*, October 1990, p. 84.
22. B. Weisendonger, "Doing the Right Thing" *Sales & Marketing Management*, March 1991, p. 83.
23. Reece and Brandt, *Effective Human Relations in Organizations*, p. 200.
24. Himstreet, *Business Communications*, pp. 159–160.
25. Ibid., p. 161.
26. D. L. Mathison, "Business Ethics Cases and Decision Models: A Call for Relevancy in the Classroom," quoted in W. C. Himstreet et al., *Business Communications* (Belmont, CA: Wadsworth, 1993), pp. 173–174.
27. M. Slayton, *Ethics Journal* (Washington, D.C.: Ethics Resource Center, May-June 1991), p. 7.
28. M. W. Drafke and S. Kossen, *The Human Side of Organizations* (Reading, MA: Addison-Wesley, 1998), p. 454.
29. B. Pachter and M. Brody, *Complete Business Etiquette Handbook* (Englewood Cliffs, NJ: Prentice-Hall, 1995), pp. 122, 129.
30. Drafke and Kossen, *The Human Side of Organizations*, p. 455.
31. M. Packard, "Telephone Courtesy Sets the Tone of Customer Relations," *RV Business*, May 1993, pp. 23–24.
32. D. Nigro, "Manners Matter," *Meetings and Conventions*, June 1995, pp. 76–80.
33. T. A. Daniel, "Electronic and Voice Mail Monitoring of Employees: A Practical Approach," *Employment Relations Today*, Summer 1995, pp. 1–11.
34. Drafke and Kossen, *The Human Side of Organizations*, p. 457.
35. J. Reese, "The Decline of Office Managers" *Fortune*, May 13, 1993, pp. 20–22.
36. Pachter and Brody, *Complete Business Etiquette Handbook*, p. 177.
37. Ibid.
38. Reece and Brandt, *Effective Human Relations in Organizations*, p. 191.

SECTION

Transition to Work Skills

CHAPTER

20

Selecting a Career and Finding a Job

Objectives

After completing this chapter, the student should be able to:

- Explain the reasons why people work
- Differentiate between work, occupation, job, and career
- Discuss career planning
- Explain how to match jobs with your personal characteristics
- Describe careers in production agriculture, agribusiness, and agriscience
- Explain the diversity of agricultural education job placement
- Describe the sources to use for finding a job

Terms to Know

career planning
work
occupation
job
career
job-related skills
self-management (adaptive) skills
transferable skills
gross national product
value-added
immediate job placement
postponed job placement
avocational (part-time) job placement
supplementary (tributary) job placement
job lead
network
classified ads

Imagine that you could do anything you wanted to do, forgetting the necessary qualifications, abilities, or money needed to do it. Suppose you could have any career that you wanted anywhere. What would that job or career be? Visualize yourself in that job or career, and hold that pleasant image in your mind.

During our youth, we usually have many ideas about what we want to be. However, as we grow older, we lose sight of our dreams due to a lack of career planning. Responsibilities arise unplanned, we tend to drift into occupations more by chance or need for a job rather than by choice.

In an earlier chapter we discussed goals and planning. People don't plan to fail; people fail to plan. Begin the process of creating your career plan by closing your eyes and imagining your dream job. Picture yourself at work in as much detail as possible. What are you doing? Where do you work? How much do you earn? Once you have visualized your ideal job, ask yourself how you can make part or all of this dream job come true.[1]

Figure 20–1 Horticulturalists Mark Roh (left) and Roger Lawson display new flowers developed for U.S. consumers. (Courtesy of USDA.)

Career planning concerns your life's mission, or what you see as the meaning of your life. Follow your dream occupation, or you won't feel you have really lived. It is better to have tried and failed than not to have tried at all. At least you can live your life without regrets of not giving it your best. After studying this chapter, you will be better able to find that dream job, as the people in Figure 20–1 have.

WHY PEOPLE WORK

Working and earning wages enables you to establish your independence and define your own life. Other people see and respond to you in new ways. You become more accepted as an adult and are granted the greater rights and responsibilities of adulthood.[2]

People's views about **work** vary greatly. Although all people do not value and enjoy their work, most people feel that work is an important part of a well-rounded life. Most people generally like what they do. Many studies have shown that most people would work even if they didn't have to. This view of work is not limited to adults. The interest in learning about and preparing for work has never been greater.[3]

People work for many different reasons. The reasons vary from person to person. Larry Bailey in his book, *The Job Ahead*, discusses seven reasons for working. They are as follows.[4]

- *Earn money.* The major reason people work is to earn money. Earnings are needed to buy food, shelter, clothing, and other necessities. Beyond meeting basic needs, money is used to purchase goods and services that provide comfort, enjoyment, and security.
- *Social satisfaction.* People are social creatures. Working gives people a chance to be with others and to make friends. In the work environment, people can give and receive understanding and acceptance.

- *Positive feelings.* People get satisfaction from their work. For instance, your work may give you a sense of accomplishment. Think of how you feel when you finish a school project or a difficult job task. Working also gives people a feeling of self-worth. The feeling comes from knowing that other people will pay you for your skills.
- *Prestige.* Some people work for the prestige or status of the position. Prestige is an admiration that society has for an occupation. Prestige is separate from how well the job is performed. Can you think of occupations that you consider to have prestige?
- *Personal development.* Many people have a drive to improve themselves. Work can provide an opportunity to learn and grow. Work can often be a great teacher.
- *Contributions to health.* Work can be very important for mental and physical health. This results from the work itself as well as the physical activity and exercise involved. People who are active and happy in their work tend to feel better.
- *Self-expression.* We all have interests, abilities, and talents. Work can be a way to express ourselves (Figure 20–2). It doesn't matter what kind of work it is as long as it suits the worker and is not illegal or immoral.

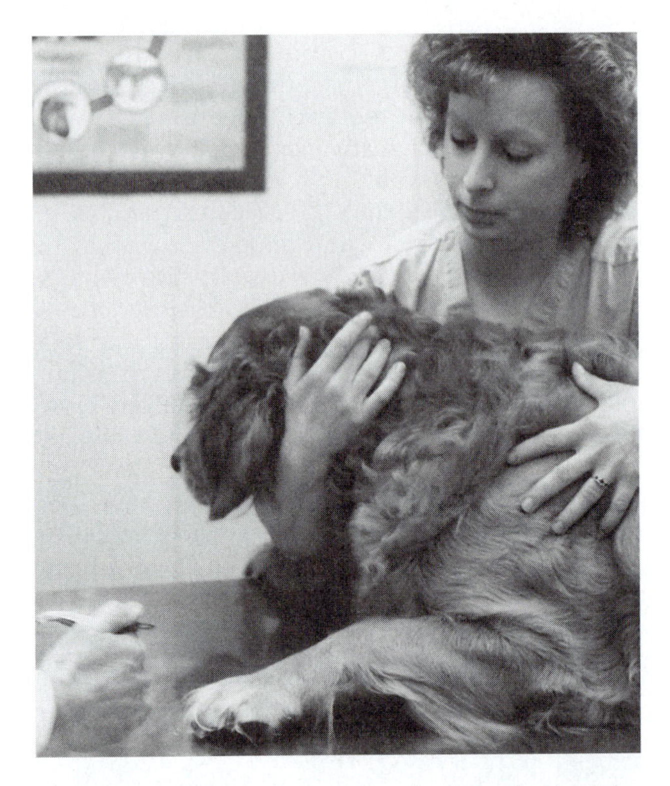

Figure 20–2 This worker expresses her love of animals by working as a veterinary technician. (Photo courtesy of Brian Yacur.)

Other reasons people work are security, success, happiness, and peers and family.

Far too many people never follow their own dreams. Their occupational choices are often shaped by what other people think they should do. They work because other people want them to work.

"You *should* go to work."
"You *should* go to college."
"You *should* serve your country."
"You *should* earn a lot of money."
"You *should* be a lawyer (or doctor, or teacher) because your parent or grandparent was."

Many people spend their whole lives doing what others think they should do rather than what they want to do. They are tyrannized by the "shoulds" of their parents, teachers, friends, social class, and other role models.[5]

TERMINOLOGY

The terms work, occupation, and job are often used interchangeably. They are similar, but consider the following differences.

Work

Work can be defined as activity directed toward a purpose or goal that produces something of value to oneself and sometimes to society. For example, work can provide you with money and a sense of accomplishment. Work by a social worker, teacher, or nurse provides benefits to society. A wage may or may not be paid, depending on the type of work, such as service or community projects.

Occupation

All occupations carry out work. An **occupation** is the name given to a group of similar tasks that a person performs for pay. For example, typing, filing, maintaining records, placing phone calls, and scheduling meetings are tasks performed by the occupation of secretary. Carpenter, salesclerk, attorney, truck driver, and chef are examples of common occupations that involve groups of similar tasks.[6]

Most occupations require specific knowledge and skills to perform. Occupations are learned on the job or in various kinds of education and training programs. A person with an occupation can work at a number of different jobs.

Job

A **job** is a paid position at a specific place or setting. A job can be in an office, store, factory, farm, or mine. For example, a nurse (an occupation) may have a job in a doctor's office, clinic, hospital, home, school, factory, or nursing home.[7]

Difference Between Job and Career

A job is a specific duty or chore performed for pay. It involves responsibilities based on an agreement between a worker and an employer about performance, hours, and other factors. A **career** is a series of jobs that is pursued in sequence to achieve the ultimate occupation that is designed by the individual. Basically, a career is something you really want to do for the rest of your life.

CAREER PLANNING

Although choosing a career is a serious step in one's life, the same holds true for planning. The more time you spend on career planning, the easier it will be to find the career of your choice. No one can foresee all the events that will occur in the future, but you can anticipate changes. When you anticipate, you use your imagination to pretend what might happen in the future.

Importance

Your career will likely be one of your most important life activities. Many years will be spent in one job or another, or in the training you need for your career. So much time will be demanded by your career that the rest of your life may revolve around it, much the way it revolves around school now. The kind of career you have will affect how much time you have for other activities, the types of people you meet, the friends you make, and the kind of lifestyle you can afford.

The amount of time you will spend working is a good reason to give careful consideration to your career. A person who works full time for 40 years, averaging 40 hours per week for 50 weeks per year, will spend 80,000 hours at work. Compare that with the 17,000 hours you will have spent in school by the time you graduate from high school!

Factors to Consider When Selecting a Career

There are many factors to consider when selecting a career. These factors include nearly every aspect of one's life. Some major factors to consider are these.

1. *Standard of living.* When selecting a career, it is important to choose one that will provide enough income to allow you to live comfortably. People view living comfortably in many ways. Some people are willing to give up things to have a career that pays more. Other people are willing to accept a lower income to live and work in a certain location. There is no right answer; the choice depends on the desires of each individual.
2. *Personal contact.* Some people are interested in careers through which they will meet and work with many different people. Others

prefer not to place themselves in that situation. It is important for each person to consider the amount of personal contact when selecting a career.

3. *Formal education required.* Almost every career requires at least a high school diploma. Many careers require a person to attend either a technical school or college.
4. *Practical experience required.* Many careers require a person to have practical experience in addition to formal education. Experience can be acquired through part-time jobs during high school and college or in full-time, entry-level jobs. Another source of practical experience is the knowledge and skills gained through a supervised agricultural experience program in agricultural education.
5. *Locations of employment.* Many people are limited in career options because of the location of the job. It is sometimes difficult for people to enter careers or to accept advancements because they are either unwilling or unable to move to a new location.[8]
6. *Whether you will like or dislike your work.* It is important to choose a career that is well suited to your own interests and capabilities so that you can be happy in your work. Sometimes, people choose a job primarily because it offers more money. These people may become disillusioned and unhappy if the job does not fit their capabilities.[9]
7. *Working conditions.* You should give careful consideration to the fact that much of your adult life will be spent on the job. The working conditions of the career should be important to you as you make your career choice. Will you be operating dangerous machinery? Handling dangerous chemicals? Breathing pollutants? All these questions should be considered.
8. *The amount of leisure time available.* Occupations vary in the amount of leisure time available. Many companies have uniform policies concerning vacations. Self-employed people, like farmers and others who own their own businesses, can set their own work schedule. Managers of restaurants, grocery, and variety stores can spend up to 80 hours per week on the job.
9. *Security during retirement.* Will your job provide you with a pension during retirement? If not, will the income be high enough while you are working that you can set some money aside for retirement? Other benefits, such as health insurance, sick leave, and disability insurance, should also be considered.
10. *Health and happiness.* The type of work you do will directly affect your health and happiness. Some occupations require a great deal of heavy physical activity; others require working in hazardous conditions.[10] Your job should also bring you satisfaction and help enhance your self-concept. Without this, happiness may not be achieved.

Many other factors are also important for career success. Much of the potential for success depends on attitude: To succeed in a career one needs a positive attitude toward one's work. How well an individual can communicate and interact with others also affects career success.[11]

Steps in Choosing a Career

Choosing your career early in life may be no better than a choice made later in life. However, an early choice, even if it changes later, can give some direction to your life and your studies. If you know where you want to go, you can plan on how to get there. Here are some logical steps to follow when choosing a career or revising your career choice.[12]

Step 1

Consider your interests, abilities, and other characteristics.

Look carefully at what you like to do. This may suggest general areas of occupations you can investigate further. Be honest with yourself about your strong and weak points. Don't forget your personality type, communication style, and learning style. All these have a correlation to the type of

Figure 20–3 Use every chance you have to explore career opportunities. Take advantage of high school career days, field trips to local businesses, and the National FFA Career Show in Kansas City. At these events, talk and ask questions, as this FFA member is doing.

job or career you select. The National FFA Career Show is an excellent place for agricultural education students to explore careers (Figure 20–3).

Step 2

Narrow the field of jobs.
After you have evaluated your own interests, you can narrow your choices. In most occupational areas, several groups of related occupations will match your interest and abilities. Your first choice is subject to change, but it can be a starting point for planning a career. If your strength is psychomotor skills, working as either a mechanic, carpenter, or building contractor could be your career choice.

Step 3

Study the requirements of the job.
Look carefully at the requirements, working conditions, and other characteristics of a job and become familiar with all the facts. Poor career choices are sometimes made because of inadequate information or misunderstanding the facts.

Step 4

Plan for alternative occupations.
Your first choice might not work out. Try to anticipate what you can do if it does not. All occupations have related occupations. Determine the occupations that relate to your choice and be prepared to enter one of those if the first choice does not work out.

Step 5

Plan ahead for career preparation.
Plan ahead so that you are ready when the time comes for you to go to work. If the job requires special training, plan to obtain that training. Important information included in a plan for career preparation could include arranging for a special course in high school, obtaining part-time work experience while you are still in high school, and planning for additional education beyond high school.[13]

Step 6

Be willing to pay the price for success.
Educational planning is an important part of the plan for career preparation. Many students decide once they enter college that they want to be veterinarians, doctors, or engineers. That decision must be made in the eighth grade or earlier. You may need to take as much biology, chemistry, and math in high school as possible. Do you want to be a sports writer or announcer? If so, you will need to work on oral or written communication skills. You can take speech courses and

English courses, as well as practicing public speaking in agricultural education and the FFA.

Step 7

Getting working experience.
Work experience is also important for career preparation. Direct work experience in the occupation you desire is best, but you may have to select another job because of insurance requirements, legal requirements, or job availability. If, for example, you cannot get a job working as an agricultural mechanic, you could gain valuable background through a related job, such as working for an auto mechanic. This job would allow you to work with motors and do auto maintenance and repair.

CHOOSING A JOB

Once you have a career path, you can determine if it is the right choice for you. Many of today's workers just "fell into" the kind of work they do either because they did not make any special plans or because it was necessary to get a job. Planning and careful evaluation can make the difference. For the next 45 years or so, you will be spending close to half your waking hours working. Much of the nonworking half will be taken up with tasks related to your job: getting ready for work, commuting, taking lunch breaks, or resting up from work.[14]

Are you looking for a job that offers good security or high risk? Perhaps you do not think much of either extreme and would prefer something in between. What you prefer is one thing; what you get depends on how wisely you choose a career. That's because the career you pick is the number one influence on the kind of life you will have. It will be a major factor in determining

- how much money you will make
- how much you will travel and where
- the type of home you will be able to afford
- if you will always be supervised by others or perhaps be your own boss someday
- what kind of car you will drive
- how much leisure time you will have and how you will spend it
- where in the world you will live
- whether you will look forward to steady paydays, an uncertain future, commissions[15]

Self-Assessment

The unique qualities of every individual should be matched to an occupation that utilizes those qualities. A career is more fulfilling and enjoyable if an individual is well suited to the job, and the employer is more pleased with the work performed.[16]

Evaluate your values and philosophy of life. You must decide if relocation is acceptable, judge your strengths and weaknesses, and acknowledge your goals. Career counselors should candidly discuss strengths and weaknesses with you. You need honest evaluations and will eventually appreciate them. You need to know the areas of your personality that need development.

Finding Your Special Abilities

Everyone is good at something. In fact, most people are good at many things and don't give themselves credit for them. You may take for granted many things you do well that others would find hard or even impossible to do. We all have special talents or abilities.

Knowing what you do best is important when you're deciding what kind of work is right for you. It makes a lot of sense to do the things you do best. If you do, you will probably be more successful. Your self-concept is enhanced and happiness is more attainable when you do the things you do best.

It is also important to do things you enjoy doing. If you enjoy what you do and you are good at it, too, your life will be satisfying.

Know Your Skills

One survey of employees found that 90 percent of the people they interviewed could not explain their skills.[17] They could not answer the question, "Why should I hire you?"

Knowing what you are good at is important in selecting a job. It also helps you decide what type of career you will enjoy and do well in. Most people think of skills as **job-related skills**. Everyone has other types of skills that are also important for success on a job. The other two important types are self-management and transferable skills.[18]

Job-related skills are skills you need for a specific job. An agricultural mechanic, for example, needs to know how to service and repair engines and machinery. A carpenter must know how to use various tools and be familiar with a variety of tasks related to that job. These job-related skills can be learned in vocational classes in high school, post-secondary schools, apprenticeship programs, and other places.

Self-management (adaptive) skills are often defined as personality or personal characteristics. They help a person to adapt to or get along in a new situation. They are some of the most important things to bring up in an interview. For example, honesty and enthusiasm are traits employers look for in a worker. Key skills are skills that employers find particularly important. They often will not hire a person who does not have most of these critical skills. They are

- Accepts supervision
- Excellent attendance record
- Gets along with co-workers
- Hard-working
- Gets things done in a timely manner
- Productive
- Punctual

Transferable skills are skills you can use in many different jobs. You can transfer them from one job to a very different one. Some are more important in one job than another. Writing clearly, for instance, is a skill you can use in almost any job.

It is important that you know and identify the skills you possess. Most potential employees think job-related skills are their greatest asset. Although these skills are important, employers often select potential employees with less experience because of their adaptive or transferable skills.[19] Therefore, you need to list your transferable skills (Figure 20–4).

Even if you do not have a specific job title, you must know the type of things you want to do and are good at before you start your job search. This means defining the specific job rather than a job in general. If you already have a good idea of the type of job you want, select the skills that will help you as you interview and then enhance those skills to increase the chance of getting the job you want.

Occupational Groups

Most jobs fit into one or a combination of categories. In *How to Choose the Right Career*, Louise Schrank divided occupations into six groups. Each group attracts different kinds of co-workers and involves a different kind of work.[20]

1. *Body workers*. These people enjoy physical activity, work with their hands, or heavy work requiring strength and endurance. Satisfaction comes from a sense of achievement at seeing the work that is completed. Body workers enjoy seeing the concrete results of their work. They may need special clothing for their work and may be dirty or physically exhausted at the end of their work day. Body workers often work with objects, machines, plants, or animals, and they often like to work outdoors. They may be production agriculturalists, athletes, physical instructors, or blue collar workers. Examples of appropriate jobs are mechanic, coach, forest ranger, lumberjack, carpenter, truck driver, brick layer, bulldozer operator, and horticulturist. These people also work in skilled technical or service jobs. Body workers are typically practical, rugged, athletic, healthy, and aggressive.
2. *Data detailers*. These people use numbers or words in their work in very exact ways. They know that being attentive to detail is important, and they like to work without errors. They often have good clerical or math abilities. They are white collar or office workers.

Transferable Skills

Key Skills (Critical)

instructing others	meeting deadlines	organize/manage projects	written communication
managing money, budget	meeting the public	public speaking	skills
managing people	negotiating		

Other Transferable Skills

Working with Things

assemble things	observe/inspect
build things	operate tools, machines
construct/repair/build	repair things
drive, operate vehicles	use complex equipment
good with hands	

Working with Data

analyze data	evaluate
audit records	investigate
budget	keep financial records
calculate/compute	locate information
check for accuracy	manage money
classify things	observe/inspect
compare	record facts
compile	research
count	synthesize
detail-oriented	take inventory

Working with People

administer	patient
advise	perceptive
care for	persuade
confront others	pleasant
counsel people	sensitive
demonstrate	sociable
diplomatic	supervise
help others	tactful
instruct	teaching
interview people	tolerant
kind	tough
listen	trusting
negotiate	understanding
outgoing	

Working with Words and Ideas

articulate	inventive
communicate verbally	library research
correspond with others	logical
create new ideas	public speaking
design	remember information
edit	write clearly
ingenious	

Leadership

arrange social functions	mediate problems
competitive	motivate people
decisive	negotiate agreements
delegate	planning
direct others	results-oriented
explain things to others	risk-taker
influence others	run meetings
initiate new tasks	self-confident
make decisions	self-motivated
manage or direct others	solve problems

Creative/Artistic

artistic	expressive
dance, body movement	perform, act
drawing, art	present artistic idea

Figure 20–4 Job skills are very important, but these transferable skills are also important to employers. In the workplace, these skills translate into productivity, whether the job is in sales or human development.

Data detailers hold jobs involving clerical or numerical tasks, such as banking, bookkeeping, data processing, accounting, and insurance adjusting. They are usually good at following instructions and attending carefully to detail work.

3. *Persuaders*. Persuaders like to work and talk with people and enjoy convincing others to see things their way. Persuaders often work in sales, law, or politics. Their success is measured by how well they influence others. In management or sales positions, these workers persuade people to perform some kind of action, such as buying goods or services. They like to lead or manage for organizational or economic gain.
4. *Service workers*. Service workers find satisfaction in helping others. They often work in schools, hospitals, or social service agencies, as teachers, nurses, or counselors. Working in education, health care, or social welfare, they teach, heal, or help people. Hairdressers, waiters, instructors, health care workers, and tour guides are part of this group. They like to inform, enlighten, help train, develop, or cure people.
5. *Creative artists*. Creative artists are people who express themselves through music, dance, drama, writing, or art. Many creative artists can only afford to work at their chosen jobs part-time because demand and pay for many creative workers' jobs is low. They may earn all or part of their income by working in related areas in which their creativity, knowledge, and skill are used in at least part of the work. These people work with words, music, or art in a creative way. Actors, musicians, composers, authors, landscapers, floral designers, and sculptors are in this group.
6. *Investigators*. These people like to observe, learn, investigate, analyze, and evaluate. Investigators enjoy asking *why* and *how* questions in their work. They work with scientific or technical information, applying it to new situations. They may work in scientific research or analysis as well as applications. Usually performing scientific or laboratory work, investigators research how the world is put together and how to solve problems.

You may notice that your six categories of work somewhat correlate with personality types, as discussed in Chapter 2. Personality type relates to communication style, leadership style, and learning style. Occupational groups correlate with personality type and learning style, as shown in Figure 20–5.

Occupational Group	USDA Categories	Personality Type	Learning Style
Body workers	Agricultural production specialist	Sanguine/choleric	Convergers/accommodators
Data detail	Managers and financial specialist	Melancholy	Assimilators
Persuaders	Agricultural marketing, merchandising, & sales	Sanguine/choleric	Convergers/accommodators
Service workers	Social service professionals	Phlegmatic	Divergers
Creative artists	Education and communication specialist	Melancholy	All four styles
Investigators	Scientists, engineers, & related specialists	Melancholy	Assimilators

Figure 20–5 Your choice of career should correlate with your personality type and learning style. Otherwise, you may choose the wrong occupation and wonder why you are not happy with your job.

CAREERS IN PRODUCTION AGRICULTURE, AGRIBUSINESS, AND AGRISCIENCE

Agriculture is the largest single industry in the United States. It provides nearly one person in five with a job. Relatively few of these jobs are in production agriculture. Most of the jobs in the agricultural industry involve activities that move farm products from the farmer's gate to the consumer's plate.[21] There are also many career opportunities providing supplies, equipment, and services to production agriculturalists, agribusinesses, and agriscience workers.

The Bases of Jobs in Agriculture

To see the impact of agriculture/agribusiness on the U.S. economy, we need to reference the yardstick that measures the value of goods and services America produces in a year, the **gross national product (GNP)**. Agriculture accounts for 17 percent of the GNP and provides more than 20 percent of all the jobs in the country. Two percent of the GNP comes from firms or people who sell goods and services to farmers and ranchers. Thirteen percent of the GNP comes from related industries, including ice cream makers, textile mills, flour mills, tanneries, breakfast food makers, and a host of others. These related industries purchase food and fiber from farmers and ranchers and then process and package it into a **value-added** product to sell to consumers.[22] This group of businesses and industries makes up the world of agribusiness and agriscience.

Over 20 percent of America's work force is employed in some phase of the agricultural industry. There are seven people working in agribusiness for every farmer. In fact, there are over 8,000 job titles in agriculture. They all work together to provide food and fiber for the planet's growing population.[23]

Virtually any career in which you may be interested can be applied to agriculture. Take engineering, for example. Today, farmers are leveling fields with lasers to decrease erosion and using robotic equipment to do dangerous or repetitive jobs. If progress is to continue, agriculture needs the best and brightest young minds working to solve tomorrow's agricultural engineering challenges.

An increasing population means a greater demand for food and fiber. It also means a growing demand for qualified people in the agricultural industry. Almost 10 percent of today's professional jobs in agriculture go unfilled simply because there are more jobs than people who understand agriculture.[24] The opportunities are increasing. Agriculture is changing rapidly and many of tomorrow's careers have not yet been imagined. It is an exciting, challenging field in which to work. Whether you are interested in business, computers, mechanics, or communications, America's largest industry has many career opportunities in the following six categories.

1. *Agricultural production specialists.* The production sector produces commodities to be processed for consumers. It includes people working with plants and animals. Production agriculture provides many raw products, such as beef, pork, grains, fruits, vegetables, ornamental horticulture, and timber.
2. *Scientists, engineers, and related specialists.* Agriscience, with its related occupations of engineering, biochemistry, genetics, and physiology, is the fastest growing area in agriculture. This is agriculture's cutting edge. If you are interested in applying scientific principles to practical situations, this may be the area that best fits your career aspirations.
3. *Communication and education specialists.* More than ever before, the agricultural industry today needs to tell its story to the rest of the population. If you are interested in sharing the news, maybe a career in education and communications is within your abilities and talents.
4. *Social service professionals.* The service area provides help to other areas so that they work more efficiently and effectively. As with most other industries, an increasing number of social service professionals are needed. If you

like working with people and filling an important role in your community, this may be the career area for you.

5. *Agricultural marketing, merchandising, and sales representatives.* This area processes, markets, and distributes agricultural products to consumers. It provides equipment and materials for production. This area is sometimes referred to as "gate to plate" because it turns the raw product from the production sector into products that are usable by consumers. There is much demand for agricultural products today. Consumers expect to walk into supermarkets and find the shelves bursting with choices. If you are interested in sales and helping people acquire the goods and services they need, a career in agribusiness or agricultural marketing could be your area.
6. *Managers and financial specialists.* For today's agricultural industry to operate, it must have management and financial specialists. From your local bank's agricultural loan officer to the USDA's economists, this is an area that demands both agricultural and business skills.[25]

Distribution of Jobs for Graduates

Potential careers for each of the six agricultural categories are shown in Figure 20–6. (Refer also to Appendixes G and H.) The percentage of employment of the six agriculture categories is also shown, along with the surplus of graduates in each category. Careers in agriculture have diversified far beyond farming. Agriculture graduates are filling jobs you may not even recognize as agricultural.

Agricultural Education Job Placement

There is often confusion on the placement rate of students coming from agricultural education programs. Many well-educated professionals do not understand the diversity of occupations in agriculture. They still see agriculture as cows, sows, and plows; weeds, seeds, and feeds. They believe that less than 2 percent of the workforce is involved in the agricultural industry, whereas, as you just learned, it is really over 20 percent. Even surveys asking for placement rates of students going into agriculture do not reflect the total diversity of occupations that were selected because of training received in agricultural education. To help understand the diversity of placement of students, consider the following four areas: immediate job placement, postponed job placement, avocational (part-time) job placement, and supplementary (tributary) job placement. A discussion of each follows.

Immediate Job Placement

Many students enter the job market immediately after high school graduation. This is called **immediate job placement**. The jobs selected may be directly related to production agriculture, agribusiness, or agriscience. On the other hand, their jobs may not be directly related to agriculture, but the skills for the job were learned through the agricultural education program. For example, indirect placement would occur when, through instruction in agricultural mechanics, a student becomes a surveyor, plumber, bricklayer, or small engine repair mechanic. Obviously, the job would not be credited as agricultural placement. Direct placement in agriculture could include the jobs of production agriculturalist, greenhouse worker, nursery worker, florist, agricultural mechanic, or produce worker. This is where many outside the field of agriculture stop with their perception of placement in agriculture, but due to the diversity, business, and science of agriculture, immediate job placement reflects only a small percentage of the placement of agriculture students.

Postponed Job Placement Due to Advanced Training

Many careers in agriculture are available only after **postponed job placement** due to advanced training. This advanced training is post-secondary

Agricultural Production Specialists	Scientists, Engineers, and Related Specialists	Communication and Education Specialists	Social Services Professionals	Marketing, Merchandising, and Sales Representatives	Managers and Financial Specialists
7.5%	**28.8%**	**7.6%**	**9.7%**	**32.4%**	**14.9%**
CAREERS	**CAREERS**	**CAREERS**	**CAREERS**	**CAREERS**	**CAREERS**
Aquaculturalist	Agricultural engineer	College teacher	Career counselor	Account executive	Accountant
Farmer	Animal scientist	Computer software designer	Caseworker	Advertising manager	Appraiser
Farm manager	Biochemist	Computer systems analyst	Community development specialist	Commodity broker	Auditor
Feedlot manager	Cell biologist	Conference manager	Conservation officer	Consumer information manager	Banker
Forest resources manager	Entomologist	Cooperative extension agent	Consumer counselor	Export sales manager	Business manager
Fruit and veg-etable grower	Environmental scientist	Editor	Dietitian	Food broker	Consultant
Greenhouse manager	Food engineer	Education specialist	Food inspector	Forest products merchandiser	Contract manager
Nursery products grower	Food scientist	High school teacher	Labor relations specialist	Grain merchandiser	Credit analyst
Rancher	Forest scientist	Illustrator	Naturalist	Insurance agent	Customer service manager
Turf producer	Geneticist	Information specialist	Nutrition counselor	Landscape contractor	Economist
Viticulturist	Landscape architect	Information systems analyst	Outdoor recreation specialist	Market analyst	Financial analyst
Wildlife manager	Microbiologist	Journalist	Park manager	Marketing manager	Food service manager
Surplus of Grads: +10.6%	Molecular biologist	Personnel development specialist	Peace Corps representative	Purchasing manager	Government program manager
	Natural resources scientist	Public relations representative	Population control	Real estate broker	Grants manager
	Nutritionist	Radio/TV broadcaster	Regional planner	Sales representative	Human resource
	Pathologist	Training manager	Regulatory agent	Technical service representative	Development manager
	Physiologist	**Surplus of Grads: +27.1%**	Rural sociologist	**Shortage of Grads: –18.2%**	Insurance agency manager
	Plant scientist		Young program director		Insurance risk manager
	Quality assurance specialist		**Shortage of Grads: –12.7%**		Landscape manager
	Rangeland scientist				Policy analyst
	Research technician				Resources and development manager
	Resource economist				Retail manager
	Soil scientist				Wholesale manager
	Statistician				**Shortage of Grads: –15.4%**
	Toxicologist				
	Veterinarian				
	Waste management specialist				
	Water quality specialist				
	Weed scientist				
	Shortage of Grads: –15.3%				

Figure 20–6 Six categories of agricultural careers, with examples of occupations, percentage of total agricultural employment they represent, and surplus or shortage of graduates in each category. (Courtesy of USDA.)

training, whether it is a two-year vocational school, a four-year college or university education, or other experience. A problem for some states with the new technical preparation movement is that there are no two-year post-secondary programs for agriculture students except for university transfer (community college) students. Other opportunities for advanced training after high school are apprenticeships, on-the-job training, and armed services. A brief discussion of the sources of advanced training follows.[26]

Some occupations require a *college* degree. Generally, a bachelor's degree requires four years of college coursework. Many legal, professional, and engineering occupations are jobs that require college training. Colleges vary in the quality of programs offered. If you decide to go to college, select a college that has a strong program in your area of interest. Size, location, and cost are other things you will need to consider. There are many sources of information on colleges and universities. Your school guidance counselor can help you get information on colleges and universities.

Community colleges offer two-year college level courses. Community colleges offer training in more than 60 occupational areas. Occupations in business, food management, and drafting can be learned at a community college. The credit you earn may be transferred to a four-year college. A major advantage of community colleges is that they are much less expensive than four-year colleges. Your school guidance counselor can help you identify area community colleges and the programs they offer.

Vocational schools provide work-related programs for high school students and young adults. They offer a variety of programs that train students in the skills they need to get a job. Auto mechanics, building trades, and electronics can all be learned in vocational schools. Vocational schools are relatively inexpensive to attend. Your school guidance counselor can help you identify area vocational schools that offer programs in which you are interested.

An *apprenticeship* program combines classroom instruction with on-the-job training. You learn a job by working with an expert in a particular trade or occupation. Common apprenticeship programs include plumbing, electrical trades, and carpentry. Apprenticeship programs last two to five years. You earn money as you learn, but the pay is low until you finish the program. Labor unions offer apprenticeship programs. To apply, check with a local union office. Competition is stiff for apprenticeship programs.

When you start a new job, you will probably work with a more experienced worker who will help you learn the job. This is called *on-the-job training*. On-the-job training may require a few days to a few years to complete. Usually, your pay is low until you complete the training program. Your state employment service can provide you with information on the type of on-the-job training you want. Company personnel offices may also tell you about opportunities available in the company.

The *military services* provide specialized training in many occupations. You can choose a job area that interests you and receive free training. If you think you would enjoy military life, the military services can be a good source of occupational training. New recruits must enlist for two years. You earn while you learn. If you want advanced training but cannot afford it, joining the military services may be the answer. The disadvantage to military service is that you cannot quit or drop out if you do not like it. You can get information on military training from the local recruiting office.

Postponed placement can include many careers. A few examples are agricultural education teachers, extension (county) agents, ASCS or soil conservationists, Farm Credit Services employees, Farm Bureau Insurance agents, veterinarians, and landscape architects.

Before you make a career choice, make sure you know the advanced training you will need. Be prepared to spend the time and money necessary to get the training. Do not commit yourself to a career if you are not prepared to train for it. When planning for advanced training, ask yourself the following questions.

- What education and training do I need for the chosen occupation?
- Where can I get the education and training needed for the chosen occupation?
- How much time and money does the training require?
- Am I willing to commit my time and money to receive the training?
- How will I pay for the training?

As a future worker, you must begin to prepare yourself for the world of work. Advanced training will give you the needed skills. Take advantage of opportunities for advanced training in your chosen occupation.[27]

Avocational (Part-time) Job Placement

Many students use the skills learned in agricultural education as **avocational (part-time) job placement** rather than full-time. A high school graduate who completes a placement survey may not indicate agriculture as an avocation if he or she has a full-time nonagricultural job. This concept is based on the economic reality that money can be made in production agriculture, but a "living" may not be possible in certain full-time production agriculture enterprises.

The student learns skills in agricultural education. However, the student makes an economic decision to use the skills in an avocation (part-time) as opposed to a vocational (full-time) career. Some examples of avocational agricultural enterprises are beef cattle, swine operation, strawberries, nursery production, and aquaculture.

Supplementary (Tributary) Job Placement

Some people work in areas that have no relation to agriculture, but their jobs were obtained because of leadership and personal development skills developed through a quality agricultural education program. This is called **supplementary (tributary) job placement**. These skills led to nonagricultural careers offering greater financial reward than some agricultural careers, but the skills were needed for success in the nonagricultural career. Some examples of such nonagricultural careers are attorneys, legislators, county officials, managers, supervisors, and salespeople.

FINDING A JOB

Once you have selected the career you want, it is time to find a job. People lacking job-seeking skills often let a job choose them instead of choosing a job. They may work at a job they dislike or for which they are overqualified. You should be aware of all available job opportunities from which to choose. By being aware of all the possible job openings, you will be able to pick the job that is right for you. The more sources you use in your job search, the better your chances of finding job openings for a job you really want. Knowing where to look can help you find a job that is satisfying and rewarding.[28]

Jobs don't come to you; you have to hunt for them. There are many sources of information about job openings. When you receive information about a job opening you have a **job lead**. By finding many job leads, you have a good chance of getting the job you want. Sources of job leads include the following: networking, direct employer contact, newspaper, classified ads, state employment services, private employment agencies, government jobs, in-school resources, trade magazines, professional associations, mailing resumes, and volunteering. Refer to Figure 20–7 for the percentage of people who got jobs using various sources.

Networking

A **network** is an informal group of people who have something in common. As a job seeker, your network is made up of all the people who can help you—and the people they know. Networking is the process you use in contacting these people. You may be surprised at how many people you

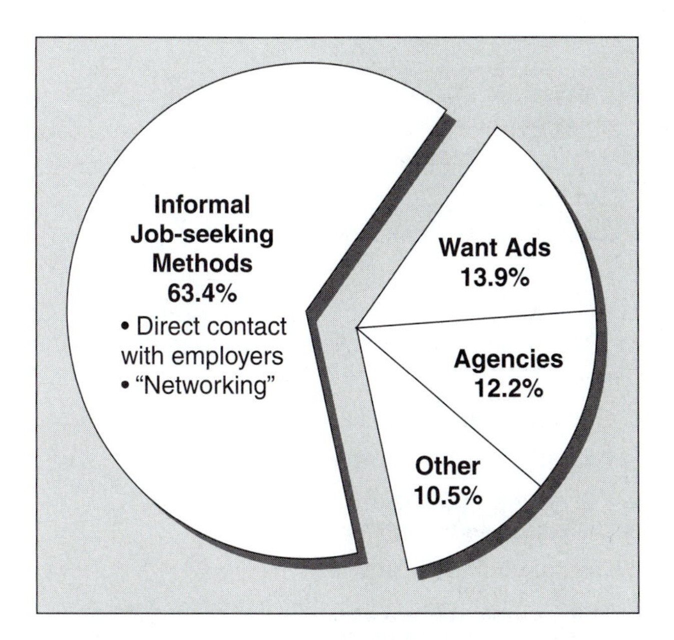

Figure 20–7 There are several ways to find a job, but networking and direct employer contact are by far the most effective. (Courtesy of U.S. Department of Labor, Bulletin 1996.)

can meet this way. Most jobs today are obtained through networking. The people in your network can do one of three things: hire you, give you job leads and information about companies, or refer you to others who can provide this kind of help.

The following is a list of potential people in your network. You may think of others to add to this list.

- Friends
- Present and former teachers
- Relatives
- Neighbors
- Former employers
- Friends of parents
- Classmates
- Former co-workers
- Members of your church
- Former classmates
- Members of sports groups
- Members of social clubs
- Your friends' parents
- People who sell you things (retail, insurance, etc.)
- People who provide you with services (hair stylist, counselor, mechanic, etc.)
- Members of a professional organization you belong to (or could quickly join)[29]

Each of these people is a contact for you. Obviously, some lists and some people on those lists will be more helpful than others, but almost any one of them could help you find a job lead. Start with your friends and relatives. Call them up and tell them you are looking for a job and need their help. Be as clear as possible about what you are looking for and what skills and qualifications you have. It is possible that they will know of a job opening just right for you. If so, get the details and follow through. More likely, however, they will not, so here are three questions you should ask.[30]

- Do you know of any openings for a person with my skills? If the answer is no, then ask:
- Do you know of someone else who might know of such an opening? If they do, get that name and ask for another one. If they don't, then ask:
- Do you know of anyone who might know of someone else who might? Another way to ask this is, "Do you know someone who knows lots of people?" If all else fails, this will usually get you a name.[30]

For each original contact, you can extend your network of acquaintances by hundreds of people. Eventually, one of these people may hire you—or refer you to someone who will!

Jobs open daily because of natural transition as people retire, quit, have children, move to another city, get fired, pass away, go back to school, or switch careers. For days or even months before these openings are posted by the resources department or advertised, they can stay in the "hidden" job market as supervisors or department heads look for someone they know or wait for approval to hire a replacement. Anyone who by search, chance, or (primarily) networking becomes aware of these openings has a golden opportunity to apply—with virtually no competition. Those who wait for the help wanted ads can get caught in a multitude of other applicants.[31]

Some experts say over 70 percent of all jobs are never advertised. Employers do not like to advertise. When employers put an ad in the paper, they have to interview all sorts of unknown people. Employers, in large part, are not trained interviewers and do not enjoy it. The interviewees do their best to create a good impression, but the interviewers eliminate most of them by finding their weaknesses.[32] It is not fair for either side. Most jobs are usually filled by networking.

Cashiers
Retail salesclerks
Machine operators, assemblers, inspectors, and tenders
Stock handlers and baggers
Cooks
Child-care workers
Janitors and maids
Food counter workers
Food preparation workers
Laborers (includes construction)
Waiters and waitresses
Secretaries and typists
Dining room attendants
Construction trade workers (includes carpenters, painters, electricians, and so on)
Receptionists and information clerks
Truck drivers
File and library clerks
Farm and nursery workers, gardeners, and grounds keepers
Stock and shipping clerks
Freight and materials movers
Garage and service station attendants
Nurse's aides, orderlies, and attendants
General office clerks
Bookkeepers and financial record clerks
Computer operators

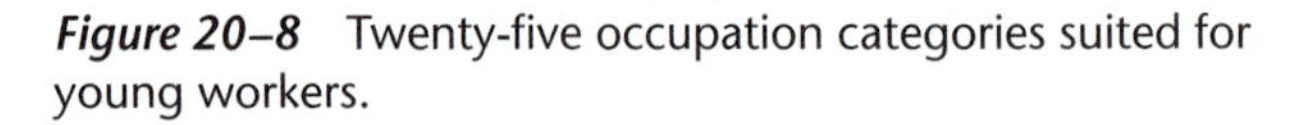

Figure 20–8 Twenty-five occupation categories suited for young workers.

Direct Employer Contact

The second most effective way to find a job is by talking directly to employers. A help wanted sign posted in a business is the oldest method of announcing a job opening. If you see such a sign, ask the employer for an application.

The Yellow Pages in your telephone book are also an important source of job information. The Yellow Pages are easy to use because businesses are listed alphabetically according to type. For example, if you are interested in working in a pizzeria, look under the heading "Restaurants" for names. Employers often have unadvertised job openings. Figure 20–8 lists the types of jobs that are suitable for young workers. You might then use the telephone Yellow Pages to make a list of companies to contact.

Frequently, employers post job openings in public places, such as grocery stores, discount stores, laundromats, community centers, and medical clinics. In addition, the local post office, courthouse, and other government buildings post dates, times, and locations of civil service examinations.

Company human resources or personnel offices can provide information on job openings in the company. You may telephone or visit a potential employer to determine if job openings exist in the company. Go dressed as you would for an interview. Be prepared to fill out an employment application form. Check bulletin boards outside the personnel office. Available jobs are often listed there. Some companies also have a separate telephone number that provides prerecorded messages about job openings.[33]

Newspaper Classified Ads

Newspapers advertise job openings in the **classified ad** (want ad) section of the paper. The best day to check the newspaper is Sunday. Only 5 to 13 percent (depending on the source) get jobs this way, but you can learn from the newspaper what kinds of jobs are available and what skills are needed for various jobs. If you see a job opening you are interested in, answer the ad quickly. Many people will apply for the same job. If you wait too long, the job will already be filled. Four types of help wanted ads are shown in Figure 20–9.

The open ad describes the job requirements, identifies the employer, and tells you how to apply for the job. This is the best type of ad. The catch-type ad emphasizes a good salary and fails to mention the qualifications and skills needed for employment. The "catch" is that the job may involve door-to-door selling or telephone sales. The agency spot ad does not include the name of the employer. This ad is used by private agencies to advertise employment available only by contacting the agency. In a blind ad the name, address, and phone number of the employer are omitted. This type of ad allows the employer to screen applications carefully, with only qualified applicants contacted for a personal interview. A list of want ad terms and abbreviations in Figure 20–10 will help you read and understand want ads.

Figure 20–9 Four common types of help wanted ads.

The State Employment Service

Both public and private employment agencies are found in most large cities. Established under federal and state laws and supported by taxes, public employment agencies help people find jobs. This service is free. Only about 5 percent of all job seekers find jobs here. Agencies are known by the name of the state where the office is located, such as Tennessee Department of Employment Security. To complete an application form, go to the employment office nearest you. After you complete the application, you will be interviewed to determine your interests and skills. If there is a job opening that seems right for you, you will be notified. The employment office will give you a letter of introduction for a personal interview.

Private Employment Agencies

Since they are not tax-supported, private employment agencies charge a fee for placement services. When you apply to a private agency, you must sign a contract agreeing to pay the agency a fee if the agency helps you find a job. Often this is a percentage of your first year's salary. In some cases your new employer will pay the fee. However, this fee payment is usually only done for high-paying professional jobs. Be sure to read the contract carefully before you sign.

adv.	advancement
aft.	after
A.M.	morning
appl.	applicant/application
appt.	appointment
asst.	assistant (helper)
ben.	benefits
bet.	between
bgn.	begin/beginning
bldg.	building
bus.	business
cert.	certified/certificate
clk.	clerk
co.	company
coll.	college
comm.	commission (pay based on sales)
cond.	conditions
const.	construction
corp.	corporation (usually a larger company)
dept.	department
dir.	director
div.	division (part of a company)
D.O.T.	Dictionary of Occupational Titles
D.O.E.	depending on experience
elec.	electric
empl.	employment
E.O.E.	Equal Opportunity Employer
eqpt.	equipment
etc.	and so on
eves.	evenings
exc.	excellent
exp.	experience
nego.	negotiable
nec.	necessary
of.	office
opp.	opportunity
pd.	paid
P.M.	afternoon/evening
pos.	position
pref.	preferred
P.T.	part-time
qual.	qualified/qualifications
refs.	references
req.	require/requirements
sal.	salary
sec.	secretary
ext.	extension (some telephones have an extension number)
F.T.	full-time
ftr.	future
gd.	good
gen.	general
grad.	graduate
hosp.	hospital
hqtrs.	headquarters
hr.	hour
hrs.	hours
hrly.	hourly
H.S.	high school
hvy.	heavy
immed.	immediate
incl.	including
ind.	industrial
jr.	junior (beginner or assistant)
lic.	license
lt.	light (a little)
mach.	machine
maint.	maintenance
manuf.	manufacturing (making things)
max.	maximum
mech.	mechanic/mechanical
med.	medical
M/F	male/female
mfg.	manufacturing
mgr.	manager
min.	minimum
mo.	month
sh.	shorthand
sr.	senior
temp.	temporary
trnee.	trainee (beginner)
typ.	typing/typist
U-W	underwriter (insurance salesperson)
w.	with
wk.	week or work
wkr.	worker
wpm.	words per minute
yr.	year

Figure 20–10 Common want ad terms and abbreviations.

Government Jobs

The U.S. government is the largest employer in the nation. If you are interested in working for the government, there are special agencies you should contact. These agencies include city civil service commissions, state civil service centers, and federal job information centers. Addresses and phone numbers are listed in the telephone book. Listed under the heading "Government," agencies are grouped by type: city, county, state, or United States government. Ask your librarian for additional information about government agencies.

In-School Resources

Three sources of job leads are available in many secondary schools. They are your teachers, guidance office, and job placement office or career center.

1. *Teachers.* Vocational education teachers and cooperative/work experience coordinators are good sources of job leads. Your own vocational teacher or coordinator is probably already helping you. Also tell other teachers that you are looking for a job. Ask for their suggestions.
2. *Guidance office.* Most schools also have a guidance office or guidance counselor. It is common for local employers to contact counselors when looking for workers. The counselor usually keeps a list of job openings or posts them on a bulletin board. Tell the counselor that you are looking for a job and ask to see any information available about job openings.[34]
3. *Job placement offices or career centers.* Not all schools have these, but for those that do, interested students generally register with the office. They may receive job counseling and other services. Job counselors help to match students with job openings and make referrals for interviews (that is, they send students to employers who are hiring).

Trade Journals and Magazines

Trade journals and magazines sometimes provide job advertisements similar to those found in newspapers. These job advertisements may be for jobs in other cities and states. You should be willing to move to another location if you consider the jobs.

Professional Associations

Many professions have special publications for people who work in that field. They are often a good source of information, and some list job openings. Local branches of national organizations sometimes list job openings, too. They are worth checking into.

Mailing Resumes

You could get lucky, but resumes sent to no one in particular will probably result in a low response rate. Expect a 5 percent or lower response rate and even fewer interviews. It is almost always better to contact the employer in person. Then send your resume before the interview. However, I have known people to get jobs with this method. It uses the principle of the law of large numbers.

Volunteering

If you lack experience or are not getting job offers, volunteer to work for free. Perhaps you could offer your services for a day or even a week to show an employer what you can do. Promise that if things do not work out, you will leave with no hard feelings. This really does work, and many employers will give you a chance because they like your attitude.[35]

A summary of how people find jobs is based on a survey of job seekers by the U.S. Department of Labor, Bulletin 1986. The percentages are as follows:

- Networking and contacting employer directly: 63.3%
- Answered want ad: 13.9%
- Referred by private employment agency: 5.6%
- Referred by state employment service: 5.1%

- Took civil service tests: 2.1%
- Other methods: 12% (examples: professional associations, placed ads in journals, went to places where employers come to hire people, trade journals and magazines, mailing resumes, volunteering)

CONCLUSION

You will face stiff competition for jobs. Once you decide what you want and what your abilities are, do not hold back. Be self-confident. As soon as you learn about a job lead, follow through with quick action. The early applicant gets the job. If you do not get the job immediately, call or go back a few days later. Let the employer know that you are really interested in the job.

The qualities of motivation and persistence, so important in finding a job, are also qualities that make you a good employee. Employers recognize this. By continuing to demonstrate motivation and persistence in pursuing a job lead, you increase the chance of such behavior being rewarded with a job offer.[36] Remember, your attitude determines your altitude.

Finding job leads may seem to take a long time. It does indeed take a lot of time to contact family and friends, search the newspaper help wanted ads, and identify other leads. Once you find a good job lead, however, things can speed up very quickly. The next step is to apply for the job. You will need to prepare a personal data sheet, complete job application forms, prepare a resume, and get ready to interview.

SUMMARY

Selecting and finding a job is one of the most important decisions of your life. Working at a paid job enables you to establish your independence and define your own life. People also work for money, social contact, positive feelings, prestige, personal development, self-expression, security, success, happiness, and to please peers and family.

The terms work, occupation, and job are often used interchangeably. However, work can be defined as activity directed toward a purpose or goal that produces something of value to oneself and to society. An occupation is the name given to a group of similar tasks that a person performs for pay. A job is a paid position at a specific place or setting. A career is a series of jobs that is pursued in sequence to achieve the ultimate occupation that is designed by the individual.

Although choosing a career is a serious step in one's life, the same holds true for career planning. The more time you spend on career planning, the easier it will be to find the career of your choice. The kind of career you select will determine the types of people you meet, the friends you make, and the kind of lifestyle you can afford. Factors to consider when selecting a career include standard of living, personal contact, formal education required, practical experience required, locations of employment, whether you will like or dislike your work, your working conditions, the amount of leisure time available to you, your security during retirement, and your health and happiness. Steps in choosing a career include considering your interests, abilities, and other characteristics; narrowing the field of jobs; studying the requirements of the job; planning for alternative occupations; and planning any career preparation.

Once you have a career path, you can determine if it is the right choice for you. In matching jobs to your personal characteristics, you must have a self-assessment, find your special abilities, and know your skills, including job-related, self-management, and transferable skills. Most jobs fit into one or a combination of categories. These six occupational groups are: body workers, data detailers, persuaders, service workers, creative artists, and investigators.

Agriculture is the largest single industry in the United States. Over 20 percent of America's workforce is employed in some phase of agricultural industry. There are over 8,000 job titles in agriculture. Agricultural careers are divided into six categories: agricultural production specialists;

scientists, engineers, and related specialists; communications and education specialists; social service professionals, agricultural marketing, merchandising and sales representatives; and managers and financial specialists.

There is often confusion about the placement rate of students coming from agricultural education programs. To understand the diversity of student placement, four areas of placement must be used: immediate job placement, postponed job placement, avocational (part-time) job placement, and supplementary (tributary) job placement. For further training after high school, consider the following alternatives: colleges and universities, community colleges, vocational schools, apprenticeships, on-the-job training, and the armed services.

Once you have selected a career you want, it is time to find a job. Sources of job leads include the following: networking, direct employer contact, newspaper classified ads, state employment services, private employment agencies, professional associations, mailing resumes, and volunteering. Networking is by far the best method, followed by direct employer contact. These two methods account for over 63 percent of all placements.

As soon as you learn about a job lead, follow through with quick action. Be motivated and persistent. The early applicant gets the job. Remember, your attitude determines your altitude.

End-of-Chapter Exercises

Review Questions

1. Define the Terms to Know.
2. What are 11 reasons people work?
3. Explain the importance of career planning.
4. List 10 factors to consider when selecting a career.
5. List seven steps in choosing a career.
6. The career you select will be a major factor in influencing the kind of life you will have in what eight ways?
7. Name the six occupational groups.
8. Name the six categories of agricultural careers.
9. List the percentage of employment in each of the six agricultural categories along with the percentage of surplus or shortage.
10. What are the four placement categories that reflect the diversity of agricultural education?
11. What are six sources of advanced training after high school?
12. Name eight agricultural careers that are examples of postponed job placement due to advanced training.
13. What five questions should you ask yourself when planning for advanced training?
14. What are five examples of avocational (part-time) placement?
15. What are six examples of supplementary (tributary) job placement?
16. Name 11 sources of job leads.
17. List 16 people or groups that could be a part of your job hunting network.
18. If one contact in your network does not know of a job lead, what three follow-up questions should you ask?

19. What are four methods of establishing direct contact with employers?
20. Name and briefly explain the four types of newspaper classified ads.

Fill-in-the-Blank

1. ____________________ skills can be learned in vocational classes in high school, post-secondary schools, apprenticeship programs, and other places.
2. ____________________ skills are often defined as personality or personal characteristics and help a person to get along in a new situation.
3. ____________________ skills are skills you can use in many different jobs.
4. One survey of employers found that ____________________ percent of the people they interviewed could not explain their skills.
5. Agriculture provides nearly ____________________ person of every five with a job.
6. Agriculture accounts for ____________________ percent of the gross national product.
7. There are ____________________ people working in agribusiness for every farmer.
8. There are over ____________________ job titles in agriculture.
9. Almost ____________________ percent of today's professional jobs in agriculture go unfilled because there are more jobs than people who understand agriculture.
10. The avocational (part-time) job placement concept is based on the economic reality that money can be made in production agriculture but a ____________________ may not be possible for certain individuals in some full-time production enterprises.

Matching

_____ 1. Love to work and talk with people and enjoy convincing others to see things their way
_____ 2. Find job satisfaction in helping others
_____ 3. Use numbers or words in their work in very exact ways
_____ 4. People who express themselves through music, dance, drama, or writing
_____ 5. Work in scientific research or analysis as well as applications
_____ 6. Enjoy physical activity, work with machines, plants, and animals
_____ 7. Cannot quit or drop out of training if you do not like it
_____ 8. Learn skills by working with more experienced worker
_____ 9. Labor unions as well as schools offer these programs
_____ 10. After successfully completing four years of study, you receive a bachelor's degree
_____ 11. Offer two years of college-level courses
_____ 12. Train students in a formal setting in the skills they need to get a job

A. body workers
B. data detailers
C. persuaders
D. service workers
E. creative artists
F. investigators
G. apprenticeships
H. college degree
I. vocational schools
J. armed services
K. community colleges
L. on-the-job training

Activities

1. Besides money, select three reasons why you would work, and briefly explain or justify each.
2. Study the factors to consider when selecting a career. Then, write a 200- to 300-word essay entitled, "The Type of Job I Want." Include or address the factors as you write the essay. Be prepared to present the essay to the class.
3. Review the "Transferable Skills" in Figure 20–4. Select 10 skills that you feel you possess and could bring out during an interview as your strengths. Select 10 skills that you feel need improvement. Write a short statement about each skill that needs improvement and tell how you can improve it.
4. Six occupational categories were discussed in the chapter. Select an occupational category and five specific jobs that could be of interest to you. Explain some general characteristics of that category that attract you to that career interest. Note: These characteristics are explained in the "Occupational Groups" section of this chapter.
5. Write down 20 jobs that may be of interest to you. Match the 20 jobs you selected with the six agricultural categories in Figure 20–6.
6. Evaluate your present situation with respect to the type of career you are considering. Select the type of advanced training you may need for that job. Justify your selection of advanced training.
7. Compile a list of people that could be a part of your network. Make sure your list has at least 10 names.
8. Locate the open ad in Figure 20–9 and answer the following questions.
   a. What is the title of the position?
   b. What are the qualifications for the job?
   c. Who is the prospective employer?
   d. What is the phone number?
   e. Is the job full-time or part-time?
   f. How much money could you make weekly?
   g. Would you have to work on weekends?
9. Using any of the 11 sources of job leads, compile a list of 5 potential employers. These potential employers can be either for part-time employment while in school or college, full-time employment after high school graduation, or full-time employment after further advanced training. Include the employer's name, address, phone number, type of work, working hours, and estimate of salary.
10. **Take it to the Net.**

    Explore careers on the Internet. Listed below are several Web sites that contain information on careers. The sites contain everything from information on what job is best for you to information on where the jobs are. Browse all the sites. Find a couple of sites that you feel are helpful and summarize them. Search terms are also listed if you want additional information.

    Web sites

    <http://www.life-career.com/consulting.html>

    <http://www.careermag.com/>

    <http://jobstar.org/tools/career/index.htm>

    <http://tbrnet.com/assessment/index.shtml>

    <http://www.mnworkforcecenter.org/cjs/cjs_site/index.htm>

    <http://www.americasemployers.com>

    <http://www.ajb.dni.us/>

    <http://www.bestjobusa.com>

Search Terms

careers
career interest testing
selecting a career
selecting right career

Notes

1. L. W. Schrank, *How to Choose the Right Career* (Lincolnwood, IL: VGM Career Horizon, A Division of NTC Publishing Group, 1994), p. 3.
2. L. J. Bailey, *The Job Ahead: A Job Search Worktext* (Albany, NY: Delmar, 1992), p. 1.
3. Ibid.
4. Ibid., pp. 2–4.
5. Schrank, *How to Choose the Right Career*, p. 4.
6. Bailey, *The Job Ahead: A Job Search Worktext*, p. 3.
7. Ibid., p. 4.
8. K. J. Bacon and R. Birkenholz, *Careers I Unit for Agricultural Science I Core Curriculum* (Columbia, MO: Instructional Materials Laboratory, University of Missouri, 1988), p. 7.
9. B. R. Stewart and B. Hunter, *Career and Personal Development for Plant Science Core Curriculum* (Columbia, MO: Instructional Materials Laboratory, University of Missouri, 1983), p. 1.
10. Ibid.
11. Bacon and Birkenholz, *Careers I Unit for Agricultural Science I Core Curriculum*, p. 7.
12. Stewart and Hunter, *Career and Personal Development for Plant Science Core Curriculum*, p. 5.
13. Ibid.
14. "Career Planning: Unit I" *Effective Employment Practices* (Stillwater, OK: Curriculum and Instructional Materials Center, Oklahoma Department of Vocational and Technical Education, 1988).
15. Ibid.
16. *Career Planning Teaching Curriculum: The Right Fit* (Lubbock, TX: Creative Educational Video, 1994).
17. J. M. Farr, *America's 50 Fastest Growing Jobs* (Indianapolis, IN: JIST Works, 1994), p. 151.
18. Ibid.
19. J. M. Farr, *Getting the Job You Really Want* (Indianapolis, IN: JIST Works, 2000), p. 14.
20. Farr, *America's 50 Fastest Growing Jobs*, p. 153.
21. Bacon and Birkenholz, *Careers I Unit for Agricultural Science I Core Curriculum*, p. 1.
22. *Employment for Agribusiness* (Stillwater, OK: MAUCC, 1992), p. 19.
23. "The Industry Too Big to Ignore," *American Careers Magazine*, Fall, 1992.
24. Ibid.
25. K. J. Coulter, et al., *Employment Opportunities for College Graduates in the Food and Agricultural Sciences* (Washington, D.C.: USDA, Published by Texas A & M University, College Station, Texas. EIS 86-174, 7/86-10M, 1986).
26. *How to Identify Sources of Advanced Training:* Skills for Success No. 7 (Nashville, TN: Tennessee Department of Education, 1993).
27. Ibid.
28. *How to Identify Sources of Job Openings:* Skills for Success No. 41 (Nashville, TN: Tennessee Department of Education, 1993).
29. Farr, *Getting the Job You Really Want*, p. 49.
30. Farr, *America's 50 Fastest Growing Jobs*, p. 156.
31. H. Gieseking and P. Plawin, *30 Days to a Good Job* (New York: Simon & Schuster, 1994), p. 15.
32. Farr, *Getting the Job You Really Want*, pp. 43–45.
33. Bailey, *The Job Ahead: A Job Search Worktext*, p. 8.
34. Ibid.
35. Farr, *Getting the Job You Really Want*, p. 41.
36. Bailey, *The Job Ahead: A Job Search Worktext*, p. 9.

CHAPTER 21

Getting the Job: Resumes, Applications, and Interviews

Objectives

After completing this chapter, the student should be able to:

- Prepare a resume
- Write a letter of application
- Complete a job application form
- Prepare for an interview
- Interview for a job
- Explain what to do after the interview
- Discuss how to accept or reject a job

Terms to Know

resume
interview
personal management skills
teamwork skills
references
letter of application
applicants
application form
personnel office
not applicable
interviewee
interviewer
hypothetical
follow-up letter

Once you have selected, searched, and located the type of job you want, it is time to get it. In all probability, you will face stiff competition for the job. Be positive and assertive. As soon as you learn about a job lead, follow through with quick action: Get the first available interview. If you do not get the job immediately, call or go back a few days later. Let the employer know that you are really interested in the job. The key is to be assertive but not so aggressive that your potential employer perceives you as obnoxious.

The qualities of motivation and persistence, so important in finding a job, are also qualities that make you a good employee. Employers recognize this. By continuing to demonstrate motivation and persistence in pursuing a job lead, you increase the chance of getting a job offer. Besides motivation and persistence, you should have a good resume, fill out applications appropriately, and make a positive impression during the interview. The purpose of this unit is to develop your skills in these three areas to further enhance your chance of getting the job you want.

PREPARING A RESUME

Purpose

It is estimated that in the course of a lifetime, a person living in the United States will change jobs an average of 8 to 10 times.[1] Therefore, it makes sense to learn what is involved in obtaining the job you want and how to exit gracefully when it is time to change jobs.

Once you have examined your capabilities, checked into employment opportunities, and identified some goals for yourself, it is time to put this information together. Getting a job consists of doing all these things, then following through with a **resume**, job application, and an **interview**.

Identify the personal characteristics, skills, education, and aptitudes that you bring to the workplace. The resume is the best tool to capture all this information in a neat, concise format. On one brief, descriptive page, a resume illustrates who you are, what you can do, and where you want to go.

Remember that the employer is using the resume to screen job applicants, and the prospective employee seeks to highlight why he or she should be considered. The process of constructing a resume begins by compiling as much information as you can about yourself in a concise manner. The student in Figure 21–1 is completing a resume as a requirement for one of her classes.

Tips for Using a Resume

A good resume will help you get an interview. Use your resume in the following ways.

- Get the interview first. It is almost always better to contact the employer by phone or direct contact. Send your resume after you schedule an interview so the employer can read about you before your meeting. Valuable interview time will be spent discussing your skills, not your education.

Figure 21–1 By completing a resume, you will be ready when you have a job offer. A resume tells who you are, what you can do, and where you want to go.

- Send copies of your resume to everyone in your job search network. They can pass them along to others who might be interested.
- If you cannot make direct contact, send your resume in the traditional way. An example would be answering a want ad with only a box number for an address, but don't expect much to happen from this approach.[2]

Resume Basics

A typical resume in American business may have only seconds to catch the eye of a potential employer. The resume must then hold that person's attention and sell the person it represents.

What makes an employer want to read a resume? Employers are all different, but some guidelines will help ensure that your resume has a chance to market you. First, think of magazine advertising you see every day: It is usually short, to the point, and covers the subject completely; it includes nothing to prevent you from wanting to read it. A resume is nothing more than an advertisement—selling your services and time to a potential employer.[3]

The following guidelines will help you prepare a successful resume.[4]

1. *Write it yourself.* Look at examples of resumes, but don't copy them. Your resume won't sound like you, and you need to be comfortable with it.
2. *Make every word count.* Try to limit your resume to one page. After you have a first draft, edit it at least two more times. If a word or phrase does not support your ability to do the job, cut it out.
3. *Make it error free.* Ask someone else to look for grammar and spelling errors. Check each word again before you print it and send it to an employer.
4. *Make it look good.* Use a good quality printer and have it copied on a good quality paper. Appearance, as you know, makes a lasting impression.
5. *Stress your accomplishments.* A resume is no place to be humble.
6. *Be specific.* Give facts and numbers. Instead of saying you are good with people, say, "I supervised and trained five people in shipping and receiving and increased their productivity by 30 percent."
7. *Don't delay.* Many job seekers say they are still improving their resume when they should be out looking for a job. A better approach is to do a simple, error-free resume first, then actively look for a job. You can always work on a better version at night and on weekends.
8. *Keep it lively.* Use action verbs and short sentences. Avoid negatives of any kind. Emphasize accomplishments and results.

Writing the Resume

Although there are many kinds of resumes and many ways to prepare them, the following information will give you the major parts that should be included in your resume.

Personal data should be at the top of the resume: your name, address, and home telephone number. Be sure the information is correct and does not contain typing errors because it is the only way the employer has to contact you.

The job objective states the type of job you want. This lets the employer see if it matches any job openings in the company. Unless you are applying for a particular job, you may want to use a broad objective so that you will be considered for openings in related jobs. You may choose to prepare different resumes that match the objective with the job opening for which you are applying.

Skills Technical, teamwork, leadership, personal management, and employability skills should be listed here. List the skills you have that can be used in the job you are seeking. Be specific and brief. Remember to list all your skills.

Personal Management Skills

Perhaps you are the type of person who can easily learn new skills and identify and suggest new ways to get the job done easier. If you are a person who can work without supervision, list it. Some people have the ability to organize; others show imagination or initiative. Whatever your personal management skills are, you need to make a potential employer aware of them. Market yourself. Other examples of **personal management skills** or assets are these.[5]

- Meeting deadlines
- Following instructions
- Following directions
- Making suggestions
- Being organized
- Being able to meet school/work deadlines
- Paying attention to details
- Being able to follow written and verbal instruction and directions

Use action words and avoid use of the word *I*. Words like *experienced*, *recognized*, *advised*, and *operate* sound energetic and dynamic. They convey a sense of expertise without being arrogant. The good resume uses action words throughout, such as the following.[6]

- Coordinated
- Managed
- Operated
- Demonstrated
- Displayed
- Formatted
- Edited
- Revised
- Constructed
- Planned
- Negotiated
- Achieved
- Accomplished
- Maintained
- Designed
- Developed
- Identified
- Completed
- Delivered
- Processed
- Consulted

Expect to rewrite your resume several times to get the results you want. Because the document is short and concise, it is important to think through each part of its construction. Build single ideas into a simple phrase, then thoughtfully choose the most appropriate words to convey your message.[7]

Teamwork Skills

Some other types of skills employers are looking for are **teamwork skills**.[8] Even if you have never had a job, you can prove your teamwork skills to potential employers by showing them that you have actively participated in a group such as the baseball team, yearbook staff, community group, choir, or band. The band members in Figure 21–2 are planning last-minute details before they march as a team. By participating in any group activity, you show that you can follow the group's rules and values and listen to other group members. Other teamwork skills are being adaptable, being a team player, following rules, and being willing to compromise.

Leadership Skills

Include leadership skills if you possess them and if they are relevant to the job for which you are applying. Leadership is a skill that all employers look for in a potential employee. However, they also want someone who can either lead or follow, depending on what is needed, and who listens to other group members.

Figure 21–2 Playing in a marching band is a good way to develop teamwork skills. These band members are learning to adapt, follow rules, and compromise.

Education List in reverse chronological order all the schools you have attended. This means you should begin with the last school attended and work backward in time. Write the full name of the school, including the city and state in which the school is located. Do not include the street address. List the years you attended the school, eliminating months or days of the month.

Programs/Courses
List the program in which you were enrolled and the courses you studied that specifically qualify you for the job. Be sure to include any high school subjects that relate to your job objective and will help you perform effectively in the job. Include special skills you have developed and other educational experiences you have had, including cooperative education, on-the-job training, and youth apprenticeship programs.

Honors
If you received any special honors or recognition, including club membership and offices held, write in that information.

Documents
Include the type of document you received for completing each part of your education or training. The document could be a diploma, degree, certificate, or license. Remember to include the year in which you received the document.

Work Experience Work experience indicates that you have participated as a productive member of the labor force. Include all work experience, especially work directly related to your career objective. You may want to include volunteer experience as well as paying jobs. List your work experience in reverse chronological order, with the most recent job first. Include the following information.

Place of Employment
List the names and addresses of each company or firm. Omit street address; just write the city and state.

Dates
State the time periods you worked from year to year. Do not include months or days of the month.

Job Title
Use a job title for each place of employment. Every job title is important, even if it is a dishwasher or cashier. The job title shows you were responsible for a particular position.

Description of Duties
Write a brief description of your duties at the job. State the main tasks without explaining every detail. The student in Figure 21–3 is gaining work experience in soil compaction.

Achievements
List any special accomplishments or honors you received at a particular job. Be sure to include the year you received the honor.[9]

Personal In this section of your resume you list personal interests that will help you stand out from other applicants with similar qualifications.

Figure 21–3 Work experience is one of the most important elements in your resume. The young woman in this picture will list the work experience she's gained on soil compaction in her resume, along with the employer's name. (Courtesy of USDA.)

By reading this section, a possible employer can see you as a total person. Include personal qualities, sports you enjoy playing or watching, and other interests.

Evaluate your personal characteristics for their selling points and job significance. Be as objective as you can. It is important to know your personal strengths and weaknesses. Decide which are pertinent to the job you are after. An honest appraisal may even help you determine where your interests lie and can save you time, energy, and frustration.[10] Some examples of useful personal characteristics are these.

- Adaptability
- Self-confidence
- Spirit of cooperation
- Assertive attitude
- Cheerfulness
- Good manners
- Good temperament
- Tact
- Sensitivity
- Commitment

References Some authorities say not to list **references**; other authorities say to list them. Do whatever you feel is the most appropriate for your region of the country. Even then, you may want to list references on some resumes for a certain job and not list references on other resumes. Always get permission from individuals before using their names.

Figure 21–4 provides an example of a good resume for a high school graduate. It is limited to one page and provides a wealth of information for a potential employer. This person chose to list his references.

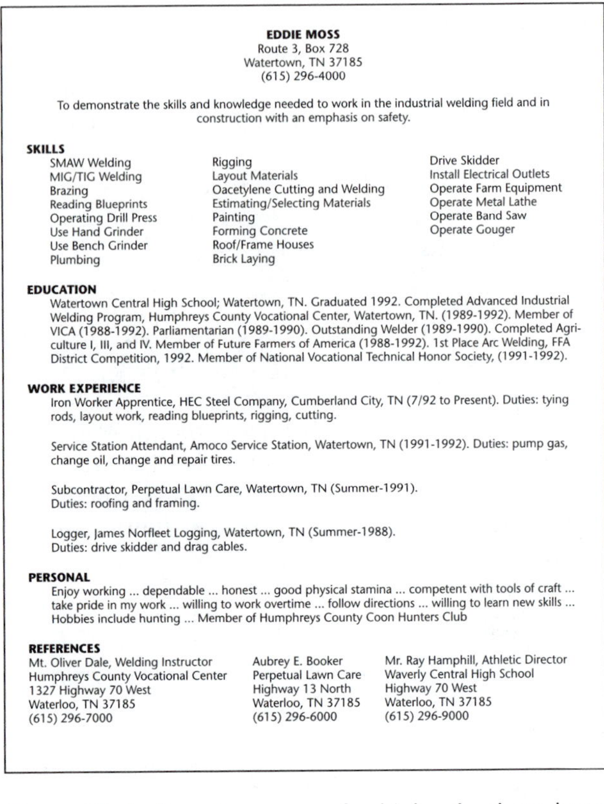

EDDIE MOSS
Route 3, Box 728
Watertown, TN 37185
(615) 296-4000

To demonstrate the skills and knowledge needed to work in the industrial welding field and in construction with an emphasis on safety.

SKILLS

SMAW Welding	Rigging	Drive Skidder
MIG/TIG Welding	Layout Materials	Install Electrical Outlets
Brazing	Oacetylene Cutting and Welding	Operate Farm Equipment
Reading Blueprints	Estimating/Selecting Materials	Operate Metal Lathe
Operating Drill Press	Painting	Operate Band Saw
Use Hand Grinder	Forming Concrete	Operate Gouger
Use Bench Grinder	Roof/Frame Houses	
Plumbing	Brick Laying	

EDUCATION

Watertown Central High School; Watertown, TN. Graduated 1992. Completed Advanced Industrial Welding Program, Humphreys County Vocational Center, Watertown, TN. (1989-1992). Member of VICA (1988-1992). Parliamentarian (1989-1990). Outstanding Welder (1989-1990). Completed Agriculture I, III, and IV. Member of Future Farmers of America (1988-1992). 1st Place Arc Welding, FFA District Competition, 1992. Member of National Vocational Technical Honor Society, (1991-1992).

WORK EXPERIENCE

Iron Worker Apprentice, HEC Steel Company, Cumberland City, TN (7/92 to Present). Duties: tying rods, layout work, reading blueprints, rigging, cutting.

Service Station Attendant, Amoco Service Station, Watertown, TN (1991-1992). Duties: pump gas, change oil, change and repair tires.

Subcontractor, Perpetual Lawn Care, Watertown, TN (Summer-1991).
Duties: roofing and framing.

Logger, James Norfleet Logging, Watertown, TN (Summer-1988).
Duties: drive skidder and drag cables.

PERSONAL

Enjoy working ... dependable ... honest ... good physical stamina ... competent with tools of craft ... take pride in my work ... willing to work overtime ... follow directions ... willing to learn new skills ... Hobbies include hunting ... Member of Humphreys County Coon Hunters Club

REFERENCES

Mt. Oliver Dale, Welding Instructor	Aubrey E. Booker	Mr. Ray Hamphill, Athletic Director
Humphreys County Vocational Center	Perpetual Lawn Care	Waverly Central High School
1327 Highway 70 West	Highway 13 North	Highway 70 West
Waterloo, TN 37185	Waterloo, TN 37185	Waterloo, TN 37185
(615) 296-7000	(615) 296-6000	(615) 296-9000

Figure 21–4 Sample resume of a high school graduate stressing skills, education, work experience, personal interests and characteristics, and references.

Computer Resumes

Many major companies and recruiters are placing resumes on computerized in-house applicant tracking systems. These systems file resumes and retrieve those that match specific job descriptions.

When retrieving resumes, these systems usually look for key words that characterize the applicant's skills and achievements. For an accounting manager, the computer may look for words such as *supervisor*, *manager*, and *BS, Accounting*. For a salesperson, *exceeded quota*, and *will travel* would be key words.

To cover all your electronic bases, especially if you know that the company to which you are applying screens resumes by computer, add a summary of key words at the top of your resume (under your name and address). Tag it: "Key Word Index."[11]

Variety of Resumes

Make your resume work for you. Although a format was presented, you can add or delete categories. If you received many honors and awards in high school, create your own category entitled

"Special Honors and Awards." If you were in several leadership events, such as parliamentary procedure or public speaking, create a section called "Leadership." Whatever your strength, if it is to your advantage, create that heading just for you.

Portfolios Many employers request portfolios. Portfolios are visual resumes used by people in creative or artistic fields to showcase their work. They are often oversized leatherbound or plastic briefcases containing pictures, photographs, articles, illustrations, and other creative products. Portfolios are usually accompanied by some printed information. Certain jobs, such as the landscaping industry, are more receptive to portfolios than others.

Sample Resumes The following resumes will assist you in presenting your resume in an organized, attractive form. If you have little or no full-time experience, they will help give you ideas as you review your summer jobs, part-time jobs, extracurricular activities, training, and job interests. The following sample resumes describe

- A recent high school graduate (Figure 21–5)
- A person with considerable experience (Figure 21–6)

SAMPLE RESUME
FOR A RECENT HIGH SCHOOL GRADUATE

JOAN SMITH
123 Some Street, City, State, Zip
Home (123) 980-2576, Work (123) 985-8973

OBJECTIVE: Seeking full-time office position involving data entry and/or computer operations.

EDUCATION: Big City High School, Big City, MO
Certificate: Computer/data entry (June, 1992)

Related course work:
-Introduction to computer literature
-Business accounting
-Computer operations, mainframe
-Records management
-Computer operations, micro
-Office machines

CAPABILITIES:
-Accurate and efficient worker.
-Work well without supervision.
-Pay attention to details.
-Computer experience: Wordperfect, Lotus 1-2-3, Q&A
-Typing 65 WPM, alpha and numerical filing.
-10 key calculator operation.

EXPERIENCE: Office worker, Ace Apple Barn, Small Town, RI, (1986-1988)
Worked in records department.
-Processed work orders.
-Checked inventory.

Switchboard Operator, Jane's Telephone Exchange, Summersville, CA (1988).
-Operator for message center.
-Answered phone system and recorded messages.

Figure 21–5 Sample resume of a recent high school graduate, including objective, education, capabilities, and experience. (Source: R. A. Wolf & T. A. Wolf, *Job Hunter's Secret Handbook for Success*. St. Joseph, MI: WINSS, 1993. Used with permission.)

SAMPLE RESUME
CONSIDERABLE EXPERIENCE

BILL ROBERTS
507 Happy Valley St., Some City, IL 44548
Home (456) 453-6432, Car (456) 412-2222

OBJECTIVE: Seeking full time position as Building Supervisor and/or Foreman

EDUCATION: University of Michigan, Bachelor of Science, Drafting Major.
Seminars Completed
"Managing Workers Today"
"Insight Seminars/Training for Construction
"Advanced Constructors Framing"

EXPERIENCE: General Builders, Detroit, MI.
-Building duties: Framing, Roofing, Trimming, Overseeing Subcontractors, Plumbers, Electricians, Masons, Painters, Overseeing up to 30 laborers.
-Arranged for purchase of all supplies and equipment.
-Conducted training of many apprentice carpenters.
-Residential and commercial experience.

CAPABILITIES:
-Attention to details.
-Excellent problem-solving ability.
-Mechanically inclined.

MILITARY: U.S. Army
Second Lieutenant.
Honorable Discharge.

REFERENCES: Available on request.

Figure 21–6 Sample resume of a person with considerable experience. (Source: R. A. Wolf & T. A. Wolf, *Job Hunter's Secret Handbook for Success*. St. Joseph, MI: WINSS, 1993. Used with permission.)

SAMPLE RESUME
REENTERING THE JOB MARKET

JUDITH G. HEAD
445 Newton Dr. Berkeley, Ca. 57650
Home (878) 453-9846, Car (133) 678-0896

OBJECTIVE: To obtain a position as Managing Editor of newspaper

EDUCATION: University of California, Los Angeles, CA (1979)
School of Journalism
Attended graduate coursework (37 hours completed)
B.A. Double Major in Journalism, English

CAPABILITIES:
-Well organized and responsible.
-Excellent verbal and written communication skills.
-Excellent leadership qualities.
-Specialize in problem solving.

EXPERIENCE: Assistant Managing Editor, *Los Angeles Times* (1984-1988)
-Managed ten reporters in city department.
-Supervised layout of paper.
-Established modern methods of conducting business.

Reporter, *Long Beach Daily* (1980-1983)
-Managed the city desk.
-Set up new procedures.
-Wrote under own byline.
-Researched leads on stories.

REFERENCES: Available on request.

Figure 21–7 Sample resume of a person reentering the job market. (Source: R. A. Wolf & T. A. Wolf, *Job Hunter's Secret Handbook for Success*. St. Joseph, MI: WINSS, 1993. Used with permission.)

SAMPLE RESUME
NO HIGH SCHOOL DIPLOMA

GEORGE McIVER
111 Mt. Curve, Atlanta, GA 78976
Home (567) 466-2548

OBJECTIVE: Seeking a full-time position as stock clerk.

EDUCATION: Essex High School, Atlanta GA
-Attended through eleventh grade —Sept. 1988 through June 1990.
-Presently studying for G.E.D.

CAPABILITIES:
-Follow written instructions and directions.
-Follow verbal instructions and directions.
-Attend school/work daily and on time.
-Demonstrate self control.
-Pay attention to details.
-Can stock 75 cases per hour.

EXPERIENCE: Piggly Wiggly Supermarket, part-time stock clerk, Summer 20__

Figure 21–8 Sample resume of a person with no high school diploma. (Source: R. A. Wolf & T. A. Wolf, *Job Hunter's Secret Handbook for Success*. St. Joseph, MI: WINSS, 1993. Used with permission.)

- A person reentering the job market (Figure 21–7)
- A job seeker who has yet to graduate from high school (Figure 21–8)

Use these examples to develop appropriate ideas and approaches in creating your own resume.

LETTER OF APPLICATION

Another way to act on a job lead is to write a **letter of application**. You might do this when acting on a suggestion from another person or responding to a newspaper help wanted ad. A letter of application is often known as a cover letter when it is mailed along with a resume. A combination cover letter and letter of application should have four parts or paragraphs (Figure 21–9).[12]

1. *Reason for writing.* In the first paragraph, you should explain your reason for writing. Name the job for which you are applying. Also, tell how you learned about the job.
2. *Point out qualifications.* Use the second paragraph to briefly point out your qualifications. Give facts, but don't boast. Employers look carefully at this paragraph.
3. *Call attention to resume.* The third paragraph calls attention to the resume. It may also be wise to give the date you are available for employment.
4. *Ask for appointment.* In the last paragraph, ask for an appointment. Tell how you can be contacted. Close the letter with a courteous comment or thank you.

6428 Valley Road
Cambden, OH 67423
April 6, 2000

Mr. Donald Young
Service Manager
Smith Auto Sales, Inc.
274 Oakland Street
Cambden, OH 67423

Dear Mr. Young:

I learned from one of your employees, Mr. Ken Jenkins, that you plan to hire a new mechanic in a few weeks. I would like to apply for the position.

I have the training and experience to do the job. For the last two years, I have worked at Goodman's Tire and Auto Center. I primarily do tune-ups, general engine repair, front wheel alignments, and wheel and brake work. I am satisfied with my present job. However, I would like to work for a new car dealership where I can better use my diagnostic and mechanical abilities. I hold a state inspection license and own my own tools.

I have enclosed a copy of my resume that provides further details about my background. I could be available for employment following a two-week notice to my present employer.

May I have an appointment to discuss the job with you? I can be reached after 4:00 P.M. at (627) 353-2761. I would appreciate being considered for the job.

Sincerely,

Ronald Fisher

Figure 21–9 A combination cover letter and letter of application should have four parts or paragraphs: reason for writing, point out qualifications, call attention to resume, and ask for appointment.

Be Brief and Specific

Notice that the sample letter is short and to the point. The purpose of the letter is to attract and hold the reader's interest. It should not attempt to give facts that are better stated in a resume and job interview. If you are qualified for the job, the letter and resume should make the employer want to invite you for an interview.

Appearance

In preparing your resume and letter of application, appearance is critical. Your resume and application letter should have a neat, error-free, professional appearance. Use a white, high-quality 8½″ × 11″ bond paper with matching envelope. Well-prepared documents say you are knowledgeable, competent, and interested in the job.[13]

You should not expect your resume and application letter to be perfect on the first try. If you want your application to make a positive impression, revise your letter and resume until they are perfect. Remember, employers are impressed by applications that say, "I cared enough to do my very best!"[14]

Helpful Hints

Consider the following helpful hints for a letter of application.[16]

- *Be assertive.* The first 20 words are the most important; they should attract the reader's interest in the letter.
- *Value to employer.* Tell your story in terms of the contribution you can make to the employer.
- *Use simple direct language and correct grammar.*
- *Let your letter reflect your individuality.* Avoid appearing aggressive, overbearing, familiar, cute, or humorous. You are writing to a stranger about a subject that is serious to both of you.[15]
- *Send it to someone by name.* Get the name of the person who is most likely to supervise you. Call first to get an interview. Then send your letter and resume.
- *Get it right.* Make sure you spell the recipient's name correctly and use the correct title. Any error in spelling or grammar will not make a good impression.
- *Be clear about what you want.* If you want an interview, ask for it. If you are interested in that organization, say so. Give clear reasons they should consider you.
- *Be friendly.* A professional, informal style is usually best. Avoid a hard-sell, "Hire me now!" approach. No one likes to be pushed.
- *Target your letter.* Typical reasons for sending a cover letter include responding to an ad, preparing an employer for an interview (the best reason!), and following up after a phone call or interview. Each of these letters will be different.

- *Follow up.* Remember that contacting an employer directly is much more effective than a letter. Do not expect a letter to get you many interviews. They are best used to follow up after you have contacted the employer.

APPLICATION FOR EMPLOYMENT
(PRE-EMPLOYMENT QUESTIONNAIRE) (AN EQUAL OPPORTUNITY EMPLOYER)

PERSONAL INFORMATION

DATE April 15, 2000

NAME Fisher Ronald R. SOCIAL SECURITY NUMBER 351-44-5751

PRESENT ADDRESS 6428 Valley Rd. (STREET) Cambden (CITY) Ohio (STATE) 67423 (ZIP)

PERMANENT ADDRESS (STREET) Same (CITY) (STATE) (ZIP)

PHONE NO. 627-353-2761 ARE YOU 18 YEARS OR OLDER? Yes ☒ No ☐

ARE YOU EITHER A U.S. CITIZEN OR AN ALIEN AUTHORIZED TO WORK IN THE UNITED STATES? Yes ☒ No ☐

EMPLOYMENT DESIRED

POSITION engine + powertrain mechanic DATE YOU CAN START May 1 SALARY DESIRED Open

ARE YOU EMPLOYED NOW? Yes IF SO MAY WE INQUIRE OF YOUR PRESENT EMPLOYER? Yes

EVER APPLIED TO THIS COMPANY BEFORE? No WHERE? WHEN?

REFERRED BY Ken Jenkins

EDUCATION	NAME AND LOCATION OF SCHOOL	NO. OF YEARS ATTENDED	DID YOU GRADUATE?	SUBJECTS STUDIED
GRAMMAR SCHOOL	Cambden Elementary School District #95	8	yes	general curriculum
HIGH SCHOOL	Cambden High School	4	yes	vocational curriculum automotive
COLLEGE	Hillside Community College	2	yes	automotive technology
TRADE, BUSINESS OR CORRESPONDENCE SCHOOL	———			

FORMER EMPLOYERS (LIST BELOW LAST THREE EMPLOYERS, STARTING WITH LAST ONE FIRST)

DATE MONTH AND YEAR	NAME AND ADDRESS OF EMPLOYER	SALARY	POSITION	REASON FOR LEAVING
FROM June 19– TO present	Goodman's Tire + Auto Center – Cambden, OH	6.85 hr.	service technician	Currently employed
FROM June 19– TO May 19–	Hunter's Auto Repair – Eastbrook, OH	4.30 hr.	general auto repair	part-time only
FROM Aug. 19– TO May 19–	Texacon Service Station – Cambden, OH	3.75 hr.	auto maintenance	co-op student learner
FROM TO				

WHICH OF THESE JOBS DID YOU LIKE BEST? Hunter's Auto Repair

WHAT DID YOU LIKE MOST ABOUT THIS JOB? engine diagnosis

REFERENCES: GIVE THE NAMES OF THREE PERSONS NOT RELATED TO YOU, WHOM YOU HAVE KNOWN AT LEAST ONE YEAR.

	NAME	ADDRESS	BUSINESS	YEARS ACQUAINTED
1	Earl Thompson	Goodman's Tire + Auto 219 E. Sycamore Cambden, OH	Service manager	1.5
2	Archie Hunter	Hunter's Auto Repair 2025 W. Walnut Eastbrook, OH	owner/manager	2.0
3	Frank Hopkins	Hillside Community College – Eastbrook, OH	auto instructor	2.0

Figure 21–10 Sample of a correctly completed job application form.

Completing a Job Application Form

The most common procedure to screen individuals for available jobs is to ask **applicants** (potential employees) to fill out a job **application form**. It is not unusual for employers to have dozens, or even hundreds, of applicants for a single position.[17] The information provided on the form helps employers sort out the best qualified person for the job. After application forms are screened, a small number of people are invited to interview.

Reasons Job Application Forms Are Eliminated

Very often, qualified job seekers "don't make the cut." In a busy **personnel office** (a company's department for handling all staffing issues), your application could be eliminated because it is messy, because you didn't list enough experience, or you showed that your last job paid more than the one now open. There are many reasons for rejecting an applicant. You may be able to do the job for which you applied, but you may never have the chance.

There is no standard form for a job application, therefore, your qualifications may not have been listed. Most employers design their own forms so that they get information most helpful to filling their job openings. By having the same information on each application, the employer can quickly and easily compare the qualifications of the applicants.

Tips for Filling Out a Job Application Form

Application forms differ from company to company (Figure 21–10). Although these forms appear simple, it is easy to make a mistake. If you take a copy of your resume with you, you will be less likely to forget important information or make mistakes. The following tips are helpful for filling out a job application form.[18]

- Before you write any answers, read the form carefully.
- Study the instructions so you will know what information to provide.
- It is a good idea to print the information. Be sure to sign your name in places where your signature is indicated. Use your correct name, not a nickname.
- A certain amount of space is allowed for each answer. Do not try to fit too much information into a space.
- Make sure your answers are in the right place.
- Read and understand the instructions. You will give the impression that you cannot follow directions if you fail to follow the instructions.

- Neatness is a must. If you erase or cross through answers, the employer may think of you as a careless or sloppy person. You may want to ask for two application forms. Use the first as a rough draft. Copy your answers neatly onto the second form.
- Answer all questions on the form. If a question doesn't apply to you, use "NA," which means **not applicable**, or draw a short line through the blank. Remember, if you leave a blank space, the employer might think you forgot to answer a question.
- Be honest. Never give false information.
- List the specific job for which you are applying. Although you may be willing to accept any job, do not write "Anything" in the blank. Indicate that you are interested in and qualified for a specific job.
- Make sure you did not make mistakes. Every blank should be filled in.
- After you have completed the form, check it carefully. Misspelled words and crossed-out responses give a poor impression of your ability. Use a dictionary if you need to check your spelling.

A job application form that is neatly and accurately completed may be the determining factor in getting an interview. Just from looking at the job application form the employer forms an impression of the applicant: neat or messy, can complete an assignment or cannot follow instructions, is thorough or careless. The impression you give with your job application form can help you or hurt you in your job search.[19]

Personal Data Sheet

A personal data sheet is useful in completing a job application and in developing a resume (Figure 21–11). It is helpful to take the personal data

PERSONAL DATA SHEET
IDENTIFICATION

Name ______ Soc. Sec # ______
Address ______ Zip ______
Birth Date ______ Birth Place ______
Telephone (__) ______ Height ______ Weight ______
Hobbies/Interests ______
Honors/Awards/Offices ______
Sports/Activities ______
Other ______

EMPLOYMENT HISTORY
(Start with present or most recent employer.)

1. Company ______ Telephone (__) ______
Address ______
Employed from Mo. ______ Yr. ______ / to Mo./Yr. ______
Supervisor ______
Position/Title ______
Last Wage ______ Reason for Leaving ______

2. Company ______ Telephone (__) ______
Address ______
Employed from Mo. ______ Yr. ______ / to Mo./Yr. ______
Supervisor ______
Position/Title ______
Last Wage ______ Reason for Leaving ______

3. Company ______ Telephone (__) ______
Address ______
Employed from Mo. ______ Yr. ______ / to Mo./Yr. ______
Supervisor ______
Position/Title ______
Last Wage ______ Reason for Leaving ______

EDUCATIONAL BACKGROUND

School Name and Address
Dates Attended: From: ______ To: ______

Elementary:

Junior High:

High School:

Courses of Study ______ Rank ______ GPA ______
Favorite Subject(s) ______

REFERENCES

1. Name ______ Title ______
Address ______ Zip ______
Relationship ______ Telephone (__) ______

2. Name ______ Title ______
Address ______ Zip ______
Relationship ______ Telephone (__) ______

3. Name ______ Title ______
Address ______ Zip ______
Relationship ______ Telephone (__) ______

Figure 21–11 Sample personal data sheet.

APPLICATION FOR HOURLY EMPLOYMENT

Personal

Name (First) (Middle) (Last) S.S.#

Address (Street) (City) (State) (Zip) Phone Number (Area Code) (Number)

Are you at least 18 years old? ____Yes ____No If No, Birthdate

Do you have the legal right to remain and work in the United States? Yes No

Have you ever been employed by this company?___Yes___No If Yes, explain________

What prompted you to apply for work here?
___Company___Image___Agency___Friend___Relative___Newspaper___Other__________

PERSON TO BE NOTIFIED IN CASE OF EMERGENCY
Name ____________________Home ____________________
Address ____________________Work Phone ____________________
Have you ever been convicted of a felony?___Yes___No If Yes, explain__________
(Conviction will not necessarily from employment?) ____________________
Personal Interest ____________________

Job related organizations, clubs, professional societies ____________________
(Omit those which indicate sex, race, religion, creed, color, national origin, ancestry, and or age)
Do you have any medical problems that would interfere with or detract from your ability to perform your job? __Yes __No If yes explain ____________________
Is any member of your family (spouse, parent, etc.) employed in this industry?
__Yes __No If yes explain ____________________

Education

	Name and location of school	Dates attended From To	Circle Highest Year Completed	Major & Minor Fields of study	Degree(s) or Diploma
High School			9 10 11 12		
Technical/ Vocational School					
College/ University			1 2 3 4		
Other					

Honors Received

Availability For Work

WHAT HOURS ARE YOU AVAILABLE FOR WORK?

	SUNDAY	MONDAY	TUESDAY	WEDNESDAY	THURSDAY	FRIDAY	SATURDAY
FROM							
TO							

Type of Schedule Desired ___Part Time___Full Time
Do you plan to work elsewhere or attend school while employed? ____________
Do you have any obligations which would affect working as scheduled? ____________
How soon after accepting an offer would you be able to start working? ____________

Experience

Employer Phone From To
Address City, State, Zip Position
Responsibilities Supervisor's Name
Starting Salary / Wages
Reason for leaving May we contact? Final Salary / Wages

Employer Phone From To
Address City, State, Zip Position
Responsibilities Supervisor's Name
Starting Salary / Wages
Reason for leaving May we contact? Final Salary / Wages

Employer Phone From To
Address City, State, Zip Position
Responsibilities Supervisor's Name
Starting Salary / Wages
Reason for leaving May we contact? Final Salary / Wages

U.S. Military Service
Veteran of U.S. Military Service?___Yes___No If yes, what branch?__________
Dates of Service?____________________Rating or rank achieved?__________
Duties and Responsibilities?____________________

Important

I represent that the information in this application is correct to the best of my knowledge and understand that any misstatement or omission of information may be grounds for dismissal. I authorize the references listed above to give you any and all information that they may have. I hereby release all parties from all liability from any statements or information provided.

In consideration of my employment, I agree to conform to all policies of this company, and that my employment can be terminated "at will," with or without cause, and with or without notice, at any time, at the option of either the Company or me. I understand that no manager, supervisor or any representative other than the President has authority to enter into any agreement for employment. I further understand and specifically acknowledge that any agreement for employment other than "at will," must be in writing and signed by the President and me.

This application was designed to comply with Federal Civil Rights Act, Title VII, The Age Discrimination Act of 1967, and State Fair Employment Practice Laws. The company therefore gives assurance that no question answered is or will be used to discriminate adversely in matters of race, creed, color, national origin, ethnic background, age, or sex.

Signature Date
685281 9/88 REV.

Figure 21–12 Sample application for hourly employment.

sheet with you when filling out an application to prevent omitting any pertinent information.

Sample Application Forms

To understand how to complete applications, study the two applications in Figure 21–12 and Figure 21–13. Figure 21–12 is an application for hourly employment. Figure 21–13 is a typical form used in hiring carpenters, custodians, cooks, drivers, nurses, secretaries, shipping clerks, and various other civil service classifications.

Preparing for an Interview

By being granted an interview, you have already achieved some degree of success in the job search. Someone in the company liked what they saw on a resume or application and decided to find out more about you. In the interview, you verify information contained in your resume or on your job application. Essentially, you are selling yourself to the employer. You have to be as good in person as you appeared to be on paper.

Purpose of the Interview

The interview also provides you with a chance to investigate further the specific job and the company. You need to know whether the job, the company's employees, its policies, and the work environment meet your expectations. This information allows you to determine whether you really want to work for the organization. Remember, your needs are as valid as the interviewer's.

APPLICATION FOR HOURLY EMPLOYMENT

Social Security Number — Last Name — First Name — Middle Name

Mailing Address (Street number and name) — City

State — County — Zip — Home Phone ( ) — Business Phone ( )

Birthdate (Respond only if you are under 18)

Have you ever taken any examinations under the State Universities Civil Service System of Illinois? __Yes __No

Are you a citizen of the United States? ___Yes ___No

If no, do you have a visa that permits you to work in the USA? __Yes __No Type Permit No. Exp.Date

Type of work desired

Type of Employment ____Full-time ____Part-time ____Either — Date Available

Circle HIGHEST grade/level completed:

GRADE SCHOOL 1 2 3 4 5 6 7 8 — HIGH SCHOOL 1 2 3 4 GRAD/GED — COLLEGE AA BA MA PhD

School	Name and Address	Dates of Attendance from / to	Major and Minor areas of study	No.of Sem./Qtr. Hr.(specify)	Type of degree	Did you graduate?
College						
College						
College						
Business/Tech. School						

Please indicate which of the following skills, experience, etc. you have:

___Typing–wpm________ ___Word processing ___Computer programming (specify): ___Foreign language(s) (specify):
___Shorthand–wpm________ ___Accounting/bookkeeping
___Use of transcription equipment ___Cashiering
___Data Entry ___Valid Drivers License, Class___

List any technical or professional registrations, certifications and/or licenses which you possess (include expiration dates):

PLEASE LIST ACCURATELY YOUR EMPLOYMENT HISTORY, INCLUDING MILITARY SERVICE. BEGIN WITH YOUR PRESENT OR MOST RECENT JOB. ALSO LIST PERIODS OF UNEMPLOYMENT OF TWO OR MORE MONTHS. RELEVANT VOLUNTEER EXPERIENCE SHOULD BE INCLUDED. USE ADDITIONAL SHEET(S) IF NECESSARY.

Current or Last Employer — **Address**

Job Title — Supervisor's Name — No. Supervised by you

Beginning Date (mo., yr) — Starting Salary $ — Ending Salary $ — Reason for leaving — May we contact Employer? ___Yes ___No

Ending Date (mo., yr) — Duties________

Full Time Years Months — Part Time Years Months

*If part time, number of hours worked per week:

Employer — **Address**

Job Title — Supervisor's Name — No. Supervised by you

Beginning Date (mo., yr) — Starting Salary $ — Ending Salary $ — Reason for leaving — May we contact Employer? ___Yes ___No

Ending Date (mo., yr) — Duties________

Full Time Years Months — Part Time Years Months

*If part time, number of hours worked per week:

Have you ever been convicted of a misdemeanor or a felony? ___Yes ___No
If yes, please explain:

A conviction record will not necessarily be a bar to employment; factors such as age and time of the offense, seriousness and nature of the violation, and rehabilitation will be taken into accounts in terms of the position applied for.

Have you ever been suspended or discharged from a position? ___Yes ___No
If yes, please explain:

SIGNED________________________ DATE__________

Figure 21–13 Typical application form used in hiring for civil service jobs.

Neither you (the **interviewee**) nor your **interviewer** knows the other, and both of you are hoping for a positive encounter. During the interview, a determination will be made about whether each of you wants the relationship to continue. If it is mutually successful, another meeting will be arranged.[20]

Purpose of Practicing for Interviews

You will hardly ever have an opportunity to meet your interviewer before the interview. Considering that interviewers look for the best and brightest candidate, you need to at least meet, and even try to exceed, their expectations of who they want to do the job. Before you start practicing for the interview, ask yourself how you can best accomplish this task. You need to look good, know what the job involves, understand how you can be an asset to the company, and be ready to explain why you would be the best choice. Each of these items needs to be addressed if your interview is to be successful.[21]

Your resume and application present a professional image that needs to be replicated in the interview. This requires that you consider how you look for this very important meeting. Since the interview is such an important step in getting the job, you should prepare carefully for it. The more you prepare, the better your chances of making a good impression on the employer.

Practicing Interview Skills

An interview for an entry-level job lasts about 15 to 30 minutes. You may be a little nervous when you think about going for a job interview; this is perfectly normal. To reduce any stress and help build your confidence, you may want to role-play some practice interviews. Something as important as a job interview deserves advance preparation. You would not go for a driver's license exam without practicing your driving skills, would you?

You may be able to set up a classroom interview situation. Arrange a desk and a couple of chairs as you might find them in an office. The instructor or a fellow student can play the role of interviewer. Take turns playing the interviewee being interviewed for a **hypothetical** (pretend) job, as shown in Figure 21–14. Try to make the interview as realistic as possible. If you have access to equipment, videotape the interviews. It can be very instructive to see yourself participating in an interview.

By practicing the interview, you will become more aware of what is involved in thinking about a question and answering it out loud. It can be a valuable learning experience to discover, for example, how much you stumble and hesitate. Do not try to memorize answers, but do practice until you respond easily. Make special efforts to rid your speech of "uhs," "you knows," and the like.[22]

Figure 21–14 Practicing your interviewing skills will help you respond more easily. These students are practicing the roles of interviewer and interviewee.

Practice Interview Questions

Think of questions the employer may ask. There are some questions you may be asked at almost every interview. Some of these questions are listed below. Study these questions before going to an interview. Be sure you can answer the questions clearly and positively. You can make a good impression by expressing yourself well. Practice answering these questions so that you will feel more confident in the actual interview. Try preparing and then rehearsing your responses to the following commonly asked interview questions.

- *What can you tell me about yourself?* Emphasize aspects of yourself that are particularly relevant to this job.
- *Why did you leave your last position?* Concentrate on your desire for new challenges. Stay positive.
- *What are your strengths?* This is the time to humbly present your strong qualities. Do not hesitate to share these with potential employers. Try to relate these strengths to the specific job for which you are interviewing.
- *What are your weaknesses?* For this question, try to frame strengths as if they are weaknesses—"I tend to work too hard" or "I try to be too much of a perfectionist"—instead of volunteering that your work is sloppy or you show up late.
- *What were your major accomplishments in each of your jobs?* Refer to your personal data sheet, if necessary.
- *Why are there gaps in your work history?* Tell the truth here, in the best possible light, but make sure employers understand that this is not a red flag with regard to your capabilities—"I needed some time to reevaluate my career direction and priorities" or "I wanted to look around carefully before taking a new job in order to find the best possible match."
- *Why should I hire you?* Reflect on your skills and accomplishments; match them with your understanding of the position's responsibilities as well as company goals and values—"I am a

highly motivated self-starter who has always taken the initiative to develop new ideas and programs." For example, you have described yourself as a very aggressive person who really wants to take the market by storm. "Given my history of accomplishments and working style, I think we would make an excellent team."

- *What was your relationship with your last boss like?* If it was not the greatest, be honest but tactful—"I respected her, but I do not feel she took advantage of my full potential."
- *How do you usually get along with your co-workers?* Try to portray yourself as a team player, if possible.
- *Describe your work personality.* Be positive, of course.
- *Which job did you like the least? Why?* Do not dwell on money here; few jobs pay as well as we would like. Instead, try to pick an experience that will show in a roundabout way what you do want—"I liked my job at XYZ Corp. the least because I was unable to utilize my broad range of abilities, and there seemed to be no room for advancement."
- *Which job did you like the most? Why?* By hearing what you valued in the job you liked the most, the interviewer can assess your potential with the company.
- *If you could do things differently, what would you change?* Emphasize something you did well and discuss how it could have been even better.
- *Have you ever been fired? Why?* Remember, frame your answer as positively as possible. Do not try to whitewash the truth; employers will see through this. If you made mistakes, take responsibility for them, but demonstrate that you learned something from these mistakes, so that you will not repeat them.
- *What do you want to be doing five years from now?* Make sure your answer reflects some career growth—"I hope to be an assistant manager by then"—and a desire to remain in the company. Employers do not want people who will stagnate or decide to leave after a year.[23]

Learn about the Company or the Job

Find out as much as possible about the job and the company before your interview. Ask people you know who might have information about the company. Personal contacts can give you inside information. For instance, you might find out about such things as the working conditions or the turnover rate of personnel. Further information may be available from the company itself. Check with the company on the availability of brochures, catalogs, annual reports, or other descriptive materials. Firsthand information may be possible if the place of employment is a restaurant, retail store, or similar public place. Visit the establishment to get a feel for the atmosphere. You can observe the type of work done and perhaps have the opportunity to ask employees a few questions.

Do not forget the library. Information is available in several directories that describe corporations by name. A librarian can help you find such references. Gather information on products or services produced, growth rate, standing in the industry, and so on.

If information about the company is not available, find out something about the company's type of industry. Let us say that you are going to interview for a job in a property management firm. Find out what services such firms provide.

When you finish gathering information about the company, write a list of questions that you would like to ask about the job or the company. For example, you might ask: "Why did the job become vacant? Will any more training be required? What are the working hours? Who will my supervisor be if I get the job?" It is generally best to avoid asking about salary or benefits. If the information is not provided by the interviewer, you can ask after you have been offered the job.[24]

Collect Materials Needed for the Interview

Collect materials you will need. You will need a copy of your resume to leave with the employer. Your resume can also be used to provide information for the job application form you may be asked to complete. Be sure you have a pen and pencil for the job application form, a personal data sheet, and a list of questions you will ask. You may want to carry your papers in a folder or briefcase.

Attend to Appearance

The first impression an employer forms of you is usually based on your appearance. Grooming and dress will influence the interviewer's final decision. In John Malloy's book, *Dress for Success*, he states that a person's reaction to styles, colors, and combinations of clothing is highly predictable. So are the tastes of hiring professionals. Therefore, a very carefully chosen interviewing "uniform" worn on a well-groomed person is very important.

Check Last-Minute Details

Write the date, time, and place of the interview on a card. Check and then double-check the information. You may want to make a trial run so that you will know where the company is located. If more than a week goes by between the time you made the appointment and the actual interview, call to confirm it.[25]

Plan to arrive at the interviewer's office 5 to 10 minutes ahead of schedule. Introduce yourself and state why you are there. Do not take anyone with you to the interview; you do not want to give the impression that you cannot do things on your own.[26]

You may have to wait a short time in an outer office or reception area. During that time, you should relax, read, or look over your list of questions. Be pleasant to others in the reception area. Do not smoke, chew gum, or do anything distracting.[27] The interviewer may later ask for the receptionist's opinion of you.

Interviewing for the Job

Interviews are the doorways to jobs. Virtually no one (with the possible exception of the boss's son or daughter) can land a job without an interview. You can look great on paper, your references can give you great support, but the person who does the hiring will want to see you and talk with you face to face before making a decision about offering you a job. People hire people; they do not hire resumes or college credentials.[28]

Link with the Interviewer

When you meet your interviewer, be friendly, firm, and direct. Smile, make direct eye contact, and use a firm but gentle handshake. "Hi, I'm John Doe. Nice to meet you." Once seated, the beginning of the conversation can sometimes be a bit awkward for both sides. If you sense this, signal that you are comfortable and ready to talk by bringing up a neutral subject. For example, compliment the interviewer about something in his office, a trophy or an interesting picture, and ask about it. If your research has turned up some facts such as the company's sponsorship of some scholarships or participation in a local charity, you might talk briefly about one of these subjects. Rapport begins when you show real interest in the other person.[29]

There are many things you could do during the first few minutes of an interview. Following are four suggestions from experienced interviewers.

1. *Allow things to happen.* Relax. Do not feel you have to start a serious interview immediately.
2. *Smile.* Look happy to be there and to meet the interviewer.
3. *Use the interviewer's name.* Be formal. Use "Mr. Smith" or "Ms. Sharpe" unless you are asked to use another name.
4. *Ask some opening questions.* After a few minutes of friendly talk, you could ask a question to get things started. For example: "I'd like to know more about what your organization does. Would you mind telling me?" or, "I have a background in ________________ and I'm interested in how these skills might be used in an organization such as yours."[30]

Body Language

It is not always what you say, it is how you say it. The importance of body language (nonverbal communication) cannot be emphasized enough. What you do with your body and how you move provide many clues to the interviewer. They tell how you feel about yourself and others. The casual movement of your shoulders or the posture you assume can sometimes provide many more signals to the interviewer than your actual words. Following are five considerations to help you project a positive image.[31]

Figure 21–15 Both interviewer and interviewee appear relaxed and self-confident. The young man is also exhibiting positive nonverbal messages, as indicated by his facial expression.

1. *Posture.* The way you stand and sit can make a difference. When you lean back, you may look too relaxed. Always project a confident attitude by leaning forward slightly. This conveys a relaxed, highly interested attitude to the interviewer. Make slow, easy gestures, let your voice have the ring of optimism and keep your chin up and your hands still. Keep your head on an even plane with the interviewer and maintain sincere eye contact and slow, regular breathing.[32] Your posture should project a high level of confidence, as shown in Figure 21–15.
2. *Voice.* You may be nervous, but try to sound enthusiastic. Your voice should be neither too soft nor too loud. Practice sounding confident. Applicants who feel at ease during an interview speak in a normal tone of voice and answer questions without much hesitation.
3. *Eye contact.* In our culture, people who don't look at a speaker's eyes are considered shy, insecure, and even dishonest. Although you should never stare, you look more confident when you look at the interviewer's eyes while you listen or speak.
4. *Distracting habits.* If the applicant has rapid eye movement, speaks with his hand over his mouth, shifts in his chair, rubs his chin, has a blank look, or looks at the ceiling, the interviewer will conclude that the applicant is very uncomfortable or is not telling the truth.[33] You may have nervous habits you do not even notice. Pay attention! Most interviewers find such habits annoying. For example, do you play with your hair or clothing? Say something like "You know?" or "Uh" over and over?
5. *Videotape.* The best way to see yourself as others do is to have someone videotape you while you role-play an interview. If that is not possible, become aware of how others see you and try to change the negative behavior. Your friends and relatives can help you notice any annoying habits you have that could concern an interviewer.

Effective Communication

Let the interviewer set the tone and pace of the interview. Adjust yourself to the style of the interviewer. For example, if the interviewer is serious and businesslike, your style should be similar. If the interviewer is cheerful and outgoing, you may need to brighten up a little. Try to establish a compatible relationship with the interviewer.

Communication skills, which are important at every step of the job search, are even more so in the job interview. Be sure to listen carefully and speak clearly. Answer each question briefly, but do not give one-word or one-line answers. If you think that the interviewer has not understood your answer or that you have not made yourself clear, try again. Stay on the topic until you are sure that the interviewer has understood your message.

Answer a question only after the interviewer is completely finished. Otherwise, you risk making a bad impression. You may also miss the exact question or important information that may be added to the question. Listening to the interviewer is as important as speaking thoughtfully and clearly. The ability to listen shows your attentiveness and reflects your interest in the job. At times, you may want to ask the interviewer the meaning of a word or phrase. Do so. You must understand a question before you can answer it.[34]

Answering Questions

If you have practiced answering questions as you prepared for the interview, you should be fine. However, be prepared for interviewers to ask almost anything. They are looking for potential problems. They also want to be convinced that you have the skills, experience, and personality to do a good job. In a recent survey, employers said that over 90 percent of the people they interviewed for a job could not answer a problem question. Furthermore, over 80 percent could not explain the skills they had for the job.[35] With proper practice, this should not be a problem.

Asking Questions

An interview involves two-way communication. Of course, the interviewer will ask you questions. The interviewer will also expect you to ask questions. It is wise to refer to a list of questions you have made beforehand. Hold the list near your lap so you can glance at it as you talk.[36]

Do not be in a hurry to ask questions. Wait until the interviewer invites them. A pause in the conversation once the interview is well under way may be the time for you to bring up your questions. Be careful, though, not to interrupt the interviewer. By all means, if you have not already been invited to do so, request an opportunity to ask your questions before the interview ends.[37]

Ask questions about the company, even if you have done some good research on the firm. There is much you need to know to maneuver toward a job. Avoid naive questions such as, "How long have you been in business?" and "What products do you make?" The interviewer will expect you to know such elementary information if you were interested enough to apply for a job.

Here are some good questions to ask. The interviewer's answers could help you sharpen your own responses during the rest of the interview and any subsequent interviews for the same position. Consider the following.

- Of course, I've read the description of this position in your ad. Could you tell me more about any other responsibilities of the job that weren't listed?
- Where do you feel this position could lead for the right person?
- To whom would I report? Could you tell me more about what he (or she) expects of the person filling this position?
- Is there any question about the company or this position that you would have asked if you were on my side of the desk?[38]

Concluding the Interview

Try to get a feeling for when the interview has run its course. The interviewer may stand or simply say right out, "Well, I think that I have enough information about you at this time." To help bring an interview to its conclusion you can ask, "Are there any more questions I can answer?"

Many job applicants fail to ask for the job. This is a big mistake. If you want the job, say so. Say something like, "I know I can do the work, Mr. Morgan, and I would like to have the job."

Seldom does an interviewer make a job offer or reject an applicant at the conclusion of an interview. Usually the interviewer wants to think about and compare all applicants before making a final decision. In some cases, the interviewer's role is to evaluate and make recommendations only. The actual employment decision may be made by another person. Nevertheless, end the interview to your advantage.[39]

- Thank the interviewer by name.
- Express interest.
- Arrange a reason and a time to call back.
- Say good-bye.

If you do learn that the company cannot use you, ask about other employers who may need a person with your skills. Thank the interviewer, shake hands, and leave. On the way out, thank the secretary or receptionist.

AFTER THE INTERVIEW

The following steps should be taken after each interview. They will help you prepare for your next interview and ensure that you leave a positive impression with each prospective employer.

Step 1

Evaluate your performance.
Take a few minutes to evaluate your performance. Could you have answered some questions better? Did the interviewer ask questions you were not expecting? If so, write these questions down so you will be ready if they are asked in another interview. Evaluate your appearance and your overall presentation. You will learn from each experience and will improve your skills in making a good impression.[40]

Step 2

Write a follow-up letter.
Within two days of your interview, write a brief **follow-up letter** thanking the employer for the interview (Figure 21–16). Mention again that you are interested in the job, referring to your job-related skills and experience. A follow-up letter will help the interviewer remember you and may be the deciding factor in whether or not you get the job.[41] Such a letter may also accomplish the following things.

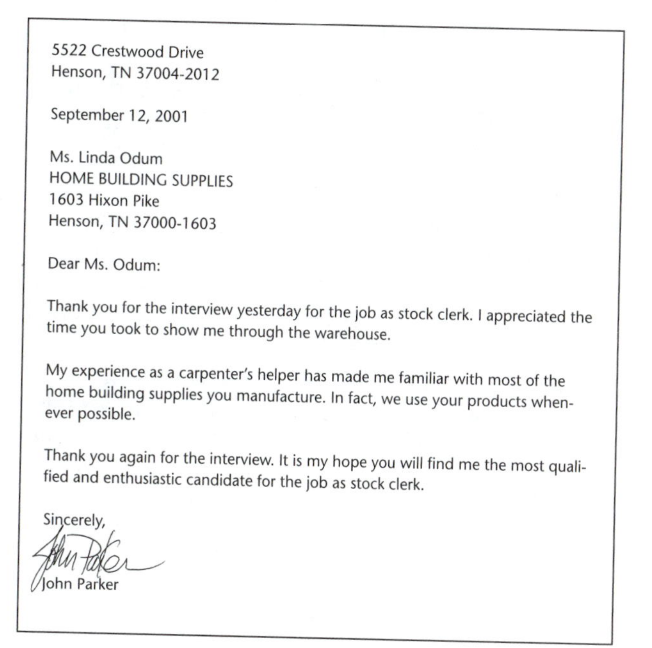
5522 Crestwood Drive
Henson, TN 37004-2012

September 12, 2001

Ms. Linda Odum
HOME BUILDING SUPPLIES
1603 Hixon Pike
Henson, TN 37000-1603

Dear Ms. Odum:

Thank you for the interview yesterday for the job as stock clerk. I appreciated the time you took to show me through the warehouse.

My experience as a carpenter's helper has made me familiar with most of the home building supplies you manufacture. In fact, we use your products whenever possible.

Thank you again for the interview. It is my hope you will find me the most qualified and enthusiastic candidate for the job as stock clerk.

Sincerely,

John Parker

Figure 21–16 Sample follow-up letter to send after an interview.

- Thanking the interviewer helps to build a courteous relationship.
- Having your letter in hand keeps your name in front of the interviewer.
- Taking time to write a letter tells the interviewer of your continued interest.
- The letter allows you to reinforce any key points you discussed during the interview.
- If you forgot to mention something important during the interview, you can put it in your follow-up letter.[42]

Step 3

Thank-you notes.
Suppose the interviewer told you that you would not be hired or that you are no longer interested in the job. Send a letter to thank the interviewer for considering you. Sending a thank-you note is a simple act of appreciation that hardly anyone ever does. They also have a practical benefit: People who receive them will remember you. Employers say they rarely get a thank-you note. They describe people who do send notes with positive terms, such as thoughtful, well-organized, and thorough. A thank-you note will not get you a job you are not qualified for, but it will impress people. When a job opens up, they will remember you. People in your job search network will also be more interested in helping you. If they know of an opening or meet someone who does, they will think of you.[43]

After completing these steps, wait and try to relax. Continue to pursue other job leads in the meantime. If you have not heard from the company, get in touch. You can do so sooner if the interviewer indicated that a decision would be made sooner.

National FFA Interview Career Development Event

The National FFA Organization has selected a new interview career development event sponsored by Tractor Supply Company (TSC). It includes the cover letter, resume, job application, interview, and follow-up letter. See Appendix I for the rules of the event and score cards.

Accepting or Rejecting a Job

You may or may not be offered the job during the interview. Usually, though, the employer makes a decision later. Employers like to interview several people for a job before making their selection.

Discuss Conditions of Employment

Job offers usually come by telephone. This gives the employer and the applicant a chance to discuss the details of the job offer. If the conditions of employment have not been discussed earlier, now is the time to ask about them. These include such things as working hours, salary, fringe benefits, and so on. You will want to know when you start work and if there is anything special that you need to bring or be prepared to do the first day. For example, you might need to pick up a uniform.[44]

What to Do if You Get Two Job Offers

It is possible to be considered for a job at different places at the same time. Let us say that you have been interviewed for jobs at both a department store and a restaurant. The department store offers you a job and you accept it. You should phone the restaurant and tell them you have taken another job.

Rejecting a Job

What if a company offers you a job you do not want? Be polite. You never know when you may be contacting them again. Give a brief explanation of your reasons. Regardless of your reasons, do not criticize the employer.

Dealing with Rejection

Not all interviews result in job offers. In fact, most of them probably will not. Dealing with rejection is something we all must learn to do. Being disappointed is normal, but do not react with anger toward an employer. By accepting rejection gracefully, you keep alive your chances for a future job. For example, what happens if the person chosen for the job turns it down? You may be next in line for it.[45]

Evaluating Job Offers

In all companies, employees come and go and new jobs open up. If you are good enough to have been invited for an interview, then you are qualified for a job. Do not get discouraged. Whether at that company or somewhere else, a job will open up for you. Finally, you get the job offer. What do you do?

In evaluating job offers, the first rule of thumb is never say "yes" immediately. Express your thanks and interest, but take a day or two to make sure it really is the best offer for you. Do the positives outweigh the negatives? Be sure to think qualitatively as well as quantitatively. For example, opportunities to do challenging work for good pay may outweigh the disadvantages of a long commute. Perhaps the money is a little low, but the opportunities for advancement and growth are great. Try to evaluate whether the long-term benefits outweigh the short-term difficulties. When all this is done, ask yourself: "Do I want to take this job?" Only you can determine the right answer to this question![46]

CONCLUSION

By properly selling yourself through your resume, job application, and interview, you will increase your chances of getting a job. The job search will be over. You will be leaving, or at least spending less time, in the familiar world of the classroom. The changes you will experience may be scary at first. Remember what it was like going from junior high to high school? A similar experience awaits you now. You are going from the known into the unknown. This can be exciting and frightening at the same time. You are going from high school into the world of work. By taking the time now to learn what to expect, you can prepare yourself for a smooth transition into your new role as a worker.[47]

SUMMARY

Once you have selected and located the type of job you want, it is time to get it. The qualities of motivation and persistence, so important in finding a job, are also qualities that make you a good employee, and employers recognize this. Getting a job consists of doing resumes, job applications, and interviews.

A good resume will help get you an interview. Write it yourself, make every word count, make it error free, make it look good, stress your accomplishments, be specific, do not delay, and keep it lively.

The resume will include personal data, job objective, skills (personal management, teamwork skills, academic, and technical), education, work experience, and references (optional). Many resumes are recorded on a computerized in-house applicant tracking system. Resumes can be prepared in a variety of ways, but remember, make your resume work for you.

Another way to act on a job lead is to write a letter of application. A letter of application is often known as a cover letter when it is mailed with a resume. A letter of application should have four parts: the reason for writing, to point out qualifications, to call attention to your resume, and to ask for an appointment. In developing your letter of application, be brief and specific, give attention to appearance, be assertive, use simple direct language and correct grammar, let your letter reflect your individuality, send it to someone by name, get it right, be clear about what you want, be friendly, target your letter, and follow up.

The most common screening procedure is to ask potential employees to fill out a job application form. Job application forms are often eliminated because they are messy, not enough experience is listed, and the last job paid more than the new one now open. Follow recommended procedures as you complete a job application form. Many people take a previously completed personal data sheet with them when filling out an application to prevent omitting any pertinent information.

By being granted an interview, you have already achieved some degree of success in the job search. An interview helps you sell yourself to the employer and provides you with a chance to further investigate the specific job and sponsoring company. To do your best in an interview, you must practice. There are 15 to 20 questions you can always expect during an interview. Practice answering these before the interview. Learn about the company, attend to appearance, collect materials needed for the interview, and check last-minute details before the actual interview takes place.

Interviews are doorways to jobs. When the interview begins, connect with the interviewer, smile, use the interviewer's name, and ask some open questions. Be aware of your body language, which includes posture, voice, eye contact, and distracting habits. Communicate effectively, answer questions, and ask questions. Conclude the interview by thanking the interviewer by name. Express interest in the job, arrange a reason and a time to call back, and say good-bye. After the interview, evaluate your performance and write a follow-up letter or a thank-you note. Evaluate the job offer and either accept it or reject it.

END-OF-CHAPTER EXERCISES

Review Questions

1. Define the Terms to Know.
2. List eight basic guidelines to help you prepare a successful resume.
3. What are the seven major parts of a resume?
4. List eight examples of personal management skills.
5. Explain what each of the four paragraphs should include in a letter of application.
6. What are 10 helpful hints to help you write a good letter of application?
7. What are three reasons your job application may be eliminated by an employer?
8. What is the purpose of an interview?
9. What are 15 questions that you will probably be asked in an interview?
10. What are five things that you need to take to the interview?
11. What are four things you could do during the first few minutes of an interview?
12. What four major areas need special attention during the actual interview?
13. List four questions an applicant could ask an interviewer.
14. What are four things to do as you conclude the interview?
15. What five things should a follow-up letter accomplish?
16. What should you do if you get two job offers?
17. Explain the procedure for evaluating a job offer.

Fill-in-the-Blank

1. At best, a resume will help you get an __________________ .
2. In preparing your resume and letter of application, __________________ is critical.
3. On a job application form, you should write __________________ if a question doesn't apply to you.
4. Before you write any answers, __________________ the job application carefully.
5. Study the __________________ so you will know what information to provide on the job application form.
6. It is a good idea to __________________ the information on a job application form.
7. Do not try to fit too much __________________ into a space on the job application form.
8. Every __________________ should be filled in on a job application form.
9. Find out as much as possible about the __________________ and the __________________ before your interview.
10. If more than a __________________ goes by between the time you made the appointment and the actual interview, call to confirm it.
11. Plan to arrive at the interviewer's office __________________ to __________________ minutes ahead of schedule.

Matching

_____ 1. Same as cover letter
_____ 2. Same as nonverbal communication
_____ 3. Lack of it during an interview can cause suspicion
_____ 4. Speaks with hand over mouth
_____ 5. Use when trying to improve interview techniques
_____ 6. Should project a high level of confidence
_____ 7. Complimenting during beginning of interview
_____ 8. Purpose is to obtain an interview with a prospective employer
_____ 9. Written to the interviewer thanking him or her for the interview
_____ 10. Gives the applicant an opportunity to present detailed information about himself or herself

A. videotape
B. linking with interviewer
C. eye contact
D. follow-up letter
E. letter of application
F. posture
G. distracting habit
H. cover letter
I. resume
J. body language

Activities

1. Write a resume using accurate facts about yourself. Type or have the resume typed or printed with a quality printer. Be neat and include all information that might help you get a job. Figure 21–4 is an example to follow.
2. Cut out a help wanted ad for a job in your occupational area from the classified ad section of the local paper. Write a letter of application to accompany the resume you prepared in activity 1. Use the example in Figure 21–9 as a guide.

3. A personal data sheet is useful in completing a job application and in developing a resume. It is also helpful to take the personal data sheet with you when filling out an application to prevent omitting any pertinent information. Complete a copied form provided by your teacher of Figure 21–11 for each item that applies to you.
4. Figure 21–12 is a form used for hiring entry-level workers for hourly employment. Complete the form provided by your teacher by filling in each item that applies to you.
5. Figure 21–13 is a typical form used in hiring carpenters, custodians, cooks, drivers, nurses, secretaries, shipping clerks, and various other civil service classifications. Complete the form provided by your teacher by filling in each item that applies to you.
6. Suppose you are an employer and you have 35 applicants for one position. Would you interview all applicants for the job? If not, how would you decide which individuals to interview?
7. If you were an employer, what questions would you ask an applicant during an interview?
8. Describe how to convince an employer to hire you for a job.
9. Ask three other classmates to volunteer to participate as a group. Select one person in the group to play the role of interviewer; other group members will play the roles of job applicants. Everyone in class should carefully observe the interviews. Following the interviews, discuss with the class the correct and incorrect behavior shown in the interviews. Part of the evaluation is appropriate dress.
10. Write a follow-up letter thanking the employer for an interview. Use any form you wish, or follow the format of Figure 21–16.
11. **Take it to the Net.**

    Explore resumes and interviewing on the Internet. The Internet is full of sites that can help you write your resume and prepare for an interview. Several of these sites have been listed below. Browse all the sites, and choose a couple that you find useful and summarize them. For additional information on resumes and interviews, use the search terms listed below.

    Web sites

    Resumes

    <http://www.l.umn.edu/ohr/ecep/resume>

    <http://www.eresumes.com/>

    <http://tbrnet.com/resumes/index.shtml>

    <http://jobsmart.org/tools/resume>

    Interviewing

    <http://tbrnet.com/interview/index.shtml>

    <http://www.sfsu.edu/%7Ecareer/planning/resume.html#interviewing>

    <http://www.smartbiz.com/sbs/arts/irish8.htm>

    Search Terms

    resumes

    interviewing

    writing a resume

    preparing for an interview

    interview questions

Notes

1. M. Gavin and G. Foster, *Preparing to Succeed in Marketing* (Austin, TX: The University of Texas at Austin Educational Resources Extension Instruction and Materials Center, 1993), p. 61.
2. J. M. Farr, *Getting the Job You Really Want* (Indianapolis, IN: JIST Works, 2000), p. 110.
3. *Employment in Agribusiness* (Stillwater, OK: MAVCC, 1992), p. 31.
4. Farr, *Getting the Job You Really Want*, p. 110.
5. R. A. Wolf and T. O. Wolf, *Job Hunter's Secret Handbook for Success* (St. Joseph, MI: WINSS, Winning International Secrets of Success, 1993), pp. 17–18.
6. Gavin and Foster, *Preparing to Succeed in Marketing*, p. 64.
7. Ibid.
8. Wolf and Wolf, *Job Hunter's Secret Handbook for Success*, pp. 17–18.
9. J. E. Hulbert, *Effective Communication for Today* (Cincinnati, OH: South-Western Publishing Co., 1991), pp. 443–445.
10. Wolf and Wolf, *Job Hunter's Secret Handbook for Success*, pp. 17–18.
11. H. Gieseking and P. Plawin, *30 Days to a Good Job* (New York: Simon & Schuster, 1994), pp. 81–82.
12. L. J. Bailey, *The Job Ahead: A Job Search Worktext* (Albany, NY: Delmar, 1992), pp. 39–40.
13. J. M. Kelly and R. Volz-Patton, *Career Skills* (Encino, CA: Glencoe Publishing Co., 1992), p. 130.
14. Hulbert, *Effective Communication for Today*, p. 452.
15. Wolf and Wolf, *Job Hunter's Secret Handbook for Success*, p. 46.
16. Farr, *Getting the Job You Really Want*, p. 45.
17. Bailey, *The Job Ahead: A Job Search Worktext*, p. 22.
18. Kelly and Volz-Patton, *Career Skills*, pp. 132–134.
19. *How to Complete a Job Application Form*, Skills for Success, No. 44 (Nashville, TN: Tennessee Department of Education, 1993).
20. Gavin and Foster, *Preparing to Succeed in Marketing*, p. 64.
21. Ibid.
22. Bailey, *The Job Ahead: A Job Search Worktext*, p. 52.
23. A. S. Hirsch, *VGM's Careers Checklist: 89 Proven Checklists to Help You Plan Your Career and Get Great Jobs* (Lincolnwood, IL: VGM Career Horizons: A Division of NTC Publishing Group, 1991), pp. 151–155.
24. Bailey, *The Job Ahead: A Job Search Worktext*, p. 53.
25. Ibid., p. 56.
26. Ibid.
27. Ibid.
28. Gieseking and Plawin, *30 Days to a Good Job*, p. 98.
29. Ibid.
30. Farr, *Getting the Job You Really Want*, pp. 85–86.
31. Wolf and Wolf, *Job Hunter's Secret Handbook for Success*, pp. 55–56.
32. Ibid.
33. Ibid.
34. Bailey, *The Job Ahead: A Job Search Worktext*, p. 68.
35. Farr, *Getting the Job You Really Want*, p. 87.
36. Bailey, *The Job Ahead: A Job Search Worktext*, p. 70.
37. Ibid.
38. Gieseking and Plawin, *30 Days to A Good Job*, p. 101.
39. Bailey, *The Job Ahead: A Job Search Worktext*, p. 71.
40. L. J. Bailey, *Working Skills for a New Age* (Albany, NY: Delmar, 1990), pp. 52–57.
41. L. A. Masters, *Finding and Holding a Job* (Cincinnati, OH: South-Western Publishing Co., 1992), p. 111.

42. Bailey, *The Job Ahead: A Job Search Worktext*, p. 72.
43. Farr, *Getting the Job You Really Want*, p. 89.
44. Bailey, *The Job Ahead: A Job Search Worktext*, p. 74.
45. Ibid.
46. Hirsch, *VGM's Careers Checklist: 89 Proven Checklists to Help You Plan Your Career and Get Great Jobs*, p. 167.
47. Bailey, *The Job Ahead: A Job Search Worktext*, p. 83.

Employability Skills: Keeping the Job

CHAPTER 22

Objectives

After completing this chapter, the student should be able to:

- Identify skills wanted by employers
- Discuss personal management skills needed by employees
- Discuss teamwork skills needed by employees
- Discuss academic and technical skills needed by employees
- Discuss the employability characteristics of a successful worker in the modern workplace
- Discuss employer and employee responsibilities
- Explain the importance of responding to authority
- Explain the importance of ethics in the workplace
- Explain how to get job promotions
- Explain the proper procedure for leaving a job
- Outline a complaint and appeal process
- Demonstrate knowledge of personal and occupational safety practices in the workplace

Terms to Know

academic skills
technical skills
employability skills
occupation-related skills
dependability
insubordination
memorandum
competence
technical knowledge
cooperative skills
gossip
compromise
trustworthiness
ambitious
capability
commitment
resignation
pride

Technology is changing at a rapid pace. The workplace is changing and the skills that employees must have in order to keep up are also changing. One thing that is not changing, however, is the fact that employers are looking for employees with leadership, a work ethic, and human relations skills.

You are developing these job skills each day of your life. The personal qualities and skills you develop today will follow you into the workplace. The kind of person you are today will affect the kind of employee you will become. Employers look for employees who have certain qualities and skills. Although **academic skills** and **technical skills** are important, employers want good, honest, hard-working people. Developing these qualities and skills will help you to be a success on the job.

Figure 22–1 Employability skills include hard work, dependability, sharing, listening, making decisions, communicating, and competing in a chosen occupation.

Skills that employers want can be divided into three major categories. Personal management skills, teamwork skills, and academic and technical skills are the foundation for gaining employability skills. Personal management skills include dependability, responsibility, setting and accomplishing goals, making decisions, honesty, and exercising self-control. Teamwork skills include organizing, planning, listening, sharing, and being flexible. Academic and technical skills include communicating, planning, understanding, problem solving, and competency in a chosen occupation. This chapter addresses these topics and others associated with **employability skills**, which must be learned to gain and maintain meaningful and productive employment. The intensity of the two workers in Figure 22–1 is a good indication that they possess many of the employability skills just mentioned.

SKILLS EMPLOYERS WANT

Some high school students find work immediately after graduation, but many others take much longer. Why? Although the educational degrees may be the same, other factors are at work. When employers were asked what they looked for in an applicant's resume, the top answer given was the person's record of achievement; related work experience was the next answer given. Ironically, further down the list was the candidate's educational achievement. The factors that influenced the final selection after an interview were level of enthusiasm, 31 percent; professional attitude, 25 percent; the right "chemistry," 23 percent; and experience level, 21 percent.[1] Another factor that nearly all prospective employers seek in candidates is individuals who have a vision of where they are going by practicing goal setting.

A partial listing of **occupation-related skills** and personal attributes that employers look for when they hire people follows. Nearly all of these factors are being observed by the interviewer in one way or another.[2]

Achiever—The internal drive to be up and doing, to get things done; energetic, competitive, on the fast track.

Activative—Presence, and the ability to have input.

Anticipative—Being able to predict consequences.

Attitude—You cannot succeed unless you think you can! Your attitude determines the outcome of any endeavor.

Commitment—To assist associates and staff in the success of the organization. You absolutely have to be committed to achieving your goals.

Command—The ability to take charge and speak out with authority. The propensity for adopting a definite stance.

Competitiveness—The desire to come out on top.

Courage—The capacity to increase one's determination in the face of resistance. The ability to carry out difficult assignments.

Credibility—Defines self as professional. Takes pride in the quality of work accomplished.

Dedication—The commitment that follows vision, empowers self, and enables others.

Dependable—Can be relied on and trusted in all situations.

Developer—The desire to help others grow and mature; the capacity for taking satisfaction from each increment of growth.

Discipline—The ability to self-structure your time and environment. A talent for getting things organized, which enhances efficiency.

Drive—The desire to accomplish, even when discouraged.

Empathy—Feelings for another person's position and needs. Being able to recognize how others think and feel about their problems.

Ethics—Having standards of right and wrong; the capacity to live by a set of principles, choosing what is morally correct.

Focus—The ability to choose a direction and maintain that direction or goal. A flair for identifying key priorities, targeting attention appropriately, and staying on track.

Ideation—The capacity to explain events; the ability to act as a problem solver and source of innovative ideas.

Knowledge—You must know something about what you are doing! Clear and certain mental perception; understanding of the problem.

Loyalty—Devotion for a person, group, or cause, which cannot be obtained by force.

Organization—System for establishing goals and objectives in a given time frame.

Responsible—Taking ownership of personal behavior, especially work and understanding how to help others feel similar about their work. The ability for seeing that directions and rules are followed.

Self-confidence—Belief in your ability; power; not afraid to make judgments; the desire to convince others of your ability.

Team player—To fit within a group of employees who set out to accomplish similar goals. Willing to do whatever is necessary for the success of the organization.

Values—Principles, ideals, and high standards of conduct.

If you are willing to adhere to the highest standards of business and professional conduct, can present yourself as accurate and honest, and convince the employer that you will conduct yourself on such a high plane, you will probably get a job if the employer is hiring. At least you'll be referred to someone who is looking for an employee with your skills. Then you can demonstrate the employability skills discussed in this chapter.

More and more, employers are looking for employees with not only specific or "technical" skills for a job but other skills, too. The Michigan Employability Skills Task Force—including leaders from business, labor, government, and education—determined the general skills that every student should have, not only for entry-level jobs but for jobs at all levels. These skills are shown in Figure 22–2, along with their subcategories, are personal management skills, teamwork skills, and academic and technical skills.[3]

Today, employers are stressing these skills because they are important in getting a job done. Besides high school training, technical skills generally can be learned through on-the-job training. However, the other skills, such as personal management, teamwork, and academic skills, will be required when hired.

PERSONAL MANAGEMENT SKILLS

These skills help a student become a responsible adult. They guide you in developing dependability and responsibility, setting and accomplishing goals, doing your best, making decisions, acting honestly, and exercising self-control. The following skills are all positive attributes that will make you a more desirable potential employee.

Attend Work Daily and on Time **Dependability** is a skill that is expected in the business world. The sooner you develop this skill, the easier it will be for you to be a mature, reliable person naturally without even thinking about it.[4] Being punctual should be every person's goal. Successful people are on time.

Meet Work Deadlines Meeting deadlines shows you are responsible enough to get work done on time. You are able to devise a plan and organize your work to complete a task on time. Meeting a

General Skills That Every Student Should Have

Personal Management Skills
- Attend school/work daily and on time
- Meet school/work deadlines
- Develop career plans
- Know personal strengths and weaknesses
- Demonstrate self-control
- Pay attention to details
- Follow written instructions and directions
- Follow verbal instructions and directions
- Work without supervision
- Learn new skills
- Identify and suggest new ways to get the job done

Teamwork Skills
- Actively participate in a group
- Know the group's rules and values
- Listen to other group members
- Express ideas to other group members
- Be sensitive to the group members' ideas and views
- Be willing to compromise if necessary to accomplish the goal
- Be a leader or a follower to accomplish the goal
- Work in changing settings and with people of differing backgrounds

Academic and Technical Skills
- Read and understand written materials
- Understand charts and graphs
- Understand basic math
- Use mathematics to solve problems
- Use research and library skills
- Use specialized knowledge and skills to get the job done
- Use tools and equipment
- Speak in the language in which business is conducted
- Write in the language in which business is conducted
- Use scientific method to solve problems

Figure 22–2 Personal management skills, teamwork skills, academic skills, and general skills every student should have for jobs at all levels. (Used with permission of the Michigan Educational Assessment Program.)

deadline may require working and coordinating work with others. Failure to make a deadline may interfere with someone else's work and may cause a breakdown in the flow of work. Failure to meet a deadline may lose a customer and that would mean loss of profit for the company. Failure to meet deadlines may cost you your job.

Develop Career Plans Career plans include setting goals and accomplishing them. Without goals, we have no focus or direction. In reality, happiness is determined by whether or not we achieve our goals. If we have no goals, how can we achieve happiness? Those who are successful in the workplace know where they are going and how to get there.

Know Personal Strengths and Weaknesses Do not accept a job if you do not have the skills, unless you know you will be trained for the job. It is best to look for a job in which you can use what you know so you can be successful and useful to your employer. Identifying your weaknesses will enable you to know where it is you need improvement. It is best to look for a job that fits your talents and abilities. One way to find out is through your school's testing program. Aptitude testing will help you identify your strengths and weaknesses. Your vocational counselor or guidance counselor can assist you with interpreting test results.

Demonstrate Self-Control Situations may arise on the job that displease you or upset you, but as an employee you will need to keep your composure. Keeping your composure requires thinking before you react. You must know when to listen and when to speak, when to act and when to pause, when to work and when to play. Thinking things out or delaying a response will help you cool down and give you time for an appropriate response. Good timing and common sense are everything when it comes to self-control. Unresolved issues must often be taken to a supervisor for a solution. Losing self-control with a supervisor is **insubordination** and may cost you your job. Most employers will give a reprimand for insubordination. All companies are different, but the first time it happens, you may receive a verbal warning, which will be noted in your employee file. The second time you may get a warning **memorandum**, a formal written communication, which is put into your employee file. The third time you may be suspended from work without pay for three to five days. The next time you may be terminated. In some companies, insubordination is grounds for immediate dismissal.

Pay Attention to Details Paying attention to details eliminates mistakes and prevents startovers. Correcting errors or starting over wastes company time and money. Doing a job right the first time is your responsibility. Being alert and asking questions will help if details are not clear. If you do not pay attention to details, you cannot produce the level of quality needed for the firm to compete in the global economy. Being observant of your surroundings and understanding how your job fits into the total organization helps you understand the importance of your job. Details of your job will then become more important and meaningful to you.

Follow Written Instructions and Directions Instructions are given for a reason: They are procedures developed for production and to get quality work done efficiently and on time. Not following instructions causes delays and slows progress from one stage to the next, eventually causing confusion and disrupting the process. Not following instructions may cost you your job.

Whether you are a construction worker, a technician, or a government worker, you must be capable of reading and following written instructions and directions. If you do not understand something, ask questions. That is how smart people got that way. There is no such thing as a stupid question.

Follow Verbal Instructions and Directions One of the most important personal management skills is the ability to communicate effectively. Most jobs require that you be able to communicate by speaking, listening, and writing. The way you communicate may influence whether or not you get a promotion. It also makes a difference in how you get along with your co-workers.

For many jobs, good communication is vital. For instance, salespeople need to be able to speak well to describe merchandise to customers. Waitresses need to be able to write down what their customers are ordering as well as talk and listen to them.

Listening skills are a very important part of communication. A good listener looks at the person who is talking, asks questions, avoids interrupting the speaker, and evaluates the message instead of the speaker. Do not let emotional words distract you, and try to control your wandering mind. Time can be wasted when you listen poorly. Employers want people who listen and follow instructions.

Oral instructions require the ability to listen and perform as requested. If an employer tells you to do something, you should follow through to the best of your ability until the task is completed. Following instructions, listening, and getting all the steps right the first time is important. If you are not sure, ask questions.

Some jobs require following detailed directions. Workers must pay close attention to the directions. Other jobs need few directions. Whether your job requires simple directions or complex directions, there are steps that can help you follow directions easily.

1. *Listen carefully.* When your employer or supervisor is giving you directions, pay close attention to what he or she is saying. Do not let your mind wander. Concentrate on getting all the directions.
2. *Write down the directions, if necessary.* You may be given a job that requires many steps to complete. When receiving lengthy directions, write them down. Do not depend on your memory. You may forget an important part of the directions.
3. *Make sure you understand the directions.* If you are not sure exactly what your supervisor means, ask questions. It is better to ask questions than to do the job wrong.
4. *Think before doing.* Go over the directions you have been given. Certain things have to be done in a certain order. Are there tools or materials you need to gather before starting? Break the directions down into steps. Decide which step should be first, then decide in what order the other steps should be done. Thinking through the steps may save you time and mistakes.
5. *Avoid shortcuts.* If your supervisor tells you to do the job a certain way, follow the directions you are given. You may know a quicker or easier way to do the job, but do not take shortcuts. Your supervisor has reasons for doing things a certain way.[5]

Work Without Supervision Working without supervision is being able to perform your job without the supervisor standing over you. Do not work to please your boss, work to please yourself, like the person in Figure 22–3. You earn the employer's trust because you show initiative, and

Figure 22–3 When a person is totally involved in his work, there is little need for supervision. (Courtesy of USDA.)

you know your job well enough to be on your own. Keep your mind on your work and do not waste company time. Stay with your job until finished. If you finish, look around for other things that need to be done or assist others if needed.

Learn New Skills It is important to know the skills needed to do your job well. However, employees also need to learn new skills to stay up to date with new technologies. Being willing to learn new skills and applying them to your work may help you keep your job.

Many employers train new employees and look for people who are willing to learn. Every company has its own way of doing things. As a new employee, you will need to learn the company's procedures and policies as well as your job tasks. An employee who has been on the job should be willing to learn more to do the job well and stay up to date. Your employer may ask you to change tasks and give you something else to do. Changes in the company structure may require you to learn new skills and responsibilities.

In today's workplace, learning is an integral part of everyday life. The skill of knowing how to learn, or learning to learn, is a must for every worker. The skill of learning to learn is the key to acquiring new skills, sharpening the ability to think through problems, and meeting challenges in the workplace.[6]

From the employer's perspective, the skill of knowing how to learn is cost effective because it can overcome the cost of retraining efforts. When workers use efficient learning strategies, they absorb, retain, and apply training more quickly, saving their employers money and time.[7]

Identify and Suggest New Ways to Get the Job Done Some jobs will grow, others will not. One thing is for certain: The skills required of workers will change over their working lives. Do not be afraid to learn, identify, or suggest new methods to help get the job done more effectively. Some companies offer bonuses or incentives for useful suggestions from workers.

Use Organizational Skills You will need organizational skills for every type of job. You must be able to organize your time, tasks, and belongings. To give yourself enough time to get all your work done, you will need to plan ahead. Breaking large tasks down into a series of smaller tasks makes jobs seem easier. For example, work supplies should be kept in order to keep you from wasting time looking for the tools you need.

Demonstrate Personal Values at Work True leaders are not afraid to demonstrate personal values in the form of examples in their everyday lives. Ask yourself: What are my values? Another way of asking this question is, in what do I really believe? What gives my life inner quality? For some people, financial success is what gives meaning to their work. For others, lending a helping hand to others less fortunate gives their efforts value.[8]

Teamwork Skills

Teamwork skills help you work effectively and efficiently within a group. They include organizing, planning, listening, sharing, flexibility, and leadership.

Teams are collections of people who must rely on group collaboration if each member is to experience optimal success and goal achievement.[9] Teamwork is a managed, planned, systematic coordination of effort by a group with a common goal in an optimally productive way.[10] Teamwork is managed because someone officially or informally exercises control. Teamwork is planned because it results from preparation and organization. In the workplace, teams are organized so that individuals' talents and skills can be directed through group effort to the accomplishment of vital tasks and goals. Here are eight aspects of teamwork.

1. *Actively participate in a group.* To gain the experience necessary to be effective in the workplace, you should start out by being

active in school or community groups like the FFA, VICA, band, athletic team, yearbook staff, youth groups, choir, and others. By participating in these kinds of groups, you practice the teamwork skills necessary to further your success in the workplace. You also learn how to organize and plan group activities.

2. *Follow the group's rules and values.* When you get involved in a group, you must become familiar with the group's rules and values. A good example of this is in basketball, in which you must not touch a person who is shooting the ball. If you do not follow this basic rule, your team will be heavily penalized, possibly causing you to lose the game. In business, just as in basketball, one must obey the rules of good business ethics and values. The alternative means suffering possible lawsuits, fines, loss of business, or the possibility of losing your job.[11]
3. *Listen to other group members.* The importance of listening has been mentioned several times previously. It is a virtue that needs to be cultivated. A young teacher observed a brilliant professor during a group meeting. He did not enter into the group discussion. Afterward, the teacher asked the professor, "Why didn't you join the discussion?" The professor answered somewhat in jest, "I know what I know, and if I listen to others to know what they know, I will become even smarter."
4. *Express ideas to other group members.* Although listening is a virtue, there is a time to speak. Your opinions are just as valid as the other person's. Be sure your ideas have been thought through before you speak. Do not just make contributions to hear yourself talk, or speak to feed your insecurities. The exchange of ideas and concepts in a group is a tremendous way of pulling together thoughts to help solve a problem (Figure 22–4).
5. *Be sensitive to group members' ideas and views.* Everyone has feelings. In a group discussion, frail egos may be open to abuse if handled the wrong way. When people share their innermost feelings or beliefs, they are allowing you to come into an area of their private emotional world. This is a trust that should not be taken lightly. Be careful to earn their trust by being kind and sensitive to their feelings.[12]

Figure 22–4 Although listening is a virtue, there is a time to speak. The young woman above is expressing her ideas to the group.

6. *Be willing to compromise in order to follow democratic majority rule.* Do not be stubborn. When a group is working, there will be times when members must bend to an opposing opinion to make the group successful in its goal. If the general consensus is the opposite of your view, state your point clearly and allow the group to judge. Work as a team! Majority rule is usually the best policy.[13]
7. *Be a leader or follower to accomplish the goal.* There are times to lead and times to follow. Knowing the right time to do one or the other is the mark of a leader. If you know more about an issue than other members of the group, you should take the initiative and show leadership. If not, allow another person to take the lead. Be your best as a leader or follower to accomplish the team's goal.
8. *Work in changing settings and with different people.* As we enter the twenty-first century, we are becoming a more diverse society. This

means that we will be working in different environments with a variety of people. We must be willing to change and experience new things. We must meet new people and gain knowledge. When your knowledge increases, your self-concept increases, and you become worth more to your employer.

ACADEMIC AND TECHNICAL SKILLS

The inability of large numbers of new work force entrants to meet the reading, writing, computational, and technical standards required by many segments of American business is fast becoming an economic and competitive issue for U.S. companies challenged by foreign enterprises.[14]

Reading has historically been considered the fundamental vocational skill. A person must be able to read to find out about available jobs, to get a job, to keep a job, to get ahead in a job, and to change jobs.

Writing skills today are important in almost every occupational field. Employers first judge a potential employee's writing ability by the quality of an application, cover letter, or resume.

The quality of letters, memorandums, progress reports, work orders, requisitions, recommendations, and instructions is regarded as an indicator of the overall quality of an employee's work.[15]

The third competence skill, computation, is no less important than the first two. There is no question that employers today are focusing more and more on an employee's ability to compute at increasing levels of sophistication.[16] The reason for this is simple: technology requires it.

Of course, an important requirement for job success is competence in occupation-related skills. Occupation-related skills are the things you need to know or do in a particular occupation. For example, an administrative assistant might need skills in typing and shorthand. **Competence** is the ability to do something well. Successful workers are the ones who do the best they can to become competent at the job. They work to improve skills and learn new skills to increase competence in occupation-related skills.[17] Ten academic and technical skills are briefly discussed.

1. *Read and comprehend written materials.* One of the first and most important skills you must obtain is the ability to read and understand what you have read. Before you continue on the road to success, you must have mastered reading as well as comprehension. Successful people spend much time reading and understanding what they have read. The more you read, the smarter you get.
2. *Interpret charts and graphs.* In today's high-tech world, some basic tools that have been around for a long time are still being used effectively to make a point. These tools, known as charts and graphs, use statistics in a graphic form to convey information easily. Displaying the information in this form enables the reader to understand the data at a glance.[18]
3. *Calculate basic math.* The ability to do simple math is a basic part of life and work. Math is used to solve problems. Cashiers and bank tellers need to be able to make change. This includes addition, subtraction, multiplication, and division. More difficult math may be used in certain jobs. You should be able to add a column of numbers in your head, have the multiplication tables one through 10 memorized, and work simple division in your head without using pencil and paper.
4. *Compute to solve problems.* Many of today's jobs require that you be able to compute cost, time, volume, percentages, and fractions. A worker should be able to add; subtract; multiply; divide whole numbers; use fractions and decimals; and calculate simple interest. You should also be able to calculate time, volume, area, weight, and distance (Figure 22–5).[19] Truck drivers calculate their mileage and time spent on the road.

Figure 22–5 Many jobs, such as that of pilot, require math and problem-solving skills. Numbers, figures, and calculations are all part of being a pilot.

Figure 22–6 Certain jobs require the use of specialized equipment. This student is gaining experience in the workplace while attending high school. (Courtesy of USDA.)

Machinists and carpenters must be able to calculate measurements.

5. *Use research and library skills.* Do not reinvent the wheel; much of what you are planning to do has already been done. Save yourself time by developing good research and library skills. Ask a librarian to assist you. In some cases, the source of information once located can be downloaded from the computer or secured from another library through an interlibrary loan. If you are having trouble locating information, ask for help.
6. *Use tools and equipment.* The tools and equipment of the modern office include computers, calculators, word processors, fax machines, telephones, and printers. Almost every business has its own high-tech tools and equipment in our highly specialized world. No matter what field you choose, you will have to learn how to use special tools and equipment. The sooner you can start to familiarize yourself with the tools of your trade, the further ahead you will be when you start your job (Figure 22–6).[20]
7. *Speak the language in which business is conducted.* Every business has its own vocabulary. Many companies use acronyms that are part of the daily terminology. You must learn the language in order to be successful. Some examples of acronyms are USDA (United States Department of Agriculture), ERIC (Educational Resource Information Consortium), DACUM (Developing a Curriculum), NASA (National Aeronautics and Space Administration), and NHL (National Hockey League).
8. *Write the language in which business is conducted.* In the workplace, formal communications are written. For example, employees receive important information through memorandums, handbooks, and manuals. Employees are expected to read and practice these instrtuctions. Workers are also asked to fill out forms; write notes, letters, and reports; and keep records. Skillful written communications enable your co-workers and supervisor to understand you.
9. *Use the scientific method to solve problems.* Problem solving and decision making are discussed at length in Chapter 13. It is discussed only briefly here to show how problem solving fits into the workplace. Problems arise on the job and, as an employee, you should be able to solve problems you encounter. Being

able to figure out the solution to a problem will make your work easier and make you more productive. In the workplace, creative thinking is generally expressed through the process of creative problem solving. Employers want employees to solve problems, to find workable solutions. If employees are involved in problem solving, they will generally feel better about their work because they were involved in the solution. The scientific method (discussed in Chapter 13) is a very effective procedure used to solve problems.

10. *Use specialized knowledge to get a job done.* Almost every job requires a certain degree of specialized knowledge. Keeping up with the latest developments in an occupation increases a worker's specialized and **technical knowledge**. Technical knowledge includes everything you need to know to do your job. It is the "why" and "how" for doing things. In our rapidly changing society, successful workers are the ones who keep up with innovations in their occupations. Successful workers keep learning to improve technical knowledge. They read and talk to people who are currently doing what they want to do.[21]

Employability Characteristics of Successful Workers

Employers look for certain qualities when hiring new employees. Understanding these qualities and their importance is a step toward job success. L. A. Liddell's book, *Building Life Skills*, addresses several things that help make a successful employee and enhance relationships. These qualities are a positive attitude, cooperation, dependability, trustworthiness, working hard, respecting others, handling criticism, exhibiting appropriate dress and grooming, showing initiative, and being diligent.

Positive Attitude

Your attitude will affect your job success. Employers look for a positive attitude when hiring and promoting workers. They want workers who can get along with others. Employers tend to hire people with friendly, outgoing personalities, especially those who must work with customers or supervise others. Your work attitude shows how you feel about your job and your co-workers. If you work hard and get along with others, you are showing a positive work attitude. If you show up late, complain, or do poor work, you are displaying a negative work attitude.

Displaying positive work attitudes on the job is very important. Your job success depends on it. Displaying the following attitudes will make your employer and fellow employees respond to you positively.[22]

- *Show enthusiasm and pride.* Be happy and positive about the work you do and the place where you work. Enthusiasm helps you tackle big problems without getting discouraged. Like any attitude, enthusiasm is contagious. If you have enthusiasm at work, you may spread it to other workers.
- *Be cheerful.* Look on the bright side. Show your co-workers and employer that you are happy to be at work and glad to do the job. Smile.
- *Be dependable.* Always come to work on time. Do what you are paid to do. Carry your own weight. Be willing to help other workers when they need help.
- *Be willing to learn.* Nobody knows everything. If you are asked to learn a new skill, look at it as a way to increase your chances for success.
- *Show initiative.* Be ready and willing to learn new skills and to help others. Do not wait for someone to tell you what to do; find work that you can do and do it without being told.
- *Use self-control.* Show patience and tolerance for your co-workers. Do not get outwardly upset when you are asked to do an unpleasant task. Do not insist on having things your way all the time.

- *Be cooperative.* Learn to work with others as a team; do not try to compete with co-workers.
- *Accept criticism.* Accept criticism and use it to help you do a better job. Be open to suggestions from others. Do not think of criticism as negative.

If you have a negative attitude, you can change it. It is not easy to change an attitude, but it is possible. You are the only person who can change your attitude. If you want to develop a positive general attitude, however, you must put some effort into it. If you need help in changing your attitude, follow these suggestions.

- *Think positive.* If you have a negative attitude about something, look for something about it to like. Try to find the good points instead of the bad points. If you have a bad attitude about school, think of the good things you get from school. It may be hard at first to think of good things. Say something positive about school to your friends every day. Soon you may find yourself liking school!
- *Avoid negative influences.* Do not let people with negative attitudes spread their attitudes to you. Although negative attitudes can be spread, you can control your own attitude. Protect your attitude against the negative influence of others. This may mean avoiding people who complain or make you feel negative.
- *Look for examples.* Think of people you know who have positive attitudes. Watch what they do and say that shows this attitude. Try doing and saying some of these things; if people respond to you in a friendly way, you have shown a positive attitude.
- *Find positive influences.* Make friends with people who have positive attitudes. Spend time with them. They will help you feel good about yourself.
- *Respect yourself.* Think of your good points. Think of yourself in positive ways. You have a lot going for yourself. You are fortunate to be you. Do not get mad at yourself when failures occur; everyone has setbacks. If you blame yourself too much, you will lack the self-confidence to succeed the next time. Learn from your mistakes and think positively about the future.[23]

Be Cooperative

Most jobs require you to work with other people. To get the work done, you must cooperate with these people. Developing good relations with co-workers makes your work more pleasant and enjoyable, and you get more work done. No matter how good you are at your job, you must get along with your co-workers to be successful. Some people are easy to get along with; others are impossible to please. When interacting with people, figure out ways to get along with them. **Cooperative skills** are ways of getting along with people. These skills help you get the most out of your interactions with others. There are basic things you can do to help you develop good cooperative skills with co-workers.[24]

- *Accept differences in others.* People have different lifestyles and different values. Respect the way others choose to live and behave, within reason. Remember, your lifestyle and values may seem strange to others. Everybody is different. Do not judge others by your standards. If you accept others as they are, they will tend to accept you.
- *Avoid assumptions.* Do not make judgments about others unless you know the facts.
- *Do your share of the work.* As a beginner, you will be taught how to do the job. Co-workers usually are happy to help you get started. Develop a good working relationship by listening carefully to instructions and asking questions related directly to the work you are doing. Once you have learned the job, do not expect others to help you get your work done. Co-workers will respect you if you do your share of the work.
- *Check your appearance.* The first thing others notice about you is your appearance. Create a good first impression by dressing appropri-

ately. Personal grooming is also important. Do your best to maintain a good appearance.

- *Have a good attitude.* People with positive attitudes are friendly and pleasant. Those with negative attitudes complain and become unpleasant. If you have a positive attitude, people will like being around you and working with you. This attitude shows that you care about your job and your co-workers.
- *Avoid gossip and disputes.* **Gossip** is talking about other people. Gossiping can hurt your co-workers and cause anger. You should avoid gossip if you want to have good relations with co-workers. If two or more co-workers have a dispute, stay out of it. Taking sides in a dispute may damage your working relationship with co-workers.
- *Control your emotions.* There may be times you feel hurt or angry with co-workers. Everybody has these emotions. You must learn not to react to these emotions on the job. If you cannot control your emotions, co-workers will avoid you.
- *Learn to compromise.* You cannot have things your way all the time. When you disagree with a co-worker, learn to **compromise**. You compromise when both workers give a little to come to an agreement. Listen to what your co-workers have to say, think about their reasoning, and decide how much you are willing to change. Nothing can be done unless you and your co-worker agree to compromise.
- *Be considerate and sensitive.* Treat other people as you would like to be treated. Consider their feelings and their points of view.
- *Choose your words carefully.* There are many ways of saying the same thing. Be clear about what you mean, but use tact to get your message across and stay positive. Think about how the other person will react to what you have to say.
- *Check your own behavior.* Is it possible that poor human relations with another person may be your fault? Have you created a problem with this person? Sometimes you can say or do things that hurt or offend others. Look at your behavior from the viewpoint of others. Is there room for improvement?

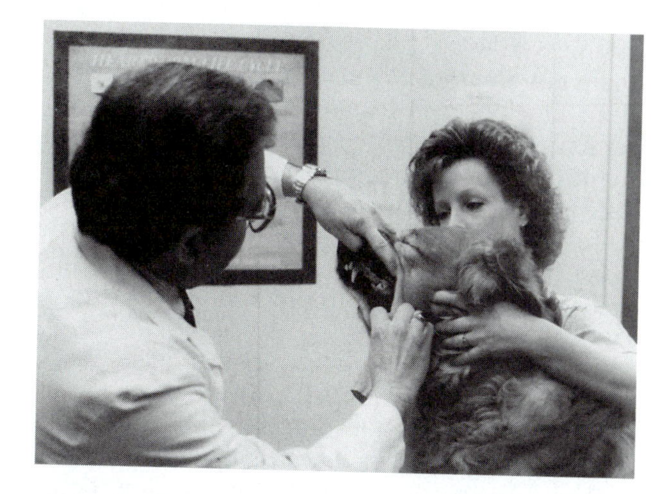

Figure 22–7 Cooperation with employers, employees, and people outside your job is necessary for success. (Photo courtesy of Brian Yacur.)

The two workers in Figure 22–7 are exhibiting cooperative skills in their professional area.

Be Dependable

As an employee, you are most likely to succeed if you show dependability. Your employer needs to depend on you to get your work done. You should not take too many breaks or expect others to do your work for you. Perform your job to the best of your ability.

Part of being dependable is being punctual, which means you are on time to work and appointments. A good employee is on time every time. Employers want to know that you will be at work when you are supposed to be because they are depending on you to get a job done. If you are not at work, you cannot complete the tasks you are assigned. If you are punctual, your boss can depend on you to be there.[25]

Reliability is also a part of being dependable. Employers will not accept many excuses for tardiness to work. Exceptions such as a flat tire or a late bus are understandable. Forgetting to set an alarm clock or oversleeping is not. Some employees have the mistaken belief that their jobs are insignificant, so their absence will not make much difference to the company. Being absent from the job causes hardship for others who must do your work. A good employee understands the importance of individual jobs to the smooth working of a company.[26]

Before you are hired, most employers check into your background. What you have done in the past is a good indication of what you will do in the future. If you are a punctual and reliable student, you are likely to be a punctual and reliable employee. You will probably have a similar record at work. Being punctual and reliable will help you be a success on the job. Your reputation precedes you.

Be Trustworthy

Another quality you will need to be a good employee is **trustworthiness**. Trustworthy people are honest, and honesty is important to every employer. Employers want to know that you will tell the truth.[27] For instance, a cashier must be trusted with the money in the cash register (Figure 22–8). Your boss will feel comfortable trusting you with the money if you are an honest person.

Trustworthiness also means doing your job and doing it right. Wasting time and materials costs money. The job should be done right if you are given proper instruction. Not having to go back and check on your work to see if you did it right will earn the employer's trust. Being honest and earning your employer's trust is a great asset in an employee.

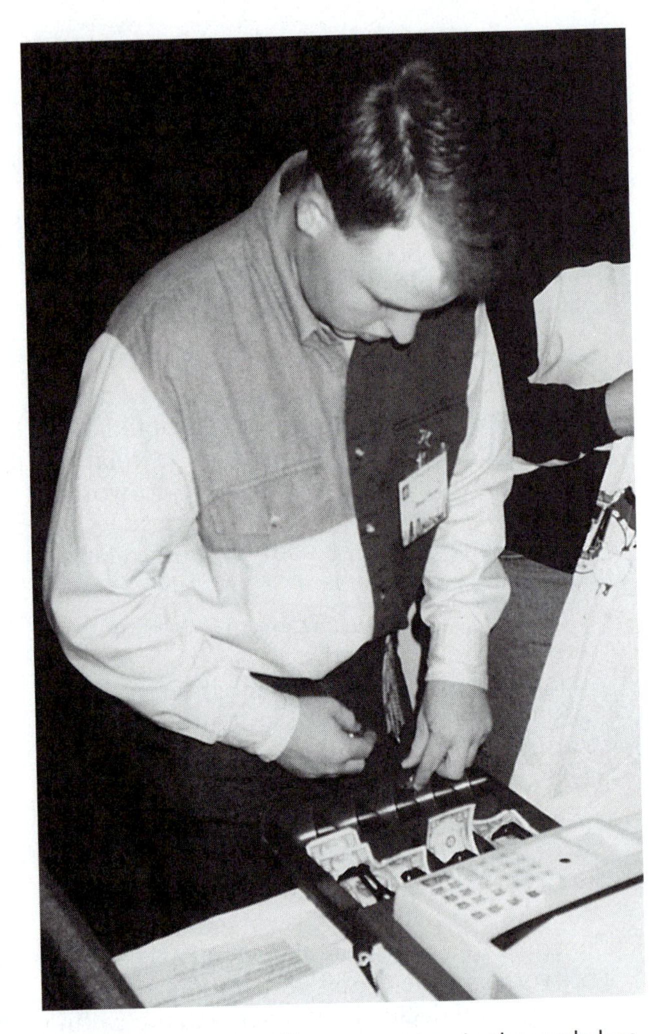

Figure 22–8 Many skills are necessary in the workplace, but none is more basic than honesty.

Work Hard

The way for you to develop the habit of working hard now is always to do your best on school assignments, being willing to put the time and effort needed into every assignment. This includes both the classes you like and the classes that are not your favorites.[28] Try to be neat and make as few errors as possible.

Fulfilling your responsibilities as a family member can help you develop a willingness to work. Being willing to help with duties around the house and other assigned tasks shows that you are **ambitious**. In other words, being alert, energetic, showing a willingness to work, and caring about what you do helps to develop a mindset that will carry through to your job.

Respect Others

To be a successful employee you must be aware that all people want to be treated well. You should show respect for others. Respect involves simple, common courtesy, staying within certain boundaries, and following the directions you are given. It involves being aware of the rights of all people with whom you work,[29] including not interfering with others while they do their jobs. Respecting others means treating others as you want to be treated.

Bosses in particular should be treated with respect. The respect you owe your boss is much the same as the respect you owe your teachers. Your teachers and boss give you direction and help you learn. You respect them for their power, experience, knowledge, and the help they can give you in learning.[30] Learning who has authority and respecting their position will be a valuable asset in your job success.

Handle Criticism

Handling criticism is difficult; no one likes to be wrong or make mistakes. Take criticism constructively instead of personally, try to keep an open mind, and always be willing to make improvements and learn from mistakes. Employers want to help you be successful because your success benefits the company.

Knowing how to give criticism can also be helpful in the workplace. For instance, you may be asked to help a new employee learn how to do a job. When you criticize this person, you want to be tactful. You want your comments to help your fellow employee do a better job. You do not want to hurt his or her feelings. Feeling empathy for the co-worker may help, so put yourself in his or her shoes. Try to say something positive before you say what or how the employee needs to improve.[31]

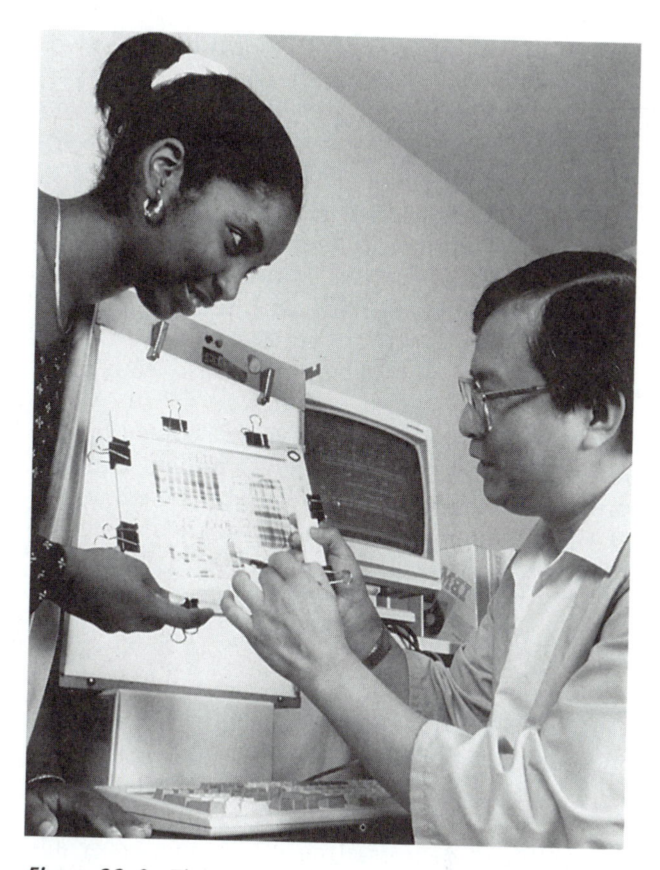

Figure 22–9 These co-workers are appropriately dressed, well groomed, and wearing smiles. (Courtesy of USDA.)

Exhibit Appropriate Dress and Good Grooming

Workers who are well groomed are likely to have a positive effect on other people at work (Figure 22–9). Well-groomed workers are clean, use good personal hygiene, and wear clean clothes that are suitable for the job. Presenting a good appearance makes you feel better about yourself and causes others to have a more positive attitude toward you.

Show Initiative

Showing initiative is being the first to start or introduce something. Good employees find something else that needs to be done when they complete a given task, and then they do it. A good employee does not wait around to be told what to do or when to do the job. Rather, the employee knows the job well and performs tasks without supervision. The employee looks for ways to improve performance and cares about the operations of the workplace. Showing initiative requires being alert, working with urgency to get a job done, and looking around for ways to assist others if needed. Good employees also learn more than what they were required to learn. Many companies encourage employees to come up with ideas that will save time, save money, and improve productivity. Employees who are successful at their jobs may be rewarded with bonuses or promoted to jobs with greater responsibility.

Be Diligent

Most employers want employees who will stay with a job until completion. Staying late to get a job completed may sometimes be needed; this is called going beyond the call of duty. Diligence means sticking with an assignment no matter how difficult or frustrating it becomes, especially if it requires your continuous attention. A good employee does not quit in the middle of a project but sees it through to completion.

Be Capable

One of your major responsibilities as an employee is to do your job to the best of your ability. The employer hired you because he or she feels you are capable of doing your job. Your **capability**, or your potential for doing the job, is important to the employer. The way your job affects the work of others in the workplace ultimately affects the success of the business as a whole.[32]

When you begin your job, follow directions and do your work exactly as you are told. You may not understand the reason for doing each task in a certain way, but your employer does have a reason for giving you those directions. After you have shown your ability, you may be able to suggest other ways of doing certain tasks, which indicates your capability as well as your initiative.[33]

Show Commitment

Similar to dependability and diligence is **commitment**. You can be counted on to attend meetings after hours or on weekends. If there is a special project, you will be there to do your best. Commitment is especially appreciated in community organizations. Nothing is more frustrating than planning an event and having people not follow through on their commitments. Commitment means really staying with a task when the going gets tough; it is not supporting the team only when the team is winning.

Employer and Employee Responsibilities

Employers Must Meet Employee Needs

The employer must attempt to fulfill the needs of every employee. In most cases, one employee needs more supervision in one area than another. The employer must lead the employees, give them responsibility, and create a feeling of pride. Employees must have a good understanding of working conditions. Motivated employees who have faith and trust in their employer can make the difference between success and failure. An employer can convey trust by delegating authority and by giving employees an opportunity to advance. Employers should take an interest in the well-being of their employees and recognize achievements. Some other aspects that are important to a good employer-employee relationship are the following.

- Correcting without embarrassing the employee
- Sharing unpleasant tasks
- Anticipating needs and problems
- Asking for feedback

Employees Should Perform to the Limits of Their Abilities

Employees should take an active role in maintaining a good relationship with the employer, recognizing that both employee and employer are part of a team. Employers expect employees to be dependable and loyal to the organization. Employees also must respect themselves and the employer. Cooperation with the employer and with co-workers is also very important. Employees cannot always expect the employer to know there is a problem. Employees should be able to discuss conflicts and should expect feedback. A positive attitude can improve every situation, and it makes for a productive work environment. A good employee shows a number of traits and qualities, which are discussed below.

Honesty

Honesty is more than just not taking things that do not belong to you. It means giving a fair day's work for a fair day's pay. It means holding up your end of the bargain. An employee's thoughts and actions should be honest. Nothing brings quicker dismissal or surer disgrace for an employee than dishonesty.

Loyalty

Loyalty means supporting a person or an organization, and it is a two-way proposition. Employees expect employers to protect their interests, to provide steady work, and to promote them to better positions as openings occur. The employer has a right to expect employees to support the company, to protect its interests, and to speak well of the firm to others. They also expect employees to confine minor problems to the workplace and to keep business matters confidential.

Willingness to Learn

Production procedures and services differ from business to business. Employees are expected to learn each company's methods and procedures. Due to the installation of new machinery and equipment, it is often necessary for experienced employees to learn new skills and operating procedures. Employees given a promotion or greater responsibilities are usually the ones who learned new skills and became more valuable to the firm.

Willingness to Assume Responsibility

Most employers expect employees to recognize the task at hand and then do it. It is tiring to have to ask repeatedly about the progress of an assigned task. When an employee receives instructions about the performance of a task, he or she is expected to follow directions. After the employer delegates responsibility, the employee should assume the responsibility and perform the duties without further instruction.

Respect for Authority

Authority is the responsibility to give commands, act, and make final decisions. Authority in agribusiness is usually entrusted to managerial personnel who are responsible for production, sales, or service. Even though many agribusinesses are relatively small and uncomplicated, employee respect for those in management positions is important.

Ability to Get Along with People

Being able to get along with others is the most highly prized of all employee traits. This ability seems natural for some people, whereas others gain the ability through thought and training. One must learn to think of people as individuals, to view a situation from another person's viewpoint, and sincerely want to get along with others. The ability to get along well with people pays big dividends in the form of profits. Employers appreciate good employee attitudes and temperaments.

Willingness to Cooperate

One way to exhibit a cooperative attitude is to offer to assist other employees after completing one's assigned tasks. In most firms there are certain tasks no one likes to do. Employees who are willing to do these tasks make a good impression on the supervisor and gain the cooperation of other workers. Employees should work as team members alongside supervisors and co-workers.

Rules and Regulations

People cannot work well together unless they understand the work. This includes knowing which tasks to do, the proper procedure for completing the tasks, the time required for completion, and who does the tasks. In any work situation, rules and regulations are a necessity.

Tardiness and Absenteeism

Since all people are creatures of habit, there is a tendency to continue doing the same thing unless the consequences are unpleasant. The habit of being late may have begun in early school days, when it was difficult to get up when called. This habit will get a person into trouble on the job, as well as at school. Clocks govern both business and industry. It is the employee's responsibility to get enough rest to be ready for the next day's work and to be at work on time. Failure to do so may prevent other employees from starting on time.[34]

Occasionally everyone has to take time off from work. Very few employers expect people to work when they are sick or when they are having a serious family problem. However, it is possible to get into the habit of letting unimportant matters keep one from work. This results in lost production and imposes a hardship on those trying to get the job done with a limited number of employees.

Employees must always notify an employer in advance of a planned absence. An employee is negligent if he or she does not notify the employer in case of illness or an important matter that keeps him or her from work. If circumstances prevent this, the employee should notify the employer as soon as possible. Failure to contact the employer is the worst way to handle an absence. It leaves the employer not knowing if the employee is merely late or whether someone else needs to do the work.

RESPOND TO AUTHORITY

No matter how much you enjoy doing your job, there will be times when you are asked to do tasks or duties you do not like. Your employers will expect you to do the task even if they know you do not want to do it. They expect you to accept your share of the work and do the job to the best of your ability. How you react to employers who assign you these tasks will affect your future. This is called responding to authority.

Appropriate Response to Authority

Your supervisor, or boss, is the person of authority who has the responsibility of seeing that the work gets done. To do this, your boss must decide which workers will do the various tasks needed to get the work done. Do not complain or try to have the job assigned to someone else. Show a willingness to cooperate with your supervisor and co-workers.

Not all bosses are kind, fair, and understanding. You may enjoy the job but dislike the way you are told to do your job. Your boss may shout at you or give impersonal orders. Developing a cooperative working relationship with this authority figure may be difficult. Try to find ways to get along with this type of person: Follow the orders you are given exactly; smile and show the boss you are willing to do the tasks. Remember, the relationship you develop with your immediate supervisor is important to your job satisfaction and advancement.

When You Disagree with Your Supervisor

There may be times when you disagree with an authority figure. Before acting on your feelings, consider the situation: If you are a beginner, you may be asked to do tasks that no one else wants to do. In this situation, the best thing to do is to accept the tasks you are assigned. Before long, you may work your way up to another position, and another beginner will be given the unpleasant tasks.

Figure 22–10 When you disagree with your employer or supervisor, wait until work hours are over and ask for a private meeting. Talk tactfully and calmly, as these two people are doing.

If you look at a situation carefully and believe you are being treated unfairly, there is a way to take care of the problem. Do the tasks as they are assigned to you during your working hours. When your work hours are over, ask to speak privately with your supervisor (Figure 22–10). Do not get angry or abusive. Explain your feelings about the situation and try to work out a solution.

Always remember that the company pays your salary and expects you to do the tasks you are assigned. Every job has unpleasant tasks. Do all the tasks you are assigned to the best of your ability. Your response to your supervisor's authority can lead to success or failure on the job.[35]

Ethics in the Workplace

In Chapter 19, we discussed ethics in the workplace. Remember, ethics refer to your standards and values—they are the basic principles you live by. Your concept of what is right or moral and what is not affects your conduct at work, at home, and in the community. Your code of ethics determines how you behave in different situations. To act ethically is to choose what is right. People who stick to their personal standards have integrity. If you choose to do what you know is wrong, you are behaving in an unethical way. You must be willing to accept responsibility for your choices. This is true at work and in other aspects of your life.

The Importance of Good Business Ethics

Your personal ethics and your employer's ethics are reflected in the community. A good worker provides worthy service for value received; otherwise, customers will take their business to someone else. For example, an auto mechanic with good ethical standards determines what is wrong with a car, makes the necessary repairs, and charges a fair price. The customer can depend on this auto mechanic and will return to the auto mechanic if other repairs are needed, knowing that the mechanic treats all customers fairly. An unethical auto mechanic may make unnecessary car repairs and overcharge the customer. Customers are often aware they have paid too much for the repairs. As a result, they will take the car to another auto mechanic the next time it needs repairs. An auto mechanic who continues to make unnecessary repairs will have a reputation for unethical business practices and will soon have no customers. Customers like to be treated fairly. If your work reflects a high ethical code, customers will keep coming back. Following are ways you can show a high ethical standard on the job.

- *Be honest.* If you sell a product, make sure it does exactly what you say it will do. Do not say it is better than it really is just to make a sale. Put yourself in the customer's place. Would you be satisfied with the product?
- *Provide worthy service.* When providing a service, do the best possible job you can. If you install air conditioners, make sure everything is in working order before leaving the job. Do not overcharge for your time or services.
- *Have a positive attitude.* More than anything else, customers respond to a warm, friendly,

and helpful worker. Although some customers are hard to like, treat them as you would want to be treated.[36]

JOB PROMOTIONS

Getting a job is the beginning of your career, not the end. One measure of success is your ability to improve your skills. A more obvious measure of success is job promotion. A promotion is an employer's way of indicating that a worker can assume increased responsibility. There are several things an employer looks for when deciding which workers to promote.

- *Job mastery.* Job mastery, or knowing how to do your job, is an important consideration in getting a promotion. Your employer will notice your skill at doing your job and your willingness to learn new skills.

  Mastering your job means your work reflects both quality and quantity. In order to be promoted you may need additional training and education to do the job. You may be able to get the training you need from the company in which you work, but you may have to attend evening classes for training. The successful worker never stops learning. Learning new skills leads to promotions.[37]
- *Initiative.* Initiative means doing tasks without being told. Initiative indicates a willingness to accept responsibility. If you show initiative on the job, your employer knows you can get the work done without much supervision.[38]
- *Good employability skills.* Along with initiative, if we took all the positive traits of an employee in this chapter and gave each a ranking of one to five, with five being the best, you would need to rank at least a four on most of them to get promoted. You need to be a good follower before you can become a good leader.
- *Understands employer or supervisor responsibilities.* If your promotion means getting into a management or supervisory position, you should learn that managers also have certain responsibilities. Sometimes these obligations are quite complex and carefully written out. The sooner you understand these, the sooner you will be in a position to get promoted. Some of the things you should be aware of are company rules, training procedures, evaluation procedures, importance of and laws on safe working conditions of employees, payroll procedure, hiring procedures, promotion procedures, dismissal and discipline, complaint and appeal procedures, and leadership techniques.

A Word of Caution: A promotion is not always the right thing for you. Just as bigger is not necessarily better, job promotions are not always an improvement in your work life. Consider your long-time goals before you make any short-term decisions. The extra responsibility, time, and commitment may not complement your life, especially if you are still in school. Refer to Figure 22–11 for techniques that lead to a more self-fulfilling job.

Techniques for a More Self-Fulfilling Job

Always be enthusiastic!
Never quit.
Use common sense.
Believe in yourself even when no one else does!
Work smart.
Smile!
Maintain a positive attitude at all times.
Think big!
Work hard.
Enjoy your life!
Plan your work and work your plan.
Stop making excuses!
Remember: When the going gets tough, the tough get going.

Figure 22–11 These techniques should help you get more out of your job and may even help you get a promotion.

Leaving a Job

A letter of **resignation** is your official notice to your employer that you intend to leave your job. The letter should state a reason for leaving (i.e., a new job or moving to another city). This is not the place to complain or get even. Never burn any bridges; you never know when you may need to go back to a job. You may wish to use this job for a reference in the future and everything, including your letter of resignation, will add to or detract from the impression you leave behind.

Always give at least two weeks' notice, more if you can. Write the letter to your immediate supervisor. If you feel that other members of the organization need to be personally informed, send copies of the letter to them. If possible, say something personal and positive in your letter and express your regret about leaving.

Complaint and Appeal Process

Most companies have employee handbooks. Most of these handbooks contain a section on the complaint and appeal process. To outline the process here would be confusing because it would probably differ somewhat from any employee handbook. All employee handbooks should clearly define the following five major areas of worker's rights.[39]

- Complaint and grievance rights
- Rights involving the procedures and nature of discipline
- Job security, including procedures and valid reasons for dismissal
- Rights concerning promotion, career advancement, training, and growth
- Rights concerning eligibility for insurance, pensions, personal time, and other benefits

The following section presents one way to handle the complaint and appeal process.

Corporate Due Process

In general, a worker's complaint is handled by using the employer's in-house complaint procedure. If that procedure fails to produce a mutually satisfactory resolution, the worker can file a claim or grievance with an arbitration board. At filing, each side is required to provide a deposit toward payment of the costs of arbitration, which are shared equally. The first and required effort at resolution is informal mediation, in which a professional mediator seeks to resolve the dispute directly and to reconcile the disputants.

If mediation fails, the grievance moves to arbitration. Each side can choose either to represent itself or be represented by any chosen agent, such as a labor relations expert, co-worker, personnel official, union representative (whether or not the grievant is a union member), or lawyer. The arbitrator is chosen by the parties themselves from lists of certified arbitrators or from rosters maintained by, for example, the American Arbitration Association. As is now sometimes the case in private arbitration, each party has the right to strike potential arbitrators from the list. Alternately, each side is granted an unlimited number of strikes, with an arbitrator being assigned by default by a labor board if all the names are struck.

The arbitrator hears the grievance, the employer's response, and any other information either party thinks is relevant. Hearings proceed under informal rules similar to those currently used in the arbitration of collective-bargaining agreements. The burden of proof is on the employee to establish the validity of the claim. After consideration, the arbitrator issues a binding judgment.[40]

Personal and Occupational Safety

Personal and occupational safety in the workplace centers around three major areas: awareness, being prepared and having a plan, and knowing your boundaries.[41] Employee handbooks include sections on personal and occupational standards. Be sure to study this information when you begin employment. The Occupational Safety and Health Administration (OSHA) also has strict safety rules to which employers and employees must adhere.

Awareness

Being aware is the most important safety skill. Practice being aware by paying attention to small details in your environment. Awareness skills can be very important in your work setting.

When you begin to work at a new job, you will very likely be in a setting different from your home or school. You may work in a community other than where you live, where the customers, co-workers, and bosses are almost strangers. You will be working with supervisors who have some authority over you and co-workers who are your peers.

On your first day of work, make a mental checklist of the physical surroundings. For example, if you working in a fast-food restaurant, note the entrances and exits. Note if there are security guards and where they stand. Is your workplace small and confined, such as a cashier's position? Is it spacious, like a sales clerk's in a department store?

Start paying attention to the environment right outside your workplace when you enter or leave. Are lots of people around, or is it quiet? Is it light or dark? What do the entrances and exits look like? Are you outside, near parking lots, alleys, sidewalks, trees, and bushes, or indoors, as in a mall? The time to take this inventory is when you start working. Once you have been there a while, the environment may start feeling comfortable because it is familiar, even though it may compromise your safety.

Being Prepared and Having a Plan

Having inventoried your work environment and determined that it is okay does not mean that you are guaranteed perfect safety. Accidents and crime occur in many settings, not just on the streets. The next step is to figure out a safety plan for your particular situation. If your workplace is a public business, robberies could be a safety problem. Many fast-food restaurants and convenience stores are targets for robberies. Most of these places have some sort of security system, including surveillance cameras or a panic button by the register that contacts the police. If your workplace does not have a procedure, ask for one. If they do not provide one for you, work out in your mind what you might do if you were confronted. If your workplace is a warehouse or factory setting, plan to avoid accidents and have a plan ready in case an accident does occur.

Know Your Boundaries

One of the first steps in learning how to avoid becoming the victim of a personal crime by someone you know is to understand the concept of *boundaries*. Your home is probably marked off by a fence—that is a boundary. The basketball court in your gym is marked off by painted lines—these too are boundaries.

People have boundaries, too. Learning to understand your boundary system is important to personal safety. We may tolerate some behavior because we grew up with it. For example, if you grew up in a home where people hugged you and pinched your cheek, you may believe that it is okay for people to touch you without asking permission.

It is important to note there is nothing "right" or "wrong" about your boundary system. It is what it is. What is important is that your beliefs about your "comfort zone" may compromise your safety.

CONCLUSION

How you feel about your job is reflected in the way you act and the willingness with which you do your duties. Successful workers show **pride** in their work. If you choose a job that matches your interests, aptitudes, and abilities, you will enjoy your work and take pride in what you are doing. This gives you a positive attitude about yourself and your work. If you enjoy your work, you will show pride and enthusiasm on the job. If you give it your best, you should be a successful employee. In closing, consider the following illustration of the importance of putting just a little more effort into whatever you pursue.

At 211 degrees, water is just hot water, powerless. At 212 degrees, water is live steam, with more inherent power than it has ever been possible to harvest. At 211 degrees, the water in a locomotive boiler exerts not one ounce of pressure. At 212 degrees, the water in the boiler has the power to pull a long train of cars across a mountain pass.

So it is with human beings. Often the difference in success and failure is just turning up the heat one degree. You are the only person who determines your destiny. If you honestly want to succeed, you will find ways to do it. Remember, if your attitude is positive and your efforts match your goals, your progress will be limitless.

Summary

Although technology is changing at a rapid pace, one thing that is not changing is the fact that employers need people with leadership, work ethics, human relations, and other employability skills.

When employers were asked what they looked for on an applicant's resume, the top answer given was the person's record of achievement. Related work experience was next, and farther down the list was the candidate's educational achievement. Factors that influenced the final selection were enthusiasm, professional attitude, the right "chemistry," and experience level.

Some of the job skills and personal attributes that employers are looking for when they hire people are achiever, anticipative, activative, attitude, commitment, command, competitiveness, courage, credibility, dedication, dependable, developer, discipline, drive, empathy, ethics, focus, ideation, knowledge, loyalty, organization, responsible, self-confidence, team player, and values.

The Michigan Employability Skills Task Force, including leaders from business, labor, government, and education, determined the general skills that every student should have, not only for entry-level jobs but for jobs at all levels. These skills are personal management, teamwork, and academic and technical skills.

Personal management skills include coming to work daily and on time, meeting work deadlines, developing a career plan, knowing personal strengths and weaknesses, demonstrating self-control, paying attention to details, following written and oral instructions and directions, working without supervision, learning new skills, identifying and suggesting new ways to get the job done, using organizational skills, and demonstrating personal values at work.

Teamwork skills include actively participating in a group, following the group's rules and values, listening to other group members, expressing ideas, being sensitive to members' ideas and views, being willing to compromise if necessary to accomplish goals, being a leader or follower to accomplish goals, and working in changing settings and with different people.

Academic and technical skills include reading and comprehending written materials, interpreting charts and graphs, basic math, computing to solve problems, using research and library skills, using tools and equipment, speaking and writing in the language in which business is conducted, using the scientific method to solve problems, and using specialized knowledge to get a job done.

You must exhibit human relationship skills with your fellow employees. Relationships are enhanced by a positive attitude, being cooperative, pride and enthusiasm, dependability, trustworthiness, working hard, respecting others, handling criticism, exhibiting appropriate dress and good grooming, showing initiative, and being diligent, capable, and commited.

Workers must also learn to respond to authority and practice ethics in the workplace. You can show a high ethical standard on the job by being honest, providing worthy service, and having a positive attitude.

Getting a job is the beginning of your career, not the end. An obvious measure of success is job promotion. There are several things an employer looks for when deciding which workers to promote: job mastery, initiative, good employability skills, and understanding a supervisor's responsibilities.

A letter of resignation is your official notice to your employer that you intend to leave your job. Never burn any bridges; you never know when you may need to go back to a job. If you honestly want to succeed, you will find ways to do it. Remember, if your attitude is positive and your efforts match your goals, your progress will be limitless.

End-of-Chapter Exercises

Review Questions

1. Define the Terms to Know.
2. List 25 job skills and personal attributes that employers are looking for when they hire people.
3. Name 13 personal management skills needed by employees.
4. What are five steps that can help you follow directions?
5. Name eight teamwork skills needed by employees.
6. What are 10 academic and technical skills that help make you a good employee?
7. What are 12 qualities that a successful employee shows with fellow employees?
8. List eight ways to exhibit or display positive work attitudes.
9. List five things you can do to change from a negative to a positive attitude.
10. What are 11 things you can do to help develop good cooperative skills with co-workers?
11. List four appropriate responses to authority.
12. How can you take care of the problem if you feel you have been treated unfairly by your employer?
13. What are three ways to show high ethical standards on the job?
14. What are four things an employer looks for when deciding which worker to promote?

Fill-in-the-Blank

1. Your attitude will affect your job __________________.
2. Part of being dependable is being __________________ and __________________.
3. Respect involves simple common __________________.
4. Take criticism __________________ instead of personally.
5. Breaking large tasks down into a __________________ of __________________ tasks will make jobs seem easier.
6. True leaders are not afraid to demonstrate personal __________________ in the form of examples in their everyday lives.

7. __________________ refer to your standards and values.
8. Employers first judge a prospective employee's writing ability by the quality of an __________________, __________________, or __________________.
9. A __________________ is an employer's way of indicating that a worker can assume increased responsibility.
10. Never burn any __________________. You never know when you may need to go back to a job.

Matching

_____ 1. Juan always starts the work day in a pleasant mood.
_____ 2. After finishing the typing Linda was asked to do, she started straightening up the office.
_____ 3. Although his boss is rather unfriendly, Arnold accepts this and tries to be friendly to others.
_____ 4. Performing your job without the boss standing over you.
_____ 5. Pedro always gets to work on time.
_____ 6. Beyond being dependable and showing initiative, you can also be counted on because of your sincere desire for the organization to succeed.
_____ 7. After Maria's employer showed her what she was doing wrong, she worked hard to improve that skill.
_____ 8. Miguel doesn't agree with the boss on how to get the job done, but he does know who the boss is.
_____ 9. Joe has been with the store for 20 years; the new system of inventory was difficult, but he worked hard to learn it.
_____ 10. Elena makes mistakes, but she knows what is right and wrong and is not swayed by negative influences at work.
_____ 11. Carmen realizes she shortchanged a customer and will be 50 cents over in the cash drawer. Instead of putting the 50 cents in her pocket, she leaves the money in the drawer.

A. honesty
B. initiative
C. willingness to accept change
D. dependability
E. willingness to accept criticism
F. cooperation
G. positive attitude
H. working without supervision
I. respond to authority
J. commitment
K. ethics

Activities

1. From the list of 25 job skills and personal attributes that employers seek in employees, select five that will present the biggest challenge to you. Write a short paragraph on each explaining why you need to improve.
2. Twelve qualities were discussed to enhance relationships with your employer and employees. Interview an employer in your community and ask what qualities they look for in employees. Compare their answers with the 12 qualities in the book.
3. You are interviewing for a job and the interviewer says, "Convince me that you can be a self-learner." Write a 100-word essay on what you will tell the interviewer.

4. From the list of 12 personal management skills, select 3 that could be a problem for you. Write a plan of action for each of the 3 explaining how to improve so you can become a better employee.
5. Review the list of 10 academic and technical skills in this chapter. Select 2 that could prevent you from getting the job that you desire. Write a plan of action on how you can improve in the 2 areas.
6. To determine local work skills needed, ask your boss or some other authority where you work the 10 most important things they look for in an employee. If you do not work, ask a person at church, in your neighborhood, or a friend of the family. Report these to your class.
7. Write a letter of resignation. Remember, this is not to complain or get even. Say something personal and positive in your letter and express your regret about leaving.
8. Write a letter of recommendation for a friend. In this case, the friend is you. State in the letter why you would hire your friend. Do not hold back. List as many positive attributes as you can.
9. Write three positive potential employability skills for each member of your class.
10. **Take it to the Net.**

    Explore employability skills on the Internet. Several Web sites containing information on employability skills are listed below. Browse the sites and find one you feel is the most useful. Summarize that site and tell why you feel it is useful. If you are having problems with the listed sites or just want more information, use the search terms.

    Web sites

    <http://www.smartbiz.com/sbs.careers.htm>

    <http://www.nwrel.org/scpd/sirs/8/c015.html>

    <http://www.edgar.k12.wi.us/edgar/emskil.htm>

    <http://www.conferenceboard.ca/nbec/eprof-e.htm>

    <http://www.employabilityskills.bc.ca/>

    Search Terms

    employability skills

    work skills

    work ethics

    job skills

Notes

1. M. Thatcher, "One Graduate in Two Still Jobless," *Personal Management* (October, 1993).
2. *Life Underwriter Training Council Course Guide*. Business Insurance Course (Washington, D.C., 1989).
3. C. Smith et al., "Michigan Employability Skills Profile," *Educational Leadership* (March, 1992), pp. 32–35.
4. R. A. Wolf and T. O. Wolf, *Job Hunter's Secret Handbook for Success* (St. Joseph, MI: WINSS, 1993), p. 74.
5. *How to Follow Directions*, Skills for Success, No. 15 (Nashville, Tennessee: TN Department of Education, 1993).
6. A. R. Carnevale, et al., *Workplace Basics* (San Francisco, CA: Jossey-Bass Publishers, 1990), p. 17.
7. Ibid.
8. Wolf and Wolf, *Job Hunter's Secret Handbook for Success*, p. 74.
9. W. G. Dyer, *Team Building* (Reading, MA: Addison-Wesley, 1987), p. 4.
10. R. E. Lefton et al., *Improving Productivity Through People Skills* (Cambridge, MA: Ballinger, 1980), p. 388.
11. Wolf and Wolf, *Job Hunter's Secret Handbook for Success*, p. 78.
12. Ibid.

13. Ibid.
14. Carnevale, *Workplace Basics*, pp. 18–25.
15. J. M. Lannon, *Technical Writing*, 2nd Edition (Boston: Addison-Wesley, 1997).
16. W. E. Brock, "Future Shock: The American Workforce in the Year 2000," *The American Association Community, Technical and Junior College Journal* 57(4): 25–26.
17. *Job Requirements of the Successful Worker*, Skills for Success, No. 2 (Nashville, TN: Tennessee Department of Education, 1993).
18. Wolf and Wolf, *Job Hunter's Secret Handbook for Success*, p. 71.
19. Ibid., p. 72.
20. Ibid., p. 72.
21. *Job Requirements of the Successful Worker*, Skills for Success, No. 2., p. 25.
22. *Developing Good Work Attitudes*, Skills for Success, No. 9 (Nashville, TN: Tennessee Department of Education, 1993), p. 163.
23. Ibid., p. 164.
24. *Working with Others*, Skills for Success, No. 12 (Nashville, TN: Tennessee Department of Education, 1993), pp. 209–211.
25. L. A. Liddell, *Building Life Skills* (South Holland, IL: Goodheart-Wilcox Co., 1998), pp. 516–517.
26. R. Busse, "The New Basics," *Vocational Education Journal* (May, 1992), p. 47.
27. Liddell, *Building Life Skills*, p. 517.
28. Ibid., p. 516.
29. Ibid., p. 517.
30. Ibid.
31. Ibid.
32. N. Wehlage, *Goals for Living: Managing Your Resources* (South Holland, IL: Goodheart-Wilcox, 2000).
33. Ibid.
34. *Entrepreneurship in Agriculture* (College Station, TX: Instructional Materials Service, 1989), 8716-B, pp. 2–4.
35. *Responding to Authority*, Skills for Success, No. 14 (Nashville, TN: Tennessee Department of Education, 1993), pp. 242–243.
36. *Ethics Associated with the Choice Occupation*, Skills for Success, No. 10 (Nashville, TN: Tennessee Department of Education, 1993), pp. 80–81.
37. *The Route to Promotion*, Skills for Success, No. 40 (Nashville, TN: Tennessee Department of Education, 1993), pp. 741–744.
38. Ibid.
39. R. Edwards, *Rights at Work* (Washington, D.C.: The Brookings Institution, 1993), p. 226.
40. Ibid., p. 218.
41. D. Chaiet, *Staying Safe at Work* (New York: Rosen Publishers Group, 1995).

Glossary

A

abilities—competence in an activity or occupation

academic skills—skills learned in a formal school situation, such as science, English, and math

accommodators—people who learn by doing and feeling

accountability—answerable or capable of being explained

acronym—a word formed from the initial letters of a phrase or title, such as FFA, NASA, USA

action—a state of motion

adjourned meeting—a continued meeting set to meet again at a certain time; similar to recess

affective attitudes—the emotional feelings attached to an attitude

affective learning—personality learning based on human relations

affective skills—human relations abilities

affinity—attraction or similarity of group members

agenda—a list, plan, or the things to be done; matters to be acted or voted on during a meeting

aggressive communication—communication with the intention to dominate, intimidate, and overpower

agricultural education—program of instruction in agriculture in high schools

alternatives—different courses of action that one might take

ambiguous—having more than one possible meaning

ambitious—showing a desire to achieve or obtain power, superiority, or distinction

amendable—when it is possible to modify or change the wording, and in some cases, the meaning of the motion to which it is applied

animate—to give life to or make lively

anxiety—uneasiness about a situation or event

apathy—lack of interest or concern

applicant—someone who applies for a job

application form—a general form that applicants fill out when seeking employment

aptitude—quickness in learning and understanding or a natural or acquired talent, ability, inclination, or intelligence

ardently—with warm or intense feeling

articulation—division into clear and distinct words or syllables

assertive communication—honestly expressing your feelings and thoughts without threatening others or experiencing anxieties

assimilators—people who learn by observing and thinking

assumption—taking a piece of information for granted without researching to see if it is true

attitude—disposition toward others and ourselves; strong belief or feeling about people, things, and situations

attributes—qualities or characteristics of an individual
authenticity—genuineness; realness
authoritarian—behavioral leadership style that leads and makes decisions regardless of the wishes of the group
autocratic leadership style—a leadership style in which the leader makes decisions independent of the group
avocational—part-time in addition to one's vocation
avocational (part-time) job placement—a job or occupation that supplements a full-time job
avoidance learning—negative reinforcement

B

barriers—obstructions; mental obstructions that keep us from communicating clearly
behavior—the actions or reactions of persons or things under specified circumstances
behavioral attitudes—the tendency to act in a particular way toward a person, object, or event
behavioral leadership—leadership according to our personality style; in the context of this book it refers to democratic or authoritarian leadership
belief—conviction or acceptance that certain things are true or real
bodily kinesthetics—control of one's body and of objects; timing; trained responses that function like reflexes; a psychomotor skill
body—in a speech, the main part, including the main points
body language—the nonverbal way one communicates
brainstorming—the unrestrained offering of ideas or suggestions
brainstorming method—the process of suggesting many alternatives, without evaluation
buzz group—a group of six to eight members, with a leader and a recorder, with the purpose of discussing a topic and arriving at answers
bylaws—standing rules governing the regulation of a organization's internal affairs

C

capability—the ability to get a job done
captivate—hold an audience's attention; fascinate or charm
career—the work, occupation, or profession followed as one's life work
career planning—to plan goals with time lines for one's life work, occupation, or profession
chairperson—also called chair; the person presiding over a meeting
channel—the means by which a sender communicates a message
chapter—a local branch of an organization or club
choleric—personality type with a bossy, quick, active, and strong-willed temperament
chronological—the time order in which things occur
classified ads—the section of the paper that advertises jobs
clone—an exact replica
cognitive attitudes—set of values and beliefs a person has toward a person, object, or event
cognitive learning—learning based on theoretical symbols (numbers, words) as well as logical reasoning; academic learning
coherence—sticking together, or an orderly or logical relation of parts that affords comprehension or recognition
cohesiveness—when groups stick or work together as one
combination leadership—leading by combining all styles of leadership
commitment—applying oneself to a task until it is completed
committee—a group within an organization that is elected or appointed to perform some task
communication—the process of sending and receiving messages in which two or more people achieve understanding

competence—the ability to do something well
compromise—settlement of differences by mutual agreement
conceit—excessive feeling of one's own worth
conceptualize—to visualize an abstract concept
conceptual leadership skills—thinking skills that can be taught, such as problem solving, decision making, and delegation
condescends—to do something one regards as beneath one's dignity
confirmation behaviors—positive behaviors that affect feelings of self-worth
confronting—approaching a problem head-on or face to face
consensus method—everyone in the group agrees with the solution
consistent style—tendency to know the appropriate amount of information to consider and evaluate before making a decision
consultative leadership style—the leader seeks information from group members before making a decision
content theories—focus of attention on the factors in an individual that cause that individual to behave in a particular manner
continuum—a series connecting two extremes with an infinite number of variations, such as the decision-making continuum
conventional method—in problem solving, when the group votes on a proposed solution and it is put in place
convergers—people who learn by doing and thinking
convey—to communicate or make known
cooperative skills—the ability to work or act together for a common purpose.
creative problem/decision—a problem that needs a plan or design to help one come to a solution
credence—belief
Creed (FFA)—philosophy statement on agriculture written by E. M. Tiffany
criteria—standards, rules, or tests by which something can be judged
culture—socially transmitted behavior patterns, beliefs, and all other products of human work and thought patterns of a society, community, or population
curriculum—course of study; in this text, agricultural education

D

daily log—record and schedule of activities and appointments throughout the day
debatable—open to discussion, as of an item being considered by a group
decision making—the process by which a new or different action is selected
decode—receive and interpret a message
decorum—propriety of debate
defamation of character—saying cruel things about a person that can damage image or status
deliberate—slow and unhurried, as in the delivery of a speech
delivery—conveying, or the method of making a speech
delphi method—a series of anonymous questionnaires repeatedly given to a group and analyzed until an acceptable position or solution is attained
democratic—behavioral leadership style that leads and makes decisions based on the input of the group.
dependable—can be trusted to accomplish a task
desire—longing or hope for something
deviant—type of behavior or attitude that differs from the norm or from accepted social and moral standards
devil's advocate method—a member suggests an alternative he or she does not support to make other think and react
diligence—constant, earnest, and persistent effort
directors—task-oriented people who enjoy telling others what to do (choleric personality type)
discreet—modest or reserved
discuss—talk about motion after being recognized properly

divergers—people who learn by observing or feeling
doodling—drawing or scribbling idly
doubt—questioning
dovetail—to connect or combine precisely or harmoniously
dunce—a dull-witted or ignorant person; not a good student

E

earnestness—sincerity or seriousness
empathize—to feel as someone else feels, as if you were that person
empathy—ability to experience the feelings, thoughts, or attitudes of another person
employability skills—human relations, personal management, and personality type (affective) skills needed to be a good employee
encode—put into a form that the sender believes the receiver will understand
energy cycle—in the context of this book, scheduling things at certain times in a meeting in order to be the most productive
equalitarian—the belief that all people are equal and should enjoy equal rights and opportunities
equity—an individual's belief that he or she is being treated fairly in relation to others
ethic(s)—rules of conduct that reflect the character and sentiment of a community
ethical—conforming to the standards of conduct of a given profession or group
etiquette—practices, values, and customs prescribed by a culture or society
exact reasoning problem/decision—a problem that usually has an exact answer
exaggeration—overstatement of the importance of an incident
executive committee—comprised of chapter officers and those who chair the major committees
extemporaneous—a type of speech in which speakers prepare ideas but do not memorize exact words
external motivation—outer force or power that helps a person achieve a goal
extinction—removal of positive reward, such as eliminating an undesirable behavior by withholding rewards when the behavior occurs

F

facilitator—someone who works with a group to make things run effectively and efficiently
fear—awareness of danger
feedback—verbal or nonverbal response to a message
fidelity—being loyal, dedicated or faithful
fixed ratio schedule—rewards behavior after a predetermined amount of actual outputs
floor—recognition from the chairperson of the right to speak
follow-up letter—a letter written to the potential employer or interviewer after the interview to let him or her know one is interested in the job
force—power and expression, as in speaking
functional groups—formal groups that exist indefinitely

G

general consent—everyone agrees and no vote is necessary
germane—closely related to or having bearing on the subject of the motion
gleaned—gathered, discovered, or found out slowly and patiently
goal—the end toward which effort is directed
gossip—idle talk or rumor about the personal or private affairs of others
governmental body—a group elected or appointed to guide, direct, and administrate policy
Greenhand degree—first of four degrees of membership in the FFA
gross national product (GNP)—the dollar amount of the products that a country produces in one year
group dynamics—pattern of interactions within a group

group think—members going along with the group and not challenging ideas and recommendations because of the fear of not being seen as a team player

H

hearing—receiving sound
holistic—emphasizing the importance of the whole and the interdependence of its parts
human relations skills—skills needed to understand and work with others.
human resource frame—democratic philosophy within an organization that relates to the needs of members without strong emphasis on procedure and policy.
hygienes—conditions and practices that serve to promote or preserve health or psychological well-being; in the context of this book, they are salary, job security, working conditions, relationships, status, company procedures, and quality of supervision
hyperbole—exaggeration for effect, not meant to be taken literally
hypothetical—a made-up situation used to convey a point or an issue

I

image—how other people feel about you or perceive you
immediate goals—goals that are set to happen within a day or week
immediate job placement—jobs entered immediately after high school graduation
immediately pending motion—a formal proposal that is under consideration on the motion being discussed
impetuousness—rushing into action with little thought
important—having great value, significance, or consequence
impromptu—a type of speech delivery in which the speaker talks "off the cuff," with no chance for preparation
incidental—a motion that arises out of other motions and takes precedence over other motions when appropriate
income strategy—a plan or budget to properly manage an income
incompetent—unable to complete a task due to a lack of ability or qualification
inculcating—to teach, impress, or instill by urging or frequent repetition
inequity—an individual's belief that he or she is being treated unfairly in relation to others
inertia—tendency of matter to remain at rest if at rest, or, if moving, to keep moving in the same direction
inflections—the tone or pitch of a voice
influence leadership—leading by convincing others of an idea, so that they will follow of their own will
initiative—taking action and getting started
innate—inborn, as of characteristics or attributes
input-based communication—when a person is receiving communication
insecure—lacking self-confidence; uncertain
instrumental values—values that reflect the way you prefer to behave, such as ambitious and hard-working
insubordination—not submitting to authority, disobeying an employer
intangible—cannot be appraised for value; having no form and substance
integrity—state of being nearly perfect or seeking perfection
intelligence—the ability to respond successfully to new situations and the capacity to learn from past experience
interference—anything that hinders the sender from making the message understood
intergroup—among groups, as in competition in which group members pull together to beat rival groups
internal motivation—an inner force or power that helps a person achieve a goal
interpersonal communication barriers—differences and personal characteristics of the sender and the receiver that hinder communication

interpersonal intelligence—the ability to understand, appreciate, and get along well with other people
interview—a meeting between an employer and an applicant with the purpose of getting a job
interviewee—the person being interviewed
interviewer—the person conducting the interview
intragroup—within the groups, as when group members compete among themselves
intrapersonal intelligence—the ability to understand ourselves, sensitivity to our own values, knowing who we are, and how we fit into the greater scheme of the universe
introduction—the beginning thought to get the audience interested in a speech
irony—humor based on words used to suggest the opposite of their literal meaning

J

job—a paid position at a specific place or setting
job lead—something that helps one find out about a job
job-related skills—the skills it takes to perform a specific job
judgment problem/decision—a problem that requires one to consider many factors, list alternatives, and evaluate the alternatives before reaching a decision
justice—fairness

K

kinesics—the study of communication through body motion

L

laissez-faire—noninterference in the affairs of others
laissez-faire leadership—in which a group makes decisions with very little, if any, input from the leader
laissez-faire leadership style—the group, not the leader, solves the problem or makes the decision
leadership—the ability to preside, guide, or conduct others, activities, or events with responsibility for the final outcome
learned needs theory—states that some needs are acquired from society and people are motivated to a behavior that satisfies that need
lectern—speaker's stand or podium
left-brain people—have characteristics of a melancholy personality, tend to be logical, detailed, active, and objectives-oriented; prefer jobs that require precision, detail, or repetition
legislative body—a group (elected or appointed) with the power to create laws
letter of application—a letter used to apply for a job; when accompanied by a resume, it is often called a cover letter
linguistic intelligence—verbal ability, sensitivity to language, meanings, and the relations among words
listening—a conscious mental effort to understand what is heard
litigious—bringing a lawsuit against someone or some organization
logical-mathematical intelligence—ability in numbers, patterns, abstract thought, precision, evaluating, organizing, and logical structure
long-term goals—goals that are set to happen approximately two or more years in the future

M

main motion—introduces new business before a meeting
maintenance role—focus on maintaining the togetherness of a group
majority vote—a decision by more than half the members of a group
mass media—any form of communication that reaches a very large audience, such as television, newspaper, radio, magazine, or outdoor advertisements
mastery—high accomplishment in a given area
matrix—a rectangular array or network of intersections
melancholy—personality type that is analytical, self-sacrificing, gifted, and perfectionist

memorandum—a formal written form of communication used in a business setting

message—whatever is intended to be communicated by one person to another

metaphors—figures of speech containing implied comparisons, in which a word or phrase ordinarily and primarily used of one thing is applied to another

minimizing approach—the problem solver/decision maker chooses the first solution available to solve the problem or make the decision

minutes—the official written record of the meetings of an organization, reflecting all actions taken and meaningful discussion on important issues

modem—device that converts data from a telephone line to a computer

momentum—the quantity of motion of a moving object, equal to the product of its mass and its velocity

monotone—speaking without fluctuations in the voice pattern

motivation—an individual's desire to demonstrate behavior and willingness to expend effort

motivation—the focus of the need or desire to act

musical intelligence—sensitivity to pitch, rhythm, emotional power, and complex organization of music

N

network—an informal group of people that can be used to help one get a job

networking—an informal way of passing information between people of similar groups and similar goals

nominal group method—the process of generating and evaluating alternatives through a structured voting method

nonverbal communication—messages conveyed by a person's behavior and the physical environment

norms—group standards expected of all members

not applicable (N/A)—what one writes on an application when a question does not apply to oneself

O

occupation—a group of similar tasks a person does that supports him or her

occupation-related skills—skills that are directly related to one's job or occupation

operant conditioning—a way to modify an individual's behavior through the appropriate use of immediate rewards or punishment

opportunist—a person who takes advantage of unplanned situations

optimizing approach—the problem solver/decision maker reviews many different solutions before making a decision

orders of the day—the items of business on the agenda

organizational structure and procedural—how and through what structure a message goes from the sender to the receiver

out of order—inappropriate parliamentary procedure

output-based communication—when a person produces the communication

P

pangs—a sudden, sharp, and brief pain, physical or emotional

paraphernalia—the gavel, owl, plow, flag, secretary's book, treasurer's book, and other items used in an FFA meeting

Pareto's principle—20 percent of tasks that are high priority can generate 80 percent of the results needed to reach goals

parliamentarian—an expert on rules governing meetings who serves as an advisor to a chairperson

parliamentary inquiry—an incidental motion that requests information on the correct or appropriate parliamentary procedure for a motion under consideration at that time

parliamentary procedure—a set of rules and procedures to follow to keep a meeting orderly and harmonious, and guarantee that all persons have equal opportunity to express themselves

participative leadership style—the leader has a tentative decision/solution in mind but goes to the group for input

participative management—having employees or followers involved in decision making

passive communication—indecisive, inability to delegate, won't take a stand, and does not give direction

peer groups—groups of people who share rank, class, or age

peers—friends, associates, and co-workers who are equal in rank and abilities

perception—way of understanding a message, based on our own beliefs, knowledge, and ways of organizing information

perseverance—steady persistence in a course of action

personal management skills—internal skills required to be a good worker, such as personality, attitude, honesty, enthusiasm, and and ability to meet deadlines; same as self-management or adaptive skills

personal time—free or leisure time

personification—a figure of speech in which a thing, quality, or idea is represented as a person

personnel office—an office that keeps records of all job applicants and employees within a company or organization

philosophy—a system of values by which one lives; also, inquiry into the nature of things based on logical reasoning rather than empirical methods

phlegmatic—personality type with a calm, cool, slow, easy-going, and well-balanced temperament

physical stroke—recognition that occurs physically, such as an acknowledging smile or a pat on the back

poise—composure under pressure

policy—a rule or action of an organization adopted for the sake of being consistent and expedient

political frame—persuasive process within groups that involves efforts by members of a group to ensure their power or to protect existing power sources

popularity (perceived) leadership—leadership that is bestowed on someone due to group perceptions and not necessarily the leader's ability

portability—capable of being carried easily or moved

portfolio—evidence of a person's performance, rather than a resume, such as samples of a journalist's articles

positive reinforcement—a method of using pleasant consequences or rewards for a desired behavior

postponed job placement—jobs taken after further training or education beyond high school, such as two-year vocational schools or college

power leadership—leading by force, with the group submitting, perhaps, against its will

precedence—the right to come before; priority

presiding officer—the officer or person designated to preside over or lead a meeting; usually the president

previous question—a formal motion requesting a vote and requiring two-thirds vote for passage

pride—a feeling of self-esteem arising from one's accomplishments

primacy effect—forming impressions quickly on first contact with a person

primary amendment—an amendment applied to any amendable motion, except the motion to amend

prime time—the time of day when you get the most done

prioritize—to arrange or deal with in order of importance

privileged motion—has nothing to do with the pending motion but is of such urgency and importance that it is allowed to interrupt the consideration of other questions

problem—difference between what is actually happening and what an individual wants to happen

problem solving—the method of arriving at a decision or answer

process theories—examine how behavior is motivated

procrastination—to put off doing something until a future time

professional time—time that relates to school, work, or employment

Program of Activities—a document within the FFA organization that records the chapter's goals and the ways and means to achieve them

pronunciation—the form and accent a speaker gives to the syllables of a word

proxemics—the science of how spacing and placement gives messages

prudence—common sense, or thinking about things before you act

psychic income—praise or positive things that happen to us that give us a psychological boost or make us feel good

psychomotor learning—learning based on manual dexterity

psychosomatic—bodily symptoms experienced as a result of mental conflict

punishment—an undesirable consequence for an undesirable behavior

purvey—to advertise or circulate

put the question—call for the vote

Pygmalion—a king from Greek mythology whose beloved statue came to life; the idea that one's expectations shapes the behavior of others

Pygmalion effect—what you believe can become a perceived reality

Q

qualifier—a word or phrase that qualifies, limits, or modifies something

qualities—characteristics, innate or acquired, that determine the nature and behavior of a person or thing

question of privilege—a privileged motion that requests immediate action on an urgent matter related to the comfort, convenience, rights, or privileges of the assembly or one of its members

quotation—in a speech, the words or passage quoted by another person

R

rapport—a positive relationship with an audience or others in a group

receiver—one for whom a message is intended

recognition from the chair—permission for a member to speak, given when the chair is properly addressed

reference—a person listed on a resume who may be contacted by a prospective employer to inquire about an applicant's qualifications

reference groups—groups such as college fraternities or sororities

reflective style—in which one identifies, analyzes, and evaluates as many alternatives to solving a problem as possible

reflexive style—in which one makes quick decisions

reinforcement scheduling—timing of rewards or punishments

reinforcement—the motivation and control of behavior by the use of rewards

relaters—people who avoid risk and seek tranquility, calmness, and peace (phlegmatic personality type)

reprimand—a severe, usually formal, reproof, rebuke, or correction for something

resignation—the termination of a job, or a letter informing an employer of such termination

resilient—able to bounce back

resources—something that can be drawn on for aid

resume—a one- or two-page description of an applicant that gives his/her educational background, experiences, and qualifications

right-brain people—have characteristics of sanguine or choleric personality types, tend to be spontaneous, emotional, nonverbal, and like jobs that are nonroutine or call for idea generation through hunches and insights

rigorous—strict or hard

role model—someone you admire who has an influence on you

role-playing—acting out a situation to illustrate a real-life problem

rudiments—basic principles

S

salutation—the beginning of the speech with a greeting, either addressing or welcoming

sanguine—personality type with a warm, lively, and cheerful temperament

scantily—scarcely sufficient

scenario—a story or outline of a hypothesized story or projected chain of events

secondary amendment—an amendment applied to a motion to amend

self-communication—the ability to answer questions that you ask yourself honestly and deal with controversial or difficult issues

self-concept—the way you see and feel about yourself

self-confidence—believing in oneself or one's abilities

self-deprecating—devaluing your own abilities

self-determination—motivation from within to achieve goals

self-esteem—belief in one's abilities or respect for oneself

self-fulfilling prophecy—people tend to behave in a way that supports what they believe about themselves

self-image—the idea, concept, or mental image one has of oneself

self-management (adaptive) skills—skills that are required to be a good worker, such as diligence, honesty, and enthusiasm

self-responsibility—taking responsibility for one's own actions

semantics—the study of the shifting meanings of language, by which a word or phrase can mean different things to different people

sender—person who sends a message to someone else

sensitive—keenly aware of another person's feelings

sensitivity—the ability to understand the feelings of others

short-term goals—goals that are set to happen immediately or very soon

similes—figures of speech in which one thing is likened to another

simultaneous—happening at the same time

situation—when and where something takes place

situational leadership—leadership style that depends on or is contingent on a situation

situational timing barriers—time and place that communication takes place

skill—proficiency, ability, or expertness

skills—proficiency or ability to do something well as a result of talent, training, and practice

slovenliness—carelessness, sloppiness

SMART—acronym for specific, measurable, attainable, relevant, trackable

socialization—learning to get along with others

socializers—relationship-oriented people who appear to need the approval of those around them (sanguine personality type)

socializing—taking part in fun, entertaining, or social activities

spatial intelligence—the ability to think in vivid mental pictures; keen observation or visual thinking; recreating or restructuring a given image or situation

special committee—a committee created for a special purpose that disbands as soon as it has completed its task

spontaneity—happening or arising without apparent premeditation

standing committee—a committee that continues to exist from one year to another

statistics—facts or data of a numerical kind used in a speech to present information

status—perceived position within a group

statute—a law enacted by the legislative branch of a government

stroke deficit—lack of positive reinforcement

structural frame—authoritarian rules set up by an organization, such as policies, procedures, and organizational charts

subsidiary motion—a motion that relates to some other motion on the floor and can change or alter the main motion

subtlety—having or showing a keenness about small differences in meaning

supplementary (tributary) job placement—jobs attained through leadership and personal development skills or other skills learned in quality agricultural education programs. Examples would be legislators, attorneys, supervisors, and salespersons

surface analysis—judging the actual appearance rather than the inner nature

surface language—relates to the exterior elements that make up your appearance

symbiotic—the relationship of two or more different organisms in a close association that may benefit each

synergy—combined action of individuals that when taken together increases one another's effectiveness

synetics—the process of generating unique alternatives through role playing and fantasizing using analyses to provide mental images

T

talent area—natural ability to learn or do something

tangible—can be appraised for value, having actual form and substance

task groups—committees; formal groups sanctioned by an organization to perform specific functions

task roles—roles members in a group play in an effort to get things done

tasks—things to be done

teamwork skills—the ability to work well with others in a group situation

technical human relations—people skills that can be taught rather easily, such as listening and team building

technical knowledge—knowledge that is needed to perform a specific job, including the "why" and "how"

technical skills—psychomotor skills or skills that involve using one's hands, such as typing, welding, or drafting

temperance—just the right amount, at the right time, of any pleasure

terminal values—goals you strive to accomplish before you die, such as security and self-respect

thinkers—people who enjoy solitary, intellectual, and philosophical challenges (melancholy personality type)

tie vote—the same number for and against a motion

time management—planning how you will control your time to do the things you need and want to do

tone—the particular or relative pitch of a word, phrase, or sentence

trade union—an organization set up to join or bind workers together for a common purpose

traditional leadership—leading a particular way because it has always been that way; sometimes referred to as cultural or symbolic leadership

trait leadership—distinctive qualities a person possesses that increase the likelihood he or she will be a leader, such as intelligence

traits—distinguishing characteristics or qualities

transferable skills—skills that can be used on many jobs, such as writing clearly

trustworthiness—deserving of trust or confidence; being honest, faithful, reliable, and dependable

truth—agreement with a standard, rule, or established or verified fact or principle

tunnel vision—a narrow perspective

two-thirds vote—67 percent of all members present must favor a motion

U

unstated subsidiary motion—a subsidiary motion that may not have been seconded and has not been stated by the chair

urgent—requires immediate attention

V

value—principle or belief you consider important

value added—processing and packaging of a product that adds monetary value

value conflict—torn between two value systems, such as trying to fulfill work and family obligations at the same time

values—social principles, goals, or standards held or accepted by an individual, class, or society

value system—an individual's set of beliefs

variable interval schedule—rewards a desired behavior after a varying period of time

verbal stroke—recognition that occurs verbally, such as words of thanks and appreciation

verbose—very talkative

violation—misuse of proper parliamentary procedure

vitality—physical or mental strength or energy

vivacity—full of life and energy; liveliness; spirited

vote—to give members the right to express approval of or opposition to a particular action

vote by voice (viva voce)—a vote in which the chairperson asks for "aye" for those supporting a motion and "nay" for those opposed to a motion

W

ways and means—a step-by-step procedure of how goals will be accomplished

winner—someone who achieves victory over others in a competition

withdraw—remove a motion from consideration before a vote is taken

work—a task that people try to accomplish that produces something of value to oneself and/or society; a wage may or may not be earned

worst-case scenario—the worst possible thing that could happen in a given situation

Appendix A

The "Shotgun Approach": Implementation of Key Principles of Learning

At the beginning of every unit, a list of terms is presented. The following explains how this material can best be taught. The purpose of presenting this information is simply to illustrate one method of teaching as teachers pursue the "educator" level of teaching. Many principles of learning also are reviewed, implemented and illustrated in the "Shotgun Approach to Teaching."

One of the older methods of teaching is rote drill. Many teachers used this in teaching spelling, states, and the state capitols. Agricultural education teachers have used this method in teaching the identification of breeds, shop tools, and other subject matter used in basic agricultural production courses. Was it effective? Generally, I'm sure that it was, depending on the teacher. The shotgun approach as a method of teaching is similar, but there is much more to this method than just rote drill. The later part of this section will attempt to explain why the shotgun method should be used more often in education today. Now, an overview of the shotgun approach will be given.

The reason for the name shotgun approach is primarily the fact that it covers everything in one class period. In other words, the lesson is introduced, the subject matter is presented, and testing occurs within the same class period. Another reason for the name is that it is hard to miss with the method if you are "aiming in the right direction."

The shotgun approach can be used in presenting almost any subject matter, but the author will use as an example teaching the definition of terms. The teaching objective will be for the students to learn the definition of twenty new terms in a one-hour class period, with 80 percent of the class scoring 100 on a multiple-choice test at the end of the class period. For the sake of clarity, the following steps will be written in the form of a "job operations sheet." The steps are as follows.

1. Write all the terms on the board before the class starts. (Tell the class that you expect that at least 80 percent of them will score 100 at the end of the class and that they will possibly do even better.)
2. Ask a student, "Do you know the definition of the first term?" If the student does not know the answer, ask another student. Draw on information from the students.
3. Thoroughly build up the interest and seek answers to questions by using problem-solving type questions. (The purpose here is to make the students think.)
4. After drawing information from the students, present the definition as it should be or restate what one of the students has said.

5. Present some extra information about the term, to make it relevant. (Present examples so that students can relate the information to something that is recorded in their perceptive field. Jesus made information relevant by using parables.)
6. Ask the question repeatedly to several students until you are sure that the information (term) has been thoroughly understood and perceived. (Ask the question in several different ways if necessary.)
7. Praise the students when they answer questions correctly. (When questions are not answered correctly, do not react with a negative comment. Thank the student for the response and make him/her feel that a positive contribution has been made.)
8. Proceed to the second term and repeat steps 2 through 7.
9. After thoroughly covering the second term, go back and ask the first term again. (Do not proceed to the third term until you are sure everyone understands and perceives what is presented.)
10. Proceed to the third term and repeat steps 2 through 7. Do not continue with the next term until the third term is thoroughly understood and you ask for and receive the answers to questions about the first and second terms.
11. Proceed to the fourth term and repeat the same procedures already mentioned.
12. Continue through term 20, remembering always to ask the preceding questions "all the way back" to term number 1 before going on to the next question. (This goes faster than one would expect once the teacher gets acclimated to the approach.)
13. After all twenty terms have been covered, ask for two volunteers. Try to select an overachiever and an underachiever without the students realizing the intended difference of selection.
14. Before accepting the volunteers, make sure that they know the conditions. (They are as follows: each student will be asked to stand in front of the room and each will be asked ten questions. If they get eight out of ten correct, the student will make a score of 100. However, if only seven questions are answered correctly, the student makes a score of zero.) Nevertheless, if the student makes a score of zero, he/she has the option of taking the test again when the other students take the test.

    Note: The point in having the two volunteers in front of the classroom is to literally get the class on the edge of their chairs.

    This can be done if the teacher manipulates the questions by asking the most difficult ones first so that the volunteer misses one or two questions out of the first six or seven. Total involvement and enthusiasm on the part of the teacher are mandatory so that the students will get excited. Also, the students are thinking through the answers as the students in front of the class are participating. An optimum situation occurs when each student answers eight of ten questions correctly. Why the overachievers and underachievers? The overachievers can be asked the most difficult questions and the underachievers are asked the less difficult questions. (Do not make this obvious to the students.)

    Furthermore, since the purpose of the students being in front of the room and answering questions is for the remaining students to have their learning reinforced, the overachiever can add clarity to difficult questions and the accomplishment of the weaker student can add self-confidence to the remaining class members.

    Note: This step is not mandatory for the shotgun approach to be effective as a teaching method. If you choose to use this step, it is not imperative that an overachiever and underachiever be selected. This step does add variety and makes the class more interesting. Moreover, it increases the grades substantially.
15. After step 14, give the remaining students the test by giving a definition and having the students select the answer from the board. (It is unfair to assume that all students can learn to spell all twenty terms correctly in this amount of time.)

What are the results? Eighty percent of the students will hopefully score 100 if the approach is done correctly. However, it should be made clear that covering twenty questions is not the real goal. Fifteen questions may be all that can be covered. Allow approximately ten minutes for the testing. Stop at whatever term you are on at that point. The important thing is that the students thoroughly learn and perceive that material that is presented.

Why the Shotgun Approach Is Successful

1. **Respect for the Student as a Self-Actualizing Human Being.** According to Earl C. Kelly (1977), "What a person feels is far more important than what he knows." Kelly believes the following to be true:

   If I could achieve bringing about confidence and reduce or abolish fear, the learner would become more and more able to take in what there was to take from his environment.

   The shotgun approach emphasizes this in step 14 and step 7. Praise, confidence, and the reduction of fear of failure are essential in order for maximum learning to take place. Kelly makes the following comments (they are used here to justify the high grades and the positive attitudes that the shotgun approach instills in students).

   a. We need people who value what a human being is above the outside values so common in our society.
   b. The life that is good to live is primarily one in which the individual is loved and is able to feel it.
   c. Each person in our world is different from any other person who has been or will be.
   d. Students need to be involved in what they are doing.
   e. In order to live the life that is good to live, one needs to be respected as a person.
   f. Don't use threat or fear as a teaching technique. Don't use rejection as a teaching technique. I don't think that we have the right to reject anyone.
   g. What we need in this land of ours is better people; and if we are to have better people, the teachers of the nation will have to help produce them. What we need is a revolution among teachers on attitudes and emotions.

   Combs (1977), another leader in education, makes the following comments relative to self-actualization.

   a. Positive self-concept calls for successful experience.
   b. Success or failure does not happen unless the individual thinks it is so.
   c. We are still providing many children with experiences of failure and self-reduction, not because we want to but because we seek to force them into a common mold which they do not fit.
   d. We must have a sensitivity to people. (Recent research has shown that good teachers are such not because of method but because of sensitivity.)

   The previous comments have been emphasized because the biggest criticism the shotgun approach encounters is the high achievement of the students. Certain clubs or sponsors may resent the high grades of the traditionally low achievers. The beliefs of Kelly and Combs are emphasized because the shotgun approach defies many laws of statistical testing. For example, many people in education and/or statisticians say that a test question is invalid if X number of people answer it correctly. These people subscribe to the norm-referenced philosophy of teaching. This author, who subscribes to the criterion-referenced philosophy of teaching, has never been able to understand this when most "methods educators" recommend the identification of objectives and the teaching toward the objectives. If the teacher is a good enough teacher so that all the objectives are accomplished and the students achieve perfect scores, does it mean that all the questions should be disregarded? It means that the teacher is teaching, not merely presenting information and hoping that the students "get it." The use of the shotgun approach works under the philosophy that the teacher should be so effective that learning is so reinforced that students score high on tests, even days after the teacher has presented the subject matter. Retention of material learned is high.

2. **Thinking, Understanding, and Perception Are Maximized.** Hammonds (1968) says that "thinking may enrich things with meaning and may give increased understanding which constitutes a large part of learning. It calls for relationship-seeing discrimination." Understanding a thing consists largely in seeing why it is true. One cannot see why a thing is true unless he sees it in relation to other knowledge which he already has and understands (Hammonds, 1968). Ultimate understanding is full perception.

   The maximum observation of perception is one of the major factors in utilizing the shotgun approach. Perception goes further than understanding. For example, nobody has been able to teach or explain to this writer how the earth can be round without someone falling off. I know the earth is round. I have no doubts. I understand it. However, it still has not been perceived. No one has been able to relate this information to me so that "I can see it." Steps 3, 5, 9, 10, and 14 stress thinking, understanding, and perceiving.

3. **Awareness of Student's Perceptive Field.** Step 5 addresses this when mentioning that Jesus' method of teaching was that of teaching parables. He told stories that his listeners could relate to, young and old alike. He spoke to their perceptive field. The shotgun approach emphasizes this, as a story or incident is told about each term so that the student can relate this to something in his perceptive field. Combs (1977) makes the following comments about a person's perceptive field.

   a. The individual in a particular culture perceives those aspects of his environment that, from his point of view, he needs to perceive to maintain and enhance his self in the world in which he lives.
   b. Low scores on intelligence tests do not mean less rich and varied fields of perception; they may mean only fields of perception more widely divergent from those of the examiner.
   c. It is conceivable that low intelligence may be, at least in part, no more than a function of the goals an individual is striving to reach in achieving his needed satisfaction.
   d. The Circular Effect—If we teach a child that he is unable and if he believes us and behaves in these terms, we need not be surprised when we test his intelligence to discover that he produces at the level at which we taught him!
   e. It is notorious that children's grades vary very little from year to year through the course of schooling.

   Dewey (1938) addresses the issue, but calls it "purposes." Dewey explains purposes as follows.

   a. Observation of surrounding conditions.
   b. Knowledge of what has happened in similar situations in the past, a knowledge obtained partly by recollection and partly from the information, advice, and warning of those who have had a wider experience.
   c. Judgment that puts together what is observed and what is recalled to see what they signify.

4. **Reinforcement and Repetition.** Reinforcement and repetition are essential for a high rate of retention. The more use of reinforcement and repetition, the more learning that takes place. These two teaching-learning principles are utilized repeatedly from steps 1 through 15. Steward (1951) made the following comments about repetition.

   a. There is no getting away from this law of repetition as a determiner of what we are to remember.

b. Scientific investigations of learning prove conclusively that in many instances economy of learning is secured through drill.
c. The principle of repetition is a good one, but like everything else, it may be easily abused, and if carried to excess it may react and prove to be of positive disadvantage.

5. **Problem-Solving Approach.** The problem-solving approach is utilized in the shotgun approach in that students are asked questions. Questions are asked that make the student think. The situations are real. According to Krebs (1967), "Each teacher will use a problem-solving approach in a slightly different manner. He will adapt the approach to his own particular personality and way of doing things. The basic principle will remain the same." The example of the definitions utilizes the problem-solving approach as questions are asked. The shotgun approach would use another variation of the problem-solving approach if questions of the students were formulated and listed on the board. All other steps of the shotgun approach would be the same.
6. **Motivation.** Another crucial reason for the success of the shotgun approach is that it is motivating. The students get excited. Learning is a challenge to them. According to Frymier (1965):

   > *A teacher should capitalize upon forces within the individual for maximum learning, but to try to do things which motiviate students is ineffective in most cases. Teachers need to arrange their activities and organize their classrooms so that their operations function in harmony with students' motivations rather than against them.*

   Notice that Frymier said "ineffective in *most* cases" when referring to things that teachers try to motivate students. "This does not mean that teachers cannot arouse interest in a subject" (Frymier, 1965). The shotgun approach is one of the exceptions that does arouse interest and cause students to be motivated. Mouly (1960) offers the following comments about motivation (they are used here as examples of the effectiveness of the shotgun approach).

   a. Motivation stems directly from the concepts of needs, the self, and the phenomenal field, and incorporates not only an *energizing* of the individual but also an *orientation* of his behavior toward the attainment of certain *goals* or potential satisfiers for his needs.
   b. Not only is learning dependent on motivation, but the effectiveness of the learning is more or less proportional to the degree of motivation of the learner.
   c. Many incentives are available to the teacher in his attempt to tap motives existing in the child.
   d. Success is important from the standpoint of motivation, particularly as it leads to the development of a positive self-concept, and hence, to further success and further motivation.
   e. The teacher is, in the final analysis, the key to motivation in the classroom.

   Motivation in the shotgun approach is used in the question-techniques, parables, reinforcement and especially step 12 in utilizing the two students in front of the room.
7. **Concept of Needs.** Among other needs, the six psychological needs listed by Mouly (1960) are: affection, belonging, achievement, independence, social recognition, and self-esteem. With the exception of independence, proper utilization of the shotgun approach is effective just as any method of teaching should be. Of course, satisfying more concrete needs is an objective of the shotgun approach, but they are mentioned indirectly throughout this paper.
8. **Attention to Talents Beyond Academic or Intelligence.** Our education system is primarily based upon the academic (cognitive) or intelligence of students. Taylor's (1974) research found "intelligence tests" encompass only about eight talents (less than one-tenth of the 98 talents now known), therefore missing more than nine-tenths of the important intellectual talents now measurable." Taylor believes that instead of conceiving of students

as merely learners and reproducers, we should esteem them more highly as thinkers and producers, decision makers, communicators, forecasters, creators, and planners. Taylor found that students who rated average in academic and planning rated high in communicating and decision making. The point here is to reemphasize a point made earlier about the shotgun approach, that is, "what a person feels is far more important than what he knows." According to Purkey (1977), "The task of the teacher is to help each student gain a positive and realistic image of himself as a learner. The prevention of negative self-concepts is a vital first step in teaching."

Many other reasons could be given for the success or effectiveness of the shotgun approach as a method of teaching. However, the purpose of these comments is to introduce a method of teaching that applies to the best principles of teaching and learning as identified by some of the foremost educators in this century. Rosenshine and Furst (1971) list the following five criteria as the most important variables between teacher behavior and student achievement: clarity, variability, enthusiasm, task-orientation and/or businesslike behavior and student opportunity to learn criterion material. You can be the judge as to whether the shotgun approach meets these criteria. Finally, Warmbrod's (1977) study of this research substantiates and summarizes some of the characteristics that have been discussed previously concerning the shotgun approach. They are as follows.

a. Learning proceeds much more rapidly and is retained much longer when that which is learned possesses meaning, organization, and structure.
b. Success in achievement is one of the strongest motivation factors. Students who are successful and who therefore derive satisfaction from a learning activity are motivated toward additional learning.
c. The most effective effort is put forth by students when they attempt tasks that fall in the "range of challenge," not too easy and not too hard, in which success seems quite possible but not certain.
d. Behaviors that are rewarded (reinforced) are more likely to recur.
e. Sheer repetition without indications of improvement or any kind of reinforcement (reward) is a poor way to attempt to learn. When students are aware of their learning progress, their performance will be superior to what it would have been without such knowledge.
f. "Problem-oriented" approaches to teaching improve learning. The important thing is that the student in his learning, and in the teaching that accompanies it, should *inquire into* rather than be *instructed* in the subject matter.

In conclusion, the shotgun approach is used with maximum student learning as its goal. Schools exist for students to learn. Teaching is not a game. Teachers should never trick students or make learning a guessing game. The teacher should tell the students what he or she expects them to know or learn from the unit (objectives), and present the materials in such a fashion that optimum learning can take place. If the students score high on an evaluation, good. If they score low, the teacher in all likelihood is not teaching effectively. He or she is only presenting information or serving as a guide. This standard may seem high, but this is the first part of the twenty-first century. It is time for teachers to improve. Teachers must strive toward the goal that their students fully perceive any knowledge presented. Needless to say, the teacher cannot cover as much subject matter using the "shotgun approach" as with most other methods. However, if motivation, instilling confidence in students, making education enjoyable and, most of all, learning is the goal of the teacher, the shotgun approach should be used as an alternative teaching method.

Appendix B

Personality Characteristics

SANGUINE

- Outgoing, friendly, cheerful
- Talkative, fluent communicators
- Like people and want to be liked in return
- Usually are enthusiastic and pleasant
- In a new group, they may hold back at first, acting and feeling shy (until they sense acceptance)
- Are eclectic
- Like to develop people, build organization
- Get things accomplished through people
- Know lots of people
- Like teamwork, will involve people
- Like group discussions, encourage participation in decision making
- Tend to have the last comment; they add a P.S.
- Persuasive
- Tend to think they have told you something they have not
- Constantly selling, often themselves
- Concerned with how others respond to them
- Naturally optimistic
- Like to be noticed, often in the latest fashion
- Good in all kinds of selling

PHLEGMATIC

- Appear cool, calm, and controlled under pressure
- Appear stable, emotionally adjusted, in harmony with the world
- Roll with the punches, take things in stride
- Noted for persistence and cooperation
- Look for peace and harmony (referees)
- Have a long fuse; when they blow up, look out
- Dependable and reliable
- Do not like to be rushed at the last minute; plan ahead
- Inclined to make every move count
- Noted for good memory and being a good listener
- Organize their time to get work done on schedule
- Time, schedules, deadlines are important
- Like to know the time frame
- Warm and friendly (may be perceived as extrovert)
- Naturally good planners
- Make great friends; take time to listen
- Approach things in a methodical way
- Tend to hold things in; don't make waves

CHOLERIC

- Act on their environment rather than reacting to it
- Venturesome
- Naturally self-confident, high-ego people
- Hard-driving and decisive
- Candid (which others may take as criticism)
- More confidence in what they can do than in what others can do
- Will do details only if they relate to their results or control
- More interested in results than people
- Authoritative
- Tellers
- Must be challenged (will challenge others)
- Attack things aggressively
- Often do not realize how strongly they come across to others
- Will delegate responsibility (details) but not authority easily
- Hate having anyone looking over their shoulder
- Outspoken and direct
- Good troubleshooters
- Thrive on solving problems; when they run out, they look for more

MELANCHOLY

- They *hate* to make mistakes
- Keep things together in the company
- Naturally good organizers
- Usually careful, accurate, precise
- Will double-check themselves and others
- Are very loyal (to circle of structure)
- Naturally good with details
- There is a right way and a wrong way—right way is the *only* way
- Don't make many mistakes, but do make nervous mistakes
- Need a thorough knowledge of product to sell it
- Often rely on tradition
- Good at developing systems
- Meticulous, can be fussy
- Perfectionist
- Like to gather many facts before making decisions
- Do *not* like criticism (proves them wrong), but are often self-critical
- Quality oriented
- Appreciate knowing the rules, expectations, instructions
- Go by the book—their book

Appendix C

PERSONALITY PROFILE

Name ______________________________

DIRECTIONS: In each of the following rows of four words, place 'M' in front of the one word that most often applies to you and 'L' in front of the one word that least often applies to you. Be sure to mark every line.

STRENGTHS

1	______ Animated	______ Adventurous	______ Analytical	______ Adaptable
2	______ Persistent	______ Playful	______ Persuasive	______ Peaceful
3	______ Submissive	______ Self-sacrificing	______ Sociable	______ Strong-willed
4	______ Considerate	______ Controlled	______ Competitive	______ Convincing
5	______ Refreshing	______ Respectful	______ Reserved	______ Resourceful
6	______ Satisfied	______ Sensitive	______ Self-reliant	______ Spirited
7	______ Planner	______ Patient	______ Positive	______ Promoter
8	______ Sure	______ Spontaneous	______ Scheduled	______ Shy
9	______ Orderly	______ Obliging	______ Outspoken	______ Optimistic
10	______ Friendly	______ Faithful	______ Funny	______ Forceful
11	______ Daring	______ Delightful	______ Diplomatic	______ Detailed
12	______ Cheerful	______ Consistent	______ Cultured	______ Confident
13	______ Idealistic	______ Independent	______ Inoffensive	______ Inspiring
14	______ Demonstrative	______ Decisive	______ Dry humor	______ Deep
15	______ Mediator	______ Musical	______ Mover	______ Mixes easily
16	______ Thoughtful	______ Tenacious	______ Talker	______ Tolerant
17	______ Listener	______ Loyal	______ Leader	______ Lively
18	______ Contented	______ Chief	______ Chartmaker	______ Cute
19	______ Perfectionist	______ Permissive	______ Productive	______ Popular
20	______ Bouncy	______ Bold	______ Behaved	______ Balanced

(*continued*)

WEAKNESSES

21	______ Brassy	______ Bossy	______ Bashful	______ Blank
22	______ Disorderly	______ Cold-hearted	______ Dull	______ Grudgeholder
23	______ Reticent	______ Resentful	______ Resistant	______ Repetitious
24	______ Fussy	______ Fearful	______ Forgetful	______ Frank
25	______ Impatient	______ Insecure	______ Indecisive	______ Interrupts
26	______ Disliked	______ Spectator	______ Impulsive	______ Frigid
27	______ Headstrong	______ Haphazard	______ Hard to please	______ Hesitant
28	______ Plain	______ Pessimistic	______ Proud	______ Permissive
29	______ Anger easily	______ Aimless	______ Argumentative	______ Alienated
30	______ Naïve	______ Negative attitude	______ Nervy	______ Nonchalant
31	______ Worrier	______ Withdrawn	______ Workaholic	______ Wants credit
32	______ Too sensitive	______ Tactless	______ Timid	______ Talkative
33	______ Doubtful	______ Disorganized	______ Domineering	______ Depressed
34	______ Inconsistent	______ Introvert	______ Intolerant	______ Indifferent
35	______ Messy	______ Moody	______ Mumbles	______ Manipulative
36	______ Slow	______ Stubborn	______ Show-off	______ Skeptical
37	______ Loner	______ Lord over	______ Lazy	______ Loud
38	______ Sluggish	______ Suspicious	______ Short-tempered	______ Scatterbrained
39	______ Revengeful	______ Restless	______ Reluctant	______ Rash
40	______ Compromising	______ Critical	______ Crafty	______ Changeable

PERSONALITY SCORING SHEET

Name ______________________________

STRENGTHS

	Sanguine ____	Choleric ____	Melancholy ____	Phlegmatic ____
1	______ Animated	______ Adventurous	______ Analytical	______ Adaptable
2	______ Playful	______ Persuasive	______ Persistent	______ Peaceful
3	______ Sociable	______ Strong-willed	______ Self-sacrificing	______ Submissive
4	______ Convincing	______ Competitive	______ Considerate	______ Controlled
5	______ Refreshing	______ Resourceful	______ Respectful	______ Reserved
6	______ Spirited	______ Self-reliant	______ Sensitive	______ Satisfied
7	______ Promoter	______ Positive	______ Planner	______ Patient
8	______ Spontaneous	______ Sure	______ Scheduled	______ Shy
9	______ Optimistic	______ Outspoken	______ Orderly	______ Obliging
10	______ Funny	______ Forceful	______ Faithful	______ Friendly
11	______ Delightful	______ Daring	______ Detailed	______ Diplomatic
12	______ Cheerful	______ Confident	______ Cultured	______ Consistent
13	______ Inspiring	______ Independent	______ Idealistic	______ Inoffensive
14	______ Demonstrative	______ Decisive	______ Deep	______ Dry humor
15	______ Mixes easily	______ Mover	______ Musical	______ Mediator
16	______ Talker	______ Tenacious	______ Thoughtful	______ Tolerant
17	______ Lively	______ Leader	______ Loyal	______ Listener
18	______ Cute	______ Chief	______ Chartmaker	______ Contented
19	______ Popular	______ Productive	______ Perfectionist	______ Permissive
20	______ Bouncy	______ Bold	______ Behaved	______ Balanced

(continued)

WEAKNESSES

	Sanguine ____	Choleric ____	Melancholy ____	Phlegmatic ____
21	______ Brassy	______ Bossy	______ Bashful	______ Blank
22	______ Disorderly	______ Cold-hearted	______ Grudgeholder	______ Dull
23	______ Repetitious	______ Resistant	______ Resentful	______ Reticent
24	______ Forgetful	______ Frank	______ Fussy	______ Fearful
25	______ Interrupts	______ Impatient	______ Insecure	______ Indecisive
26	______ Impulsive	______ Frigid	______ Disliked	______ Spectator
27	______ Haphazard	______ Headstrong	______ Hard to please	______ Hesitant
28	______ Permissive	______ Proud	______ Pessimistic	______ Plain
29	______ Angered easily	______ Argumentative	______ Alienated	______ Aimless
30	______ Naïve	______ Nervy	______ Negative attitude	______ Nonchalant
31	______ Wants credit	______ Workaholic	______ Withdrawn	______ Worrier
32	______ Talkative	______ Tactless	______ Too sensitive	______ Timid
33	______ Disorganized	______ Domineering	______ Depressed	______ Doubtful
34	______ Inconsistent	______ Intolerant	______ Intolerant	______ Indifferent
35	______ Messy	______ Manipulative	______ Moody	______ Mumbles
36	______ Show-off	______ Stubborn	______ Skeptical	______ Slow
37	______ Loud	______ Lord over others	______ Loner	______ Lazy
38	______ Scatterbrained	______ Short tempered	______ Suspicious	______ Sluggish
39	______ Restless	______ Rash	______ Revengeful	______ Reluctant
40	______ Changeable	______ Crafty	______ Critical	______ Compromising

TOTALS:

M	______	______	______	______
L	______	______	______	______

COMBINED TOTALS:

M	______	______	______	______
L	______	______	______	______

PERSONALITY PROFILE

1. Animated	I am vigorous and lively.
Adventurous	I love adventure.
Analytical	I examine the nature of things.
Adaptable	I'm able to change without difficulty.
2. Persistent	I'm stubborn and will keep on.
Playful	I'm fond of play and fun.
Persuasive	I can convince you of most anything!
Peaceful	I'm calm, quiet, and tranquil.
3. Submissive	I will yield without resistance.
Self-sacrificing	I'll sacrifice myself and my interest usually for others.
Sociable	I'm friendly and love and company of others.
Strong-willed	I have a strong, unyielding mind.
4. Considerate	I'm thoughtful of others.
Controlled	I'm usually under someone else's direction.
Competitive	A good contest or rivalry interests me.
Convincing	I'll persuade you one way of the other.
5. Refreshing	You'll feel renewed after being around me.
Respectful	I show consideration.
Reserved	I keep my thoughts to myself and I'm rather withdrawn.
Resourceful	I am able to deal promptly with problems.
6. Satisfied	I'm content.
Sensitive	I'm touchy and easily hurt.
Self-reliant	I'm comfortable with my own judgment.
Spirited	I'm lively and energetic.
7. Planner	I'm structured and scheduled.
Patient	I'm tolerant and can endure without complaining.
Positive	I may be overconfident and opinionated.
Promoter	I'm good at getting things started.
8. Sure	I am certain without doubt.
Spontaneous	A lot of the time, I act on impulse.
Scheduled	I'm a real planner.
Shy	I'm timid and bashful.
9. Orderly	I am neat and tidy.
Obliging	You can count on me to help you out.
Outspoken	I'm very frank and candid.
Optimistic	I look on the bright side.
10. Friendly	I am supporting and helpful.
Faithful	You can rely on me.
Funny	You'll enjoy my sense of humor.
Forceful	I have the power to control, persuade, and influence.

(*continued*)

11. Daring	I am fearless and bold.
Delightful	I am a very pleasing and charming person.
Diplomatic	I'm very tactful in dealing with people.
Detailed	I'll deal with every item.
12. Cheerful	I'm joyful and glad.
Consistent	I hold to principles.
Cultured	I'm fairly refined.
Confident	I'm sure of myself.
13. Idealistic	I look at things the same way they should be.
Independent	I am self-confident and free from the rule of anyone else.
Inoffensive	I'll cause no harm or discomfort.
inspiring	I'll give you guidance and motivation.
14. Demonstrative	I show my feelings openly with frankness.
Decisive	I show determination and firmness.
Dry humor	I'm amusing in a low-key way.
Deep	I'm a serious person and may be hard to understand.
15. Mediator	I'll intercede for you.
Musical	I'm fond of and usually talented in music.
Mover	I won't be in a the same place for long.
Mixes easily	I'll fit in with most anyone.
16. Thoughtful	I show consideration for others' feelings.
Tenancious	I'm persistent and stubborn.
Talker	I love to talk.
Tolerant	I'll allow you to have your views and practices.
17. Listener	I will pay attention while you talk.
Loyal	I'm faithful.
Leader	I have the ability to lead or direct.
Lively	I'm never still a minute.
18. Contented	It doesn't take a lot to keep me happy.
Chief	Leave it to me to be in command.
Chartmaker	I plan and schedule almost everything.
Cute	I'm rather clever and sharp.
19. Perfectionist	Everything has to be done correctly.
Permissive	I allow others to control me.
Productive	I turn out a lot of work.
Popular	I am well liked by most people.
20. Bouncy	I'll be jumping around everywhere.
Bold	I am daring and fearless.
Behaved	You won't find me raising a ruckus.
Balanced	I am a well-rounded person.

21. Brassy	I'm a bit of a show-off.
Bossy	I'll tell you what to do and when to do it.
Bashful	I'm shy.
Blank	I'm not a very interesting person.
22. Disorderly	Everything around me is a mess.
Cold-hearted	I am unsympathetic and unfeeling.
Dull	I'll bore you to tears.
Grudgeholder	I hold resentment.
23. Reticient	I'm usually silent and uncommunicative.
Resentful	Sometimes I feel offended.
Resistant	I will not yield; I will withstand.
Repetitious	I do or say the same thing over and over again.
24. Fussy	I worry over the smallest details.
Fearful	I'm afraid
Forgetful	Count on me to forget!
Frank	I'm outspoken and candid.
25. Impatient	I have very little patience.
Insecure	I am not secure in my feelings.
Indecisive	I have a hard time making a decision.
Interrupts	I'd rather talk than listen.
26. Disliked	Many people do not like me.
Spectator	I'd rather watch than participate.
Impulsive	I may act on impulse without much thought.
Frigid	I may be cold.
27. Headstrong	You'll have trouble changing my mind.
Haphazard	I'm very casual and don't make plans.
Hard to please	Yes, I'm very hard to please.
Hestitant	I'm unsure of myself.
28. Plain	I don't call attention to myself.
Pessimistic	I usually look on the dark side of things.
Proud	I show pride in myself; I'm rather arrogant.
Permissive	I'll allow almost anything.
29. Anger easily	I get mad easily.
Aimless	I don't have a lot of goals.
Argumentative	I am apt to argue about anything.
Alienated	I'm unfriendly.
30. Naive	I'm simple and childlike.
Negative attitude	I don't think things wil work out.
Nervy	I'm fairly bold and daring.
Nonchalant	I show very little enthusiasm.

(*continued*)

31. Worrier	I worry about almost everything.
Withdrawn	I'm shy and reserved.
Workaholic	You will usually find me working overtime.
Wants credit	I'll do it but I want the credit for it!
32. Too sensitive	My feelings are hurt very easily.
Tactless	I may be offensive.
Timid	I'm pretty shy.
Talkative	I love to talk—all the time.
33. Doubtful	I'm uncertain and unsure.
Disorganized	I don't keep anything in order.
Domineering	I'm overbearing and arrogant.
Depressed	I feel gloomy, dejected.
34. Inconsisent	I'm changeable.
Introvert	I direct my interests toward myself.
Intolerant	I will not put up with much.
Indifferent	I don't show my feelings.
35. Messy	I'm unorganized and disorderly.
Moody	I'm prone to get into moods.
Mumbles	You may not be able to hear or understand me.
Manipulative	I can control you by my influence.
36. Slow	I don't hurry.
Stubborn	I'm hard to change and refuse to yield.
Show-off	Look at me, look at me!
Skeptical	I'm doubting and hard to convince.
37. Loner	I'd rather spend my time alone.
Lord over	I'll be the boss.
Lazy	I'm not eager or willing to work.
Loud	I'm noisy and clamorous.
38. Sluggish	I tend to be lazy and slow.
Suspicious	I can believe something is bad without much evidence.
Short tempered	I get angry easily.
Scatterbrained	I forget things; I'm not very responsible.
39. Revengeful	I hope to hurt you in return for what you have done to me.
Restless	I'm readyy to move on to where the action is.
Reluctant	I am unwilling to do things.
Rash	I act and speak with haste.
40. Compromising	I'm will to give up some of my demands.
Critical	I am a fault-finder.
Crafty	I may be deceitful and cunning.
Changeable	I may act differently tomorrow.

Appendix D

Communication Styles

	Socializer	Director	Thinker	Relater
Behavior Pattern	Open/Direct	Self-contained/ Direct	Self-contained/ Indirect	Open/Indirect
Performance Equivalent	Influences others	Dominance	Compliance	Steadiness
Appearance	• Fashionable • Stylish	• Businesslike • Functional	• Formal • Conservative	• Casual • Conforming
Work space	• Stimulating • Personal • Cluttered • Friendly	• Busy • Formal • Efficient • Structured	• Structured • Organized • Functional • Formal	• Personal • Relaxed • Friendly • Informal
Pace	Fast/Spontaneous	Fast/Decisive	Fast/Systematic	Slow/Easy
Priority	Relationships: Interacting	The Task: The Results	The Task: The Process	Maintaining relationships
Fears	Loss of prestige	Loss of Control	Embarrassment	Confrontation
Under Tension Will	Attack/Be Sarcastic	Dictate/ Assert	Withdraw/ Avoid	Submit/ Acquiesce
Seeks	Recognition	Productivity	Accuracy	Attention
Needs to Know (Benefits)	How it enhances their status/ who else uses it	What it does/ by when/what it costs	How they justify the purchase logically/how it works	How it will affect their personal circumstances
Gains Security By	Flexibility	Control	Preparation	Close relationships

(*continued*)

	Socializer	**Director**	**Thinker**	**Relater**
Wants to Maintain	Status	Success	Credibility	Relationships
Support Their	Ideas	Goals	Thoughts	Feelings
Achieves Acceptance By	• Playfulness • Stimulating environment	• Leadership • Competition	• Correctness • Thoroughness	• Conformity • Loyality
Likes You to Be	Stimulating	To the point	Precise	Pleasant
Wants to Be	Admired	In charge	Correct	Liked
Irritated By	• Boredom • Routine	• Inefficiency • Indecision	• Surprises • Unpredictability	• Insensitivity • Impatience
Measures Personal Worth By	• Acknowledgment • Recognition • Compliments	• Results • Track record • Measurable progress	• Precision • Accuracy • Activity	• Compatibility with others • Depth of relationships
Decisions Are	Spontaneous	Decisive	Deliberate	Considered

Appendix E

Working with the Four Personality Types and Communication Styles

Sanguines/Socializers

Co-Workers Be active with them and do not slow them down. Be energetic and ready to go. Be adventurous, optimistic, spontaneous, and fun. Compete in fun when appropriate.

Leaders Because they like co-workers to solve problems and to move quickly, be open and ready to wing it. Be willing to work on your own in hands-on activities. Respect their changes of direction and fast pace. Get actively involved in the workplace.

Parents/Adults Compliment their generosity and sense of humor. Use a direct, right-to-the point approach. Get involved in physical activities with them. Respect their lack of structure and need for spontaneity.

Cholerics/Directors

Co-Workers Try to be organized, efficient, dependable, and loyal. Remember to be on time. They are generous and like things to be returned. Respect their need for security and do what you say you will do.

Leaders Pay attention to details and be neat and orderly. Respect their need for rules and regulations. They value their positions as leaders, so follow directions carefully. Make an extra effort to be on time.

Parents/Adults Respect their need for tradition and stability. Be loyal, dependable, and truthful. Be up front with them and understand their desire for structure and security. Be clean and neat in appearance.

Phlegmatics/Relaters

Co-Workers Spend quality time, one-on-one, with them. Be aware that they wear their hearts on their sleeves. Listen to them as they listen to you and be supportive. Share your thoughts and feelings and praise their imagination and creativity.

Leaders Respect their concern for feelings of co-workers. Get along with other co-workers on the job. Offer your ideas and feelings and give the leader positive feedback. Appreciate their warmth and caring attitude. Be dramatic and expressive.

Parents/Adults Respect their need to know about you. Be truthful, sincere, helpful, open, and communicative. Take a creative approach to problem solving and cooperate with other family members. Show that you value them through thoughtfulness.

Melancholy/Thinkers

Co-Workers Be aware of their curiosity about life. Respect their need for independence and know that they are caring even though they may not show their feelings easily. Reinforce their new ideas and concepts.

Leaders Be curious, observing, and ask lots of questions. Be open to their ideas and praise their competence and knowledge. Because they like to say it once, pay attention the first time.

Parents/Adults Respect their preoccupation with wisdom, knowledge, ideas, and logic. Help them with the day-to-day details and praise their ingenuity and intelligence. Think ahead because they are future-oriented.

Appendix F

Parliamentary Procedure Career Development Event

IMPORTANT NOTE: *Please thoroughly read the Introduction Section at the beginning of this handbook for complete rules and procedures that are relevant to all National FFA Career Development Events.*

I. PURPOSE

To encourage students to learn to effectively participate in a business meeting and to assist in the development of their leadership skills.

II. OBJECTIVES

Students will be able to:

1. Use parliamentary procedure to conduct an orderly and efficient meeting.
2. Demonstrate knowledge of parliamentary law.
3. Present a logical, realistic and convincing debate on motions.
4. Record complete and accurate minutes.

III. EVENT RULES

1. **Team Make-up**- A team representing a state will consist of six members and one designated alternate listed at the state and national levels from the same chapter. The alternate is not permitted to observe competing teams, but may observe their own team. The alternate may replace a regular team member prior to the start of the event.
2. The event is open to one team per state as certified by the State Supervisor to the Teacher Services Specialist in charge of National Career Development Events.
3. The event will have four phases: written examinations, a ten minute team presentation of parliamentary procedure, oral questions following the presentation, and minutes prepared by the team secretary in consultation with the team chair.
4. The advisor shall not consult with the team after beginning the event.
5. Official FFA dress is highly recommended for participation in the parliamentary procedure career development event. Official FFA dress is required for the awards banquet.

IV. EVENT FORMAT

A. EQUIPMENT

Materials student must provide- Each participant must bring a minimum of two sharpened No. 2 pencils.

B. TEAM ACTIVITY

Presentation (750 points)

1. The national event will have three rounds, a preliminary round, a semi-final round and a final round. The preliminary round will have four to six sections.A section shall be made up of six to nine teams. Two teams from each of the sections, for a total of twelve teams will advance to the semi-final round. The semi-final

round is composed of two sections with six teams in each section. Two teams in each semi-final section will advance to the final round of four teams.

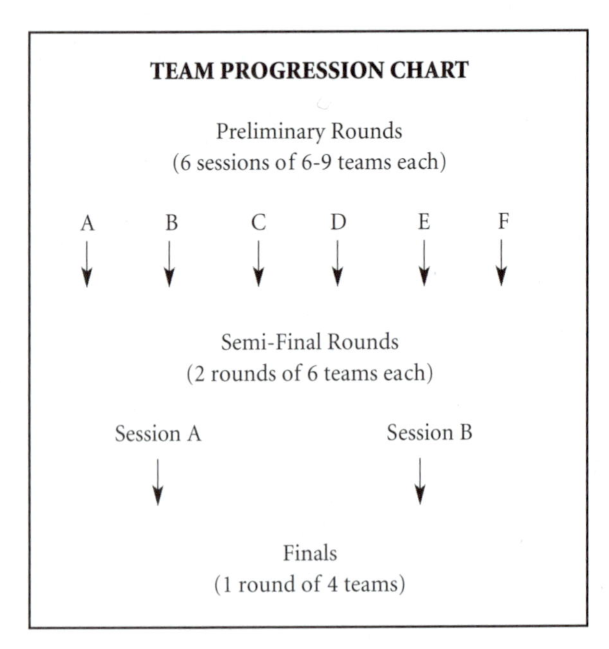

2. Teams will be placed into preliminary and semi-final rounds based on a procedure determined by the officials in charge of the event.

3. Each team will address a local chapter item of business which would normally be a part of a chapter's Program of Activities (e.g. Food for America, Project PALS, WEA, fundraisers, recreation, etc.) Consult the Official FFA Manual and Student Handbook for specific activities. The motion will be specific and must be moved as it is written on the card.

4. The team demonstrating shall assume that a regular chapter meeting is in progress and the chair shall start the presentation by saying, "Is there any new business that should be presented at this time?"A team member will then move the main motion assigned to the team.

5. The event official will assign the main motion on a 3"x 5"card. This is to be the first item of business presented. All teams in each session will use the same main motion. It is suggested that this main motion should be the first motion presented, unless orders of the day, take from the table, reconsider or rescind are required.

6. The event officials will select two subsidiary, two incidental and one privileged or unclassified motion from the list of permissible motions. These motions will be on a 3"x5"card and one will be randomly assigned to each team member. All teams in each session will be assigned the same motions. Team members will have one minute to review the main motion, the motions to be demonstrated and to identify his/her motion (which may be noted by bolding, underlining or highlighting). Members may not confer during the one-minute time period.

SAMPLE CARD

Main Motion:
I move to sell citrus as a fundraiser.

Required Motions:
Lay on the Table
Amend
Suspend the Rules
Appeal
Reconsider

7. There shall be no limitation to the number of subsidiary, incidental, privileged and unclassified motions demonstrated except that the team must demonstrate two subsidiary, two incidental and one privileged or unclassified motions designated by the officials in charge. The team may use more than one main motion as long as it pertains to the assigned main motion. While acceptable, this practice is strongly discouraged.

8. An alternate main motion not pertaining to the main motion may be used to facilitate the correct demonstration of the motion, "Call for the orders of the day," should that privileged motion be designated as one to be demonstrated by the officials in charge.

9. If the officials in charge designate "rescind, reconsider or take from the table", as a motion to be demonstrated, you could assume that you would rescind an action taken, which cannot be reconsidered, or take from the table a motion or reconsider a motion you did earlier in the present meeting. Example:"I move to rescind the motion that was passed at our last meeting about having an FFA hayride." These motions shall not be used unless they are a required motion. Unrealistic or canned debate on rescind or reconsider may be penalized at the judges discretion.

10. The top four debates per member will be tabulated in the presentation score. No more than two debates per member per motion will be tabulated.

11. A member's required motion will not be counted as an additional motion for another member. The person who makes the assigned main motion will be given credit for an additional motion (20 pts). Credit for an additional motion will only be given one time (Example: Division of the Assembly can only be used once for credit). If an alternative main motion is used, the member will be given credit for an additional motion.

12. A team shall be allowed 10 minutes in which to demonstrate knowledge of parliamentary law. Thirty (30) seconds past 10 minutes will be allowed without penalty. A deduction of 50 points will be made for every additional 30 seconds or major fraction thereof. A timekeeper will furnish the time used by each team at the close of the event.

Minutes	Points Deducted
10:00 - 10:30	0
10:31 - 11:00	50
11:01 - 11:30	100
11:31 - 12:00	150
12:01 - 12:30	200
12:31 - 13:00	250

Oral Questions (100 points)

Each of the six-team members will be asked a planned question relating to their assigned motion. No one may step forward to help correct answers on the first six questions at any time. Following these six questions,the judges will have two additional minutes to ask questions for clarification of the presentation, after which, time will be called.

Presentation Minutes (50 points)

Each team will have a secretary take minutes of the presentation. A possible score of 50 points will be allowed for the minutes. Pencils and paper will be supplied to take notes during the presentation. Following the presentation, the secretary in consultation with the chair, will have thirty minutes to prepare the official minutes. Notes taken by the secretary during the presentation must be turned in with the official copy of the minutes on Form 1. (The lowest possible score for the section is zero (0).) Event officials shall use Form 3 to score the official minutes of the presentation.

Instructions on Minutes

1. Use the example of proper minutes as illustrated in the Official FFA Secretary's Book and/or outlined in Robert's Rules of Order Newly Revised.

2. A dictionary will be permitted for writing the official minutes of the presentation.

3. The minutes will begin by recording the first item of business presented. Opening ceremonies and other preliminary information will not be used. Example:"It was moved by John Smith and seconded by Jill Jones to start the Food For America program on December 1."

4. The chair and the secretary may consult in preparing the official minutes of the presentation.A total of 30 minutes will be allowed to prepare the minutes.

5. A judge will read, review and grade the official minutes of the presentation after completion of each round of the event. The scores will be provided to the presentation judges for use in computing final scores.

B. INDIVIDUAL ACTIVITY

WRITTEN TEST (100 POINTS)

A written test will consist of 25 objective-type multiple choice questions covering basic parliamentary law and information pertaining to minutes. Thirty minutes will be allowed to complete the test. Each participant may score a maximum of 100 points. Alternates do not take the written test. The average score of the six-team members will be used to compute the total team score in each round.

V. SCORING

Guidelines for Scoring Discussion

1. It is essential that each judge observes and maintains consistent criteria in scoring discussions for the duration of the event.

2. Judges must overlook personal opinions and beliefs and score discussion in an unbiased manner. All discussion should be scored at the time it is delivered.

3. Characteristics of effective discussion include a) completeness of thought, b) logical reasoning, c) clear statement of speaker's position, d) conviction of delivery, and e) concise and effective statement of discussion.

4. A suggested grading scale is as follows:

   Excellent ....16-20 points
   Good..11-15 points
   Average ....6-10 points
   Poor ....0-5 points

5. An excellent discussion would be extremely unusual and would be characterized by a truly stirring delivery and brilliant in terms of information provided and/or suggestions for action offered. Poor discussion would be characterized by a lack of effective delivery, poor grammar, reasoning and substance. An example might be:"I think this is a good idea."

6. Most discussion would fall in the range of 8-15 points. An example of a discussion might be:"I think this is a very significant motion which should be passed for the following reasons (new, informative and logically related)." Each debate should have a logical conclusion. Good discussion would be characterized by effective delivery, substance,creative and visionary thought delivered in a convincing and compelling manner.

7. Each time a participant in the presentation discusses any motion, they may earn a score. However, an individual may never earn more than 60 points in a given presentation. Furthermore, no more than 20 points may be earned during one recognition by the chair.

8. The top four debates per member will be tabulated in the presentation score. No more than two debates per member per motion will be tabulated.

Guidelines for Scoring the Chair

1. Ability to preside – handling of motions, keeping members informed,use of the gavel, distribution of discussion. (80 points)

2. Leadership – stage presence, poise, self-confidence, politeness and voice. (20 points)

The judges will use Form 2 to score the event. The top two teams will be ranked first and second based on the judges' lowest combined rank. The remaining teams will be designated silver or bronze awards.

SCORING

Phase	Breakdown of Points	Section Points	Total Points
Written Test (average of 6 members scores)			100
Presentation			750
Total of 5 members on the floor		500	
Required motion	20		
Discussion (max. of 4 debates @ 20 pts. each)	60		
Additional motion (incl. main or alternate main motion)	20		
Chair		100	
Ability to preside	80		
Leadership	20		
Teams's General Effect		150	
Conclusions Reached by Team (Team's use of motions and discussion support disposal of the main motion)	50		
Team Effect (Degree to which discussion was convincing, logical, realistic, orderly, and efficient)	50		
Team's voice, poise, expression and appearance	50		
Oral Questions			100
Total for members' questions (6x12 pts)	72		
Additional clarification questions (2 min.)	28		
Minutes of Presentation			50
Completeness and Accuracy	25		
Format	10		
Grammar, style,legibility	15		
Deductions			
Deductions for parliamentary mistakes	5-20 pts/minor mistake		
Deductions for omitting assigned motion	50		
Deductions for going overtime	50 pts for every 30 seconds over 10:30		
TOTAL			1,000

VI. TIEBREAKERS

Tiebreakers for teams will be:

1. the total presentation score,
2. the team's average score on the written test
3. the total score for questions.

VII. AWARDS

Awards will be presented at an awards ceremony. Awards are presented to teams as well as individuals based upon their rankings. Awards are sponsored by a cooperating industry sponsor(s) as a Special Project, and/or by the general fund of the National FFA Foundation.

Each state will be provided a plaque for their state winning team. The first place national team will be presented a trophy plaque. Each member of the first place team will be presented an individual team member plaque.A national gold plaque and individual medals will be presented to the top 12 teams competing in the event; silver plaques and individual medals to the middle 18; and remaining teams and individuals competing will receive bronze. The top four teams will each receive a designated gold plaque.

Scholarships may be awarded as funded by special project sponsors. Collegiate scholarships awarded to FFA members competing at or above the local level in parliamentary procedure events may be available.

VIII. REFERENCES

This list of references is not intended to be inclusive. Other sources may be utilized and teachers are encouraged to make use of the very best instructional materials available. The following list contains references that may prove helpful during event preparation.

The official text will be the latest edition of *Robert's Rules of Order* Newly Revised (currently 1990 edition).

Additional references may include *FFA New Horizons* magazine, the Official FFA Manual, the FFA Student Handbook and the Official Chapter Secretary's Book.

National FFA Parliamentary Procedure CDE

Chapter: ______________________ Team No.: ______________

State: ______________________

Team Score Sheet

Contestants	Required Motion [20 pts. max/mber]	Discussion [60 pts. max/mber] [20 pts. Max/item]	Additional Motion [20 pts. max/mber]	Individual Questions [12 pts. max/mber]	Individual Total 112 pts. max/mber
1					
2					
3					
4					
5					
Chair	Chair's Question ~ [12 points maximum]				
	Ability to Preside ~ handling of motions, keeping members informed, use of the gavel, distribution of discussion. [80 points maximum]				
	Leadership ~ stage presence, poise, self-confidence, politeness and voice [20 points maximum]				
Team's General Effect	Conclusions Reached by Team ~ [50 points maximum]				
	Team Effect ~ [50 points maximum]				
	Team's voice, poise, expression and appearance ~ 50 points maximum				
General Questions	Questions for clarification of the presentation (2 minutes allowed) [28 points maximum]				
Written Test	Average of the six individual test scores [100 points maximum]				
	Subtotal				
Deductions	Deductions for parliamentary mistakes ~ 5 to 20 points per mistake, omitting assigned motion - 50 points				
	Deductions for overtime ~ 50 points for every 30 seconds or major fraction thereof over 10:30				
	Total Team Score				

Appendix G

Careers in Agriculture

AGRICULTURAL PRODUCTION

- Agronomist
- Aquaculturist
- Certified Seed Grower
- Dairy Farmer
- Farm Manager
- Fruit Grower
- Kennel Operator
- Mushroom Grower
- Pet Shop Operator
- Shrimp Farmer
- Turkey Producer
- Animal Breeder
- Beekeeper
- Citrus Grower
- Dairy Herdsperson
- Farmer
- Horse Rancher
- Livestock Producer
- Nut Orchardist
- Potato Grower
- Tobacco Grower
- Vegetable Grower
- Animal Keeper
- Cash Grain Farmer
- Cotton Farmer
- Dairy Management Specialist
- Fish Farmer
- Horse Trainer
- Milking Machine Operator
- Orchard Supervisor
- Poultry Hatchery Manager
- Tree Farmer
- Animal Trainer
- Cattle Rancher
- Custom Operator
- Diversified Crop Farmer
- Fish Hatchery Manager
- Hydroponics Grower
- Mink Producer
- Peanut Producer
- Rice Farmer
- Tree & Vine Fruit Grower

AGRICULTURAL SUPPLIES AND SERVICES

- Aerial Crop Duster
- Health Products Distributor
- Artificial Inseminator
- Dog Groomer
- Field Inspector
- Harvest Contractor
- Poultry Field Service Technician
- Service Technician
- Agricultural Aviator
- Animal Inspector
- Chemical Applicator
- Farrier
- Field Sales Representative, Animal Health Products
- Insect & Disease Inspector
- Poultry Inseminator
- Sheep Shearer
- Ag Chemical Dealer
- Artificial Breeding Distributor
- Chemical Distributor
- Feed Mill Operator
- Field Sales Representative, Crop Chemicals
- Ova Transplant Specialist
- Sales Manager
- Veterinarian
- Artificial Breeding Technician
- Dairy Nutrition Specialist
- Fertilizer Plant Supervisor
- Grain Elevator Operator
- Pest Control Technician
- Salesperson
- Animal Groomer

AGRICULTURAL MECHANICS

- Ag Equipment Dealer
- Ag Plumber
- Field Sales Representative, Agricultural Equipment
- Land Surveyor
- Safety Inspector
- Ag Construction Engineer
- Ag Safety Engineer
- Heavy Equipment Operator
- Machinist
- Soil Engineer
- Ag Electrician
- Diesel Mechanic
- Hydraulic Engineer
- Parts Manager
- Welder
- Ag Equipment Designer
- Equipment Operator
- Irrigation Engineer
- Research Engineer

AGRICULTURAL PRODUCTS

- Butcher
- Food & Drug Inspector
- Fruit Distributor
- Hog Buyer
- Meat Cutter
- Quality Control Supervisor
- Wool Buyer
- Weights & Measures Official
- Cattle Buyer
- Food Chemist
- Fruit Press Operator
- Livestock Commission Agent
- Milk Plant Supervisor
- Tobacco Buyer
- Cotton Grader
- Food Processing Supervisor
- Grain Broker
- Livestock Yard Supervisor
- Produce Buyer
- Federal Grain Inspector
- Fruit & Vegetable Grader
- Grain Buyer
- Meat Inspector
- Produce Commission Agent
- Winery Supervisor

(continued)

Careers in Agriculture (continued)

HORTICULTURE

- Floral Designer
- Golf Course Superintendent
- Hydroponics Grower
- Plant Breeder
- Floral Shop Operator
- Greenhouse Manager
- Landscape Architect
- Turf Farmer
- Florist
- Greenskeeper
- Landscaper
- Turf Manager
- Flower Grader
- Horticulturist
- Nursery Operator

AGRICULTURAL RESOURCES

- Animal Behaviorist
- Environmentalist
- Game Farm Supervisor
- Range Conservationist
- Water Resources Manager
- Animal Ecologist
- Fire Warden
- Game Warden
- Resource Manager
- Wildlife Manager
- Animal Taxonomist
- Forest Firefighter/Warden
- Ground Water Geologist
- Soil Conservationist
- Environmental Conservation Officer
- Forest Ranger
- Park Ranger
- Trapper

FORESTRY

- Christmas Tree Grader
- Logging Operations Inspector
- Forester
- Lumber Mill Operator
- Forest Ranger
- Timber Manager
- Log Grader
- Tree Surgeon

OTHER

- Ag Accountant
- Ag Corporation Executive
- Ag Extension Specialist
- Ag Market Analyst
- Agronomist
- Animal Physiologist
- Biochemist
- Botanist
- Computer Specialist
- Electronic Editor
- Equine Dentist
- Farm Broadcaster
- Horticulture Instructor
- International Specialist
- Limnologist
- Marketing Analyst
- Nematologist
- Parisitologist
- Plant Geneticist
- Pomologist
- Publisher
- Scientific Artist
- Soil Scientist
- Viticulturist
- Ag Advertising Executive
- Ag Economist
- Ag Journalist
- Ag Mechanics Teacher
- Animal Cytologist
- Animal Scientist
- Bioengineer
- Computer Analyst
- Credit Analyst
- Embryologist
- 4-H Youth Assistant
- Farm Investment Manager
- Foreign Affairs Official
- Hydrologist
- Invertebrate Zoologist
- Magazine Writer
- Media Buyer
- Organic Chemist
- Photographer
- Plant Nutritionist
- Poultry Scientist
- Reproductive Physiologist
- Scientific Writer
- Vertebrate Zoologist
- Agricultural Education Instructor/FFA Advisor
- Ag Association Executive
- Ag Educator
- Ag Lawyer
- Ag News Director
- Animal Geneticist
- Avian Veterinarian
- Biophysicist
- Computer Operator
- Entomologist
- Farm Appraiser
- Feed Ration Developer & Analyst
- Graphic Designer
- Ichthyologist
- Lab Technician
- Mycologist
- Mammologist
- Meteorological Analyst
- Ornithologist
- Plant Cytologist
- Plant Pathologist
- Public Relations Manager
- Rural Sociologist
- Silviculturist
- Veterinary Pathologist
- Ag Consultant
- Ag Extension Agent
- Ag Loan Officer
- Agricultural Attaché
- Animal Nutritionist
- Bacteriologist
- Biostatistician
- Computer Salesperson
- Dendrologist
- Environmental Educator
- Farm Auctioneer
- Fiber Technologist
- Herpetologist
- Information Director
- Land Bank Branch Manager
- Marine Biologist
- Microbiologist
- Paleobiologist
- Plant Ecologist
- Plant Taxonomist
- Publicist
- Satellite Technician
- Software Reviewer
- Virologist

Adapted from: *Think About It,* National FFA Center, Alexandria, VA 22309-0160.

Appendix H

Jobs in Agriculture

- Agricultural Advertiser Account Executive
- Agricultural Agency Employee
- Agricultural Attaché
- Agricultural Chemical Salesperson
- Agricultural Chemist
- Agricultural Commodity Grader
- Agricultural Commodity Inspector
- Agricultural Commodity Warehousing Examiner
- Agricultural Consultant
- Agricultural Broker
- Agricultural Business Administrator
- Agricultural Economist
- Agricultural Writer, Editor
- Agricultural Educator
- Agricultural Engineer
- Agricultural Geographer
- Agricultural Instructor
- Agricultural Journalist
- Agricultural Machinery Serviceperson
- Agricultural Management Specialist
- Agricultural Marketing Specialist
- Agricultural Missionary
- Agricultural Program Specialist
- Agricultural Researcher
- Agricultural Market Reporter
- Agricultural Statistician
- Agricultural Trade Magazine Editor
- Agriculture Advertising Writer
- Agriculturist (many countries)
- Aerial Plant Nutrient Applicator
- Agronomist
- Animal Behaviorist
- Animal Breeder
- Animal Husbandperson
- Animal Specialist
- Aquatic Weed Specialist
- Archnologist (zoology of spiders)
- Area Economist
- Associate Buyer
- Avian Specialist
- Bacterial Pesticide Specialist
- Bacteriologist
- Baker Scientist Technologist, Manager
- Bank Agricultural Representative
- Banking Official
- Beekeeper
- Biochemist
- Biological Pesticide Specialist
- Biologist
- Biophysicist
- Biostatistician
- Botanist
- Breeding Technician
- Cattle Manager
- Cereal Chemist
- Clay Mineralogist
- Climatologist
- College Agricultural Researcher
- College Faculty Member
- Commercial Cattle Feeder
- Commodities Broker (Ag)
- Conservationist—forest, range, soil, water, wildlife
- Consumer Marketing Specialist
- Cooperative Manager
- County Extension Agent
- County Extension 4-H Agent
- Crop Insurance Specialist
- Crop Physiologist
- Crop Protection Specialist
- Crop Researcher
- Crop Specialist
- Cytogeneticist (heredity through cells and genetics)
- Dairyperson
- Dairy Plant Manager
- Dairy Technologist
- Dendrologist (tree rings to date past events)
- Director of Ag Research
- Earth Scientist
- Ecologist
- Electric Co-op Manager
- Elevator Manager
- Entomologist
- Environmental Scientist
- Extension Specialist
- Exterminators—Insect
- Farm Appraiser for a bank or agricultural lending institution
- Farm Building Designer
- Farm Credit Manager
- Farmer
- Farm Equipment Mechanic

Jobs in Agriculture (continued)

Farm Machinery Dealer
Farm Machinery Designer
Farm Manager
Farm Planner
Farm Realtor
Farm Store Manager
Farm Superintendent
Farm Supply Cooperative Salesman
Feed Dealer
Feedlot Manager
Feed Mill Manager
Feed Salesperson
Feed Technologist
Fertilizer Industry Employee
Field Crop Grower
Field Representative of: Beef Breed Association
Beef/Pork Packing or Processing Company
Dairy Breed Association
Feed Company
Fruit Packing or Processing Company
Fuel Company
Insurance Company
Lumber Processing Company
Machinery Company
Plant Nutrient Company
Poultry Packing or Processing Company
Veterinary Supply Company
Fish Culturist
Fish and Wildlife Specialist
Fishery Biologist
Fishery Manager
Floral Designer
Floriculturist
Florist
Food Inspector
Food Processor
Food Retailer
Food Scientist
Foreign Agricultural Affairs Officer
Forester
Forest Products Technologist
Forest Ranger
Forestry Aid and Technician
Fruit Grower
Game Warden
Garden Center Retailer
Geneticist—plant, animal
Geochemist
Geomorphologist (genesis of earth forms)
Golf Course Superintendent
Grain Buyer
Grain Miller
Grain Processor
Greenhouse Grower
Greenhouse Owner
Groundwater Geologist
Horticultural Supplies Worker
Horticultural Therapist
Horticulturist
Hydrologist
Ichthyologist (zoology of fish)
Industrial Agriculturist
Information Specialist
Inspector—food, feed
Insurance Broker
International Agriculture Specialist
Irrigation Engineer
Irrigation Manager
IVS Volunteer
Laboratory Technician
Land Appraiser
Land Economist
Landscape Architect
Land Surveyor
Land Utilization Specialist
Life Scientist
Livestock Breeder
Livestock Buyer
Livestock Feeder
Loan Specialist
Market Analyst
Market Research Workers
Meat Cutters
Meat Department Manager
Meat Grader
Meat Inspector
Microbiologist
Microscopist
Molecular Biologist
Mycologist (botany of fungi)
Naturalist
Nematologist (zoology of nematode)
Nursery Owner or Operator
Nurseryperson
Nutritionist—plant, animal
Organizational Fieldperson
Ornamental Plant Specialist
Ornithologist (zoology of birds)
Outdoor Recreation Specialist
Packinghouse Manager
Parasitologist
Park Manager
Park Naturalist
Park Ranger
Park Superintendent
Pathologist—plant, animal
Peace Corps Administrator
Peace Corps Volunteer, especially in Africa, Asia, or Latin America
Pest Control Manager
Pesticide Residue Analyst
Pet Food Processor
Plant Breeder
Plant Pathologist
Plant Physiologist
Plant Propagator
Plant Quarantine Inspector
Pest Control Inspector
Poultry Inspector
Poultryperson
Poultry Scientist
Produce Department Manager
Produce Development Specialist
Production Credit Fieldperson
Public Relations Manager
Purchasing Agent
Quality Control Specialist
Radio Farm Director
Rancher
Ranch Manager
Ranger—forest, park
Range Conservationist
Range Manager
Range Scientist
Range Specialist
Recreation Development Planner
Researcher, Specialist
Resource Economist
Rural Sociologist
Salesperson (of any one of many agricultural outputs or inputs)
Sales Representative
Sanitarian
Securities Salesman
Science Editor
Seed Broker
Seed Grower
Seed Technologist
Soil Analyst
Soil Chemist
Soil Conservationist
Soil Fertility Scientist
Soil Physicist
Soil Scientist
Soil Surveyor
Statistician
Taxocologist (classifies plants, animals)
Taxonomist
Technical Editor
Technical Writer
Textile Researcher
Textile Technologist
Timber Manager, Specialist
Transportation Manager
Turf Producer
Turf Specialist
TV Farm Director
Vegetable Grower
Virologist (viruses)
VISTA Volunteer
Vocational Agriculture Teacher
Waterfowl Specialist
Water-Life Management Specialist
Water Economist
Water Resources Administrator Engineer
Water Resources Devlopment Official
Weed Science Specialist
Wildlife Administrator
Wildlife Biologist
Wildlife Specialist
Wildlife Writer, Editor
Youth Corps Conservation Director
Zoologist

Appendix I

NATIONAL FFA JOB INTERVIEW CAREER DEVELOPMENT EVENT

IMPORTANT NOTE: *Please thoroughly read the Introduction Section at the beginning of this handbook for complete rules and procedures that are relevant to all National FFA Career Development Events.*

I. PURPOSE

The National FFA Job Interview Career Development Event is designed for FFA members to develop, practice and demonstrate skills needed in seeking employment in the agricultural industry. Each part of the event simulates "real world" activities that will be used by real world employers.

II. EVENT RULES

1. The National FFA Job Interview Career Development Event will be limited to one participant per state.

2. The National FFA Job Interview Career Development Event will only be for students who are regularly enrolled in agricultural education during the calendar year, have a planned course of study, or who are still in high school, but have completed all the agricultural education offered. When selected, participants must be active members of a chartered FFA chapter and the National FFA Organization. A member representing a state association may participate in the National Job Interview CDE only once.

3. It is highly recommended that participants be in official FFA dress in each event.

4. Each participant's cover letter, résumé and application will be the result of his or her own efforts.

5. Participants will submit a signed statement of originality on the certification form provided through their state FFA association.

6. Participants shall be ranked in numerical order on the basis of the final score to be determined by each judge without consultation. The judges' ranking of each participant then shall be added, and the winner will be that participant whose total ranking is the lowest. Other placings will be determined in the same manner (low point score method of selection).

7. The National FFA Officers and National Board of Directors will be in charge of this event.

III. EVENT FORMAT

A. EQUIPMENT

Materials student must provide- Students must provide their own writing utensils.

B. ACTIVITIES

1. The event is developed to help participants in their current job search (for SAE projects, part-time and full-time employment). Therefore, the cover letter, résumé and references submitted by the participant must reflect their current skills and abilities and must be targeted to a job for which they would like to apply. In other words, participants cannot develop a fictitious résumé for a fictitious job. Instead, they are expected to target the résumé towards a real job that they can qualify. **By September 15th of the year that the participant**

is competing they will submit the following:

a. **Cover Letter (Points - 100)**
   1. Ten copies of a single spaced 8 1/2"x 11" white bond paper letter of intent. The paper is to be single sided only, typed with no more then ten characters per inch and block justified.
   2. Letter is to be addressed to the Superintendent of the Career Development Event and dated for the first day of the event.

b. **Résumé (Points - 150)**
   1. Ten copies of a single spaced 8 1/2"x 11" white bond paper. The résumé is to be single sided only, typed not to exceed two pages total. Suggested formats can be found in the Greggs Manual.
   2. Résumé must be non-fictitious and based upon their work history.
   3. Students are to **submit** three letters of reference.
   4. **Cover letter, résumé and references must be submitted to the National FFA CDE office by September 15th of the year that participant is competing.**

2. At the National FFA Career Development Event the following will be completed:

a. **Application (Points - 50)**
   1. Students will complete a standard job application on-site, prior to the personal interview.

b. **Telephone Interview (Points - 150)**
   1. Students will interview with one of the following three people:
      a. Human, Fiscal and Resource personnel director
      b. Employer's Assistant
      c. Employer themselves
   2. The telephone interview will last a maximum of three minutes.
   3. Students are to position themselves so that they obtain a personal interview with the company they are applying. Student should interview with the thought that the company has already received their cover letter, résumé and three letters of reference.
   4. Telephone interview will be conducted in both the preliminary and final rounds.

c. **Personal Interview (Points - 450)**
   1. The preliminary round will consist of an interview in front of a panel of judges. Each state with a contestant is to provide a person from their state to assist.
   2. Students participating in the final round will interview with three separate judges. Each interview will last twenty minutes.

d. **Follow Up Letter (Points - 100)**
   1. Participants will submit a follow up letter after each round of interviews. Students will be provided computers with word processing applications to compose and type a follow up letter. Thirty minutes will be given.
   2. Letter is to be addressed to the Superintendent of the Career Development Event, and should be a response to their most recent interview.

IV. TIEBREAKERS

Ties will be broken based on the greatest number of low ranks. Participant's low ranks will be counted and the participant with the greatest number of low ranks will be declared the winner. If a tie still exists, then the event superintendent will rank the participant's response to questions. The participant with the greatest number of low ranks from the response to question will be declared the winner. If a tie still exists then the participant's raw scores will be totaled. The participant with the greatest total of raw points will be declared the winner.

V. AWARDS

Awards will be presented at an awards ceremony. Awards are presented to individuals based upon their rankings. Awards are sponsored by a cooperating industry sponsor(s) as a Special Project, and/or by the general fund of the National FFA Foundation.

VI. REFERENCES

This list of references is not intended to be inclusive. Other sources may be utilized and teachers are encouraged to make use of the very best instructional materials available. The following list contains references that may prove helpful during event preparation.

Greggs Manual
Elements of Style - Strunk and White
Microsoft Word résumé templates

101 Toughest Interview Questions...and Answers That Win Jobs
Daniel Porto, Daniel Porot / Paperback / Published 1999

25 Reasons Why I Won't Hire You! What You Did Wrong Before, During & After the Interview!
Zenja Glass / Paperback / Published 1998

Best Answers to the 201 Most Frequently Asked Interview Questions
Matthew J. Deluca, Mathew J. DeLuca / Paperback / Published 1996

The Complete Job Interview Handbook
John J. Marcus / Paperback / Published 1994

Job Interview

Name: ______________________________ Chapter: ____________________

State: ______________________________ Member No.:____________________

Cover Letter

Composition	Possible Points	Score
Correct format and stationary	10	
Punctuation	10	
Grammar	10	
Spelling	10	
General appearance	10	
Composition Sub Total:	50	
Content		
Career goal specified	15	
Proper qualifications	35	
Content Sub Total:	50	
Composition Sub Total:	50	
Content Sub Total:	50	
Grand Total:	100	

Job Interview

Name: ______________________________ Chapter: ______________

State: ______________________________ Member No.:______________

Employment Application Scorecard

	Possible Points	Score
Legible	15	
Neat • Grammar • Punctuation	10	
Completed accordingly	10	
Consistent with resume	15	
Grand Total	50	

Job Interview

Name: ______________________________ Chapter: ______________

State: ______________________________ Member No.:______________

Final Round Follow-up letter Scorecard

	Possible Points	Score
General appearance	5	
Composition	10	
Express appreciation	10	
Comments on interview activities	10	
Express interest in position	15	
Review of relevant qualification	15	
Is requested information addressed	15	
Provisions for follow-up stated	20	
Grand Total	100	

Job Interview

Name: ______________________________ Chapter: ______________

State: ______________________________ Member No.: ______________

Personal Interview Scorecard

	Possible Points	Score
Appearance and courtesy	45	
Greetings and introduction	45	
Speech • Grammar • Vocabulary • Volume • Enunciation	45	
Attitude and personality • Forcefulness • Poise • Temperament • Sincere	45	
Ability to convince or impress interviewer • Persuasiveness • Self-confidence	45	
Knowledge and presentation of abilities • Educational experience • Occupational experience	45	
Reliability • Frankness • Consistency • Accuracy	45	
Poise • Tact • Discretion • Questions asked of interviewer	45	
Career Objective • Degree to which the contestant had determined career objective	45	
Conclusion of interview	45	
Grand Total	450	

Job Interview

Name: ______________________________ Chapter: ______________

State: ______________________________ Member No.:______________

Preliminary Round Follow-up letter Scorecard

	Possible Points	Score
General appearance	5	
Composition	10	
Express appreciation	10	
Comments on interview activities	10	
Express interest in position	15	
Review of relevant qualification	15	
Is requested information addressed	15	
Provisions for follow-up stated	20	
Grand Total	100	

Job Interview

Name: ______________________________ Chapter: ______________

State: ______________________________ Member No.: ______________

Resume Scorecard

General Appearance	Possible Points	Score
Presented in proper format	15	
Pleasing to the eye • Captures interest • Layout • Easily read	40	
Grammar • Punctuation • Typing • Spelling	20	
General Appearance Sub Total:	75	
Composition		
Personal data	10	
Career objective	10	
Educational background	20	
Work experience/skills	20	
Special experiences,activities,honors	10	
References	5	
Composition Sub Total:	75	
Composition Sub Total:	75	
Content Sub Total:	75	
Grand Total:	150	

Job Interview

Name: ______________________ Chapter: ______________

State: ______________________ Member No.: ______________

Telephone Interview Scorecard

	Possible Points	Score
Introduction	22	
Initiative	22	
Communicated effectively	26	
Exhibited ambition and efficiency	30	
Diplomatic and courteous	26	
Gathered appropriate information • Contact name • Address • Date • Time	24	
Grand Total	150	

Appendix J

Entrepreneur in Agribusiness—Is it for Me?

The American economy is based on the free enterprise system, or capitalism. This simply means that we have the right to own our own business and to make a profit. It also means that we can lose money, perhaps even all the money that was invested. There is a big risk factor for those who start their own business. Therefore, entrepreneurship may not be for everyone.

General Characteristics of Entrepreneurs

Entrepreneurs are people who have the initial vision, diligence, and persistence to follow through. An entrepreneur is a person who accepts all the risks pertaining to forming and operating a small business. This also entails performing all business functions associated with a product or service and includes social responsibility and legal requirements. There are many reasons why people are willing to take the risks of starting a business.

- Entrepreneurs work for themselves, are independent, and make their own business decisions.
- Whatever income they earn above their financial obligations is theirs to keep.
- They can test their own theories and ideas on how to run a business.
- They set their own working hours.
- They themselves set prices, determine production levels, and control inventory according to the market.
- They determine the product or service offered and control its quality as well as the overall reputation of the business.
- They solve the problems.
- They perform all the human resource functions such as hiring, training, and firing.
- They set company policy.

Not everyone satisfies the qualifications to be an entrepreneur. It takes a lot of hard work, and there may be little job security. Many small businesses fail each year for a number of reasons. You should thoroughly investigate all the aspects of starting your own business. Make sure it is what you really want and can actually handle.

Index

Note: Page numbers followed by an "f" indicate figures.